KU-252-005

1939 1973

Powering the Hydrocarbon Revolution, 1939-1973

A History of Royal Dutch Shell

Stephen Howarth & Joost Jonker

Published under licence from Boom Publishers, Amsterdam, initiating publishers and publishers of the Dutch edition.

OXFORD
UNIVERSITY PRESS

2007

fill up with SHELL

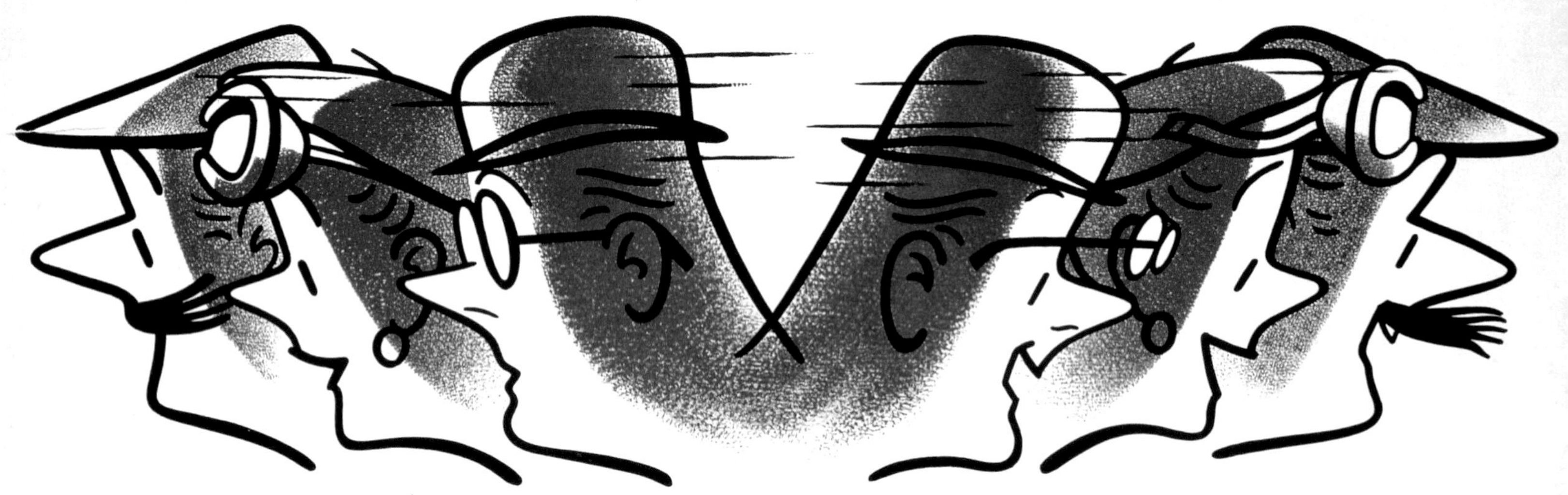

and feel the difference

[1]

Powering the Hydrocarbon Revolution, 1939-1973 is one of four volumes of the work entitled *A History of Royal Dutch Shell*, written by a team of four authors associated with Utrecht University, Jan Luiten van Zanden, Stephen Howarth, Joost Jonker and Keetie Sluyterman. It is the result of a research project which was supervised by the Research Institute for History and Culture and coordinated by Joost Dankers.

The other volumes are:

Joost Jonker and Jan Luiten van Zanden
From Challenger to Joint Industry Leader, 1890-1939

Keetie Sluyterman
Keeping Competitive in Turbulent Markets, 1973-2007

Jan Luiten van Zanden
Appendices. Figures and Explanations, Collective Bibliography, and Index, including three DVDs

Contents

5 Introduction

Chapter 1
Transformation under fire, 1938-1948

10 Quarrelling in a conflagration
12 Preparations for war
26 Defending the business
34 Becoming a crude-short business
43 Stark contrasts in the downstream business
49 The war at sea
63 The Japanese offensive in Asia
65 Targeting oil installations
70 Manufacturing for victory
78 Under the shadow of the Nazis
86 The battle of the boards
97 Back together again
103 Conclusion

Chapter 2
Remodelling the Corporation

106 Loudon takes the lead
110 The listing of Royal Dutch on the New York Stock Exchange, 1954
115 Learning to live with the shareholders
121 Towards a closer unity: simplifying and harmonizing
137 Inventing the matrix organization
140 Perfecting the matrix: the McKinsey review, 1957-1960
149 Shell Oil and the matrix model
156 Staff matters
159 The evolution of training
169 Conclusion

Chapter 3
Into ever deeper waters

172 A new order for the international oil industry
180 Expanding operations in a competitive environment
185 The Group's supply pattern
195 The offshore challenge
202 Shell Oil's overseas ambitions
207 Strategic partnerships: NAM and Gulf
220 Confronted by nationalism in Egypt and Iraq
225 Struggling to keep operations in Indonesia
241 Coming to terms with OPEC
249 Conclusion

Chapter 4
Creating competitive advantages downstream

253 The elusive ideal of Shell Oil
258 Seeking synergies in manufacturing oil
281 Transport: economies of scale afloat and ashore
298 Integrating fragmented markets
329 Conclusion

Chapter 5
Chasing the rainbow

332 Growing a second leg
342 From interface to battleground
346 Products and processes
354 Debating the chemical sector's performance
370 Beyond the core business: venturing into other industries
381 Conclusion

Chapter 6
Discovering tomorrow's world

385 The evolution of Group research
401 From cost conscious to environmentally conscious
402 Discovering the environment
405 Seeking industry support
419 Combating sulphur emissions
423 Leaded, unleaded – or something else?
426 Labelling dangers
427 The double-edged drins
441 Conclusion

443 Conclusion
449 Notes
481 List of tables and figures
483 Abbreviations
485 Bibliography
489 Illustration credits
493 Index
514 Colophon

Great Clarendon Street, Oxford OX2 6DP

Oxford University Press is a department of the University of Oxford.
It furthers the University's objective of excellence in research, scholarship,
and education by publishing worldwide in

Oxford New York

Auckland Cape Town Dar es Salaam Hong Kong Karachi
Kuala Lumpur Madrid Melbourne Mexico City Nairobi
New Delhi Shanghai Taipei Toronto

With offices in

Argentina Austria Brazil Chile Czech Republic France Greece
Guatemala Hungary Italy Japan Poland Portugal Singapore
South Korea Switzerland Thailand Turkey Ukraine Vietnam

Oxford is a registered trade mark of Oxford University Press
in the UK and in certain other countries

British Library Cataloguing in Publication Data
Data available

ISBN: 978-0-19-929879-2

Introduction

The opening of Volume 2 finds the Royal Dutch/Shell Group in a position of undiminished power, being (together with Standard Oil of New Jersey, Jersey Standard) the acknowledged joint leader of the industry, at a considerable distance from all other oil companies. Having started operations during the 1890s as potential challengers to the monopoly of the Standard Oil Trust, in 1902 Royal Dutch and Shell Transport had merged their Asian marketing operations with those of the Paris Rothschilds to form Asiatic Petroleum Company. Financial mismanagement had then pushed Shell Transport to the brink of bankruptcy, from which the company was rescued by its merger with Royal Dutch in 1907.

This had created an unusually complicated business. The two holding companies both needed to retain their national identity, Royal Dutch in order to comply with concession requirements in the Dutch East Indies, and Shell Transport so as to give the Group a British stature and corresponding access to the corridors of world power. The original intention had been to form a single, integrated operating company under the two holding companies. This was thwarted by the Rothschilds, who insisted on keeping their share in the very profitable Asiatic, so two further operating companies were formed instead, namely Bataafsche Petroleum Maatschappij for exploration and production, and Anglo-Saxon Petroleum Company for shipping. As shown in the diagram, the whole structure was held together by exclusive service agreements and overlapping directorships between the companies. As general manager of Royal Dutch, a manager of Shell Transport and of the three operating companies, Henri Deterding became in all but name the chief executive officer of the business.

From 1907 the Group expanded at a prodigious rate, acquiring production properties around the world and building up a marketing network using Shell as its main brand name. Repeated price wars with the Standard Oil Trust failed to halt this expansion, and in 1911, when the Trust was broken up following legal proceedings, the Group was already firmly established on all inhabited continents. By 1920 it had overtaken Jersey Standard, the Trust's main successor, to become the world's biggest oil enterprise. However, its precocious expansion exposed serious shortcomings in its management structure, which the strains of the First World War did little to alleviate. During the early 1920s, matters came to a head in a protracted power struggle at the top of the Group. Deterding emerged as the victor, but his defeat of proposals for a thorough managerial overhaul prevented a necessary adaptation of the Group's management structure to its greatly enlarged size. The business continued to depend more on managing directors' personal interests and interventions, and not on formal procedures, the delegation of responsibilities, and centralized administrative controls. Moreover, Deterding's increasingly volatile personal behaviour estranged him from his colleagues.

The structure of the Group in 1907.

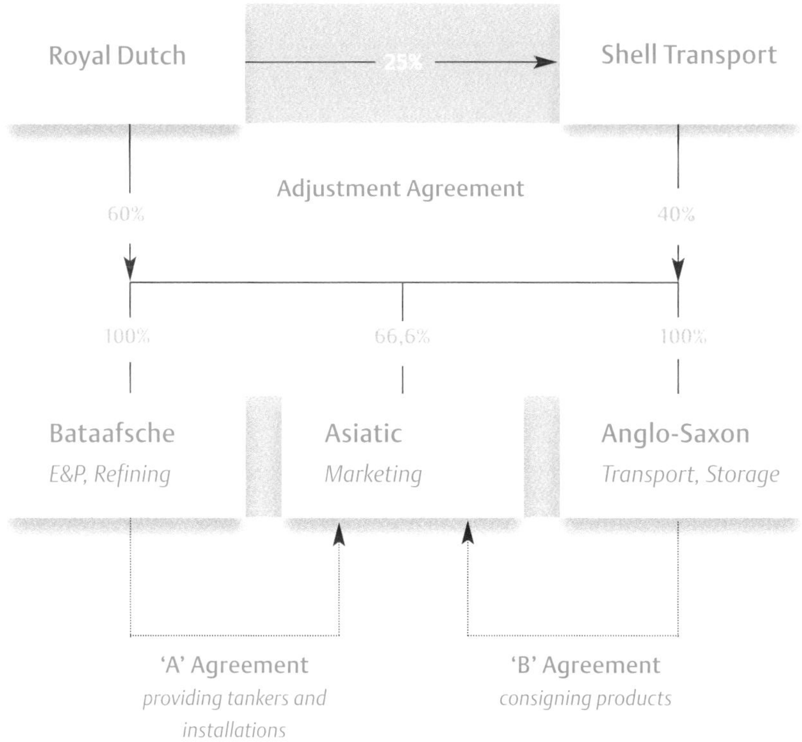

Partly because of this power struggle, Shell in the 1920s was less dynamic than in the period immediately preceding the First World War. Faced with rapidly rising world production, it entered no major new production areas and did not pursue an aggressive sales policy, but instead strove to maintain market share. The culmination came in 1928 when Deterding masterminded the Achnacarry conference, at which the major oil companies agreed to respect each others' respective positions. At the same time, however, the Group made significant forward strides in several other fields. Firstly, the business showed its resilience in the conspicuous ease with which it weathered major blows such as the loss of one-third of its production through nationalization of the oil industry by the Soviet Union in 1918. Secondly, the Group continued to grow, notably in the United States, where Shell Union expanded into a nationwide, integrated oil company. Thirdly, a new marketing style evolved using striking visual designs which simultaneously positioned the highly recognizable Shell pecten as a hallmark for reliable and high-quality products around the world, and as a brand with a distinctive local character in each of the countries served. Fourthly, research graduated from product-testing into more fundamental analyses focused on finding the best processes for making high-quality components from the cheapest crude available. Finally, in 1927 the Group took the momentous decision to diversify into petrochemicals with a view to obtaining technologies deemed vital for the future development of the oil industry. Combined with the progress made by Group research, this diversification produced some key innovations.

Overproduction in the United States following the discovery of huge new fields there pushed the oil industry towards a crisis before the Wall Street crash of 1929 marked the end of the Roaring 1920s. During the ensuing global economic Depression the Group was hit particularly badly in the Dutch East Indies and in its American operations, the latter suffering from a clear overexpansion. Against this, the market stabilization achieved by the Achnacarry agreement helped markedly to dampen the impact of the relentless fall in oil prices. Shell also sought to stem the tide with swingeing staff cuts and with new administrative policies which included comparative cost analyses.

Meanwhile economic nationalism had become an increasingly formidable opponent for the oil industry. Governments around the world took measures ranging in severity from taxation, through the formation of national oil companies and official monopolies, to the outright nationalization of foreign companies, as happened in the Soviet Union. Initially Shell reacted indignantly to any such measures. Deterding did great damage to his own public standing with excessive outbursts of rage in specific cases. However, all efforts to take a principled stand of resistance against nationalist policies foundered on the ready availability of oil from rival companies, so gradually the Group adopted a more realistic attitude, trying to get the most out of a given situation by negotiating for as long as possible. This policy failed both in Germany and Mexico, however. After coming to power in 1933, the Nazis took progressive control of the German oil industry, to the extent that by 1938 Shell had effectively lost its subsidiary Rhenania-Ossag, and in that year too Mexico nationalized its oil industry.

The Hitler regime also provided the immediate cause for Deterding's resignation in 1936. His detestation of communism led to a fascination with the New Order in Germany and acrimonious boardroom rows, which persuaded both him and his fellow directors that it was better for him to step down. The selection of his successor showed how fragile the Group's top management structure really was. Conflicting national interests and the majority share of Royal Dutch immediately acquired an importance which they had not possessed before, and the new de facto CEO J. E. F. de Kok needed all his considerable tact to hold the team of managing directors together. More than ever the Group needed a fundamental modernization of its management structure, but with Dutch and British managing directors deadlocked this appeared a remote perspective.

The first chapter of this volume shows how, during the Second World War, the managerial conflict escalated, only to be resolved in a compromise which inaugurated a renaissance. Thereafter, armed with a new purpose and conviction, the Group expanded rapidly during the hydrocarbon revolution of the 1950s and 1960s. Chapters 2, 3, and 4 tell this story, moving from exploration and production downstream, via manufacturing and transport to marketing. Chapter 5 relates how the Group pursued a vigorous diversification into petrochemicals and, a little later, also into other business sectors. In Chapter 6, the focus is first on research, which continued to yield valuable new products and processes, though it became gradually more difficult to recoup the high costs involved. The chapter then shows how Shell reacted to the environmental concerns which confronted the petrochemical industry from the 1950s.

Like its companion volumes, and as detailed in the general introduction in Volume 1, this study of Royal Dutch Shell's history was researched and written with five themes in mind: the geographical spread of operations; the Group's internal organization; its performance in relation to the competition; innovations; and the role of politics. It has likewise been based on unrestricted access to the available company records, including the minutes and other documents of the CMD and the Conference. Additional research in the archives of the Bank of England, the National Archives in Kew, De Nederlandsche Bank, the Dutch Ministry of Finance and the Beyen papers yielded further vital information. The authors wish to record their gratitude to Mr. C. Fieret at the Dutch Ministry of Finance for his assistance with our research in the department's records; to Joke van der Hulst for her always kind help with the archives of De Nederlandsche Bank; to Dr. Wim Weenink for giving access to the Beyen papers; and to Hayley Wilding for laying open the Bank of England records to us. Maurits van Os shared with us his insights about the McKinsey exercise drawn from an interview with Hugh Parker; interviews with former Group employees Bill Bentley, Prof. Dr. Frits Böttcher, Dr. H. J. Kruisinga, and Dr. E. G. G. Werner gave us valuable insights; and at a very late stage, Barbara Gamalski helped us to find additional material at Shell Deutschland. We offer our thanks to them all.

[1]

Chapter 1

Transformation under fire, 1938-1948

The threat of war in Europe found the Group in a difficult position. Deterding's retirement in 1936 had left it with a brittle top management structure, which was now increasingly exposed to dangerous chauvinist tensions. Subsequently the war dealt serious blows to the business, sharply reducing production, manufacturing capacity, and tanker capacity, and leading to heavy loss of life. Yet the Group succeeded in transforming itself during the war and emerged strengthened overall, with both operations and managerial structure geared for achieving new purposes.

The petroleum harbour at Amsterdam: the result after parts of it were set ablaze by British experts on 14 May 1940

Quarrelling in a conflagration On 13 May, 1940, Fred Godber wrote an urgent letter to his fellow Group managing director, J. C. van Eck. Since the previous September, Britain and France had been at war with Germany after that country had attacked Poland. In April, Germany had conquered Denmark and Norway, and on 10 May, it had launched a surprise attack on the Netherlands, Belgium, and France. This Blitzkrieg was irresistible. The Dutch defences were close to collapsing; the German army had advanced deep into Belgium, threatening Antwerp; an armoured column was piercing defences at Sedan in northeastern France. In Britain, Prime Minister Neville Chamberlain had just resigned after a prolonged political debate over his leadership in the war, handing over to Winston Churchill.

But it was not these momentous events that exercised Godber's mind in writing to his colleague. Instead, he presented Van Eck with a summary of the steady deterioration over the previous two years of relations between the British and Dutch managers who formed the Group's top management. Dutch managers, he wrote, had several times asserted Royal Dutch's control over the business, something which under Deterding had never happened. When in 1938 the British managers had suggested the drafting of contingency plans for the eventuality of war in Europe, their Dutch colleagues had made arrangements for securing the safety of central offices in The Hague by creating the possibility of transferring the seats of the companies resident in the Netherlands, without consulting the British, and without responding to their suggestion to reinforce the management of Bataafsche, the Dutch operating company, by giving the British directors executive powers. On putting the transfer scheme into effect, the Dutch managers had made new appointments to operating company boards without discussing them with their

[2]

British colleagues, who disagreed with some of the appointments and wanted an immediate reconsideration. Godber's letter drew acerbic comments from Guus Kessler, the senior Dutch manager in London. The British managing directors had been involved in the contingency planning all along, he noted in the margins of the letter, but they had contributed nothing substantial at all except the suggestion that Shell Transport should assume control, as in the First World War. They had also been consulted about the appointments, which only concerned some most unimportant subsidiaries anyway.[1]

Thus, as in August 1914, an external crisis exposed a serious disagreement between the Dutch and the British in the Group's top management. The fact that they were now Allies opposing a common enemy failed to stop the managerial infighting which had developed after Deterding's resignation in 1936. Some form of reconciliation on basic principles appears to have taken place

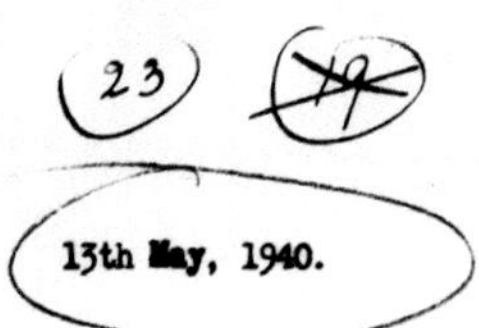

13th May, 1940.

My dear Van Eck,

I conceive it desirable to try to explain to you what has so profoundly disturbed the Shell Managing Directors and your British colleagues recently.

Deterding (as well as his and our colleagues in Holland) during all the years of his administration, took the view that the ramifications of our Group were so vast and the responsibility of managing it so great and serious that one partner alone could not take that responsibility and that the only way to keep an even keel was to operate as a Group and avoid any attempt at management control by weight of shareholding. He did this, not only because he felt that the very magnitude of the task necessitated it, but also because he was unwilling to take the responsibility without the full support of his British partners. He frequently expressed himself in these terms, and collaboration had been developed to such a fine point that consultation and agreement on all important matters became a daily routine. In all the 35 or so years of my connection with the Group, I never once heard him mention control, or claim control.

-4-

to tell you that we disagree with some of these appointments and I must ask for their immediate re-consideration.

Legh-Jones is in full agreement with the foregoing.

(Sgd) F.GODBER.

On 10 May 1940, the same day that the Netherlands were overrun by Germany, Winston Churchill became Prime Minister of Great Britain, heading a coalition government after the resignation of Neville Chamberlain.

Simultaneously, Royal Dutch/Shell's joint leadership was under great stress, as shown by Kessler's angry comments scribbled in the margins of Godber's letter of 13 May.

following Godber's letter. In June 1940, he and George Legh-Jones were appointed managing directors of Bataafsche, eliminating the anomaly laid down in the 1907 Adjustment Agreement, that Dutch directors had executive power at Anglo-Saxon and Asiatic, the two London operating companies, while the British had no executive power at Bataafsche. At last the team of directors, which had started to refer to itself as Group managing directors in 1938 became equal.[2]

The appointment of British executives at Bataafsche did not succeed in bridging the rift within the managerial team, as we shall see. However, having British directors available to represent Bataafsche also gave the company better credentials in London. At the same time, the arrangement helped to alleviate the consequences of administrative dislocation for the Group, which followed the dispersal of company departments from central offices in The Hague and London.

Preparations for war In November 1934, almost two years after Hitler came to power, the British government asked the Oil Board, an interdepartmental committee set up in 1925 but which had lain dormant since, to draft supply plans for a war against 'a European enemy'. The Board immediately started collecting data on estimates for oil supplies and tankers.[3] The first ripple from this tide reached the Group during 1936, when the oil companies agreed to raise their stocks of products for civilian uses to three months' peacetime use, the British government bearing half the cost of the extra tankage required.[4] In December of that year, the Air Ministry enquired whether the Group would consider building an iso-octane plant in the UK. Just four months earlier the production of this vital component for high-performance aircraft fuel had started at the Pernis plant, near Rotterdam, but the department's officials thought Britain could not 'rely on obtaining supplies of iso-octane from Pernis in the event of an emergency.'[5] The Group responded a few months later by promising to supply 32,000 tons of iso-octane a year for five years, to be doubled in case of war, from a plant in the UK. Construction of the unit started the following spring at Stanlow and it came on stream in the summer of 1939. The two high-pressure steel reaction vessels at the core of the installation had come from the German Krupp works.[6] By the time the plant came on stream, however, the Air Ministry had substantially increased its projections of the demand for 100-octane aviation fuel and decided to build three additional installations. These plants were to produce iso-octane by a new and untried process, based on dehydrogenating isobutane into isobutylene. At the Ministry's request, Imperial Chemical Industries (ICI), Trinidad Leaseholds Ltd., and the Group formed a joint venture, known as Trimpell, to construct and manage the installations on behalf of the British Government. One plant was sited in the UK, at Heysham in Lancashire, the other two in Trinidad.[7] The particular process which Trimpell pioneered did not become a success, because alkylation, another process for making iso-octane, proved to be far superior.[8] Before the Heysham plant had reached its design capacity, ICI took over some of the premises for making synthetic ammonia.[9]

The motives which inspired the building of the Stanlow iso-octane plant also made the Group invest heavily in other new facilities there. In 1938, the Anglo-Saxon board accepted proposals to construct plants for making alcohols, ketones, and ester salts, from slack wax, a by product of the Stanlow lube oil factory.[10] A few months later the board decided to replace the lube oil plant itself with new and greatly expanded installations, including a Duo-Sol extraction unit following the Rectiflow principle, a recent innovation developed by the Group's Amsterdam laboratory. This plant would both reduce the heavy dependence on imports from Rhenania-Ossag, Shell's German operating company, and fulfil the Air Ministry's increased requirements for special aircraft engine lubricants. The installation came on stream in July 1940, the entire output being earmarked for the Air Ministry.[11]

[4]

Below left, Shell's Heysham refinery (Lancashire, UK) displays Crude Distilling Unit no 3, with Vis-breaking Units 1 and 2 in the background. The latter improved upon the Trumble process by breaking down the viscosity of heavy fuels. The diagram (top right) shows the Rectiflow process, also known as the backwash process – another Shell discovery which enhanced the Edeleanu process by adding benzene during the manufacture of lubricants, and enabled the removal of more aromatics. Below right shows the principle of the alkylation process; and below, in contrast to such high science, a road tanker at Heysham uses the simplest method possible to take on its load – gravity.

[6]

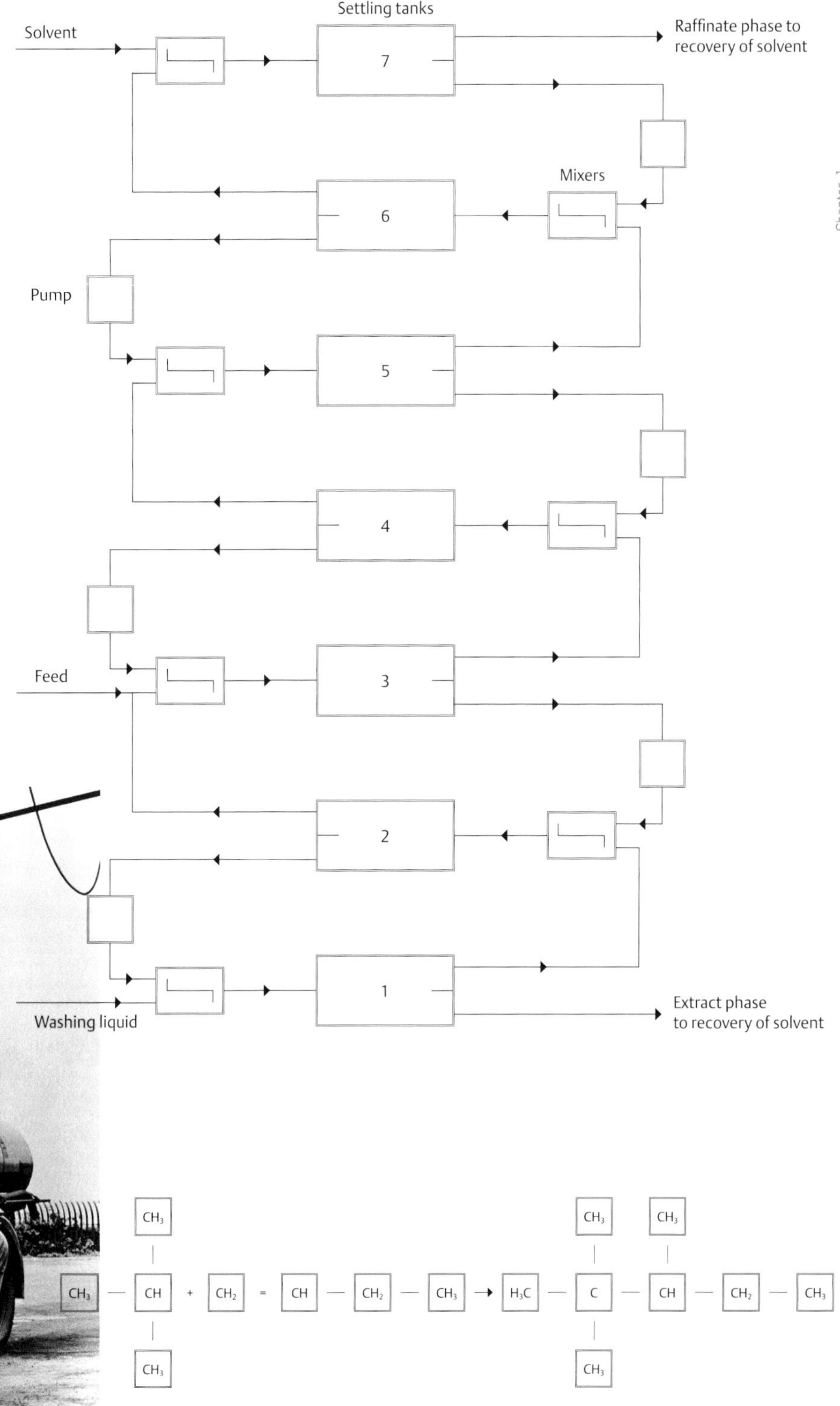

[5]

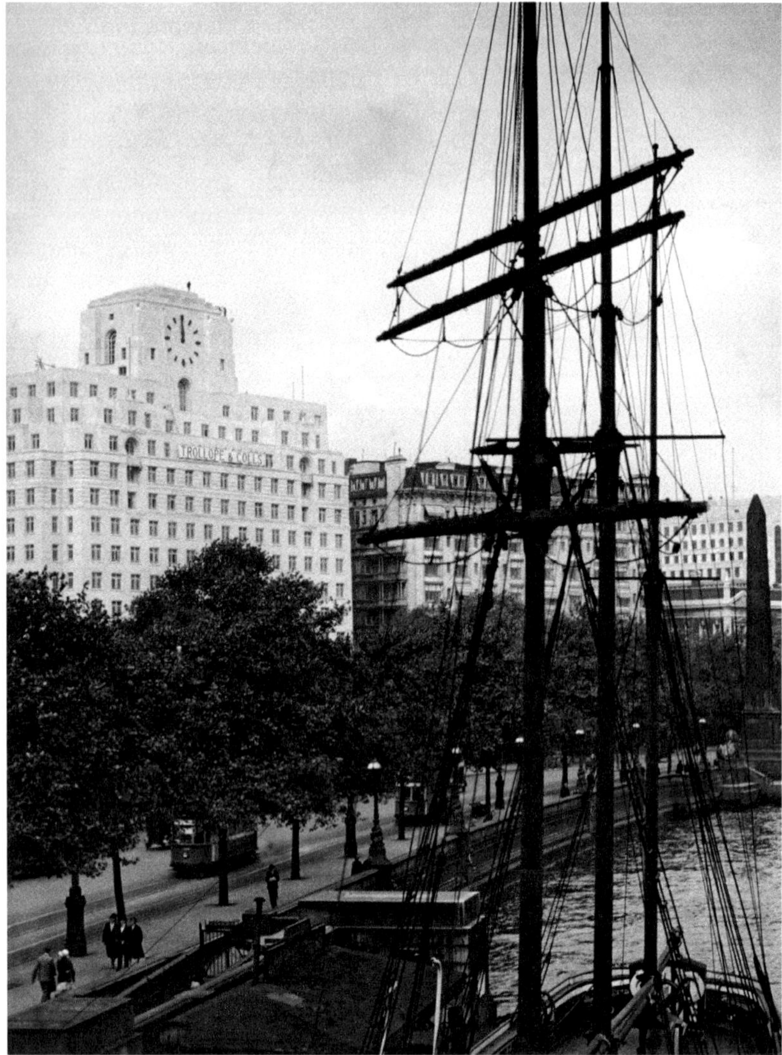

[7]

Charged with total coordination of Britain's wartime petroleum supplies and led by Andrew Agnew, the Petroleum Board was headquartered in Shell-Mex House on the north bank of the Thames.

Right, the ARP (Air-Raid Precautions) control room; the main telephone switchboard never closed, day or night, throughout the war.

The government-inspired preparations also led to the oil companies setting up, in 1938, the Petroleum Board, a joint organization to coordinate supply and distribution in wartime. Operating under the directions of a War Cabinet subcommittee, the Oil Control Board, the Petroleum Board worked as a pool running its members' installations, stocks, vans, etc., and selling unbranded oil products from unmarked outlets in a limited number of standard grades, members sharing the trade in proportion to their market share. As a consequence of this pooling, marketing became relegated to a subordinate role during the war, in Britain and elsewhere, resulting in a sharp reallocation of advertising expenses. From 1939 to 1943, the advertising budget for Europe, Africa, and Asia was cut by four-fifths; only Shell Union's expenses remained level.[12] Andrew Agnew became the Petroleum Board's chairman. He had resigned as Group managing director in 1937 but, as managing director of Shell Transport and a director of the three operating companies, he remained a key figure within the Group.[13] To a considerable extent the close collaboration between the oil companies in wartime was built on foundations laid by the 1928 Achnacarry talks and subsequently elaborated in various agreements during the 1930s (see Volume 1, Chapter 7). Product swaps between companies helped to save tanker capacity, for instance, and the Group refineries on Aruba and Curaçao collaborated closely with Jersey's Aruba plant to maximize aircraft gasoline output.[14]

The British preparations for oil supplies in wartime were remarkable for their efficiency and for their informality. The government could rely on private business to provide the installations, stocks, product supplies, and administrative arrangements required and it achieved its aims without legislation or other forms of compulsion. Indeed, the Petroleum Board, which operated in effect as a commercial company with a semi-official monopoly, lacked a proper legal basis.[15] The United States likewise succeeded in mobilizing its huge resources by combining central coordination with voluntary cooperation from the industry, but then the country also started with a fairly comfortable lead over all others. The oil companies had already taken the initiative to build substantial facilities for manufacturing special products such as iso-octane. Shell Union's aviation manager, Jimmy Doolittle, had succeeded in persuading the board to follow up the company's

[8]

technical lead in developing iso-octane production with the construction of substantial manufacturing capacity, even though the market prospects remained dim at the time. In 1936 Shell Union already operated four units, at Houston, Martinez, Wood River, and Wilmington, capable of producing more than 6 million US gallons of iso-octane a year.[16] Other oil companies followed suit and by May 1939, the US oil industry possessed a total iso-octane production capacity of 5,790 barrels a day or about 300,000 tons a year, with Shell Union having a quarter.[17] America also possessed an enormous volume of crude oil in wells kept shut to restrict the overproduction that had dogged the industry since the late 1920s, plus a framework of administrative controls with which the Roosevelt government had effected the production restraints. Having fallen into disuse following a dispute about the legal basis of intervention under the National Recovery Act, President Roosevelt reinstated the framework of controls in May 1941, after declaring a national emergency for the United States. The Secretary of the Interior, Harold Ickes became Petroleum Coordinator for National Defense.[18]

However, with one notable exception, about which more below, the governments of other major countries did not take sufficient precautions to meet their projected wartime oil demands, whether they resorted to legislation or not. The French government, for instance, repeatedly imposed minimum stock requirements on the oil companies, without attaining the targeted volume deemed necessary for war. Nor did France possess an adequate administrative infrastructure to coordinate oil policy.[19] For too long the need to safeguard vital supplies such as iso-octane went unrecognized. Though the Group had considered building an iso-octane plant in France as early as 1936, government officials started pressing for such an installation only two years later. Construction at the Group's Petit-Couronne works began in 1939, too late to be of use.[20] The Mussolini government's tight regulations concerning the Italian oil trade, introduced in 1934 with the aim of gaining control over strategic resources, led to a prolonged stalemate with the oil companies which did little to improve the country's basic supply position (see Volume 1, Chapter 7). Italy also strove to limit its dependence on oil imports by acquiring the technology to make gasoline by hydrogenating coal or heavy oil, i.e. by adding hydrogen to them, without however achieving the desired end.

Japan passed legislation similar to the Italian measures in 1934, initially with the economic aim of stabilizing the economy by tight regulation of industrial sectors, but military reasons soon gained the upper hand.[21] As a consequence of the country's aggressive foreign policy towards China, the resulting stand-off between the two most important foreign oil companies active in Japan, the Group operating company Rising Sun Oil Company and Standard Vacuum (Stanvac), turned into a struggle over access to strategic resources such as oil and more specifically to high-octane aviation gasoline. More than three years of negotiations between company managers, diplomatic representatives from the US, the UK, and the Netherlands, and Japanese government officials failed to resolve the basic issues. When Japan moved to a full-scale war

[10]

One of Shell's wartime aviation heroes, Major General James Doolittle, is seen (right) just after the war's end with a group of US Army Air Force crew, and (below) starting his take-off run from USS *Hornet* on 18 April 1942, at the beginning of the historic raid on Tokyo by 16 B-25 bombers, which decisively affected Japanese naval strategy. Doolittle began working for Shell early in 1930, selling aviation fuel, and after the war was a vice president and director of Shell Union Oil Corporation (renamed Shell Oil Corporation in 1949).

[9]

Japan began the undeclared Second Sino-Japanese War (1937-45) by invading North China. Below, anxious Chinese people are seen watching a battle between Japanese bombers and Chinese fighter planes in 1938. As the conflict escalated, western governments began to fear the possibility of Japanese military action elsewhere, and (bottom) America stockpiled ammunition at Pearl Harbor in Hawaii.

against China in 1937 and rapidly conquered large parts of that country, the western governments began to worry in earnest about the likelihood of military action against other countries. Economic sanctions were considered, including an embargo on oil supplies, but postponed for the time being as likely to push Japan into securing her supplies by attacking the Dutch East Indies. As a precautionary measure, however, the State Department and the Foreign Office asked the owners of the worldwide hydrogenation patents, including the Group, to deny Japan access to this technology for making oil from coal.[22] When the United States imposed an embargo on the export of high-octane aviation fuel to Japan, the Japanese government began to put pressure on the Dutch government to increase oil supplies from the Dutch East Indies, including 100-octane gasoline, which the Group had just started manufacturing at its Pladju installation. In August 1940, talks opened in Batavia between a Japanese delegation and representatives from the Dutch colonial government, assisted by senior oil company managers. Van Eck attended for the Group, Fred Kay for Stanvac, and Agnew arrived in the US to represent the Group's position in the heated policy debate there concerning Japan: whether to continue supplying a country with hostile intentions in Asia and a potential enemy, or whether to provoke war by cutting supplies. In a subtle move designed to show resolution with understated force, Washington ordered its Pacific fleet, engaged in exercises around Hawaii, not to return to its Californian home base but to remain posted at Pearl Harbor.[23]

By contrast, Germany chose a radically different approach to the problem of oil supplies in wartime. The Hitler government which came to power in 1933 had given a powerful boost to the production of synthetic gasoline by hydrogenation, which IG Farben had championed since the 1920s (see Volume 1, Chapters 6 and 7).

[11]

[12]

Like Shell, Esso in wartime rapidly lost control of its Romanian assets. Below, on 1 January 1940, a consignment of oil bound for Germany is awaiting departure from the marshalling yards near Ploesti.

[13]

Another of Shell's wartime aviation heroes, Douglas (later Sir Douglas) Bader, lost both legs in an air crash late in 1931. Invalided out of the RAF in 1933, he joined Asiatic Petroleum's Aviation Department, and was able to rejoin the RAF in 1939. His wartime exploits earned him many awards and an international reputation: as a prisoner of war he was treated with great respect by his captors. Afterwards he returned to Shell, becoming manager of Group Aviation Operations in 1952 and first managing director of Shell Aviation in 1958. He retired in 1969.

As a result, synthetic gasoline already covered nearly a third of oil consumption in 1939, oil production from indigenous wells bringing self-sufficiency up to 49 per cent. Four years and a massive investment programme later, the percentage had risen to an impressive 75 per cent.[24] The rest came from Romania and from oil stocks captured in conquered countries. Germany did not adopt iso-octane for raising the octane number of synthetic gasoline, presumably because the country lacked the necessary installations to produce it, but relied on adding benzol and tetraethyl lead (TEL), produced by IG Farben under licence from Jersey Standard, to counter engine knock.[25] This restricted the octane number of gasoline produced to 87, because of a technical limit to the addition of lead, which in turn curtailed the performance of *Luftwaffe* combat aircraft.[26] As early as 1938 high-octane gasoline and anti-knock additives were in such short supply that the government lowered the standard for gasoline used by private motor cars to 74 octane.[27] Hermann Goering, who was both commander of the air force and in charge of the four-year plan to raise synthetic gasoline output, accepted the importance of 100-octane gasoline only in 1938, when it was already too late to start producing it in sufficient quantities.[28]

Interestingly, Germany did possess the means to overcome the structural disadvantage suffered by the *Luftwaffe* in the form of the turbojet engine, but failed to capitalize on it. Conceived almost simultaneously by four different engineers independently of each other, jet engines reached the stage of static running tests in Britain and in Germany during 1937, within months of each other. Backed by the Heinkel aircraft works, jet development at first proceeded rapidly in Germany. In 1939 the world's first jet-powered aircraft took to the air and in 1941 Heinkel began testing a prototype twin-engined jet fighter, which proved to be far superior to any aircraft the Third Reich possessed. Bottlenecks in bringing jet engines into series production then slowed down development for some three years. The first operational jet fighter, the Messerschmitt Me 262, began entering Luftwaffe service only from July 1944, too late to turn the tide.[29] In Britain, the Group provided a crucial breakthrough when, in 1940, jet engine development stalled on interminable difficulties with fuel and combustion. A combustion chamber designed by the Fuel Oil Technical Department of Shell Mex & BP, the UK marketing joint venture with Anglo-Iranian, solved the problem, enabling the first British jet-powered aircraft to fly in 1941. The RAF started flying Gloster Meteor jet fighters also in July 1944.[30]

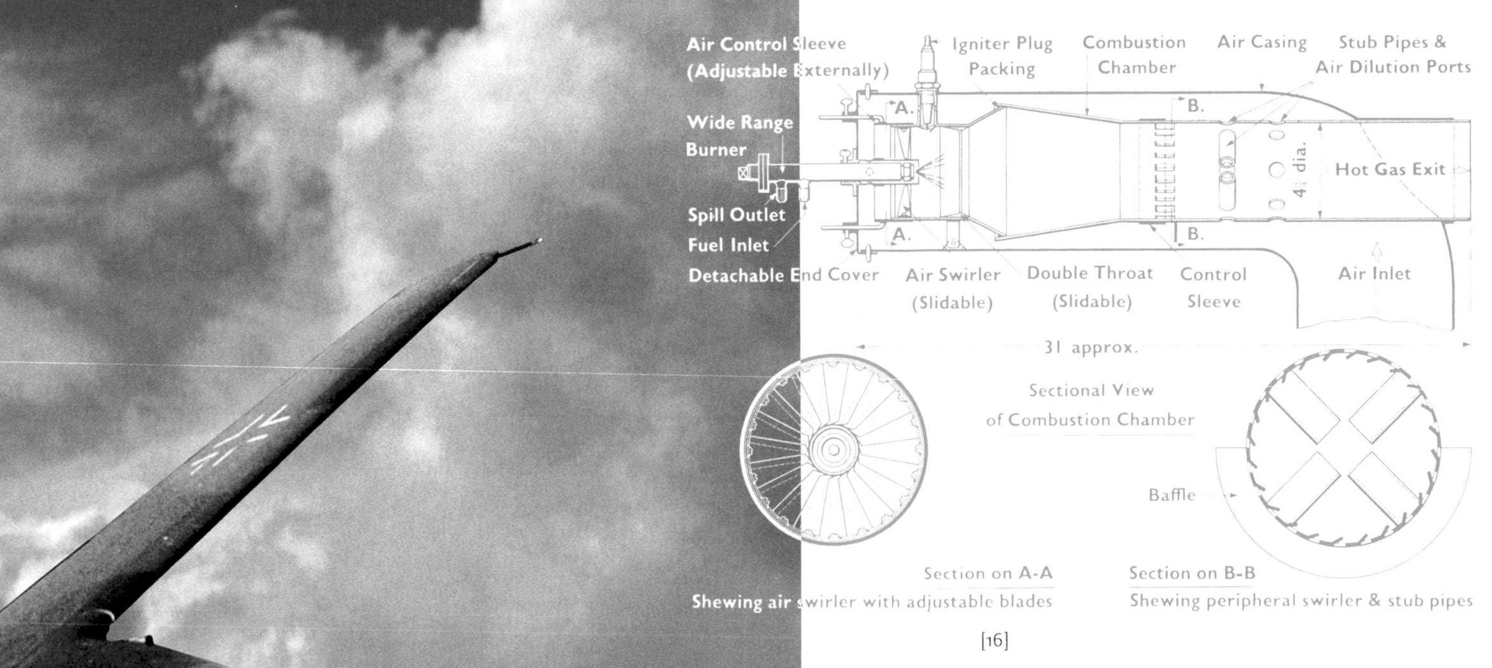

[16]

Clockwise from left, the main picture shows a Messerschmitt Me-262 jet fighter. In the race to develop effective jet engines, the Shell-designed combustion chamber (above) overcame problems experienced by the British designer Air Commodore Frank (later Sir Frank) Whittle, seen below right explaining a model engine. The inset shows one of the RAF's Gloster Meteors being serviced.

[17]

[18]

Finally, coercion loomed large in Nazi oil policy. If private business resisted official policy, for instance the building of hydrogenation works, officials resorted to force. When, in 1938, the Group refused to raise its participation in the Pölitz hydrogenation works to accommodate a substantial expansion of capacity ordered by the *Reichswirtschaftsministerium*, ministry officials threatened to take Rhenania-Ossag, which held the shares, into administration if managers did not comply (see Volume 1, Chapter 7). Other companies had to be goaded into taking part in hydrogenation ventures by similar means.[31] The government also imposed strict controls on the retail trade in oil products, setting standards, laying down numbers and locations of service stations, restricting the private use of fuel oil, and finally binding all oil companies to a semi-public body, all before the war had broken out.[32] In January 1940, the Group lost control over its German business when the government appointed a *Verwalter* or caretaker manager to take charge of Rhenania-Ossag.

These very different starting positions determined to a large extent the outcome of what followed. The Allied countries had ample access to oil and could bring an overwhelming superiority of manpower and resources to bear on getting supplies of the kind and in the volumes required. To them, oil posed a logistic problem, as it had during the First World War. This could be formidable at times; as the American General George Patton famously put it to the Supreme Allied Commander General Eisenhower when he ran out of fuel in 1944, 'my men can eat their belts, but my tanks have gotta have gas.'[33] Even so the problem was not an impossible one. By contrast, two of the three Axis powers, Italy and Japan, would face a critical situation if war cut their overseas supply lines. The oil supply position of the third, Germany, looked far better, but it was

[19]

[20]

actually rather stretched from the very beginning. For all the impressive speed of a few armoured columns during the *Blitzkrieg*, the German army remained partially mechanized and heavily reliant on railways and horse-drawn transport.[34] Oil production continued to lag badly behind the ludicrously optimistic projections and never came close to meeting the actual demand of the armed forces, thus posing a check on further mechanization. As the war progressed and mobility became the decisive factor, the German army came to rely more and more on horses, not on motorized transport.[35] Any cuts in production would immediately have disastrous repercussions, as the future was to show.

Early in 1944 the Allies began an all-out effort to destroy the Luftwaffe. Below left, a British bomber crew approach their Armstrong Whitworth Whitley aircraft; centre, on 12 May 1946, after the war, this HE-162 factory – underground in a salt mine – was found with aircraft still lined up, abandoned during construction. Right, from its bases in southern Italy, the US 15th air force destroyed this German factory manufacturing ME-262s in a raid in February 1944.

Supplies of motor fuels and lubricants were a chronic problem for Germany during the Second World War, and led to a constantly increasing use of animal power. Even so, an armed man on horseback could easily intimidate unarmed pedestrians. In October 1938, in Waldheusel in the annexed Sudetenland, German-speaking Czechs salute Nazis riding by. They might almost have been saluting medieval warriors rather than 20th-century soldiers.

[23]

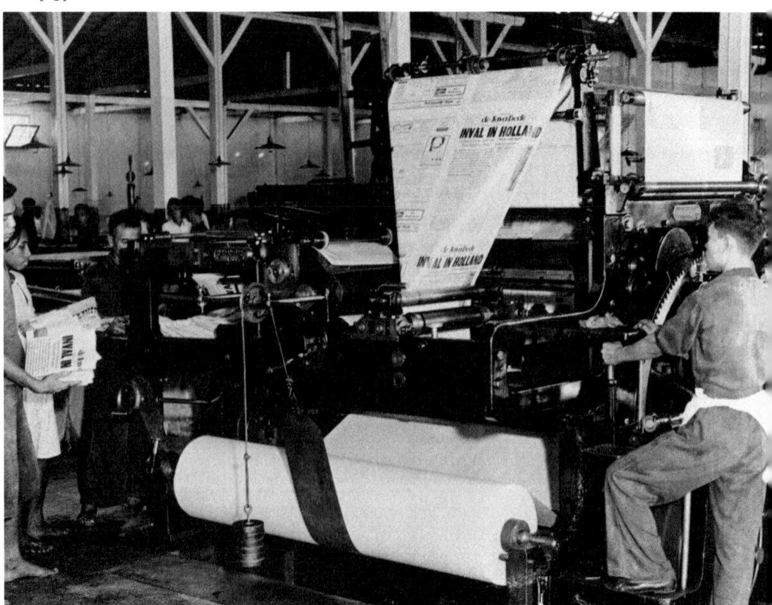

Defending the business As for the Group's own war preparations, Bataafsche had been involved with defence issues in the Dutch East Indies from the early 1920s, when the colonial government started worrying about threats to the outlying districts of the vast archipelago, such as Tarakan and Balik Papan.[36] The discussions with the military authorities about drafting plans for destroying oil stocks and installations led Bataafsche to take a decidedly contradictory stand. The company wanted a strong colonial defence, but refused to face the consequences, i.e. prepare plans for the evacuation and demolition of its installations to deny the enemy the use of vital resources. The Hague permitted the Batavia general manager to give the authorities oral information, but nothing written, about general arrangements on the various installations, so the army could draft its own plans, but no more. Similarly, Bataafsche managers militated against the deleterious effects of deep spending cuts on the readiness of army, navy, and air force, but fought tooth and nail against higher taxes to pay for increased defence spending; they pressed for vigorous action, and complained when government departments received comprehensive powers to deal with economic and military emergencies. The arrangements at the Pernis installation also reflected Bataafsche's general attitude towards economic defence. Company employees were not to engage in any demolition activities but were to leave this to army troops admitted to the installation for the purpose on giving a special password. When the signal came, the Pernis staff would do no more than mix the available product stocks together so as to make them unsuitable for direct use.[37]

The principled refusal to collaborate originated from four considerations. First, the issue of information and command. Staff would not necessarily have the military intelligence required to assess the situation correctly and proceed with the demolitions at the right time, that is to say, neither too early nor too late. This required setting up a chain of command including provision for a rupturing of the chain. Second, soldiers engaged in demolition works had the cover of the Geneva Convention, but company staff ran the risk of being shot as saboteurs. Third, cost, presumably reinforced by doubts about the validity of compensation claims for any damage inflicted if the company willingly collaborated in planning destruction; and finally, an aversion to endangering the continuity of the business under any circumstances.[38]

As regards cost, in preparing the defence of the Dutch East Indies, the cradle of the business, Bataafsche took the same commercial attitude as in discussions with the government officials in Japan, Italy, France, or Germany about increasing stocks. This was summarized in a January 1940 instruction: 'the company cannot be expected to bear the financial burden of maintaining non-commercial stock and cannot be expected to make non-commercial investments without proper remuneration'. Or, as one telex from the same month put it, 'we are anxious to assist the government whenever possible without great inconvenience or expense to ourselves', in other words, the company came first, the country second. The other main producer in the Dutch East Indies, Standard-Vacuum's Nederlandsche Koloniale Petroleum Maatschappij (NKPM), had a similar attitude to making defence arrangements.[39] The reader may remember from Volume 1, Chapter 3, that the Group nursed this kind of pragmatism concerning wartime operations during 1914-18 as well. Inflicting lasting damage on fields and installations would be too

[24]

A way of life is about to end. Anti-clockwise from below, BPM's head office at the Koningsplein in Batavia, about 1940; left, Dutch diners eating out in Java, January 1940; far left, a newspaper printing press in Batavia churns out the shocking news of the invasion of the Netherlands.

[25]

[26]

<u>Urgent</u>

In view of difficult communication to and from Holland you are undoubtedly considering advisability moving offices of Hague to London at appropriate moment

Please confirm and advise so that we can take all necessary steps here to facilitate such a transfer stop

Colleages would all welcome you here

KESSLER VAN ECK

inconvenient and too expensive. Consequently, when the oil companies, including Bataafsche, discussed the destruction of oil and oilfields with the government's *Staatsmobilisatieraad* or Mobilization Council in July 1940, they persuaded the council to set a limited target, putting wells and installations out of action for a period of six months, instead of for a year or even indefinitely, as the council had originally wanted. The demolitions would now be carried out by company teams, assisted if and when necessary by army explosives experts, a change of mind perhaps influenced by the experiences being gathered in Europe.[40]

A direct involvement with defence policy such as Bataafsche had in the Dutch East Indies remained very much the exception, however. Group companies tended to concentrate on passive rather than on active defence. The boards of the main operating companies must have debated taking precautions at least from 1936, when the Air Ministry approached Anglo-Saxon about the iso-octane plant, but practical consequences began to appear only from 1938. The measures taken included bringing as many Group tankers as possible under the British flag, strengthening the upper deck of tankers to carry guns, and building air raid shelters at installations and office buildings.[41] To minimize the likely impact of bomb damage on staff and operations at central offices, the departments of St Helen's Court were split into three groups, the essential ones remaining there, others being transferred to Teddington, west of London, near the Group's Lensbury Club sports facilities. Some departments were moved further out; the Marine and Marine Accounts departments went to Plymouth, for instance, and Anglo-Saxon also bought flats in Bournemouth for unspecified purposes. From September 1939, Anglo-Saxon board meetings took place at the Lensbury Club. The Shell Transport board relocated to a house in Newbury, Berkshire. Telexlines connected the various sites with each other.[42]

To ensure continuing close coordination between central offices, The Hague prepared plans for a gradual transfer to London of core staff with the papers needed to keep the various business functions going. This operation started in April 1939 with the selection of a core staff from the total of some 900 working at The Hague central office, who would form what became known as the Hague Party. Led by W. H. Oosten, this team came over in four groups, one in August 1939, the next in February 1940, when a German invasion seemed imminent, the third one in April, and the last one on 14 May 1940. During June and July, the team was reinforced by Dutch staff from Belgium and France, who had managed to escape to the UK. In January 1941, the total staff numbered around a hundred people. Royal Dutch and Bataafsche set up offices in Teddington, in and around the Lensbury Club.[43] The two boards also made arrangements to keep the companies and their main effects out of enemy hands should Germany invade the Netherlands. During 1939, all shares in operating companies and other paper effects were transferred from the Netherlands to safekeeping with London or New York banks.[44]

TRANSFERS TO CURACAO

Royal Dutch Petroleum Reports 'Freedom' in Operations

The Royal Dutch Petroleum Company yesterday notified shareholders here that the domicile of the company, together with all its Netherlands subsidiaries, had been transferred from The Hague to Willemstad, Curaçao, Netherlands West Indies. The company made known that "it is conducting its business free from all control by the enemy and all its resources are on the side of the battle for freedom."

The articles of association of the Royal Dutch Company and these subsidiaries have been altered to authorize the change in domicile according to the provisions of a law promulgated in Holland on April 26 last year.

[27]

On 14 May 1940, four days after the invasion of the Netherlands, Kessler and Van Eck urged De Kok and other colleagues to evacuate to Britain (far left). But De Kok remained in The Hague until his premature death, and as readers of the *New York Times* learned belatedly in April 1941 (left), Royal Dutch's legal seat was carefully kept out of Britain.

Together with other companies facing the same problem, such as Philips, Unilever NV, and the big international shipping lines, the Royal Dutch and Bataafsche boards sought ways to bring the companies themselves into safety. Philips and Unilever NV transferred the ownership of their foreign assets to overseas trusts set up for the purpose, but the shipping companies, Royal Dutch, and Bataafsche preferred to use special legislation enacted on 8 May 1940 to transfer the legal seat of companies to the overseas territories of the Kingdom of the Netherlands in case of an emergency.[45] Two days later, on 10 May, Royal Dutch, Bataafsche, and a string of other Group companies officially moved from The Hague to Curaçao, where A.S. Oppenheim, the head of the legal department, had prepared a formal company office in February. On that fateful day he sat at his desk, together with a notary public, knowing that the Netherlands were at war and waiting for the telegram instructing him to effect the transfer.[46]

Why did the Dutch Group directors opt for this particular solution and why did they choose Curaçao? With an annual processing capacity in 1940 of 8 million tons, 56 million barrels, in 1940, the refinery there was the Group's biggest overall and the third biggest in the world, after Anglo-Iranian's Abadan plant and Jersey's installations on Aruba; but from the viewpoint of management the island remained an outpost. As Oppenheim himself acknowledged, his Curaçao office was a façade.[47] There was no intention to let it do more than to convene shareholders' meetings and perform other acts required under the law and under the articles of association. Such acts assumed a purely ritual character, because under the circumstances no shareholders, directors, or managers could think of travelling thousands of miles to attend meetings or sign financial statements. In July 1940, Group managers sent Van Eck to the United States so the company would have a senior manager outside of the combat zone should Britain also be invaded by Germany. Van Eck set up a representative office for the Group in New York and remained stationed there until the end of the war, as if to highlight the fact that Curaçao was no more than a paper presence.[48] Batavia might have been a more logical choice. The shipping lines which had their main operations in Asia transferred their seats to Batavia and for some time Bataafsche had faced calls to show a greater commitment to the Dutch East Indies, for instance by raising the position of the general manager to board level.[49]

Moreover, the solution chosen left the Group's assets controlled from the Netherlands dangerously exposed to enemy action until the very last moment. By contrast, Philips and Unilever NV had the trusts for their overseas assets safely in place during the summer of 1939. Managers of the big Dutch multinationals held regular meetings to discuss joint political, legal, and financial problems and such an important subject as the legal preparations for war will have figured at those meetings. Both Philips and

Royal Dutch Petroleum Company

N. V. KONINKLIJKE NEDERLANDSCHE MAATSCHAPPIJ

Tot Exploitatie van Petroleumbronnen in Nederlandsch-Indie

THE ROYAL DUTCH AND ALL ITS DUTCH SUBSIDIARIES

including the

N. V. DE BATAAFSCHE PETROLEUM MAATSCHAPPIJ

transferred their domiciles from The Hague to Willemstad, Curacao, in May 1940 on the invasion of Holland by Germany. Shareholders will be glad to know that the Company is conducting its business free from all control by the enemy and that all its resources are on the side of the battle for freedom.

The transfer of domicile, enabling the Company to escape all enemy influence, was made in virtue of the law promulgated in Holland on April 26th, 1940, the relevant clauses of which read as follows:—

ARTICLE 1.

1. By making a corresponding amendment in the Memorandum of Association the place of domicile of a limited company domiciled in the Netherlands, Netherlands-Indies, Surinam or Curacao may be transferred to another part of the territory of the Kingdom of the Netherlands.

2. The management of the company is authorised to amend the Memorandum of Association accordingly: the consent or cooperation of the general meeting of shareholders or another body of the company is not required for this purpose. If one or more managers—respectively two or more managers jointly—are entrusted with the management of the company, each of these managers—respectively the managers jointly—are authorised to amend the Memorandum of Association.

The Articles of Association of the Royal Dutch and the B.P.M. have been altered according to the provisions of the law.

The official announcement of the transfer of Royal Dutch's legal seat to Curaçao.

Bataafsche formed part of a commission which prepared draft legislation for the transfer of company seats. The Dutch Group managers will therefore have been aware of the trust construction chosen by Philips and Unilever NV. Why then did they not do the same and invest the Royal Dutch and Bataafsche assets in a British trust managed from London? After all, for both Royal Dutch and Bataafsche, St Helen's Court and Teddington became the centre of all operations. Moreover, the Group had made an arrangement with the British Treasury, probably in September 1939, to have Bataafsche treated as a resident in the UK and thus subject to that country's foreign exchange regulations.[50] This was a logical consequence of Anglo-Saxon's position as Group treasurer; it meant that all Bataafsche's foreign currency earnings flowed to the Treasury, which in return undertook to supply the Group with any currency needed, for instance dollars to buy equipment in the US. The arrangement also covered Curaçao's unusual position, earning mostly sterling from oil exports, but needing dollars for its purchases. Keeping the Group within the sterling currency area was a vital interest for the UK, which had great difficulty in finding dollars to pay for essential imports. For the Group the arrangement was beneficial overall, though disadvantageous for some parts of the business, notably those in the Dutch East Indies.[51]

The answer probably lies in issues of management and control. The trusts effectively made the Philips and Unilever NV assets controlled by them subject to British and, in one case, American law, with the trustees exercising the management over them according to the best interests of the shareholders in those assets. When moving the company seat, the assets remained subject to Dutch law, however, and managers did not have to contend with outsiders who might have a different opinion about the interests of shareholders. In addition the move to Curaçao was probably intended as a statement, to emphasize the Dutch character of Royal Dutch and Bataafsche. Unilever NV made no bones about making Unilever Ltd. in London the focus of all its operations; Royal Dutch's choice for Curaçao suggests that managers felt uncomfortable at the prospect of relying too much on London and sought leverage by establishing an Archimedean point on Dutch territory. This must have been what annoyed Godber in writing to Van Eck: not the transfer itself, but the choice of destination. In the overcharged atmosphere of the war, such questions of nationality would assume ever greater proportions, only to suddenly melt away in 1945, as we shall see.

The transfer was not an unqualified success, either. The *Reichskommissar* A. Seyss-Inquart, the highest civil authority in the Netherlands during the German occupation, picked on a technicality to declare it null and void. Consequently the *Verwalter* appointed to manage Royal Dutch and Bataafsche had full powers to act on behalf of the concern in occupied Europe and in the countries allied to Germany, such as Romania. The *Verwalter* lost no time in appointing a new general manager, more amenable to the German demands, at Astra Romana, the Group's operating company there and that country's biggest oil producer.[52] The evacuation of the management and core staff from The Hague also had the unfortunate side effect of leaving the office and the rest of the employees there without proper guidance. It had apparently been agreed that the Royal Dutch Director-General and de facto Group CEO Frits de Kok would accompany the last staff and papers to leave the country, leaving fellow director J. M. de Booy in overall charge of central offices in The Hague, but on the afternoon of 14 May De Kok, already terminally ill, persuaded De Booy to go in his place. That evening De Booy caught a fishing boat leaving Scheveningen for Britain, one of the very last escape opportunities.[53]

[29]

Part of the crowd that gathered outside Royal Dutch's head office in a spontaneous display of sympathy at De Kok's funeral.

In September 1940 De Kok slipped discreetly out of the offices, telling only one person that he needed to rest. He went directly into hospital and did not return. De Kok died on 28 October at the age of only fifty-eight, and the news of his death created a spontaneous show of public sympathy in The Hague, best understood as an expression of mourning both for the man and for a national past, brutally ended by the German occupation, in which his long-distance flights had captured the imagination. A large, silent crowd gathered in front of the Royal Dutch office in the Carel van Bylandtlaan to pay its last respects, and an even larger number of people attended the funeral a few days later.[54] Following De Kok's death N. van Wijk, the Royal Dutch company secretary, became the senior manager of the rump in The Hague and the *Verwalter* Hauptmann Eichardt von Klass appointed him a director.[55] The Dutch companies in exile now faced the very difficult problem of appointing a successor, with all the normal rules suspended. As in 1936, Kessler was the obvious candidate, and again he did not make it. To avoid antagonizing the British managing directors further, no successor was appointed.[56] However, as we shall see, the Group managing directors soon clashed over other appointments.

Another measure taken following the German invasion of the Netherlands was the dismissal of German staff and other employees considered to be of doubtful political allegiance, such as Dutch Nazi party members, Austrians, and Romanians. Whether Anglo-Saxon and Asiatic had dismissed Germans and other enemy aliens from their overseas operating companies in September 1939 we do not know; at any rate Bataafsche had not done so. In April 1939, Bataafsche suspended some people with Nazi sympathies working at central offices in The Hague, presumably to quell unrest amongst the rest of the staff.[57] No further steps appear to have been taken until mid-May, when reports began reaching London about employees being arrested in the Dutch East Indies, Surinam,

The invasion of the Netherlands forced the Group to consider how to treat its employees of enemy nationality. Below, its Argentine company Diadema requests advice; below right, the reply from Kessler and Van Eck; right, confirmation from De Booy.

FILE Personnel 5

19 IX 0

L 11

CEP

Telegram SENT to NLT PERSONAL VANGOETHEM BUENOS AIRES
from MR.DEBOOY
No. of Words No. 18323 Sent on 20.5.40

See L. 10 21/5/40 from Buenos Ayres.

OOSTEN L 11 View present conditions we have decided to suspend immediately from all duties and notwithstanding any inconvenience all employees of enemy birth whether naturalised or not and you are in addition carefully to scrutinise all European and native staff other nationalities in line with our L 8 whether they are suspect of enemy sympathies stop

In case you wish to propose exception to decision relating to naturalised employees of enemy birth cable fully stop

Your recommendation as to temporary pay or compensation must be forwarded soonest but your action must be immediate

DE BOOY

B - re German Staff

---

2 L.4.

VS Telegram RECEIVED from BUENOS AIRES
to LC MR.KESSLER
No. of Words 45 No. 18754 Despatched at on 15.5.40
Received on 16.5.40

L-4 It becomes very difficult to keep several employees of German nationality or descent in our service fullstop

Can we dismiss when according to our judgement no objection from Argentine point of view

VANGOETHEM

See L.6 17/5/40 to Buenos Aires.

O.F

4

COPY

MC Telegram SENT to NLT PRIVATE VANGOETHEM DIADEMA ARGENTINA BUENOS AIRES
from MR KESSLER MR VANECK
No.of Words No.18122 Sent on 18.5.40

L 8 Experience Netherlands has shown presence NSB members of former members caused unrest in organisation and unless Government raise objection wish discharge all such members of former members NSB of whose loyalty you not absolutely assured beyond any possible doubt telegraph us names of employees grouped under both categories

KESSLER VANECK

file G 8 D.P. / G 1 D P.

[30]

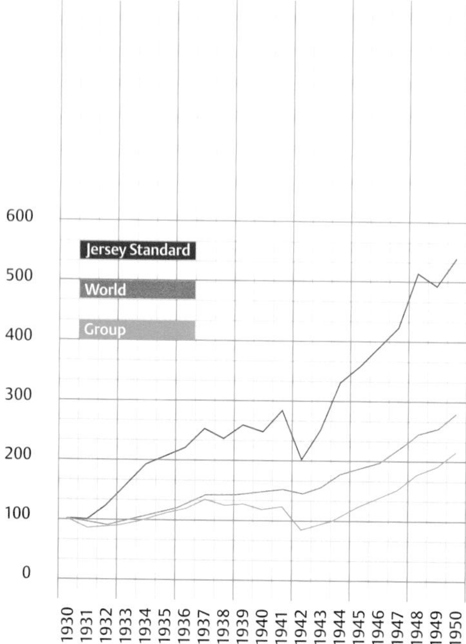

Figure 1.1
Crude oil production indices, World, Jersey Standard and the Group, 1930-1950.

and Curaçao, and about serious unrest amongst the staff of Diadema Argentina caused by the presence of twelve Germans working for that company. On the 20th of May, London central offices instructed all operating companies to dismiss employees of enemy birth and to scrutinize all other nationalities for their political allegiance.[58] The instruction appears to have been interpreted very differently around the world. Managers in the Dutch East Indies took a rigid line, no doubt because the recent defeat by Germany rankled; others acted more cautiously. When Italy entered the war in July 1940, a new complication arose. Dismissing Germans had already raised a small storm for Diadema and the company could simply not start to consider similar action against staff of Italian descent. In the end nothing was done. The Group fired a total of seventy employees following the instruction from London, from a total of around 190,000 staff.[59]

Becoming a crude-short business The war changed the Group's position in one fundamental respect: from being able to generate sufficient production to meet its needs, the Group turned into a crude-short business. In other words, the company consistently sold more oil than it managed to produce, the difference being made up by buying crude or products as needed. This situation developed from the drop in production which started with the Mexican nationalization in 1938, followed by the loss of control over Astra's Romanian production in 1940 and the Japanese occupation of the Dutch East Indies, Sarawak and Brunei two years later (Figure 1.1). By then the Group had lost a total of 37 per cent of its 1937 output. In 1943 its production picked up, but it did not regain the 1937 mark until 1946. As a consequence, the Group's share in world production fell markedly, and Jersey Standard, despite suffering setbacks of its own, took a huge lead. In 1937, the companies had both produced slightly over 30 million tons (210 million barrels), but in 1945 Jersey produced 52 million tons, (365 million barrels) while the Group remained on 30 million tons. The gap between the two companies started narrowing again only during the later 1940s.

Jersey's surging crude production did not translate into a proportional lead in output, however, because refinery runs did not rise to the same degree, so during the war the company simply switched from purchasing crude to producing from wells shut in during the Depression. Conversely, the Group increased its purchases of crude and products, but in the absence of figures for crude runs or for oil purchased we do not know by how much. The initial volume at any rate was considerable; during 1939 Asiatic Corporation in New York, the Group's American trading arm, bought 3.5 million tons of products, mostly gasoline, for sale to Group companies, a rise of 36 per cent over 1938. Purchases in the US appear to have stabilized after that.[60] The overall supply situation must then have eased considerably, the German advance on the European continent leaving the Group with some 6 million tons outside Europe which it could no longer sell to the countries now under the Nazi regime.

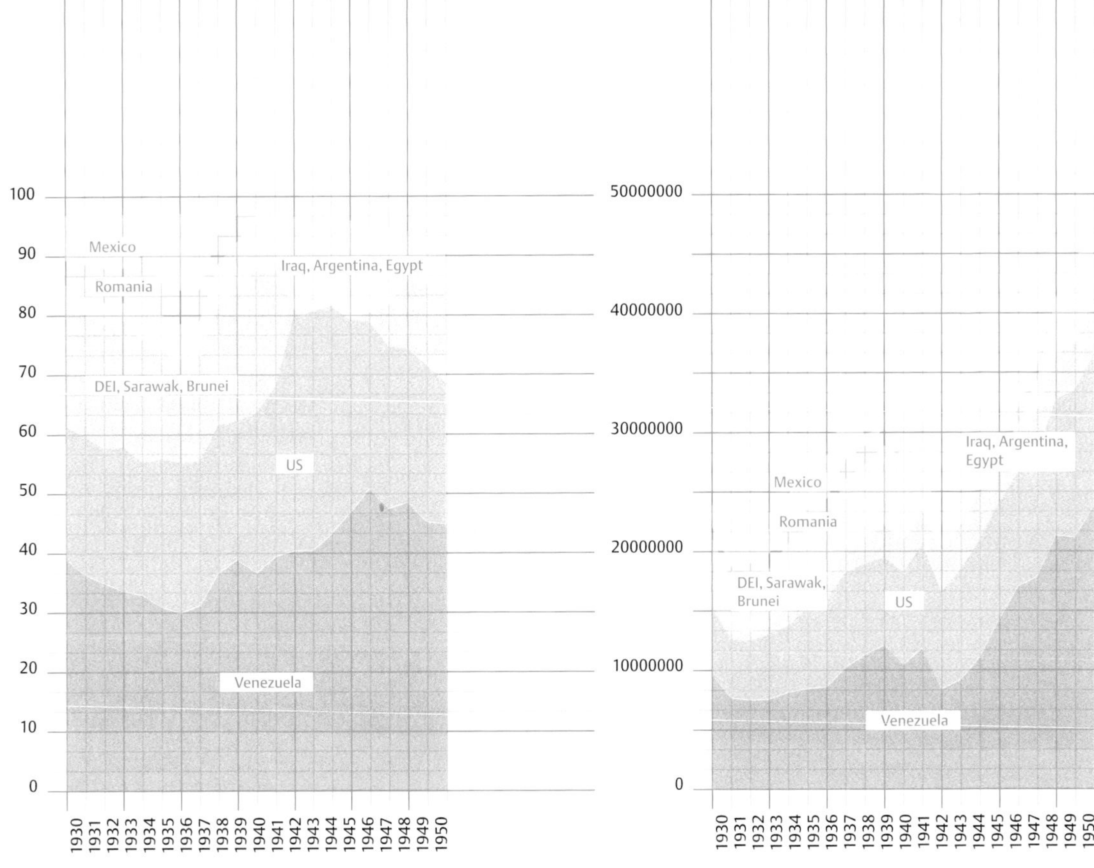

Figure 1.2
The Group's main areas of production in percentages of total production, 1930-1950.

Figure 1.3
The Group's production by main area in metric tons, 1930-1950.

Thus, during the Second World War the Group to some extent suffered from the same problem as during the First, that is to say it lost a very large share of its production, but with a very different outcome (see Volume 1, Chapter 3). Firstly, the availability of supplies previously sold on the Continent meant that the Group did not have the acute crunch from which it suffered during 1914-18; secondly, after the war managers accepted the crude shortage as a continuing fact of life. Though the Group embarked on a vigorous expansion of production, as had happened following the First World War, it did not regain a similar level of self-sufficiency as during the 1920s. Consequently, the old competitive positions were reversed. Shell had derived its original strength from a global spread of production, which Walter Teagle set out to match for Jersey Standard during the 1920s and 1930s. Jersey had now overtaken the Group in that respect, but this did not turn into a competitive disadvantage for Shell because the opening of new production areas in the Middle East coupled with the development of global crude and product markets offered plentiful supplies. The war marked a break with the past in other respects as well. Production at Astra Romana was restored quickly, only to be lost again through nationalization in March 1948. It took a very long time before the operations in Indonesia had regained their former importance, the four-year struggle for Indonesian independence between nationalists and the Dutch colonial government blocking access to oil fields and hampering reconstruction work. Brunei and Sarawak recovered much faster and together produced more oil than Bataafsche's old home ground throughout the 1940s and 1950s. So between 1938 and 1948 the Group lost two very valuable production areas from its pioneering days, Mexico and Romania, and struggled to hang on to a third, the Dutch East Indies.

The loss of supplies from these three areas increased the Group's already heavy dependence on its two main producing areas, Venezuela and the United States, which together supplied about 80 per cent of its crude between 1942 and 1946 (Figure 1.2). Percentages are slightly misleading here, however (Figure 1.3). In 1945 Venezuelan production was only 200,000 tons higher than it had been in 1939. Supplies from Venezuela fluctuated heavily and actually fell in 1940 and in 1942, largely as a consequence of German submarines threatening the ships ferrying crude oil to Curaçao. The characteristics of Venezuelan crude, predominantly a very heavy type, also limited the scope for increasing production there. The war demanded gasoline and most of all aircraft fuel, one ton of bombs dropped on Germany requiring three tons of high-octane gasoline for delivery.[61] Processing Venezuelan crude into gasoline yielded large quantities of fuel oil and asphalt, which piled up in Curaçao because the demand for these products lagged far behind that for gasoline. Pits dug to hold the surplus asphalt contained 2 million tons of it when the hostilities ended. As a result of these two bottlenecks, production in Venezuela and manufacturing in Curaçao both remained below their capacity throughout the war.[62]

By contrast, the Group's production in the United States between 1939 and 1945 increased from 6.8 million tons to 9.8 million, or from almost 48 million barrels to 70 million. This rise of 45 per cent comfortably exceeded the overall rise in US production of 30 per cent.[63] Most of the new crude supply came from Texas, which in 1944 overtook California as the Group's biggest producing area in the US. Moreover, production capability in some of the Californian fields, notably those around Los Angeles, was diminishing. Intense E&P operations did not result in significant discoveries there during the war, whereas Shell Union did find new and important crude sources in Texas and Kansas.[64] Jersey's surge

Shell Oil's processing plant in Sheridan, Texas, received wet gas from nearby wells. The columns from left to right respectively extracted the usable hydrocarbons – propane, butane mixture, isobutene, isopentane and ethane. The ethane was not for use, but had to be separated to allow the fractionation of some of the other gases. Along with some reject methane, it was then mixed with the dry gas and pumped back in, re-pressurizing the field. In the still at extreme right, natural gasoline was separated from the absorption oil.

[31]

in production stemmed almost entirely from the US, where its production almost doubled during the war. The US share in Jersey's total production rose from about a third during the 1930s to nearly half by 1945.

Supplies from four other, smaller, producing countries, Iraq, Egypt, Argentina, and Trinidad, helped to lessen the impact of lost production elsewhere. Their contribution to Group supplies almost doubled between 1937 and 1945, from 2 million tons (14 million barrels) to 3.7 million tons (26 million barrels), adding up to more than 12 per cent of the Group total in that last year. Of these four, Iraq had the most potential, but this failed to materialize. The Iraq Petroleum Company (IPC), in which the Group had a 23.75 per cent stake, had finally started production in 1934; by the end of the 1930s, the Group received just over 900,000 tons of crude (6.3 million barrels) annually from the company. Two pipelines connected IPC to the Mediterranean, a northern one to a refinery near Tripoli, then in Syria (now Lebanon), and a southern one to a refinery built at Haifa, Palestine (now Israel). This latter one, jointly owned by Anglo-Iranian Oil Company (AIOC) and the Group, came on stream in December 1939 with an annual capacity of 2 million tons (14 million barrel), which was doubled during the war.[65] When Italy entered the war in June 1940, IPC curtailed production because its oil could no longer reach European markets. The Mediterranean was now closed to Allied shipping and the trip around the Cape made the oil uncompetitive, forcing the IPC shareholders to develop outlets for their products close by. At first both pipelines were shut, and though the southern one reopened after five months, political turmoil in Iraq then almost halted production. During 1941, the Group received no more then 320,000 tons (2.2 million barrels) from its IPC share. Production regained its prewar level only in 1944.[66]

Egypt, on the other hand, showed a remarkable resurgence. Production there had declined throughout the 1930s, but in 1937 Anglo-Egyptian Oilfields Ltd., in which the Group had a 32.2 per cent stake, discovered the Ras Gharib oilfield, which came into production the following year. As war approached, production at Ras Gharib and at the Group's other main production site, Hurghada, was increased to the maximum possible, from 675,000 tons (4.7 million barrels), in 1939 to 1.3 million tons (9 million barrels) in 1945. The Suez refinery had up to date installations; just before the war, an old Dubbs cracking unit had been converted into a gasoline reforming installation and the plant also had facilities for making bottled butane gas.[67]

Trinidad was more favourably located than Curaçao to supply the UK, and became an important provider for the Allied war effort. In response, the Group operating company there, United British Oilfields of Trinidad, trebled production, from 250,000 tons (1.8 million barrels) in 1937 to 760,000 tons (5.3 million barrels) in 1945. The Royal Dutch report on operations during the war noted with regret that this had been considerably higher than would have been desirable for a rational development of the fields. The release of wartime pressures also manifested itself in a strike in the Trinidad refinery in 1946, which took a long time to settle.[68] The Group's production in Argentina did not rise nearly the same degree as in Trinidad, only increasing from 560,000 tons to 610,000 in 1945, which did little apart from enabling a reduction of imports and a consequent saving on transport.[69] Finally, exploration carried out by Bataafsche in the Netherlands under German occupation resulted in oil being discovered in the Schoonebeek field near Coevorden, which towards the end of the war was said to produce 25 tons per day.[70]

[32]

[33]

Oil fields in Trinidad (left) became an important provider for the Allied war effort, and to a European eye had a certain charm, with their storage tanks (pictured in 1952, far left) being roofed with palm thatch.

One particular problem for the Group in Latin America was the still smouldering dispute over the nationalization of the Mexican oil industry in 1938 (see Volume 1, Chapter 7), the outcome of which looked likely to determine the future shape of relations between governments and oil companies in the entire region and beyond. Bolivia had nationalized Jersey Standard's operating company there in 1937; Iran, Chile, Ecuador, Colombia and, more alarmingly, Venezuela, all appeared to be heading in a similar confrontational direction.[71] The foreign oil companies affected therefore wanted to press firmly for an annulment of the nationalization and property restitution. This point of principle covered a question of money, that is to say the valuation of oil reserves in the concessions held by the companies. The Group's operating company El Aguila, by far the biggest with nearly two-thirds of Mexican output in 1937, also had the largest concessions, including rights to the rich Poza Rica field, just on the verge of production. Considering it impossible to get what they considered a fair valuation for their concessions, the companies insisted on restitution; somewhat conceitedly confident of their superior know-how, they hoped that by holding out long enough, Mexico, unable to run the oil industry itself, would sooner or later beg them to restore production.

With concerted diplomatic support, perhaps the strategy might have worked. The British government protested immediately and vigorously, with the result that diplomatic relations between the two countries were broken off in May 1938. The American government, however, did not want to antagonize Mexico and would only support claims for compensation, not for restitution. The British protests and efforts to organize an embargo on oil from Mexico had driven the Mexican government into the arms of Germany, Italy, and Japan, all keen to get independent supplies; Japanese companies had already started exploration there. The outbreak of war in Europe further reinforced the Roosevelt administration's desire to remain on friendly terms with Mexico and keep the Axis powers out. The Dutch government took an intermediate position, recognizing the fundamental justice of the Group's claim for restitution, but unwilling to press it too hard and cause a rupture in relations. The Group disliked what it considered a weak attitude and preferred the staunch support of the British Foreign Office, but Dutch diplomats were in fact more realistic in their assessment that Mexico would never return the nationalized properties.

In the absence of united support, the oil companies made little headway in their talks with the Mexican government and their common front collapsed in November 1941, when the US signed a settlement with Mexico on the basis of a compensation payment to be determined by a joint commission of valuation experts. Protests from British and Dutch diplomats in Washingon that this set a dangerous precedent for Venezuela went unheeded. The

[34]

[35]

commission's report in April 1942 set the amount at a total of $30 million. Under pressure from the US government, Jersey Standard grudgingly accepted the settlement. The company was under heavy fire at home because of accusations that its pre-war involvement with the German chemical giant IG Farben had delayed access to the technology to make synthetic rubber; antagonizing the government over Mexico would mean being left out on a limb over Venezuela, where Jersey's interests were far greater than they had been in Mexico.

The Group's position now became untenable. Mexico had meanwhile joined the war on the Allied side and the US appeared likely to underpin the improvement of the bilateral relations with a comprehensive programme of economic aid, including an expert mission to report on the state of the oil industry and the possibilities for its reconstruction. There even appeared to be a chance that the American companies might be allowed back. In November, the Group accepted the principle of compensation as a stratagem, to create goodwill in the US and in Mexico, while privately considering the likely amount resulting from a valuation on the basis of the American settlement, estimated at $80 million, as beyond Mexican means and thus an insurmountable barrier to reaching an agreement. This was hardly an attitude conducive to smooth talks, and because the Mexican government showed a similar unwillingness to come to a quick solution, negotiations dragged on for another five years. In August 1947, the two parties finally reached agreement on an amount of $82 million plus interest since March 1938, making a total of $130 million, as compensation payment for the El Aguila shareholders. The Group considered the settlement 'very satisfactory'.[72]

The Venezuelan case was different in two respects. First, the government did not threaten outright nationalization, merely litigation over the alleged doubtful status of concessions in support of two demands: a comprehensive revision of all concession conditions including a substantial increase of royalty payments, plus a shift of refining capacity from Aruba and Curaçao to Venezuela itself. Second, diplomatic pressure played only a minor role the oil companies had to fend for themselves, mainly because the US government did not want to intervene. During the summer of 1941, the three major foreign companies operating in Venezuela, Jersey Standard, the Group, and Gulf Oil, began talks with government officials about a new petroleum law then being prepared which would enshrine the demands mentioned. As usual the

[36]

The opening of Shell's refinery at Cardón, Venezuela, on 7 May 1949. From left to right, the Minister of Labour Ruben Corredor (right) is pictured chatting with another guest, Graterol Roque; staff help the lady guests to choose gift scarves as souvenirs; and a group of guests leave the control room of Crude Distillation Unit no. 1 during their tour of the refinery area.

companies agreed to hold out together insisting on the inviolability of existing contracts; only a few years before, they had gone through a very similar cycle of legal threats and negotiations terminating in higher payments. The Group's general manager in Venezuela, B. Th. W. van Hasselt, knew that game intimately, having been the El Aguila manager until 1939.[73]

The oil companies were outflanked, however, when, after a year of negotiations, Venezuela approached Washington directly with an outline of the demands and the progress made so far. For the same geopolitical reasons as in the Mexican case, the State Department told the American companies to accept the proposed terms and to work out the details directly with the Venezuelan government. The Group had no choice but to follow suit, attempts to change individual clauses particularly onerous for its companies foundering on the State Department's push for a settlement. The agreement finally reached in 1943 awarded payments in royalties and taxes to the Venezuelan government amounting to a fifty-fifty split of net profits with the oil companies. Though a landmark in its later ramifications, this level was not entirely new; as we have seen in Volume 1, Chapter 7, NIAM, the joint venture with the Group set up to exploit the Jambi fields, gave the colonial government in the Dutch East Indies a profit share of 60-80 per cent. The new petroleum law did incorporate one innovation in the form of royalty payments in crude, enabling the Venezuelan government to start its own downstream operations. In addition, the foreign companies undertook to build export refineries in the country, for which the Group bought a site on the Paraguana peninsula. Construction of the Cardón refinery started in 1945 and it came on stream four years later. As part of the settlement, the Group agreed to limit the processing capacity of Curaçao to 200,000 b/d or 10.4 million tons a year.[74]

01|42

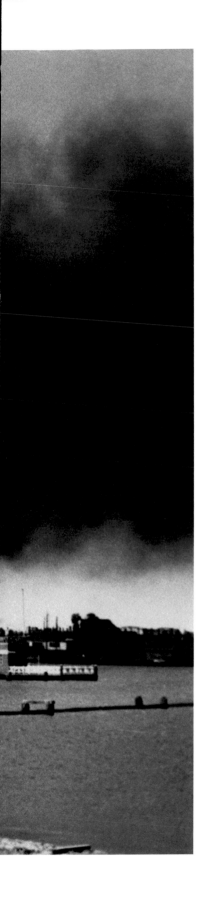

The conflagration on 14 May 1940 at the Badhuisweg in Amsterdam Noord, the location of Shell's laboratory, with the distinctive spires of the Posthoornkerk across the harbour in the background. On the same day the petroleum harbour was set on fire.

Stark contrasts in the downstream business The impact of war on the Group's manufacturing, transport, and marketing presents stark contrasts; between the destruction of installations, ships, and equipment in combat zones, and a powerful expansion of manufacturing capacity in the US and in the Caribbean; between attempts to keep going by the companies in occupied Europe, devastation for the operations in the Dutch East Indies, and a fast growing array of new, specialist products and components elsewhere; between the temporary suspension of marketing following the spread of pools and other forms of trade controls, and a strong rise of product sales; and finally between a disintegration through the loss of control over staff and assets in occupied countries, and a closer integration between operating companies in the Allied and neutral countries. Unfortunately a lack of data greatly handicaps giving a balanced assessment of the war's impact on the business. The two parent companies published very rudimentary Annual Reports during the war. Because the Group only started consolidating its accounts in 1949, these reports present no more than a superficial impression of the business.

Moreover, the depreciation policy applied by both parent and operating companies obscures a view of the underlying realities. Under normal circumstances, operating companies depreciated their installations generously. Between 1926 and 1936, the French operating company Jupiter annually wrote off on average 9 per cent of capital, leaving net profits of only 4.5 per cent. At Rhenania-Ossag, depreciations averaged 16 per cent and net profits no more than 6 per cent during the same decade, but then the company wrote off excessively because profits could not be repatriated under the restrictive exchange regime in force since 1931.[75] The war put a premium on progressive depreciations, both to build reserves for loss of assets and to limit the profits which a seller's market inevitably created. The First World War had taught a useful lesson here (see Volume 1, Chapter 3). Shell Union, the only company for which we have detailed figures, provides a good example of what is likely to have happened. Between 1939 and 1945, refinery runs rose by 27 per cent, net profits by over 140 per cent, capital expenditure by almost 40 per cent, depreciations by 80 per cent, but total assets

[38]

May 1940: the inferno in Amsterdam's petroleum harbour is extinguished, leaving smoking ruins.

[39]

[40]

[41]

Above, Shell tanks ablaze at Pernis in July 1940. Below, Britain was not immune, as the shattered remains of the London and Thames Haven Oil Wharves (later called Shell Haven) showed in September 1940.

by only 6 per cent. Dividends increased by 140 per cent and war-related taxes claimed a huge share of revenues indeed, but the disparity between growing investment and depreciations on one hand, and stagnating assets on the other shows that Shell Union, while apparently standing still in terms of assets, channelled its rising revenues into replacing and upgrading installations and into building financial reserves in the form of low book values for installations. Other operating companies and the two parent companies will have done the same. We do not know to what extent, but we may assume that the Group's customary prudence greatly reduced the financial impact of wartime destruction, and helped to build reserves for the post-war reconstruction. The opaque financial reporting also prevents us from gauging the Group's profits during the war. Shell Transport paid a 5 per cent dividend every year, but Royal Dutch suspended such payments from 1939 to 1945, when the board awarded shareholders a 25 per cent dividend to bring them up to the same level as the Shell Transport shareholders. In addition to these dividends, both companies gave their shareholders bonuses during 1947-48 by letting them subscribe to new share issues at par when the shares were trading above par value, considerably so in the case of Royal Dutch.[76] This suggests that the Group was rather more profitable than the modest dividends suggest.

The trail of destruction started with the German advance through the Low Countries and France. The efforts to demolish installations and oil stocks in the Netherlands were as spectacular as they were ineffective overall. A small British naval task force landed at the Bataafsche laboratory in Amsterdam harbour on the morning of 14 May to destroy nearby oil tanks, for which apparently no demolition preparations had been made. The tanks were hacked open with pickaxes and the contents ignited, setting off a fire which lasted four days, sending huge columns of thick black smoke into the air. The laboratory staff used the fire to burn sensitive documents. However, the tank park largely survived. Once the fires had been put out, only a third of it had been destroyed, with another 20 per cent damaged; about half remained unscathed. Rotterdam and Vlaardingen, the two other big tank parks, suffered even less damage, with the result that 85 per cent of all tankage emerged intact from the fighting in May 1940. Most of the stocks survived as well. Fire destroyed no more than 13 per cent of all oil products in the Netherlands.[77]

LES RAFFINERIES NORMANDES RENAISSENT DE LEURS CENDRES

Normandie 25.2.46

JUPITER célèbre son rallumage

La salamandre pourrait être l'emblème des raffineries géantes de Normandie. Comme l'animal fabuleux, insigne de François Ier et de la ville du Havre, elles renaissent tour à tour de leurs cendres.

Première dans cette course à la reconstruction, la Société anonyme des Pétroles Jupiter a célébré samedi le rallumage des feux de son usine de Petit-Couronne.

Plus qu'un succès technique, c'est une victoire humaine que l'on fêtait là. L'histoire vaut d'en être contée.

rent remplies avec tous les appareils délicats, toutes les pièces de bronze. Ils n'en sont sortis qu'à la libération. Les occupants n'ont jamais pu obtenir d'accord amiable pour l'enlèvement du matériel. Ils s'emparèrent quand même de l'outillage des gaz, mais ne purent jamais l'utiliser. On vient de les retrou-

The war is over, and in February 1946 'The Normandy refineries are reborn from their ashes – JUPITER celebrates its re-ignition'.

The demolition squad of British and Dutch soldiers sent to Pernis did not succeed in doing much damage there either. When the Bataafsche staff, sent away by the demolition experts, re-entered the grounds after the capitulation, they found some sheds roofless and a few tanks pierced by anti-tank rifle fire, but only the Dubbs cracker had some superficial damage. The rest of the plant was still in full working order and began reprocessing the stocks mixed up during the early days of the war. From May to December 1940, Pernis manufactured about 300,000 tons of products, largely for use by Germany. Repeated bombing by Allied aircraft also failed to have much impact on the Pernis installations.[78] With the stocks processed, Pernis fell idle and the management gave staff permission to start growing vegetables on the grounds to alleviate the food shortage. The many rabbits on the site provided a welcome source of extra protein. The German *Verwalter* sold some installations, but the plant remained otherwise intact until September 1944, when successive rounds of looting by bands of Germans stripped it of transportable equipment, cars, pumps, spare parts, measuring instruments, etc. Part of the site was then cleared to serve as a launching ground for V-1 flying bombs targeted at the UK. German army demolition troops moved in during the spring of 1945 and blew up some of the tanks, again without touching the main processing installations, which were found to have only superficial damage when the war had ended. On regaining control, Bataafsche launched a determined and concerted drive to retrieve looted equipment. Using the surviving detailed card-index inventory termed Annex C with which the staff had kept track of German requisitions and looting, the campaign succeeded in recovering a remarkably large part of the pre-war installations from locations scattered across the Netherlands and Germany. In July 1945, two months after the liberation of the Netherlands, the bulk installation handled 120,000 barrels of gasoline a month. A year later the Dubbs cracker was back on stream, returning the refinery to its pre-war capacity of processing one million tons of crude annually.[79]

The Group's Belgian bulk installations did not sustain significant damage at all during the war.[80] Destruction at the Petit-Couronne refinery near Rouen (France) appears to have had more success than similar efforts in the Netherlands, perhaps because the staff were allowed to help; two-thirds of the stock went up in flames and some of the auxiliary installations were also destroyed. As with Pernis, the main installations remained in perfect order, but the Germans preferred stripping the plant of equipment for use elsewhere to extensive repairs. The dismantling was done so thoroughly that it was eighteen months after the war before Petit-Couronne resumed processing. Stocks at several bulk installations and at the Pauillac refinery near Bordeaux were also destroyed. In 1941 German inspectors considered the plant capable of operating at no more than 30 per cent of capacity. During August 1944, Pauillac suffered two Allied air attacks which devastated the refinery installations, buildings, and residential staff accommodation. It took more than a year after the liberation of France before Pauillac resumed operations as a bulk import station and the refinery only came back on stream in April 1948.[81] The destruction of Pauillac testifies to the increasing impact of bombing as the war progressed. Initially, air raids had very little effect. Neither the Allied attacks on Pernis, nor the German bombing of oil installations and tanks in the Thames estuary during the Battle of Britain (July-September 1940) inflicted more than superficial damage.[82]

[43]

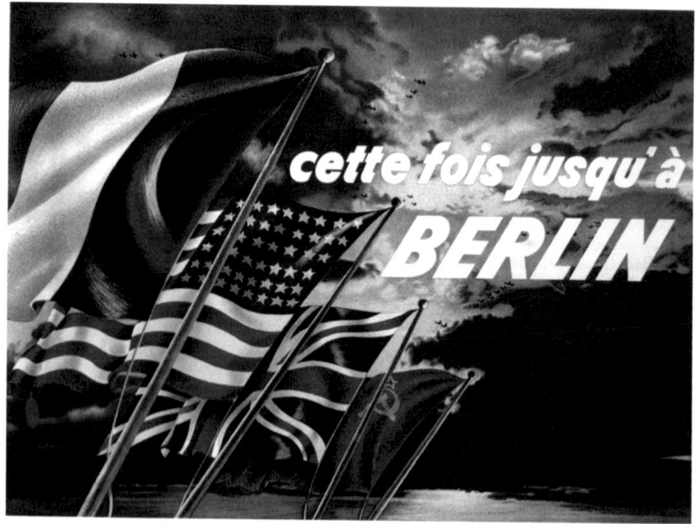

The war at sea Of all the Group's operations, its tanker fleet was by far the most exposed to enemy action. During what became known as the Battle of the Atlantic (1940-43), Germany attempted to cut British supply lines by mounting an intensive war on shipping, first with surface ships and later primarily with submarines or U-boats. Directed at cargo ships and preferably at oil tankers, these attacks had a profound impact. As early as July 1941, Britain faced a crisis in oil supplies, which was overcome with American help in the form of naval ships provided under the Lend-Lease Act; but when the United States entered the war in December 1941 and the U-boats started an offensive on shipping in American coastal waters, the situation became even worse. The number of ships sunk far outstripped the capacity of yards to replace them. By December 1942, Britain had only two months fuel oil left, which imposed a serious check on any plans for an invasion of Europe.

Three months later stocks had dwindled even further, but then the tide turned as an Allied counter offensive brought results. First, the intensification of the anti-submarine warfare curtailed the shipping losses. Second, the introduction of a new type of so-called Liberty ships, built to a uniform design and welded rather than riveted for speed of production, greatly increased the number of ships available. The oil tanker of this design, known as T2, remained the standard reference of tanker size for decades to come. Third, a pipeline network laid across America came on stream, reducing the need for tankers to carry supplies from the Gulf to the east coast.

[44]

[45]

On 8 December 1941, the day after Japan's assault on Pearl Harbor, an American workman reads that his country is at war. Three days later Hitler declared war against the US as well.

Built in four months over 1941-42, the Plantation products line (24 per cent Shell) used pipes of 12-inch and 10-inch diameter and initially ran 812 miles from Baton Rouge, Texas, to Greensboro, North Carolina. Branch lines of 8-inch and 4-inch diameter fed it from Alabama, Tennessee, and Georgia and brought its total length at that time to 1,261 miles. For a short period it was the largest and longest pipeline ever laid, and as an emergency measure, using second-hand 8-inch pipe, an additional 178 miles were completed in April 1943, all the way to Richmond, Virginia. Before long two other pipelines, both constructed by another consortium in which Shell took part, had outstripped it in length and size. Big Inch, 24 inches in diameter, was by far the widest pipeline the world had seen. Partly functioning by the end of 1942, from its completion in August 1943 it carried crude oil 1,478 miles from Longview, Texas, via Norris City, Illinois, to a terminus in New Jersey, close to New York City. The Little Big Inch products pipeline, so named because it was 'only' 20 inches in diameter, was even longer. Authorized in February 1943 and completed in a year, it ran 1,714 miles from the Houston-Beaumont refining area in Texas to Linden, New Jersey; and between them, from their completion to the end of the war, Big Inch and Little Big Inch delivered 380 million barrels of crude and products from Texas to the Atlantic seaboard. Some of this went to supply the east coast population, but most of it was shipped direct from the east coast to the UK.[83]

[47]

The miracle of Liberty ships: seen here (left and right) under construction in Portland, Oregon, are just a few of the 2,770 Liberty ships totalling 29.3 million deadweight tons built in the US in 1941-45. Based on a British design dating from 1879, they were all-welded, uniform in design, fast to build, simple to operate, and capable both of carrying large cargoes and of withstanding war damage to a remarkable degree.

[49]

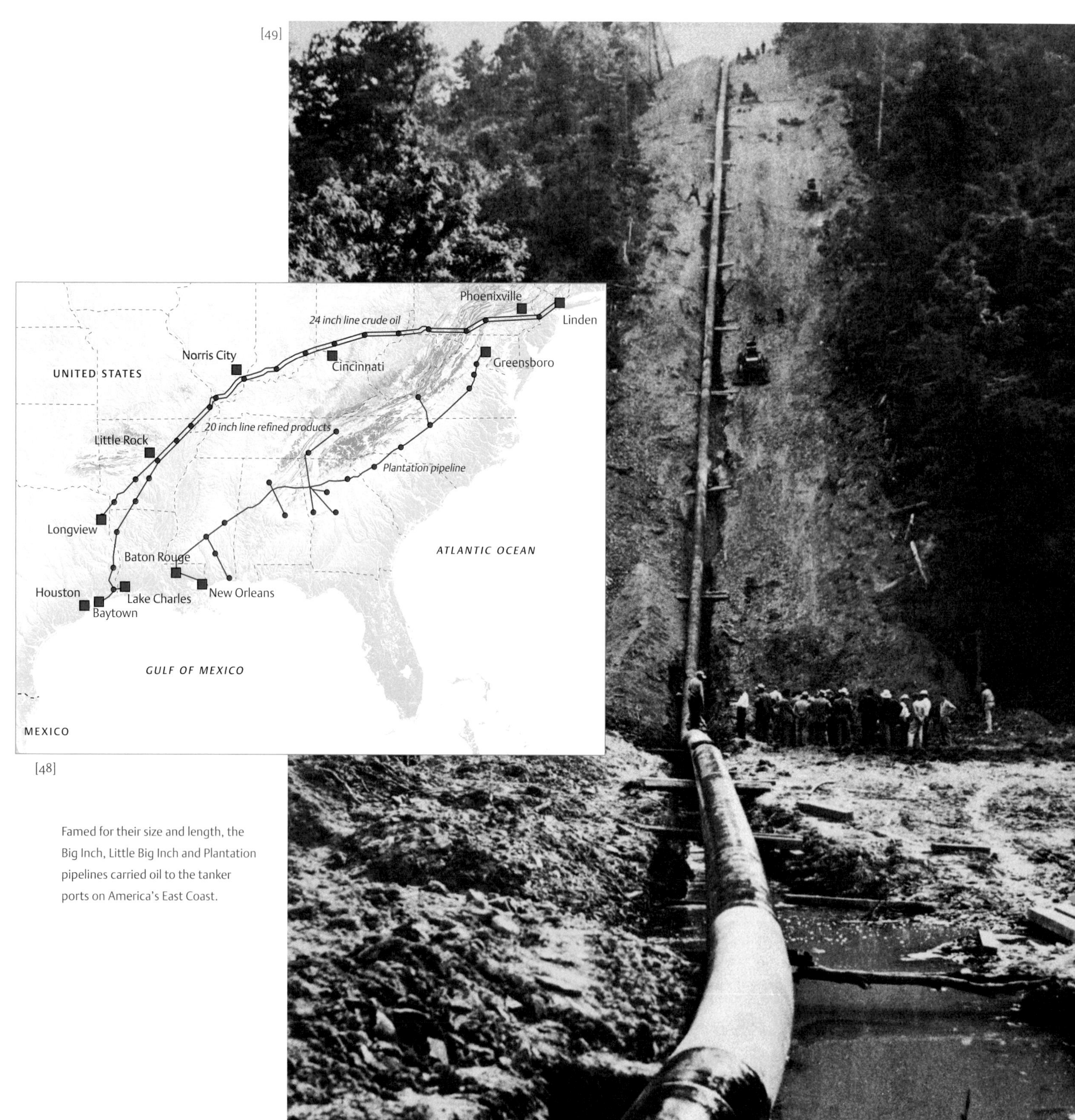

[48]

Famed for their size and length, the Big Inch, Little Big Inch and Plantation pipelines carried oil to the tanker ports on America's East Coast.

[50]

[51]

[52]

The vicissitudes of the Group's fleet fully mirror the heavy toll exacted from Allied shipping. In September 1939 the Group still had four separate fleets: British-flag ships belonging to Anglo-Saxon, aggregating 1,525,000 deadweight tons (dwt), and Dutch-flag ships belonging to the Nederlandsch-Indische Tank-Stoomboot Maatschappij (NIT), NV Petroleum Maatschappij La Corona, and the Curaçaosche Scheepvaart Maatschappij (CSM), operating small tankers on the short stretch between Curaçao and Venezuela. Dutch-flag ships aggregated 594,000 dwt, and together with Anglo-Saxon's they formed the world's largest private fleet in numbers and tonnage.[84] In addition the Group managed and part-owned the British-registered fleet of the Eagle Oil and Shipping Company, whose twenty-six tankers aggregated 317,000 dwt, making a grand total of 2,436,000 dwt.[85] Tankers had little automation; their running was a very labour-intensive operation, each typical 12-15,000 dwt ship requiring between forty and fifty officers and men, and the Group's owned and managed fleets immediately before and during the Second World War required somewhere between six and seven thousand people to operate them.

The main picture (left) shows the Big Inch pipeline crossing the Allegheny mountains. At times the line had to climb almost vertically.

Right, at Rock Hill near Glenmoore, Pennsylvania, work continued over the winter of 1943; centre, the last sections were laid at Phoenixville, Pennsylvania, on 19 July 1943; below, a whimsical signpost at Phoenixville. The entire pipeline was laid in only 350 days.

[53]

Routes of the English pipeline system.

[54]

After D-Day, the Allied invasion of Europe on 6 June 1944, it was hoped that PLUTO, the Pipeline Under The Ocean, would supply the advancing armies in France with 50 per cent of their fuel. Above left, 200 miles of flexible pipe is lying in storage, ready to be wound onto giant drums (above) which were towed out across the Channel, unwinding the pipes as they proceeded. Taking its name from the Ancient Greeks' god of the underworld, the first PLUTO carried gasoline from the Isle of Wight to Normandy. A second PLUTO later ran from Dungeness, Kent, to Calais in France. But they proved to be a serious technical disappointment, and for four crucial months in the summer of 1944 they supplied on average only 150 barrels a day – one-sixth of 1 per cent of Allied needs.

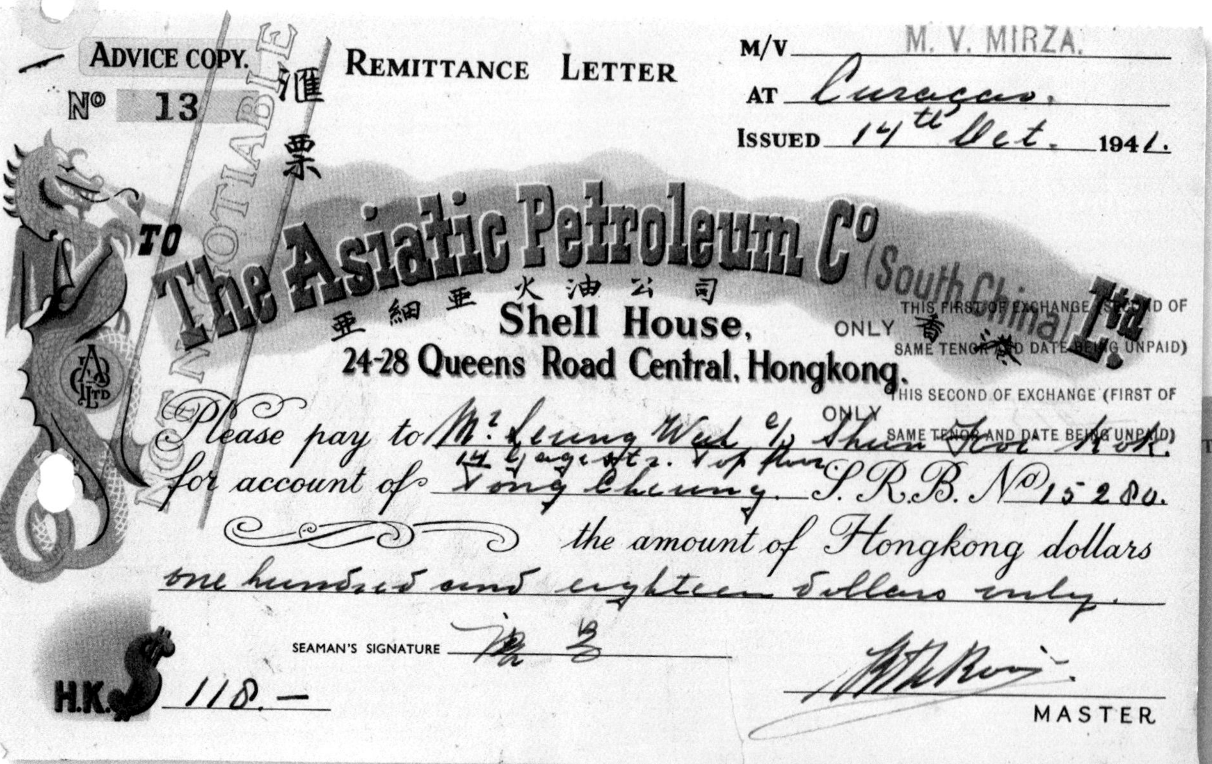

ADVICE COPY.

No 13 匯票

REMITTANCE LETTER

M/V M. V. MIRZA.

AT Curaçao.

ISSUED 14th Oct. 1941.

NOT NEGOTIABLE

TO The Asiatic Petroleum Co (South China) Ltd.

亞細亞火油公司

Shell House,

24-28 Queens Road Central, Hongkong.

ONLY 香港 THIS FIRST OF EXCHANGE (SECOND OF SAME TENOR AND DATE BEING UNPAID)

ONLY THIS SECOND OF EXCHANGE (FIRST OF SAME TENOR AND DATE BEING UNPAID)

Please pay to Mr. Leung Wai c/o Shun Hoi Hok, 14 Yaug Sts. Top Floor

for account of Tong Cheung. S.R.B. No. 15280.

the amount of Hongkong dollars one hundred and eighteen dollars only.

SEAMAN'S SIGNATURE

H.K.$ 118.–

MASTER

[56]

[57]

ne 8935 = rij-2/quarles = 1.8.47 = ps =

= marine crew dept =

"sunetta"

we confirm telephonic conversation 31.7.1947

killed

s.r.b.	20044 tan ah chwee	sailor
	19832 ho siew	carpenter

seriously injured (admitted havenziekenhuis, rotterdam)

s.r.b.	1424 chan poh	q'master
	7495 chai kin ling	q'master
	6705 siek heck see	sailor

slightly injured (returned to vessel)

s.r.b.	18471 ah poh	sailor
	5542 ah say	sailor
	11848 wong siang mew	sailor
	20042 koh meek ke	sailor
	0921 chen ah kae	pumpman
	20047 shah chen chang	no. 1 fireman
	20051 ng kwoh	donkeyman
	01363 lee yiu	fireman
	20049 wong you teng	fireman
	3227 cheong thim	fireman
	2231 seng on	fireman
	18938 yeong kew	fireman's cook

+++

Sunetta crew

Telegram RECEIVED for ASPCO MARINE LONDON

despatched UNKNOWN received

MARISA survivors landed here Datema Zwaa

Ling Che Ming Siong Ah Moy Ping Ah Kong

Lew Chung Tong Poon Lee Sang Wee Wang See

hospital

Crompton

[58]

Shell's international fleets were frequently crewed by Chinese. Left, a wartime 'remittance letter' guarantees a man's payment. Issued on board MV *Mirza* at Curaçao on 14 October 1941, the payment would be drawn against Asiatic Petroleum in Hong Kong. Below, the physical risks of wartime continued with a deadly legacy in peacetime. On 30 July 1947, the 12,000-ton tanker *Sunetta* from the Group's La Corona fleet hit a mine, resulting in the deaths of a sailor and a carpenter, serious injuries to three of the crew, and slight injuries to a further twelve.

The Dutch censor allowed the CSM management to publish a defence of its handling of the labour conflict, but prohibited a local paper from reporting about the grievances of the officers on strike. The paper's editor then left a blank space on the page under the headline over the article, filled only with a banner 'Expunged by the censor' and the text of article 4 from the Curaçao constitution outlawing slavery on the island. The censor reacted by suspending publication for three days.

[58]

DAGBLAD

AMIGOE DI CURAÇAO

59ste Jaargang, No. 3555

Het C.S.M.-Conflict

Door den Censor geschrapt.

Door den Censor geschrapt.

Manning oil tankers was extremely risky work. The ships were small targets, but slow and with a distinctive outline, with large and highly inflammable cargoes, each with a large number of souls on board, and the preferred target of U-boats. It remains one of the wonders of the Second World War that anyone on either the Axis or the Allied side was willing to crew an oil tanker at all. As one of Anglo-Saxon's officers observed, 'Floating around minefields on top of 11,000 tonnes of motor gasoline is not exactly a rest cure.'[86] In the course of the war Shell Transport alone lost sixty-six ocean-going ships totalling 632,000 dwt – over 40 per cent of its deadweight tonnage at the outbreak.[87] Including vessels taken over by Axis forces, the Dutch-flag fleets lost thirty-four out of the ninety-one ships that were in them at the beginning of the war, while the Eagle fleet lost thirteen of twenty-six.[88] There were further losses, too, of ships bought or built in wartime. The Group ended the war with its total fleet capacity considerably below the 1939 figure. Anglo-Iranian and Jersey Standard suffered similarly.[89] Some of the Group's lost ships were bombed, others mined, and a few were sunk by gunfire; however, the great majority were victims of torpedo attacks. The loss of life in these sinkings was not total in every instance, but it was severe enough overall. The Group fleet lost a total of 1,506 men, i.e. up to a quarter of the total operating crew. More than two-thirds of them, 1,025 crew members in all, were Chinese or Asian, 481 British or Dutch.[90]

The hazardous work on the tankers did lead to serious outbreaks of labour unrest in the Group's fleet. In March 1942, the Chinese crew of the tanker *Ovula*, moored in Alexandria harbour,

[61]

447.-J.M.B. 40824

MUTATIONS, m. /S. "APOLLONIA".

Du[illegible]t .... Sailed ....

STAFF ON BOARD Rank	Name	Due for Leave	Reason of Leaving	Paid off		Relieved by	Signed On	REMARKS
Act. Captain	P.Schol							
" Ch. Officer	W.Buninga							
2nd Officer	E.J.C. v/d Wiel (Wounded)							
3rd Officer	a Koppier							
Apprentice	–		M.S.G. (P.Kraus (wounded)					
Apprentice	–		(A.Dowse (~~1 British Machine Gunner (?)~~)					
Ch. Engineer	J.E.Jongejan							
2nd Engineer	A.Klijn							
Ext. 2nd ~~3rd~~ Engineer	K. de Jong (Missing)							
4th Engineer	G.Colijn							
5th Engineer								
5th Engineer								
5th Engineer								
Supernumeraries —								
Radio Officers 1st	F.A.Ahlberg (wounded)							
Radio Officers 2nd	–							
Crew	30 Chinese Crew							

On the 25th November 1940 whilst in ballast from Plymouth to Avonmouth and when between Lizard Light and Wolf Rock Light "APOLLONIA" was attacked & subjected to heavy gunfire. The attack lasted from 12.45 a.m. until 1.30. a.m. As the vessel was sinking rapidly Captain & surviving Officers, Engineers and Crew took to the only remaining lifeboat. At 2.00 a.m. Captain returned to ship in endeavour to bring vessel to port. As they approached vessel enemy craft commenced shelling ship and lifeboat, shortly afterwards "APOLLONIA" began to founder and Captain made for coast as soon as possible by sailing lifeboat arriving Penzance at 10.00 a.m. on the 25th November. Casualties:- K. de Jong, extra 2nd Engineer & 11 Chinese killed or missing, 8 Ratings unaccounted for.
Wounded.: E.J.C. v/d Wiel, 2nd Officer,
F.A.Ahlberg, Radio Officer,
P.Kraus, Gunner, and 1 Chinese.

m. /S. "APOLLONIA".

Nederlandsche Telegraaf Maatschappij
„RADIO-HOLLAND" N.V.

No

RADIO-NAVIGATIEBER[ICHT]

ONTVANGEN VAN: NMN
A/B M/S.: Vistula
TE HEFFEN KOSTEN: FL.
D.D.: 23·4 1945
DOOR: A.E.[illegible]

SSSS SSSS SSSS de VASTL –
Torpedoed position 35 56 North · 7452 We[st]
ar de NMN – 231552 ar

[62]

[63]

Nederlandsche Telegraaf Maatschappij
„RADIO-HOLLAND" N.V.

No

RADIO-NAVIGATIEBERICHT

ONTVANGEN VAN: WSL
A/B M/S.: VISTULA –
TE HEFFEN KOSTEN: FL.
D.D.: 14 4 1945 TIJD: 0300
DOOR: Anthony Rauan Bergh

To BAMS – 2A – 2B – from WSL. CK 7
Operational priority –

Submarine attacked
3604 NORTH 7405 WEST
at 142020z

150157z

A thundering explosion and at the same time burning gasoline covered the deck and bridge and the whole vessel, except the stern, was aflame. I saw the second officer leave the bridge, burning, followed by the helmsman - the look-out on the bridge was never seen again.

[60]

[64]

PARTICULARS OF CASUALTY.

1. Name of Vessel. MARISA
2. Date of Loss. 16/5/41
3. Reason of loss. Torpedoed
4. Casualties: C.M. Kandel + one Chinaman
5. Date of Survivors' arrival ~~in London~~
6. Red Cross advised (in case of relatives in Holland).
7. Families or friends in U.K. advised. none in U.K. as far as known
8. Minister of Interior advised. 8/7/41.
9. Radio Holland advised (in case of Radio Officer).
10. Disbursements Dept. advised (-do-).
11. Steamers A/cs. Dept. advised.
12. Curacao advised (if C.S.M).
13. Insurance Dept. advised.
14. Compensation for loss of effects. Advances to all
15. Estate of Deceased/s credited with Compensation for loss of Effects. 17/7/41
16. Staff advised of wages arrangements. verbally 8/7/41
17. Shell Magazine advised. (death of Kandel.)
18. Aliens' and Immigration Officer visited.
19. Passbooks and Monster books sent to Consulate. not available
20. Mr. Rowbottom's Secretary advised Captain and Chief Engineer available for interview. 7/7/41
21. Captain and/or Chief Engineer interviewed by Mr. Rowbottom. 17/7/41.
22. Master visited office. 7/7/41
23. " " Mr. de Booy. 9/7/41
24. " " Receiver of Wrecks. 18/7/41
25. Master made deposition at Consulate. 10/7/41.
26. Master appeared before Notary Public. Not for Dutch Vessel.
27. Master visited Consulate to certify death Certificate/s. 10/7/41.
28. " " Admiralty.
29. Letter informing N.S.T.C., Baltic House (Mr. Toeters) re new diplomas.
30. Provident Fund advised. verbally — 10/7/41.
31. Tombstone/s arranged.
32. Coupons 7/7/41
33. Uniform fittings.

At 2.45 a.m., our estimated position being 48°36′ N. Lat. and 8°52′ W. Long., there was a violent explosion amidships aft of the bridge. The vessel was found to have a large hole in the ship's side on starboard. The deck was completely torn open on starboard, with pieces standing upright. The vessel was also cracked across the deck to portside and from there outboard to below the water-line on portside.

The vessel immediately listed heavily to starboard, making it impossible to walk on board. She also made alarming move-ments, as stem and stern were continuou-sly moving towards each other.

[65]

At about 9.20 p.m. on May 16th 1941, while the vessel was approximately 340 miles south west of Freetown, there was a terrible explosion, which caused the entire ship to vibrate and whip violently, and it goes without saying that everyone on board then realized that the vessel had been torpedoed. It was soon appa-rent that the 'MARISA' had been hit by a tor-pedo in way of the engine-room on the port side aft.

[66]

mutinied after some months of increasing strife on the ship. When British and Dutch naval ratings boarded the ship to restore order, the mutineers attacked them with hammers and files and the ratings opened fire, killing two sailors and wounding four. Twenty-five mutineers were taken ashore and detained in Egypt.[91] That same spring the suspension of oil shipments between Venezuela and Curaçao because of U-boat threats created rising tension amongst the 420 Chinese crew of CSM ships, whom local regulations did not allow onto Curaçao. Having called a strike, the men were then taken ashore and housed for several weeks in a temporary camp on the island. In April, a group attacked police officers who had entered the camp to take some of them away. The police then fired on the crowd, killing twelve Chinese crew and wounding thirty-eight.[92] These very grave incidents formed part of a broader protest by Chinese seamen on Dutch and other Allied ships against the various forms of discrimination from which they suffered: while sharing the same hazards, they had much lower pay and poorer conditions than other crew members, notably the Europeans.[93] And the submarine threats were sufficiently frightening for the European officers of the CSM fleet also to strike during April-May 1942, forcing Van Eck to come down from New York and sort things out.[94] The justified grievances of Chinese sailors would contribute to the fundamental reorientation of labour relations after the war. One effective measure already considered in 1943 was the lowering of the retirement age for fleet staff from 55 to 50.[95]

Doing all they could to lessen the dangers at sea, Group engineers and technicians came up with five valuable inventions. The first of these, in January 1940, was the use of compressed air in keeping a torpedoed vessel afloat. The system was developed by W. Lynn Nelson, Eagle Oil's chief of technical staff, and by mid-1941 was standard equipment on all Allied tankers. It proved remarkably versatile: as well as the uses to which it was first put – forcing water out of a ruptured tank and restoring buoyancy – a vessel so equipped could use compressed air to power seawater fire-fighting pumps, the transfer of cargo, and even steering. The other inventions included the 'Eagle hood', developed by Eagle's technicians to protect against the effect of immersion in oil; a soap which worked in fresh water or salt, so a man emerging from an oil-covered sea could clean his eyes, ears, nose and mouth; and fireproof lifeboats capable of surviving at least 2,400 degrees Fahrenheit (1,315 degrees Centigrade) without the internal temperature exceeding 116 Fahrenheit (47 Centigrade).[96]

But since the best protection was to keep attackers at bay and since the surest defence of a convoy was air escort, probably the most important Group innovation at sea during the Second World War was the class of vessel known as MAC ships or Merchant Aircraft Carriers. Conceived more or less simultaneously by Shell

[67]

Far left: after conversion into a Merchant Aircraft Carrier, able to accommodate Swordfish aircraft as well as oil, Shell's Dutch-flag tanker *Gadila* became the first aircraft carrier in the Royal Dutch Navy and was closely followed by her sister *Macoma*. Left, a returning Swordfish lands on *Gadila*'s deck, and below, three Swordfish take off from one of the Dutch flag MACs in the North Atlantic, 1943.

[68]

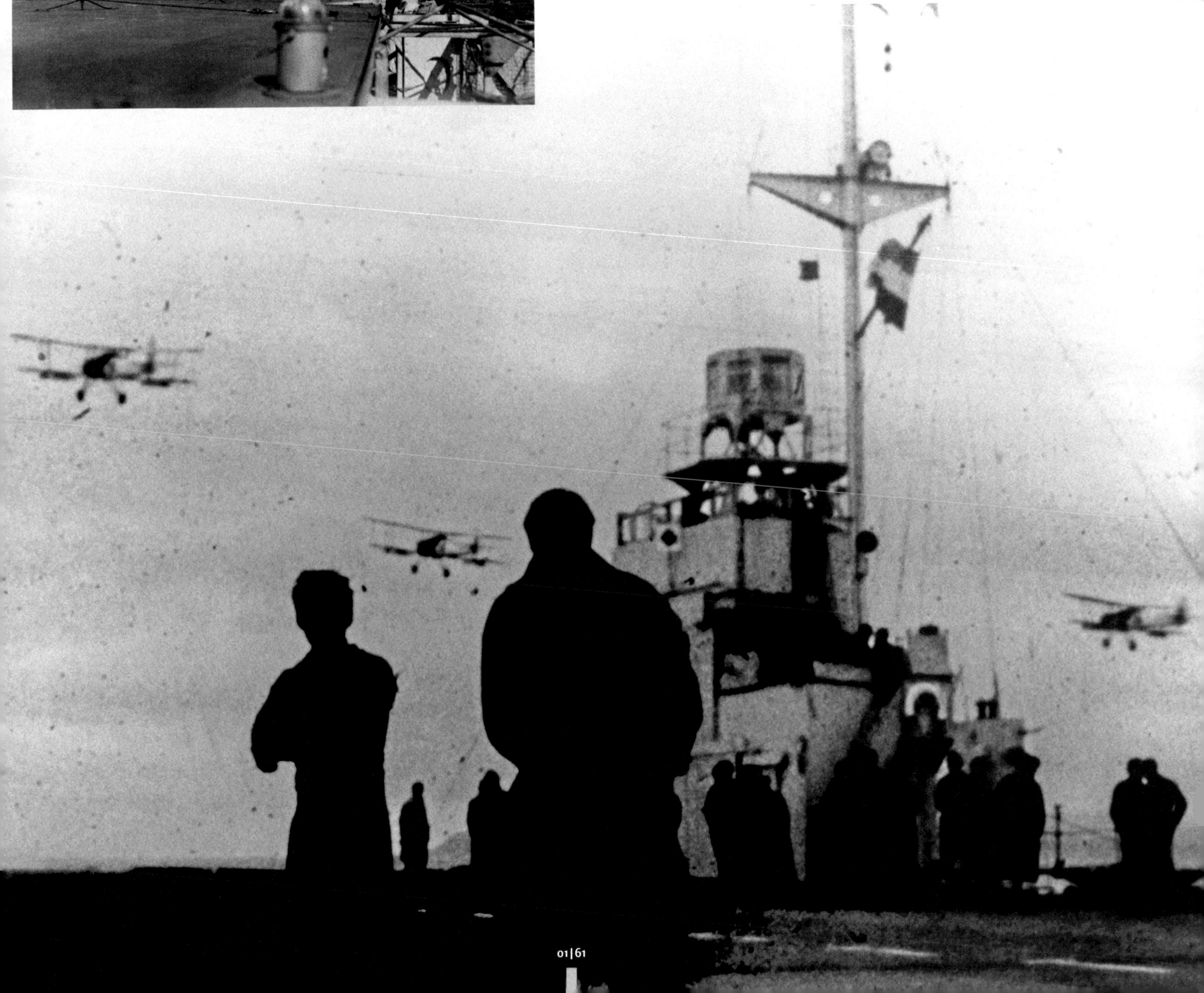

and the British Admiralty as a short-term solution to the lack of aircraft carriers, these were grain carriers (or in Shell's case, oil tankers) which after major modification were able to continue carrying their normal cargoes, reduced by only 1,000 tons, and four Fairey Swordfish anti submarine biplanes armed with torpedoes and rockets. Nine Shell ships were converted in this fashion, including two which later became the first aircraft carriers in the Royal Netherlands Navy, and all contributed to the turning of the tide in the Battle of the Atlantic. Serving for up to two years, they collectively escorted 217 convoys, making 323 transatlantic crossings; their aircraft flew 4,177 sorties, and only one of those convoys was successfully attacked.[97]

Overall the Group's pattern of loss reflected the pattern of the U-boat war, with 1942 marking the peak. In that year alone, fifteen of Anglo-Saxon's ships were sunk, eleven by torpedo in the Battle of the Atlantic, along with seven of Eagle's and six of La Corona's, while nine NIT vessels were captured by Japanese forces.[98] To relieve the pressure on British and American shipyards and to prevent damaged vessels having to travel long distances for repair, a new dry dock, the Princess Beatrix dock, was built at Curaçao, capable of repairing ships up to 20,000 dwt or 600 feet in length, and during 1944, for example, large and small repairs were made to 287 vessels in the Group's docks on the island.[99]

[69]

THE SUNDAY EXPRESS November 10 1985

ON THIS REMEMBRANCE DAY, WHO HAS HEARD OF THE HE

The British ship ro the Japanese gi like a snarling t

ON the morning of November 11, 1942, while flags around the cenotaphs in Britain dipped in homage to those who had made the supreme sacrifice, two ships, one British, one Dutch, prepared for battle in the lonely reaches of the South Indian Ocean.

Theirs was not to be a day of glory; of creaming bow-waves and thundering broadsides, but a desperate struggle to survival against impossible odds.

by CAPTAIN BERNARD EDWARDS Master Mariner

The 14,000-ton Dutch tanker *Ondina*, escorted by the 733-ton Royal Indian Navy minesweeper *Bengal*, had left Freemantle in Western Australia on November 5 bound for the island of Diego Garcia, 2,800 miles away in the Indian Ocean.

Commanded by Captain William Horsman and manned by Dutch officers and Chinese ratings, the *Ondina* was armed with a single four-inch gun mounted on her poop.

Her gun's crew, led by Able Seaman Herbert Hammond, RANR, was made up of four Royal Navy ratings, three Royal Artillery gunners and a Dutch Merchant Navy gunner.

The *Bengal*, dwarfed by her huge charge, was commanded by Lieutenant-Commander W. J. Wilson, RINR, and carried British officers and Indian ratings.

She had a top speed of around 12 knots and her sole armament consisted of a small 12-pounder gun mounted forward. For this she had only 40 rounds of ammunition.

Lt. Cdr Wilson was not unduly concerned with his lack of firepower as he set off from Fremantle with the *Ondina* dutifully following. By the autumn of that year the Indian Ocean had been largely cleared of German surface raiders and only a few U-boats were reported to be operating in the southern approaches to the Mozambique Channel, more than 2000 miles to the west of the small convoy's proposed route.

BLOODY

As for the Japanese, after a protracted and bloody foray against Allied shipping in the early summer, using armed merchant cruisers and sub-

The ship was heading straight for the *Bengal* and *Ondina* and closing rapidly. At eight miles Wilson's worst fears were confirmed. The short, unraked funnel, clipper bows and cruiser stern were all unmistakably Japanese. A few minutes later he was able to make out the funnel markings of the Nippon Yusen Kaisha (N.Y.K. Line).

Wilson's next moves were quick and decisive. Sounding Action Stations, he altered course 90 degrees to starboard away from the enemy, at the same time signalling the *Ondina* to take up station on his starboard beam.

No sooner had the *Bengal* put herself between the enemy and the tanker than a second ship came over the horizon, hard on the heels of the first. Wilson's face went white under his tan as he once again lifted his binoculars and picked out the funnel colours of the N.Y.K.

Although he was not aware of it at the time, he was facing the Japanese raiders *Hokoku Maru* (10,438 tons) and *Aikoku Maru* (8,000 tons). Both vessels carried ten 5·9-inch guns, torpedo tubes and spotter aircraft.

Wilson ordered the *Ondina* to reverse her course and make off to the south at maximum speed, giving her a rendezvous to keep 24 hours later. Then, at full speed, the minesweeper steered straight for the oncoming *Hokoku Maru*; the diminutive David rushing to challenge a Goliath more than 14 times his size.

Perhaps unable to com-

to the south as instructed by the *Bengal*, Horsman came around only enough to bring his stern mounted, and already manned, four-inch gun to bear on the *Hokoku Maru*. He did not increase speed.

When the oncoming *Hokoku Maru* had decreased the distance to 8,000 yards Horsman gave the order to open fire. On the poop, Second Officer Bartele Bakker, officer in charge of the gun, tapped the shoulder of Able Seaman Hammond, who was already crouched over his sight. The four-inch barked and recoiled.

only one option: Sending below for two white bedsheets, he ordered them to be hoisted on the flag halyards. The *Ondina* was offering surrender.

The *Aikoku Maru* ignored the white flags and continued firing on her helpless quarry. Horsman now had no alternative but to stop his engines and order his crew to abandon ship. No sooner had he given the last order than a Japanese shell burst on the bridge of the *Ondina* and Horsman was killed by flying shrapnel.

The remaining 56 men of the *Ondina's* crew, most of them unaware of their captain's death, set about abandoning ship. Although the shells continued to burst all around them, there was no panic and in less than three minutes all lifeboats and rafts were in the water and moving away from the ship.

Ironically, as the *Ondina's* men were thus engaged, they witnessed the last moments of the *Hokoku Maru*, which sank stern-first, leaving many men struggling in the water.

What little satisfaction the men gained from the demise of their erstwhile enemy was soon dissipated by the action of the *Aikoku Maru*. At 400 yards, the raider fired two torpedoes, which passed under the lifeboats and struck the *Ondina* in way of her after tanks. She immediately took a heavy list to starboard and began to settle by the stern.

For the grieving survivors of the *Ondina* there was worse to come. In a vile demonstration of man's inhu-

manity to man, th *Maru* now closed th and opened up machine guns. Fiv survivors were inju

Five shells succession hit the raid

of them fatally, Chief Engineer Jan

Mercifully, the raide revenge was short-li she suddenly sheered steamed back to pick vivors from her siste

[70]

War Stories of the Shipping Lines

THE N.V. Petroleum Mij. "La Corona" of Holland began its tale of losses at sea with the sinking of the 8,000 tons "Mamora." A few weeks later the crew of the 6,000 tons "Eulota," sailing in two open boats 10 miles away from their vessel, were sighted by a French reconnaissance 'plane which quickly brought a British warship to their rescue. The "Eulota," broken in two after striking a mine, was later sunk by gunfire from the rescue-ship as a danger to shipping.

RÉCITS DE GUERRE DE LIGNES DE NAVIGATION. La compagnie hollandaise Petroleum Mij. "La Corona" perdit d'abord le "Mamora" (8.000 t.). Quelques semaines plus tard l'équipage de l' "Eulota" (6.000 t.) qui se trouvait dans deux embarcations ouvertes à 15 [illegible] *de son bateau abandonné, fut aperçu par un avion de reconnaissance français qui* [illegible]*dement à son secours un navire de guerre britannique. Ce dernier coula par le* [illegible] *de canon l'épave de l' "Eulota", coupé en deux après avoir heurté une mine,* [illegible]*it un danger pour la navigation.*

ΠΟΛΕΜΙΚΑΙ ΠΕΡΙΠΕΤΕΙΑΙ ΕΦΟΠΛΙΣΤΙΚΩΝ ΕΤΑΙΡΕΙΩΝ. [illegible]

The Japanese offensive in Asia Of all the Group's land-based operations, those in the Dutch East Indies suffered most heavily from enemy action. Following the outbreak of the European war, Bataafsche invested heavily in expanding manufacturing capacity there, especially for the production of aircraft gasoline. Pladju received an alkylation plant and an isopentane installation, and at Balik Papan a start was made on the construction of an entirely new factory complex. In December 1941, projects for a casinghead gasoline plant at Pladju, for an installation to make aircraft gasoline components at Pangkalan Brandan by the separation of pentane, isopentane, butane, and isobutane from light crude fractions, and for a synthetic ammonia factory at Tjepu were at an advanced planning stage.[100]

At the Batavia talks in the late summer of 1940, Japan demanded a huge increase in oil supplies from the Dutch East Indies, to a total of 3.2 million tons a year, double the volume obtained until then and nearly half of the total production. With this request the American policy dilemma returned through the back door. Giving in meant supplying a potential enemy, Japan, which during the talks signed an alliance with Germany and Italy; refusing might provoke war. While Van Eck and Kay negotiated with the Japanese representatives, Agnew and Stanvac CEO George Walden worked closely with the State Department and the Foreign Office to draft a compromise deal based on the maximum capacity of their Dutch East Indies operations, reduced by present commitments. This resulted in a proposal to supply Japan with a total of 1.8 million tons annually, free on board so that Japan would have to provide the necessary – scarce – tankers, for an initial six months. The supply schedule fitted the US reservations on oil exports to Japan in that the Group undertook to deliver the 400,000 tons of high-octane aviation fuel, enabling Stanvac, which had also just started manufacturing 100-octane gasoline in the Dutch East Indies, to comply with the American embargo on that product. In October, the Japanese delegation accepted the proposal in its entirety, much to the surprise of the Dutch representatives. As it turned out, the tanker shortage prevented Japan from taking up the full quantities contracted with the oil companies.[101]

Meanwhile the British and American governments quietly imposed restrictions on Japan's ability to stockpile oil products by working on shipping lines to withdraw tankers from the Pacific and by reducing the export of steel barrels. When the Batavia supply agreement came up for renewal in April 1941, Shell and Stanvac received permission to renew for another six months, provided the volume carried over as undelivered under the previous contract was kept to a minimum. Two months later, Japan sent troops to Indochina and revealed its intentions to attack the Dutch East Indies in diplomatic code telegrams, which the State Department could read. The United States responded by effecting an informal embargo on oil exports through the freezing of foreign exchange transactions; the British and Dutch governments followed suit. The last tankers destined for Japan left the United States and the Dutch

Shell's La Corona fleet suffered its first loss on 11 March 1940 – two months before Germany invaded the Netherlands – with the sinking of *Eulota*, torpedoed by the German submarine *U-28*, 120 miles south-west of Land's End, England, while en route from Rotterdam to Curaçao.

Far left, a tale of courage retold: on 11 November 1942, the Shell tanker *Ordina* sank a Japanese raider in the Indian Ocean with gunfire.

Main picture: the devastating explosion in the destroyer USS *Shaw* during the Japanese attack on the US Pacific Fleet at its base in Pearl Harbor, Hawaii, 7 December 1941; inset, headline from the *Los Angeles Times* the following day; below right, report by the *New York World Telegram*.

IT'S WAR

Hostilities Declared by Japanese
350 Reported Killed in Hawaii Raid

East Indies in early August. With oil stocks for eighteen months of war, the Japanese government decided to embark on a desperate bid to gain economic freedom, by first knocking out the American fleet at Pearl Harbor, and then moving south to capture the Dutch East Indies. On 7 December 1941, the Japanese naval air force attacked Pearl Harbor and the United States entered the war. The next day, troops landed in Thailand and on the Malaysian peninsula, a week later on Sarawak, one of the Group's key production areas, and on 11 January 1942 in the Dutch East Indies itself.[102]

[72]

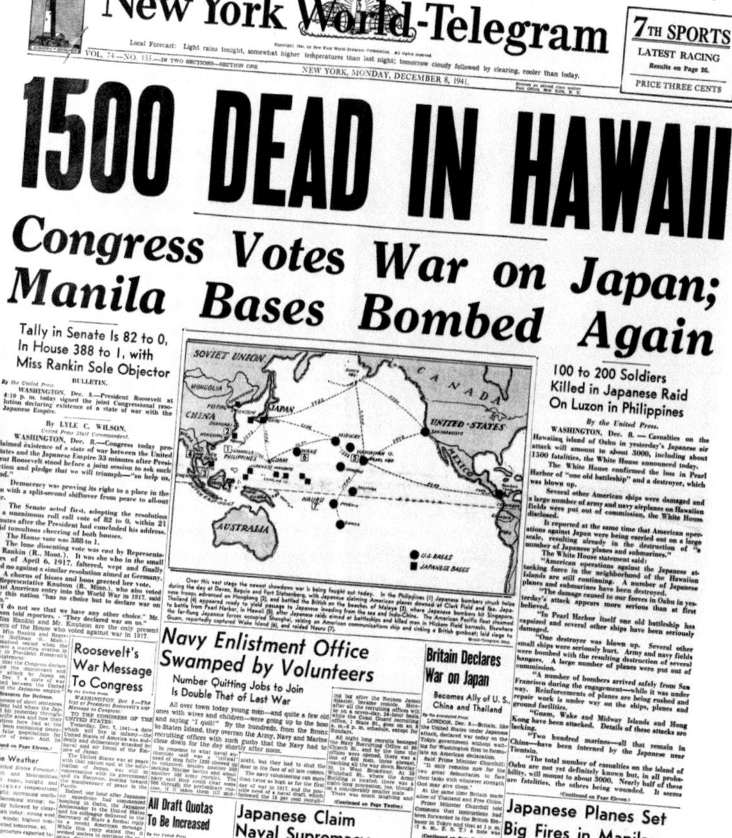

New York World-Telegram

Local Forecast: Light rains tonight, somewhat higher temperatures than last night; tomorrow cloudy followed by clearing, cooler than today.

VOL. 74.—NO. 135.—IN TWO SECTIONS—SECTION ONE

NEW YORK, MONDAY, DECEMBER 8, 1941.

7TH SPORTS
LATEST RACING
Results on Page 26.
PRICE THREE CENTS

1500 DEAD IN HAWAII

Congress Votes War on Japan; Manila Bases Bombed Again

Tally in Senate Is 82 to 0, In House 388 to 1, with Miss Rankin Sole Objector

BULLETIN.

By the United Press.

WASHINGTON, Dec. 8.—President Roosevelt at 4:10 p. m. today signed the joint Congressional resolution declaring existence of a state of war with the Japanese Empire.

By LYLE C. WILSON.
United Press Staff Correspondent.

WASHINGTON, Dec. 8.—Congress today proclaimed existence of a state of war between the United States and the Japanese Empire 33 minutes after President Roosevelt stood before a joint session to ask such action and pledge ...

The Senate acted first, adopting the resolution by a unanimous roll call vote of 82 to 0, within 21 minutes after the President had concluded his address amid tumultous cheering of both houses. The House vote was 388 to 1.

The lone dissenting vote was cast by Representative Rankin (R., Mont.). It was she who in the small hours of April 6, 1917, faltered, wept and finally voted no against a similar resolution aimed at Germany. A chorus of hisses and boos greeted her vote.

Representative Knutson (R., Minn.), who also voted against American entry into the World War in 1917, said this nation "has no choice but to declare war on ..."

"I do not see that we have any other choice," Mr. Knutson told reporters. "They declared war on us." Miss Rankin and Mr. Knutson are the only present members of the House who voted against war in 1917.

Over this vast stage the newest showdown war is being fought out today. In the Philippines (1) Japanese bombers struck twice during the day at Davao, Baguio and Fort Stotsenberg, with Japanese claiming American planes downed at Clark Field and Iba. Japanese troops advanced on Hongkong (2), and battled the British on the beaches of Malaya (3), where Japanese bombers hit Singapore. Thailand (4) appeared ready to yield passage to Japanese invading from the sea and Indo-China. The American Pacific fleet steamed to battle from Pearl Harbor, in Hawaii (5), after Japanese raids aimed at battleships and killed men in Hickam Field barracks. Elsewhere the far-flung Japanese forces occupied Shanghai, seizing an American communications ship and sinking a British gunboat; laid siege to Guam, reportedly captured Wake Island (6), and raided Nauru (7).

World-Telegram Map.

100 to 200 Soldiers Killed in Japanese Raid On Luzon in Philippines

By the United Press.

WASHINGTON, Dec. 8. — Casualties on the Hawaiian island of Oahu in yesterday's Japanese air attack will amount to about 3000, including about 1500 fatalities, the White House announced today.

Several other American ships were damaged and a large number of army and navy airplanes on Hawaiian fields were put out of commission, the White House disclosed.

It reported at the same time that American operations against Japan were being carried out on a large scale, resulting already in the destruction of "a number of Japanese planes and submarines."

The White House statement said:

"American operations against the Japanese attacking force in the neighborhood of the Hawaiian Islands are still continuing. A number of Japanese planes and submarines have been destroyed.

"The damage caused to our forces in Oahu in yesterday's attack appears more serious than at first believed.

"In Pearl Harbor itself one old battleship has capsized and several other ships have been seriously damaged.

"One destroyer was blown up. Several other small ships were seriously hurt. Army and navy fields were bombed with the resulting destruction of several hangars. A large number of planes were put out of commission.

"A number of bombers arrived safely from San Francisco during the engagement—while it was under way. Reinforcements of planes are being rushed and repair work is under way on the ships, planes and ground facilities.

"Guam, Wake and Midway Islands and Hong Kong have been attacked. Details of these attacks are lacking.

"Two hundred marines—all that remain in China—have been interned by the Japanese near Tientsin.

"The total number of casualties on the island of Oahu are not yet definitely known but, in all probability, will mount to about 3000. Nearly half of these are fatalities, the others being wounded. It seems ...

(Continued on Page Eleven.)

Roosevelt's War Message To Congress

By the United Press.

WASHINGTON, Dec. 8.—The text of President Roosevelt's war Message to Congress:

TO THE CONGRESS OF THE UNITED STATES:

Yesterday, Dec. 7, 1941—a date which will live in infamy—the United States of America was suddenly and deliberately attacked by naval and air forces of the Empire of Japan.

The United States was at peace with that nation and, at the solicitation of Japan, was still in conversation with its government and its Emperor, looking toward the maintenance of peace in the Pacific.

Indeed, one hour after Japanese air squadrons had commenced bombing in Oahu, the Japanese Ambassador to the United States and his colleague delivered to the Secretary of State a formal reply to a recent American message. While this reply stated that it seemed useless to continue the existing diplomatic negotiations, it ...

(Continued on Page Two.)

Navy Enlistment Office Swamped by Volunteers

Number Quitting Jobs to Join Is Double That of Last War

All over town today young men—and quite a few old ones with wives and children—were going up to the boss and saying "I quit!" By the hundreds, from the Bronx to Staten Island, they overran the Army, Navy and Marine recruiting offices with such gusto that the Navy had to close down for the day shortly after noon.

(Continued on Page Twelve.)

Britain Declares War on Japan

Becomes Ally of U. S., China and Thailand

By the Associated Press.

LONDON, Dec. 8.—Britain, like the United States under Japanese attack, declared war today on the Tokyo government without waiting for Washington first to formulate an American declaration.

Said Prime Minister Churchill: "It only remains now for the two great democracies to face their tasks with whatever strength God may give them."

(Continued on Page Twelve.)

All Draft Quotas To Be Increased

Japanese Claim Naval Supremacy

Four Battleships, a Carrier ...

Japanese Planes Set Big Fires in Manila

Gasoline Dumps Are Believed ...

World-Telegram Index

Churchill Tells ...

The Weather

Targeting oil installations Experiences from the European war clearly had had an impact on the economic defence preparations in the Dutch East Indies. After its early reluctance to collaborate with the *Staatsmobilisatieraad*, Bataafsche drafted detailed plans for the demolition of oil wells, stocks, and installations on all its many sites in the archipelago. The execution was now entrusted to company staff, who received formal military rank and uniforms to give them protection under the Geneva Convention. Weekly exercises made sure the squads were thoroughly trained in their tasks. Though none of the destruction plans have survived, the general plan of action at all sites appears to have been the same, targeting stocks, storage tanks, and auxiliary equipment such as power generators, stationary engines and electric motors, compressors, pumps, and pipelines, leaving the main installations intact. Wells were blocked in a variety of ways, by destroying casings, cementing shafts top and bottom, forcing pieces of equipment down the hole, and finally by dynamiting them. As the Japanese army advanced through the archipelago, the Group's producing and manufacturing sites underwent the same fate, one after another. In effecting the demolitions, staff at times appear to have gone considerably beyond the original directive of making installations unfit to use for a period of six months. We do not know how all individual sites fared. At Balik Papan, the Group's second biggest manufacturing complex in Indonesia, the demolition schedule had a group of 120 employees working five to six hours, which in the event took longer. The installations appear to have been quite effectively destroyed; the Japanese did not attempt to rebuild them, presumably because the navy, to which the Borneo fields and plants had been assigned, could use the heavy oil from Borneo and Tarakan without refining.[103] Pangkalan Brandan, Tjepu, and possibly Wonokromo seem to have received less damage. When the remainder of the colonial army in the Dutch East Indies capitulated on 28 March, only Pladju, Bataafsche's biggest and most modern refinery in the country, had escaped almost unscathed, because parachute troops captured it before destruction could begin. The refinery did suffer from a heavy bombardment by Allied aircraft in 1945, however. Despite their militairy uniform and the protection due under Geneva Convention, many Group staff were executed by the Japanese forces for having participated in destroying oil stocks and installations.[104]

Once again, the overall damage from demolitions appears to have been somewhat less than one would have thought given the drama with which contemporary accounts describe them. The Group lost about one million tons of stocks, the equivalent of slightly more than two months' production.[105] The Japanese occupying forces succeeded in restoring oil production and manufacturing with remarkable speed. In 1943, crude production in the Dutch East Indies reached 75 per cent of the 1940 level, and refinery output just over 40 per cent, sufficient for Japan's needs.[106] Output from Group properties averaged more than 55 per cent of the 1941 level, ranging from 32 per cent in Tarakan to 60 per cent in Pladju and 72 per cent in Tjepu.[107] The destruction of properties at Stanvac's subsidiary NKPM also appears to have varied greatly.[108] By contrast, the human sacrifice from the war and the occupation was very great. In 1941, the Group employed almost 1,800 Europeans and probably around 20,000 Asians in the Dutch East Indies.[109] Of the European staff, 317 employees and thirty of their wives died during the war, either killed in the fighting, executed for sabotage, or perishing during evacuation or in captivity.[110] We do not know how many of the Asians in Group employment died. The occupying forces would have used a large number of labourers to get the oil industry operating again. Given the fact that the Japanese

[73]

[74]

The war reaches Balik Papan with 'the disconsolate sight of one of millions of installations' and (main picture) 'irreparably devastated tank parks'.

[77]

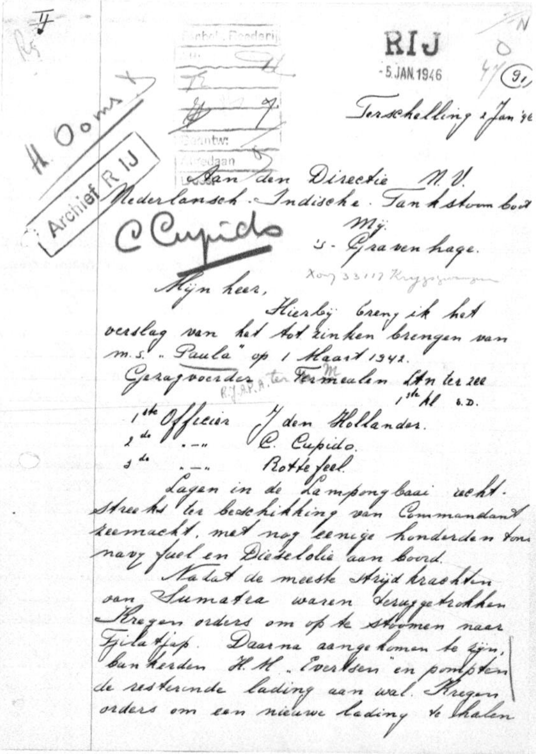

RIJ
-5 JAN 1946

H. Ooms
Archief RIJ

Terschelling 2 Jan 46

Aan den Directie N.V.
Nederlandsch-Indische Tankstoomboot Mij.
's-Gravenhage.

C Cupido

Mijn heer,

Hierbij breng ik het verslag van het tot zinken brengen van m.s. "Paula" op 1 Maart 1942.

Gezagvoerder Fermeulen Ltn ter zee 1ste kl.

1ste Officier J. den Hollander.
2de " C. Cupido.
3de " Rottefeel.

Lagen in de Lampongbaai rechtstreeks ter beschikking van Commandant Zeemacht, met nog eenige honderden tons navy fuel en Dieselolie aan boord.

Nadat de meeste strijdkrachten van Sumatra waren teruggetrokken kregen orders om op te stoomen naar Tjilatjap. Daarna aangekomen te zijn, bunkerden H.M. "Evertsen" en pompten de resterende lading aan wal. Kregen orders om een nieuwe lading te halen

[78]

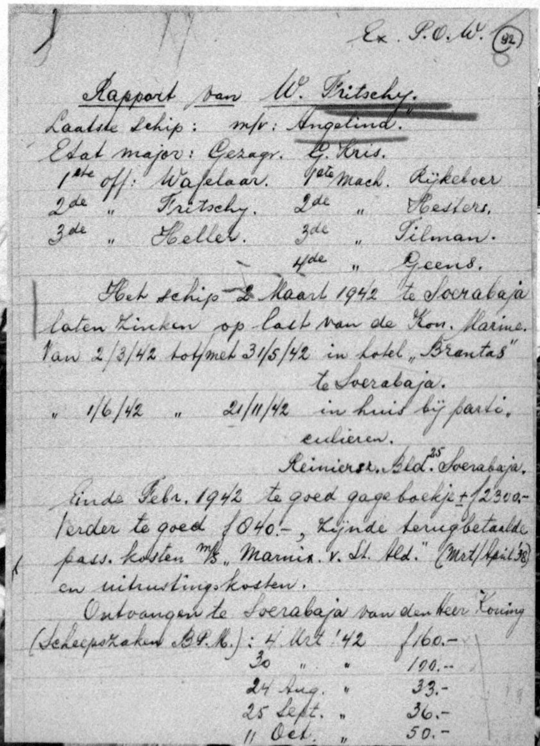

Ex. P.O.W. (92)

Rapport van W. Fritschy.

Laatste schip: m/v: Angelina.
Etat major: Gezagr. G. Kris.
1ste off: Wafelaar. 1ste mach. Rijkelvoer.
2de " Fritschy. 2de " Hesters.
3de " Heller. 3de " Tilman.
4de " Geens.

Het schip 2 Maart 1942 te Soerabaja laten zinken op last van de Kon. Marine.
Van 2/3/42 tot/met 31/5/42 in hotel "Brantas" te Soerabaja.
" 1/6/42 " 21/11/42 in huis bij particulieren.
Reiniersz. Bld. 25 Soerabaja.

Einde Febr. 1942 te goed gageboekje ± f 2300.-
Verder te goed f 840.-, zijnde terugbetaalde pass. kosten m/s "Marnix v. St. Ald." (Met/April 38) en uitrustingskosten.

Ontvangen te Soerabaja van den heer Koning (Scheepszaken B.P.M.):

4 Aug '42	f 160.-
30 " "	190.-
24 Aug. "	33.-
25 Sept. "	36.-
11 Oct. "	50.-

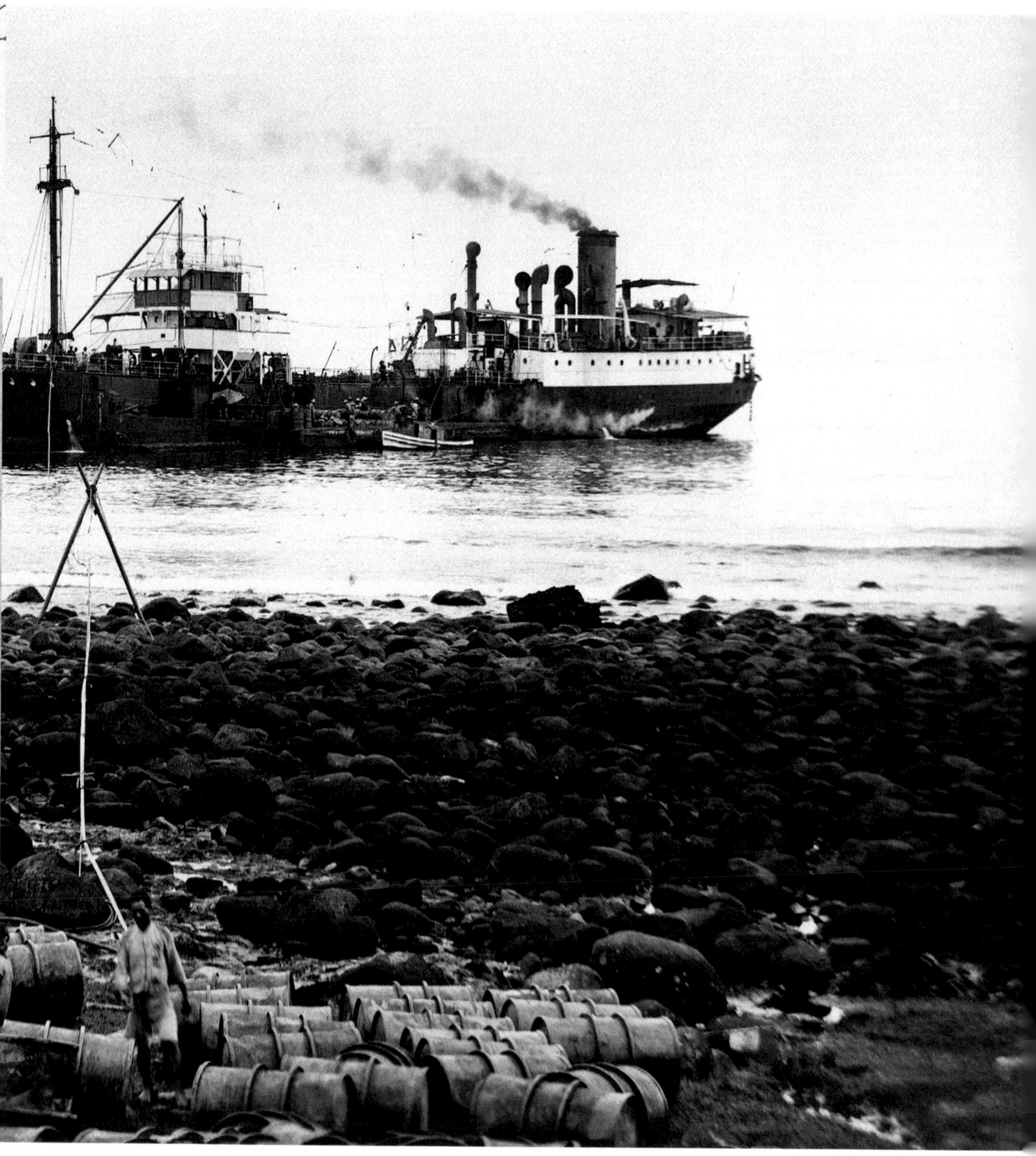

[76]

In the Far East, Shell's sea-captains received orders to sink their own ships. Top left, at the harbour of Tandjong Priok, *Paula* was ordered to be sunk behind the mine field in order to create an obstacle for enemy shipping. Firstly secret papers were destroyed, then instruments and arms; then at 9 p.m. on 1 March 1942 the ship was sunk. The whole crew was taken ashore safely and its Chinese members were paid off. The next day (bottom left) at Surabaya, *Angelina* was the last of the Royal Dutch ships in the region to be scuttled. The main picture shows *Angelina* in happier days, anchored of Formosa on 23 February 1937.

treated the indigenous people forced to work for them far worse than they treated the Europeans, we must assume that upwards of 3,200 Asian Group employees died during the war as well.

In the end, Japan's ability to sustain the war foundered on the Allied efforts to cut its oil supplies by targeting the tankers transporting them from the Dutch East Indies. Oil imports to Japan fell from 1943 and by late 1944 the supply position had become critical, imposing heavy restrictions on all military operations.[111] Though it was hard fighting followed by the two atom bombs which secured the Allied victory over Japan in August 1945, it was the stranglehold on oil which prepared the ground.

The Allies applied the same strategy in Europe, and with very similar results, although achieved over a much longer period. The first target in denying oil to Germany was of course Romania. From the outbreak of war in 1939, Britain and France attempted to lock up supplies by purchasing oil on long-term contracts, Shell Union taking up Astra Romana's output. The two Allies also tried to deny Germany oil by tying down transport facilities. As Romania started moving from neutrality to an alliance with Germany during 1940, this became more and more difficult. In September a pro-German general forced the King's abdication and installed himself as dictator, after which the Romanian industry was increasingly drawn into the Axis war effort.[112]

During the autumn of 1940 Astra Romana was brought into the German fold when Bataafsche's *Verwalter* appointed a new pro-German board and a Dutch Nazi general manager, J. H. W. Rost van Tonningen. Rost had had a good career with Bataafsche as an engineer, rising to the position of visiting technical inspector of the Group installations in Italy, Austria, Hungary, Yugoslavia, and Romania. During the 1930s he developed a keen interest in fascism, joining the Dutch Nazi party, NSB, probably in 1935. A month before the German invasion of the Netherlands, Bataafsche suspended Rost from work because of his dubious political allegiance. Reinstated in July 1940, presumably through the influence of his brother, a prominent leader of the Dutch Nazi movement, Rost fitted the German intentions perfectly. With the Nazis' domination of Europe seemingly secured, he regarded it as his task to steer Astra into a strong position within the new economic order, independent of, yet closely allied with, German interests. Consequently, he took the initiative to build an alkylation plant for making 100-octane gasoline at Astra's Ploesti works and he accepted a deal which gave Kontinentale Öl AG, which the Nazi government wanted to nurse into a big international oil company, a half share in Astra's concessions. Construction on the alkylation plant started in 1941, but the war circumstances prevented its completion.[113]

During the war, Romania's crude production declined, but deliveries to Germany climbed from 1.4 million tons in 1940 to 3 million tons three years later, about one-third of oil consumption.[114] Once Romania had joined the Axis in November 1940, the Allies started planning to knock out the oil industry by bombing. In June 1942, Allied aircraft mounted a first attack on Ploesti, the main manufacturing centre for all companies producing in Romania, and despite sometimes appalling losses in men and machines the raids continued at a rising pace until August 1944, when Romania switched to the Allied side and concluded an armistice with the Soviet army. The Germans had driven huge numbers of forced labourers to rebuild the oil refineries around Ploesti.[115] The scale of the devastation is not exactly known, but Astra's plant emerged from the war in a surprisingly good condition overall. Some of the installations had been knocked out, others operated at a reduced capacity; major units such as three of the four Dubbs crackers, the Trumble plant, the vacuum distiller, and the gasoline reformer were still in full working order. Within months the refinery was back at 77 per cent of its pre-war processing capacity.[116] The Allied bombing offensive against oil installations in Germany itself, which started in

[79]

[80]

May 1944, was rather more successful. Within four months, the output of synthetic gasoline had dropped from 92,000 to only 5,000 barrels per day. The production of aircraft fuel had dwindled to almost nothing, effectively grounding much of the *Luftwaffe* and giving total air superiority to the Allies. The Pölitz hydrogenation works, in which the Group's German operating company had a one-third stake, shared in the destruction. The company's other main installations appear to have escaped the worst. The Hamburg installations are said to have been heavily damaged; the plants in Monheim and in Reisholz, and the Regensburg bulk installation, received little damage, and the Voltol lube oil works in Freital near Dresden none at all. Oral Group tradition has it that managers supplied the Allied air forces with the plans of installations in Germany, thereby helping to destroy them more effectively. We have found no material in Group records to corroborate that story, but it is probably true because there existed a committee of four members, including an oil industry expert from Shell and one from Jersey Standard, which assisted the British Air Ministry in interpreting bomb damage to oil installations. However, the varying level of destruction at the German plants suggests that any information passed through that committee about them had little practical effect on the extent of the damage inflicted.[117]

Manufacturing for victory

During the course of the war the Group not only lost a third of its crude output, but also nearly a third of its primary processing capacity.[118] As a consequence, Jersey Standard built up a lead in this respect as well, its refinery capacity increasing by over 50 per cent between 1940 and 1945, nearly all of it in the United States. At Shell Union, the only Group company for which we have comparable figures, refinery capacity rose by a comparatively meagre 27 per cent.[119] However, the essence of Group wartime manufacturing lay not in the overall growth of operations. Presumably because of its relative crude shortage, Shell gave precedence not to simply raising volume, but to upgrading plants, and to developing and making a remarkable variety of sophisticated hydrocarbon components and specialist hydrocarbon derived products. This policy was at the same time the direct result of, and a further stimulus to, the Group's deepening commitment to research and chemicals since the 1920s (see Volume 1, Chapter 6). One simple figure, again from Shell Union, highlights the preference for product quality and product range over volume: with only 6 per cent of America's refinery capacity, the company produced more than 13 per cent of the country's aviation gasoline.[120]

Indeed, the search to maximize the production of aircraft fuel boosted many if not most of the manufacturing innovations adopted or developed by the Group during the war. A very

Aimed at cutting off German oil resources, the Allies counterattack included the bombing (far left) of Ploesti, Romania, by US forces, the destruction (left) of Harburg refinery, Germany, and (right) the colossal blaze at Harburg, June 1944.

[81]

important breakthrough was the introduction of fluid bed catalytic cracking, developed by a consortium of oil companies, including the Group. The process involved aerating a fine powder of alumina and silica compounds and circulating it at such high velocities that it took on the flow characteristics of a liquid, enabling it to mix thoroughly with and crack the incoming feedstock. After passing through a cyclone separator, the cracked vapours passed into a fractionating tower to be condensed into the desired products; the catalyst was recycled by burning off the coke deposits, and then used again. Catalytic cracking marked a breakthrough in the manufacturing of high-octane gasoline, because it produced a base stock of 80 octane, much higher than that of conventional thermal cracking, thus reducing the quantity of tetraethyl lead and iso-octane needed to bring gasoline up to the 100-octane level. After the formation of a patent pool in 1942, the process came into general use in the United States as part of a government-sponsored aviation gasoline programme, launched to provide incentives for the construction of the massively expensive cat crackers. Shell Union built its first cat cracker under this scheme at Wilmington; it came on stream in November 1943, followed a few months later by two more at Wood River. The installations were combined with plants to manufacture alkylates from the offgases from cracking, and with special hydrogenation units to apply Shell's Base Stock Hydrogenation Process, which eliminated the tendency of cracked gasoline to form deposits during storage and use, an old problem with which Group scientists had wrestled since the mid 1930s.[121]

Another important step in expanding the production of high-octane gasoline consisted of taking the alkylation process for making iso-octane further by another process, namely isomerization. Like alkylation, this had been developed simultaneously by different companies using different methods: the Group's first isomerization unit, designed to convert readily available butane into its less available isomer, isobutane, came into operation in June 1941, and another Shell process converted pentane and hexane into their isomers, all of which could then be alkylated. Alkylates, that is to say, the octanes created by alkylation, could now be produced in practically unlimited quantities, and this and their high anti-knock value made them the absolute bedrock of wartime aircraft fuel production.[122]

The Curaçao refinery did not get a cat cracker during the war, but the Group built another Dubbs cracker there, plus two alkylation units, a high-vacuum distiller, an isomerization installation and a cumene plant.[123] As discussed above, Curaçao's capacity for manufacturing gasoline depended on the limited storage facilities for fuel oil and asphalt, but the refinery had no such limits when it came to making gasoline components. In 1941, the Group also started making 'avaro' at Curaçao. Avaro, short for aviation aromatics, was a good example of Shell's ability to maximize the yield of 100-octane without undue waste or expenditure. Made by blending benzene, cumene, toluene and similar aromatics, it could include fractions left spare from other processes, and it did not need new technology – rather the opposite: it was thermally cracked in units normally used for civilian gasoline production which, because of the new emphasis on military products, would otherwise have been standing idle.[124] Its drawback was low volatility: it could only be used mixed with highly volatile isomers, which were scarce.[125]

Nonetheless, avaro formed another strand in meeting the 100-octane production challenge, as did cumene or isopropyl benzene, an aromatic hydrocarbon with a very high anti-knock value. Building on its long familiarity with the Edeleanu extraction process, the Group occupied a strong position in the manufacturing of aromatic hydrocarbon mixtures, used in Britain for mixing special performance boosting fuels. With two of the most desirable aromatic hydrocarbons toluene and benzene in short supply, Group scientists chose cumene as an alternative. In February 1942, Shell Union's Emeryville laboratory had succeeded in developing a process for producing synthetic cumene by alkylating benzene with propylene and four months later manufacturing started in the US and at Curaçao. Having a very high boiling point, no more than 10 per cent of cumene could be added to a given volume of aviation gasoline, but cumene made a disproportionate contribution. Shell offered the formula free of charge to all other oil companies, and US production of 100-octane gasoline rose from approximately 2.5 million tons a year at the beginning of 1942 to 7 million by the end of 1943; and it has been calculated that 23 per cent of the increase was because of cumene. Annual production continued to rise to 16 million tons by mid-1944 and over 20 million by the start of 1945.[126] Meanwhile in 1943 the Group had taken the development of performance boosters one stage further with the production of xylidine, also known as cumene substitute or CS. Added to 100-octane fuel, 3 per cent of xylidine plus a very small additional amount of tetraethyl lead created a 'super-fuel', an extremely rich blend rated at 100/150-octane. The development of this began with Shell in the UK and then continued at Emeryville, the result being a fuel which gave fighter planes an additional 15 per cent emergency acceleration at full throttle. In 1943, Shell Union undertook to build and manage a xylidine plant for the US army at the Cactus ordnance works in the Texas Panhandle district. It came on stream within seven months, but was shut again after a year of operations, because the Allied air forces had achieved air superiority in all theatres and no longer needed performance boosters.[127]

As during the First World War (see Volume 1, Chapter 3), the Group devoted great efforts to relieve the scarcity of toluene, the vital ingredient for high explosives. Already before the war Bataafsche's Amsterdam laboratory had begun looking for cheaper alternatives to the then standard method of extracting toluene from crude oil by the Edeleanu process. During the war Shell Development's Emeryville laboratory continued this research and found a cheap and efficient extractor in carbolic acid (phenol), which produced toluene that was at least 99 per cent pure, and on 2 December 1940 a unit for the purpose was completed at Houston. Extractive distillation builds elegantly upon the principle that, when a mixture of two components cannot be separated by normal fractionation, a third may be added to lower the vapour pressure of one of the two and dissolve it out. In the case of toluene, by adding carbolic acid, the toluene was dissolved out while the other fractions were distilled off as usual, and the resultant mixture was redistilled to separate the toluene and the phenol, which could be used again. But the elegance of the method was marred, at least as

[82]

The Allied counterattack also included the construction of numerous oil installations to enhance their own resources. Bottom, photographed in 1954, the cat cracking unit at Wood River was built during the war. Three more were built at Houston, Pernis and Stanlow after the war was over.

Centre, Shell Chemical's Houston plant was already under construction in the summer of 1941, before the US came into the war. Top, taken in the 1940s, is a general view of the gasoline hydrogenation unit at the Dominguez refinery.

[83]

[84]

regards toluene, by the fact that even in the richest toluene-bearing fractions, the yield was only about 1.5-2.2 per cent of the volume processed. Foreseeing this, the Emeryville scientists sought to make synthetic toluene by too from other hydrocarbons – specifically and successfully, by using a combination of catalysis and dehydrogenation to remove six atoms of hydrogen from each molecule of methyl cyclohexane; and to ensure supplies of the latter, they also applied an isomerization process to dimethyl cyclopentane. Shell Union immediately started building synthetic toluene at Dominguez and at Wood River, which came on stream in December 1942 and January 1943 respectively. However, Jersey Standard's alternative method of producing synthetic toluene with the Edeleanu process proved superior and, after overcoming some early problems, provided a greater percentage of wartime toluene needs than did the Group's process.

Aromatics such as toluene and cumene of course belonged to the core business, but the Group ventured far beyond that in the search for products to support the Allied war effort. In a perverse transposition of Kessler's 1927 wish to turn offgases into 'something we can ship', managers in May 1940 even considered a scheme to manufacture poison gas from ethylene in the UK and in France.[128] Vanadium provides a better example of the Group extending its product range. Tiny amounts of this soft silver-grey metal, which adds strength, heat-resistance and toughness to steel, appear in Venezuelan crude. Group engineers found a way to extract it, resulting in enough supplies for the treatment of 85,000 tons of steel.[129] The most spectacular extension, into synthetic

When sources of natural rubber were cut off by the war, butadiene became vital as an important constituent of synthetic rubber. With spherical storage tanks such as these at Signal Hill, California, the high internal pressure of the liquefied gas was resisted equally in every direction and the gas remained liquid. Inset, the formula for butadiene.

beta-butylene		2-3 dichlorobutane		1-3 butadiene
CH_3		CH_3		CH_2
|		|		‖
CH	Chlorine →	CHCl	HCl → split off	CH
‖		|		|
CH_3		CHCl		CH
|		|		‖
CH_3		CH_3		CH_2

rubber, laid the groundwork for a full-scale diversification. Synthetic rubber had been manufactured in Germany as early as 1901, and received a good deal of attention there in the years between the World Wars. As with alkylation and isomerization, different companies experimented with different methods: the German method made butadiene, the principal constituent of synthetic rubber, from acetylene gas, but this was expensive.

American experiments focused on the possible use of cheaper hydrocarbons. In 1938, Emeryville had a pilot plant on stream for making synthetic butadiene from butylene, using a process of chlorination. Three years later Shell Chemical was first to open a commercial plant, adjacent to Shell Union's refinery at Houston. Initially this process seemed a great success, and in normal peacetime commercial terms it was: in 1942, Houston produced 4,000 tons of butadiene, enough for over 1 million car tyres. Yet the system could not meet the huge demands made on the industry in war, and, as with synthetic toluene, an alternative process worked out by Jersey Standard proved better geared for high demand; so it was not until the companies were allowed to pool their knowledge that real success in this regard was found.[130] This pooling formed part of an American government programme designed to overcome the serious shortage of natural rubber. Under this programme, the Group agreed to build and manage a factory producing 25,000 tons of butadiene a year using Jersey's butylene dehydrogenation process. The site chosen was Torrance, near Los Angeles, in close proximity to suppliers of hydrocarbon gases. The plant, which employed special purification facilities to upgrade the products of other manufacturers engaged in the programme, started production in July 1943; by the end of 1944, the rubber shortage had turned into a surplus and synthetic rubber had become commercially viable.[131] In butadiene production as in other manufacturing areas such as 100-octane gasoline and toluene, US government funding provided crucial support. In 1942 the Anglo-Saxon board estimated the cost of the Group's five current war-related projects in the US 122 million dollars, of which the government had financed almost two-thirds, with the rest coming from the Group. With the output contracted to government agencies at fixed terms, the pay-out times were three years or better. Thus by providing management, know-how, entrepreneurship, and a minimum of finance, the Group could explore various manufacturing processes at close range, low risk, and very little cost.[132] However, one immensely valuable opportunity went for nought. In 1937, Shell Development had nursed a process for making synthetic glycerine through the pilot plant stage, but subsequent discussions with the chemical company Du Pont about a joint venture failed to reach a conclusion. Wartime construction priorities then intervened and it would take another decade until the Group launched synthetic glycerine production at Houston.[133]

Group employees and laboratories demonstrated their ingenuity not only in manufacturing, but also in many other areas of vital wartime research. They assisted in the development of radar, devised a chemical additive which jellified ordinary gasoline and made it into a devastatingly efficient fuel for anti-personnel flame-throwers, and developed phenolic inhibitors to enable the long-term storage of aviation gasoline. They were constantly testing fuels for automotive and aviation engines, including jet engines; they developed substances to inhibit the formation of rust; they invented a new method of making penicillin and developed an infra-red spectroscope for analyzing complex hydrocarbon streams. Another interesting innovation was the portable pipeline, designed by Sydney Smith of Shell Union and of great value in assuring the

[87]

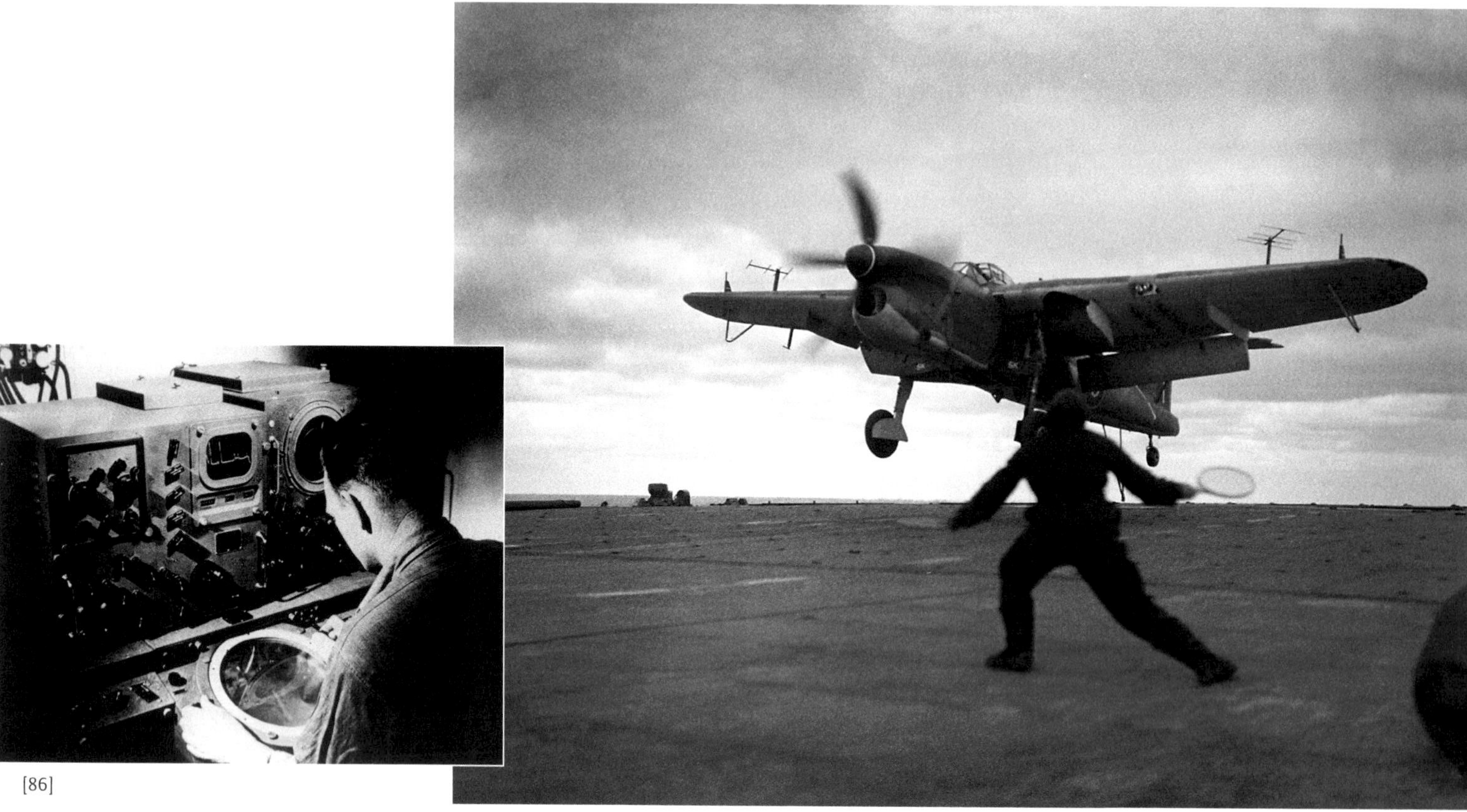

[86]

ability of oil supplies to keep up with fast-moving armies.[134] These innovations did not lead to fully-fledged commercial operations; only the Group's chemical operations came into their own. During the late 1930s the contours of a tentative programme for chemicals encompassing butadiene, fertilizers and pesticides, detergents, and resins had begun to show (see Volume 1, Chapter 6). However, in 1940 the main products were still fertilizers, alcohol, and solvents, all produced on a commercial if modest scale at Stanlow, Pernis, Dominguez, and Martinez. The start of solvents and butadiene production at the Houston chemical plant in 1941 and of ester salts production at Stanlow in 1942, followed two years later by the decision to build a new solvents plant at Stanlow, raised chemicals from an interesting offshoot to a business function in its own right, closely integrated with manufacturing oil.[135] Chemicals proved to have a ready market and seemed to have a bright future, Shell Chemical Corporation and individual products such as Teepol making handsome profits.[136] Pesticides, which the Group had already sold before the war, increasingly attracted attention as a bright prospect. In 1944, managers intensively studied a proposal to set up an agricultural research station in Argentina with a view to making pesticides there, only to drop the idea because of the unsettled conditions in the country.[137]

Indeed, chemicals achieved such a scale that managers faced issues of control and priority. In February 1943, a memo written for the Anglo-Saxon board spelled out the difficulty of drawing a boundary between oil and chemicals, using the production of alcohols, ketones, and ammonium sulphate as an example: 'Shell Chemical produces secondary butyl alcohol from a butane-butylene fraction which it acquires from the refineries of Shell Oil. In the preparation of the feed for this process isobutylene is removed and returned as polymer to Shell Oil for use in the production of aviation gasoline. Shell Chemical in turn converts

During the Second World War the Allies were aided by many new inventions. Far left, a member of the US Navy is running a check on some new experimental radar equipment, and centre, radar antennae are visible on the aircraft's wings. One of Shell's developments – macabre but effective – was the jellification of petrol for use in flame-throwers, enabling US Marines in the Marshall Islands (left) to clear buildings held by Japanese forces.

secondary butyl alcohol to methyl ethyl ketone, and the hydrogen thereby produced is sold to Shell Oil where it is also used in the manufacture of aviation gasoline. Acid used in the manufacture of alcohol is later used by Shell Oil in gasoline treating and is then returned again to Shell Chemical for the production of ammonium sulfate.'[138]

From this position, the paper went on to argue that the operations of the two companies were so intertwined that it would be better to merge them which, it was hoped, would result in 'a greater mutual appreciation of the problems of each business and a closer coordination between their respective operations. This would eliminate any possible conflict in interest and would be of particular value in the period of expansion and development that lies ahead.'[139] The London managers were not immediately enthusiastic. In their view, the long-term value of Shell Chemical to the Group might well be higher than the price now to be obtained; having Shell Union purchase Bataafsche's half-stake in Shell Chemical might attract undesirable press publicity about the relations between these companies; keeping Shell Chemical independent might be wiser considering the possibility of the US government obtaining a tighter grip on the oil companies.[140] These objections did not convince the American managers and in July 1943, Shell Union bought out Bataafsche, reorganizing Shell Chemical into a company division. Anglo-Saxon clearly preferred chemicals to remain stand-alone. In February 1945, the chemical operations in Britain were lifted out of Shell Refineries and Marketing and transferred to the newly formed Shell Chemical Company Ltd.[141]

In allen Betriebsstellen unserer Organisation, in den Werken und Lagern, in den Laboratorien und wissenschaftlichen Versuchsabteilungen wird mit größter Anspannung für Aufgaben geschaffen, die heute vor allem wichtig sind. Was dabei an Erfahrungen gewonnen, an Verbesserungen erzielt wird, dient heute schon der Kriegswirtschaft und wird morgen unserer gesamten Kundschaft zugänglich sein.
RHENANIA-OSSAG

An advertisement from the March/April 1941 issue of *Der Ring*, house magazine of the Group's company Rhenania-Ossag. Though under German control since January 1940 and thus formally no longer part of the Group, Rhenania-Ossag still used the pecten logo, at least in in-house publications such as this.

Under the shadow of the Nazis

The German occupation of most of the European continent effectively took the operating companies there outside the Group's control. We have already noted the Nazi government's appointment of a *Verwalter* for Rhenania-Ossag in January 1940; the Bataafsche *Verwalter* subsequently assumed formal control over the companies in countries under German occupation or in the German sphere of influence, such as Hungary.

Of the Group's companies under Nazi control only Astra, Rhenania-Ossag, and Nafta Italiana continued operating at their former levels. As we have already seen, Astra was drawn into the German war effort. As one of the two biggest German oil companies and the main lube oil manufacturer, Rhenania-Ossag was an industry leader in the country. Following Hitler's annexation of Austria and Czechoslovakia, Group managing directors sanctioned Rhenania-Ossag taking over the Shell companies in those countries.[142] With the rupture of overseas supplies, Rhenania-Ossag turnover plummeted, but the company formed part of the official oil cartel and thus had a share in the processing and distribution of any oil coming in, which assured a steady, if meagre, flow of revenues. In December 1940 the *Verwalter* activated the hidden financial reserves built up during the 1930s to raise the company's capital from 75 million to 120 million Reichsmarks. A year later Rhenania-Ossag floated a bond loan of RM 60 million to pay off an old loan from Bataafsche and finance some new installations.[143] Meanwhile the relationship between parent company and subsidiary had to some extent been reversed by the appointment of Rhenania-Ossag's research director as *Verwalter* over Bataafsche's Amsterdam laboratory, to ensure that it would contribute to the German war effort.

The Group similarly lost control over its subsidiary Nafta Italiana in July 1940. After Italy's entry into the war, the Mussolini government sequestered the company, putting the Group general manager, W. de Graan, in prison and appointing the president of AGIP as caretaker manager. Since Bataafsche held 80 per cent of the Nafta shares and Anglo-Saxon only 20, von Klass vigorously protested and attempted to get the sequestration overturned. However, in 1942 the Italian government fulfilled a long-standing ambition to create a national oil company and merged Nafta into AGIP.[144]

[90]

[91]

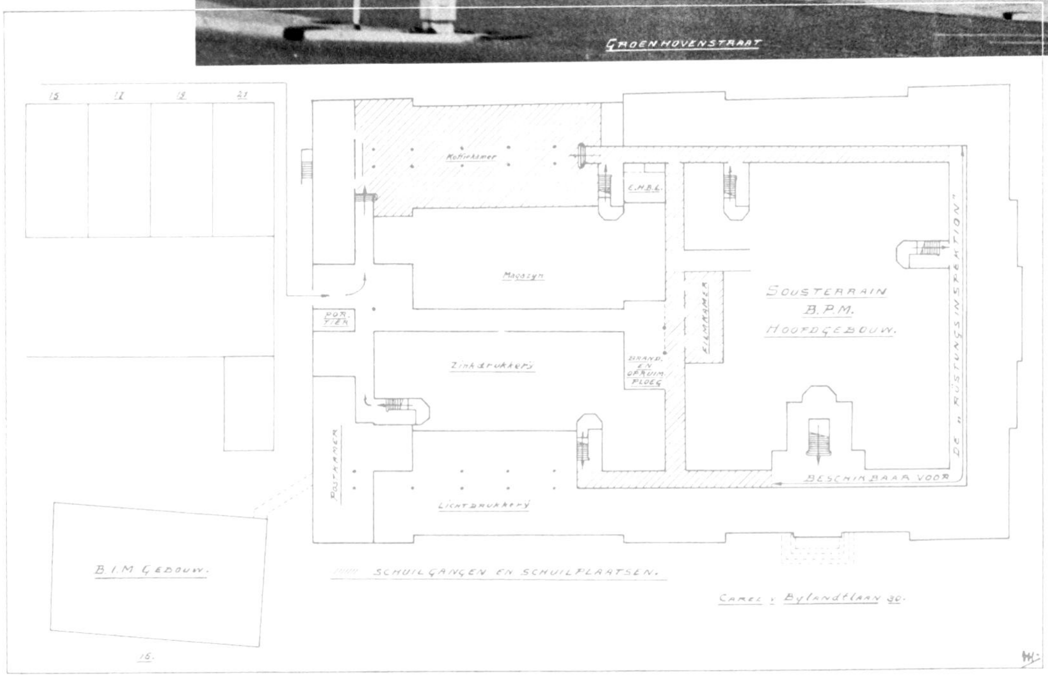

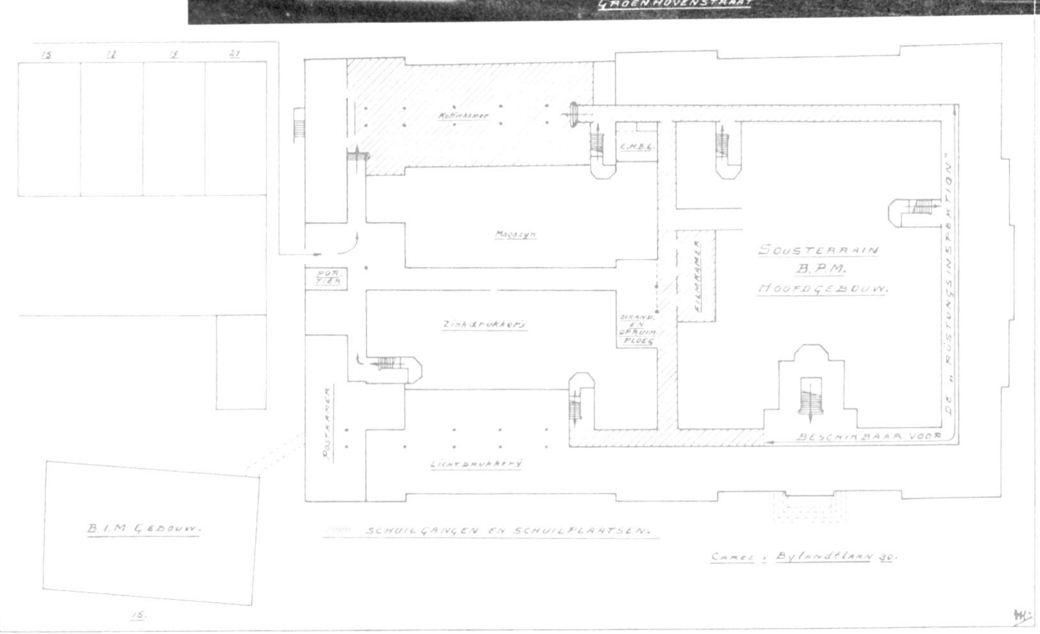

Above, as an epitome of the Nazi occupation of the Netherlands, the Swastika flag flies on the classic Dutch façade of Royal Dutch's head office at 30 Carel van Bylandtlaan in The Hague. In the offices' ground plan, left, preparations for their defence include emergency exits and shelters.

[92]

[93]

All other European companies had to reduce their operations sharply following the suspension of overseas imports. Norske Shell dismissed 30 per cent of its staff, and Danske Shell a full 50 per cent.[145] In July 1940, Bataafsche put 2,376 of its 5,373 employees in the Netherlands on half pay. The sale of remaining stocks from Pernis and elsewhere plus various other sources of income generated sufficient revenue to pay the skeleton staff and the company ended the war with a small surplus.[146] By December 1943, Pernis still had some stocks left, a little gasoline, some white spirit, substantial quantities of lube oils and fuel oil, and a surprisingly large volume of butane gas, more than ten times the volume present in May 1940.[147] Some employees found work in the government-organized distribution of oil products. Bataafsche kept other staff in work by forming a joint venture with the German company Gewerkschaft Elwerath for E&P operations in the eastern regions of the Netherlands and along the border with Germany.[148] The Amsterdam laboratory and the Delft engine-testing station, which together employed over 1,000 staff, received instructions to start generating their own income by attracting outside work. Amsterdam succeeded so well in getting commissions that in 1943 a memo to von Klass complained that the laboratory did not contribute any serious work to the German war effort. One of the commercial activities undertaken was the modification of motor vehicles to run on butane/propane and gas made from coal and charcoal. Following the growing interest in pesticides, Amsterdam laboratory started investigating DDT, producing small quantities and preparing plans for a pilot plant. Another field of investigation was polyvinylchloride or PVC. Researchers discovered a way for making this plastic from the coke oven gases at MEKOG's IJmuiden plant, which was later to enable Bataafsche to bypass existing patents based on using offgases from cracking oil.[149] Following the same line of thinking, the French Group company Jupiter embarked on large-scale charcoal-burning activities, buying woodlands and setting up a processing and distribution network.[150]

As Bataafsche *Verwalter*, von Klass had the difficult task of running a fairly large and international company which had lost its purpose. He strove to maintain the integrity of Royal Dutch as a holding company with an eye to keeping the business strong so it could serve Germany after the war. To that end he travelled around to exercise supervision over Group companies in occupied Europe, negotiated successfully to acquire the companies in Hungary, Yugoslavia, and Greece, but failed to get hold of companies in neutral countries because he could not show title to the shares, which were in London. When Germany invaded the Soviet Union, von Klass immediately contacted Berlin to resuscitate the Group's claim on its nationalized properties in that country.[151] To make himself agreeable, he had Bataafsche sell idle installations and equipment to German companies and tried to mobilize employees to work in Germany, but apparently without much success.

[94]

Passierschein Nr. I 48559
Laissez-Passer No.

Name: / Nom: Eckhardt von Klass
Beruf: / Profession: Direktor
Staatsangehörigkeit: / Nationalité: Dt. Reich
Wohnsitz: / Domicile: Berlin
Geburtsdatum und -ort: / Date et lieu de naissance: 8.9.97 Greifswald
ist berechtigt, unter Vorlage von / est autorisé présentant: Pass Nr. 164 R/1283/37
in der Zeit vom / à partir du 9.2.41 bis / jusqu'au 23.2.41
einmal und zurück / une fois aller et retour à franchir
die / la frontière belg.franz. Grenze
in Richtung von / dans la direction de Berlin nach / à Frankreich
(bei Rückreise in umgekehrter Richtung) zu überschreiten.
(au voyage de retour dans le sens contraire)
Reiseziel: / Lieu de destination: Paris
Durchreise durch das Gebiet / le voyage par le territoire
ist / sind genehmigt. / est / sont permis.
Bemerkungen: / Observations:
Berlin, den 6.2.41
Für den Generalquartiermeister

Nichtzutreffendes ist zu streichen. / Rayer la mention inutile.

Berechtigt nicht zur Einreise nach und Durchreise durch Holland. / Ce Laissez-Passer n'autorise pas l'entrée en Hollande ni sa traversée.

[95]

1

Hiermit bestätige ich die Uebernahme des Flugbenzins empfangen per Leichter ..Delft..................

Die Siegel (Marke B.P.M.L.A.) waren unverletzt. (12 Stück)

Eine Durchschnittsprobe der Ladung war beigefügt.

Ort........................ Datum.....28.Mai 1940.......

Der Kommandant des Flugplatzes

The *Verwalter* Eckhardt von Klass, caricatured (far left) in 1945 by Johannes Maat (presumably a Group employee), and (above left) Von Klass in his office at 30 Carel van Bylandtlaan. Above right is one of Von Klass's travel documents, and right, a receipt for aviation fuel, both prominently stamped with the Nazi eagle.

[96]

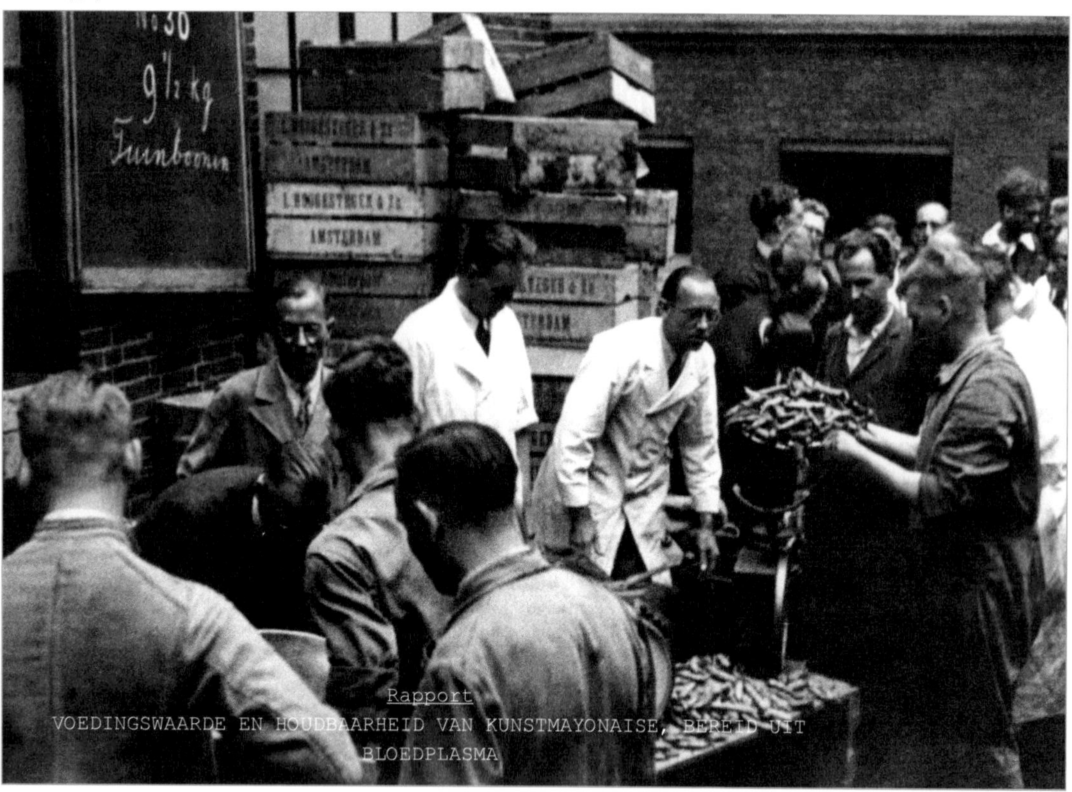

As the war dragged on, the Group's Amsterdam laboratory set up its own food distribution system (above left), and became the site of strange activities forced by shortages of every kind of necessary item. From inside the laboratory came a detailed report (below left) into the best way to make synthetic mayonnaise, and outside, the laboratory's grounds (right) were excavated to considerable depths by employees in search of coal.

Rapport
VOEDINGSWAARDE EN HOUDBAARHEID VAN KUNSTMAYONAISE, BEREID UIT BLOEDPLASMA

De Kunstmayonaise werd bereid volgens het recept ons door den heer Mulder verstrekt.
3 gram gelatine werd opgelost in 30 cm³ water en deze oplossing met 30 cm³ kruidenazijn (4%ig) aan 250 cm³ plasma onder roeren toegevoegd. Het mengsel werd vervolgens ca. 1/2 uur op een waterbad onder intensief roeren verwarmd.

Dit product werd gehomogeniseerd door het tusschen de walsen van een verfmolen te laten vloeien en vervolgens geanalyseerd. Het geanalyseerde product bevat dus niet de voor de practijk gekozen toevoeging van smaakstoffen, n.l. 4 gram mosterd, 1/2 gram keukenzout en 0,2 gram peper.

TOEVOEGING	1½ dag	2½ dag	7 dagen	14 dagen
0.1% salicylzuur in alcohol	goed, reuk naar blik-groente	reuk schimmelig	schimmel gegroeid	schimmel groeit nog steeds
0,[illegible] [illegible]zuur (in alcohol)	goed, reuk naar blik-groente	onveranderd	onveranderd	onveranderd
dubbele hoeveelheid azijnzuur	goed, eenige reuk naar blikgroente	onveranderd	reuk schimmelig	schimmel groeit nog steeds
0,2 Kaliumnitraat, opgelost in water	goed, reuk naar blik-groente	onveranderd	reuk erg schimmelig	schimmel groeit nog steeds
5-voudige concentratie aan azijnzuur	goed	goed	goed	goed

[97]

Von Klass also forced Bataafsche to comply with the Nazis' anti-Semitic policy. In general, the rump organization left in The Hague showed little enthusiasm for his initiatives and seems to have harboured comparatively few Nazi sympathizers. After the war, out of a pre-war total of more than 5,000 staff in the Netherlands, Royal Dutch dismissed fifty for collaboration with the Germans.[152] The managers tasked with implementing anti-Jewish measures delayed their implementation for as long as possible, presumably because they understood the dangers to Jewish colleagues. During 1938 and 1939, various Group companies had exerted themselves to find shelter and work for employees and other refugees from Germany, creating an awareness of the very real threat posed by the Nazi regime.[153] In October 1941, the German authorities in the Netherlands issued a decree for the dismissal of Jewish employees from private companies. The commercial banks fired their Jewish staff that same month, as did Unilever; Philips set up a special department for its Jewish employees in December, hoping to keep them out of harm's way by offering them work under the company's protection.[154] Bataafsche responded rather more

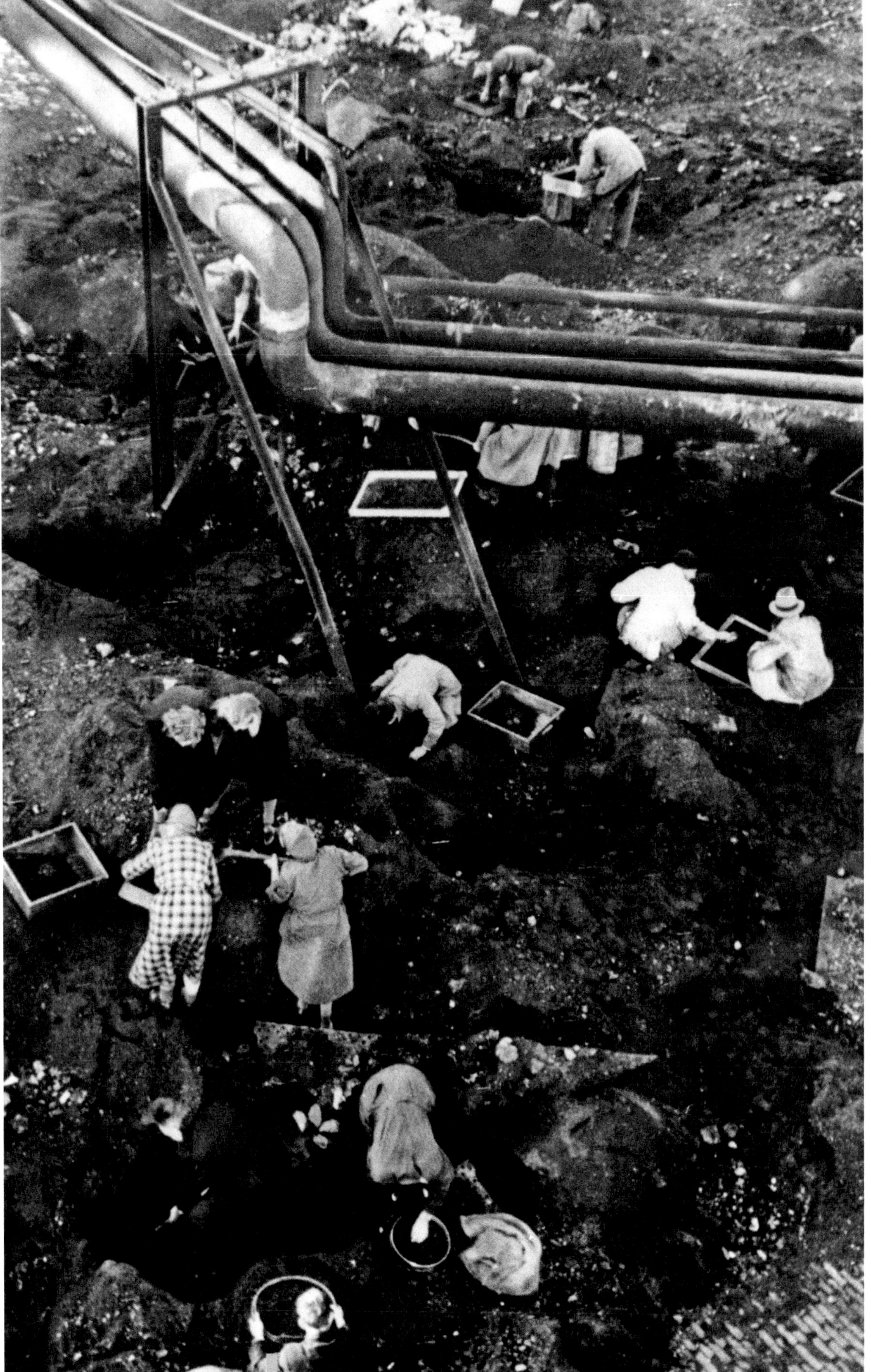

Necessity is the mother of invention, and in Birmingham, UK, taxis were converted to run on gas rather than gasoline. But with steel urgently needed for the armed forces, there was no chance of using pressurized tanks, and the fuel had to be carried in gaseous form, not liquid. One bagful lasted 15-20 miles. In Amsterdam (main picture), not even that opportunity existed, and another solution was found.

[99]

slowly, in February 1942 sending a circular to its Dutch offices instructing them to have the staff complete an *Ariërverklaring*, a form giving particulars about their descent. This survey resulted in forty people being classified as Jewish under the German laws. In March, von Klass ordered them to be put on half pay and then dismissed with effect from the end of April. At least twenty of them did not survive the war.[155]

In September 1942 von Klass volunteered to send men and installations from Bataafsche to help with the planned rehabilitation of the Soviet oil industry, now that the German army had reached the Caucasus and looked likely to conquer the great production regions there. An inspection team toured Group installations to select suitable ones and two Bataafsche engineers, both NSB members, were appointed to head a department entrusted with preparing the mission to the Caucasus. The initial plans envisaged a team of 200-300 men able to operate as an independent unit, which would first dismantle the Petit-Couronne refinery and then rebuild it in the Soviet Union. A call for volunteers resulted in 228 applications, of which 106 came from Bataafsche staff and the rest from outsiders with friends or family members in the company. However, Kontinentale Öl had already contracted a company to dismantle Petit-Couronne and did not want help. The two prospective leaders showed themselves incapable of forming a proper team and, because no senior staff had applied, the mission remained seriously underqualified for any work. In the end, nothing happened; during November 1942 the German push towards Grozny and Baku was beaten back, ending all prospects of

capturing the oil industry.[156] In the autumn of 1943 von Klass halted the sales of idle plant and equipment to Germany and ordered the recommissioning of the Dubbs plant at Pernis as part of a plan to crack Romanian oil there. The order was eagerly taken up; it helped to keep staff busy and thus free from the danger of being sent to work in Germany, and Bataafsche managers reckoned that sending crude from Romania to Rotterdam would prove impossible anyway, as was indeed the case.[157]

The Caucasus project and the attempted restart of Pernis underline the structural weakness of von Klass's position at Bataafsche.[158] The company contributed very little to the German war effort, while the Nazi regime's designation of Kontinentale Öl as the spearhead of its oil policy relegated Bataafsche and its *Verwalter* to the sidelines. Von Klass tried to get back into the game by coaxing Astra into an alliance with Kontinentale and by devising useless projects. For everything he did, von Klass remained heavily dependent on a handful of Bataafsche middle managers who sympathized with the Nazis, but who themselves did not have much grip on the organization either. A March 1943 memo to von Klass summarized the problem concisely. The installations managed by Bataafsche all lay idle and would remain so for the rest of the war. The employees had found a variety of non-essential occupations, studying to improve their job qualifications, or preparing reports on useless subjects such as the geology of Mexico. Even those who were sympathetic to the Nazis found the situation intolerable and attempted to find productive jobs elsewhere.[159] During the last phase of the war the offices ground to a halt. Public transport stopped, preventing people coming into work and as food became ever scarcer, all efforts were concentrated on providing staff and their families with means for survival.[160]

The battle of the boards If Godber had hoped that his May 1940 letter to Van Eck might clear the air between the British and Dutch managing directors, he was disappointed. During the first years of the war, relations between them went from bad to worse until, at an unknown point in time, the two sides finally found each other again in a compromise solution. As Godber's letter showed, the friction originated in a struggle for power which had started after Deterding's retirement as de facto Group CEO in 1936 (see Volume 1, Chapter 7). Deterding had truly acted as a bridge between the parent companies, his very personal leadership style and unquestioned authority preventing differences of opinion between British and Dutch managing directors becoming a wedge.

Once he had gone, a wide crack opened. The central issue was always the question of ultimate control over the Group, a confrontation between insensitive Dutch assertions of their majority and a British failure to accept that the business was not simply a British business. If Dutch directors were too quick to see threats to their position of power, the British directors were prone to present their own interest as the self-evident Group interest, and anything that went against that as harmful to the integrity of the business. Conversely, the Dutch appear to have been more imaginative in devising managerial changes to eliminate bottle-necks in the organization, with the British keen advocates of the status quo, as often as not interpreting proposals for change as Dutch attempts to take control. And of course, each side considered its own contribution to the business as more important than the other, the Dutch likely to emphasize technology and the Dutch East Indies, the British the brand and the power of the British Empire. De Kok had done much to bridge the gap with his tactful personal diplomacy, and the emergence of the term Group managing directors for the five top managers showed him trying to build a team, but the bonding between them remained tentative.

The Group's planning for war immediately widened the crack to a chasm. British managers used the threat to the Group's assets managed from The Hague to press for greater British control over them by the appointment of Godber and Legh-Jones as managing directors of Bataafsche, so they could represent the company in

With his tactful personal diplomacy, J. E. F. de Kok (in the centre of this photo from 1938) did much to heal the managerial rift between Royal Dutch and Shell Transport at that time.

Britain should the Netherlands suddenly be overrun by Germany. Agnew even proposed altering the ownership ratio from 60:40 to 50:50. He did this presumably with the argument that it would reinforce the Group's position with the British government. From time to time in the past, officials had declined to give political support because of the majority foreign ownership. Though the Foreign Office had, as we have seen, pursued the Group's claims in Mexico with vigour, the Treasury probably raised the point again in discussions about currency arrangements and Bataafsche's prospective status as a company resident in the UK. The Dutch directors interpreted the British proposals as driven by a desire to take control and accordingly resisted them. Coupled with the Dutch directors' brinkmanship in assuring the safety of the Group's companies and assets in The Hague, this attitude considerably vexed their British colleagues.[161]

The delicate managerial balance was then upset by the German occupation of the Netherlands and the transfer of Bataafsche and Royal Dutch to Curaçao. The Netherlands was no longer a country, but just one of the many governments in exile, with much dignity but very little power, and probably quite unable to defend the Dutch East Indies against the imminent Japanese attack. These circumstances fanned the chauvinism of the Dutch Group directors, who suddenly found themselves holding a priceless asset which might one day help to ensure the resurgence of the Netherlands. As Kessler and De Booy put it, 'After the war the importance of the Royal Dutch assets relative to the total assets is likely to be still larger than it has ever been in the past. We (we feel) are the Trustees for the Nation in that matter and whether we like it or not we must defend the property.'[162] But they could no longer support the force of their arguments with the power of numbers. In addition to De Kok, all the Group's Dutch non-executive directors remained in the Netherlands, except for one Royal Dutch *commissaris*, Daan Crena de Iongh, who happened to be visiting the Dutch East Indies in May 1940. Following the seat transfer, the Royal Dutch and Bataafsche directors in enemy territory had been suspended from their office, leaving the three managing directors out on a limb facing the full complement of directors on the British side. Shell Transport's formidable team was led by the chairman Walter

Samuel, and the managing directors Bob Waley Cohen and Andrew Agnew. Samuel (1882-1948), the second Viscount Bearsted, had succeeded his father Marcus Samuel in 1921 as head of the family banking firm M. Samuel & Co. and as chairman of Shell Transport. Samuel & Co. was Shell Transport's merchant bank and a respected firm in the City of London, and Bearsted was consequently a person of authority and wide influence. Waley Cohen (1877-1952) had the closest and longest experience in running the Group, having been part of Deterding's original team from 1907 until 1928. He then resigned from active management, but kept non-executive directorships in the principal operating companies, and pursued other wide-ranging interests in business and society. Agnew (1882-1955) had resigned as managing director of Anglo-Saxon in 1937, but he remained managing director of Shell Transport and non-executive director of the principal operating companies. His chairmanship of the Petroleum Board made him the informal leader of the industry in Britain, with access to the corridors of power in both Whitehall and Washington. The two Group managing directors on the Shell Transport side, Fred Godber (1888-1976) and George Legh-Jones (1890-1960), knew each other well from the central period of their career in the United States. Godber became president of Roxana in the same year, 1922, as Legh-Jones assumed the presidency of Shell Oil in California. Whereas Godber returned to London in 1929 to take up his appointments to the boards of Shell Transport and the principal operating companies, Legh-Jones, a tough Welshman with a reputation as a brilliant negotiator, stayed in America until 1934, when he became manager at Asiatic. He joined the Shell Transport board in 1937 and succeeded Agnew as managing director of Anglo-Saxon at the end of that year.[163]

Guus Kessler (1888-1972) was the youngest of the three Dutchmen, but he had the longest experience as a managing director, having been appointed to the boards of Royal Dutch and the principal operating companies in 1922-24. Moreover, he also had a seat on the Shell Transport board since 1929 and so (apart from Deterding and, after Deterding's retirement, De Kok), he was the only director to sit on all five main boards. After a few years cutting his teeth in Russia, Kessler had spent his career based at the London office, where his deep understanding of the business, combined with a thorough grasp of detail and a wide knowledge of operations on the ground acquired by frequent travelling, had

Walter Samuel

[102]

Robert Waley Cohen

[103]

Fred Godber

[104]

George Legh-Jones

[105]

made him a pivotal figure. The long experience of working abroad did not lessen Kessler's feelings for the Netherlands, however, and during the war his British colleagues discovered a fervent chauvinist in him. By contrast, Jan van Eck (1880-1965) or, to give him his full name and title, J.C. van Panthaleon Baron van Eck, had a far more cosmopolitan, businesslike and relaxed attitude towards questions of nationalism than Kessler. Van Eck accepted an apprenticeship with the Group in 1908, following his graduation as a lawyer from Utrecht University. He passed through various departments in The Hague and London before being sent to the United States in 1911 with Frank Harris, manager of Asiatic's gasoline department, to set up a marketing organization there. Van Eck acted as president of Shell in California until 1923, when he moved to New York in order to set up the new holding company for the Group's American business, Shell Union, of which he became president and, from 1933, vice-chairman. In the managerial changes following Deterding's retirement, Van Eck was appointed managing director of Royal Dutch and the principal operating companies, and stationed in London. After four years in the job he had wanted to retire in 1940, considering sixty a proper age to do so, only to relent at the urgent request of all his colleagues. Van Eck had worked closely with both Godber and Legh-Jones during the Group's formative years in America, which made him ideally suited to be the Group's representative in the US during the war, and to act as a broker between the British and Dutch managing directors. James de Booy (1885-1969) was in many respects the odd man out amongst Group managing directors. A former naval officer, he had resigned his commission to join Bataafsche in 1919, working his way up from site manager in the Dutch East Indies to become general manager of Caribbean Petroleum Company in 1926. He then returned to The Hague and became company secretary at Bataafsche in 1932, obtaining the same position at Royal Dutch two years later. He became managing director of Royal Dutch and the operating companies at the same time as Van Eck, but stationed in The Hague. During the war, De Booy combined his work for the Group with various official functions for the Dutch government in exile, which led him to resign as director and accept a ministerial post in 1944.[164]

Guus Kessler

[106]

J. C. van Panthaleon van Eck

[107]

James de Booy

[108]

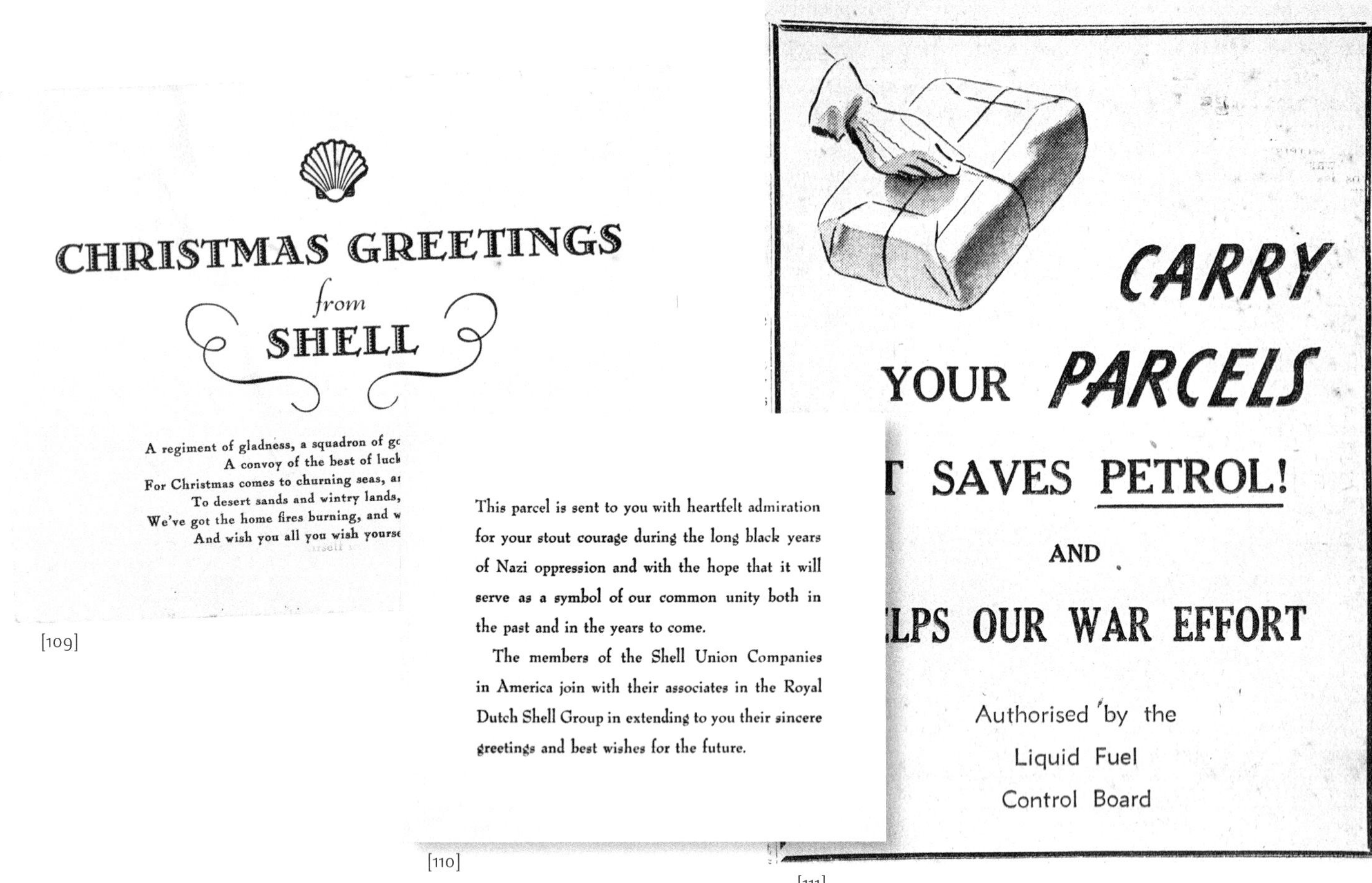

CHRISTMAS GREETINGS
from
SHELL
A regiment of gladness, a squadron of
A convoy of the best of
For Christmas comes to churning seas,
To desert sands and wintry lands,
We've got the home fires burning, and
And wish you all you wish
This parcel is sent to you with heartfelt admiration for your stout courage during the long black years of Nazi oppression and with the hope that it will serve as a symbol of our common unity both in the past and in the years to come.
The members of the Shell Union Companies in America join with their associates in the Royal Dutch Shell Group in extending to you their sincere greetings and best wishes for the future.
CARRY
YOUR PARCELS
SAVES PETROL!
AND
LPS OUR WAR EFFORT
Authorised by the
Liquid Fuel
Control Board

[109]

[110]

[111]

Van Eck's secondment to New York reduced the number of Dutch directors in London to only two, and De Kok's death had created a vacancy for the position of Director-General of Royal Dutch, the informal Group CEO. In November 1940 Kessler and De Booy presented their British colleagues with a memorandum announcing their intention to have Royal Dutch appoint new Dutch non-executive directors; they had probably already decided to leave De Kok's position open for the duration of the war, judging this to be too important a decision to be taken by them alone.[165] To reinforce the depleted boards at a time of crisis for the business made sense of course, and there was also a practical reason for doing so in the case of Bataafsche. The Dutch East Indies Mining Law required the boards of companies holding concessions in the colony to have a Dutch majority and with the suspension of the directors in occupied territory Bataafsche no longer met that condition. There was no pressing need for appointments, however; in the end Kessler and De Booy themselves woke up the sleeping dog of the colonial government to Bataafsche's non-compliance, months after the row over the appointments had started.[166] The two directors can scarcely have been motivated by a wish to reinforce the operating companies' boards either, for most of the candidates lived either in the US or in the Dutch East Indies, so they could not contribute materially to the running of the business.

The board appointments were therefore a ploy to assert Royal Dutch's control over the Group and this is exactly what the British directors took them to be.[167] They reacted indignantly against the proposals, to the point of threatening to challenge any appointments in court as unfair impositions by the majority shareholders on the minority. For a period of three to four months in the middle of the war, the Group's normal managerial process

Though the Allied war effort had no fuel shortage, economising on oil by reducing the need for transport was a common theme of many official campaigns (left).

At Christmas 1945, Shell Union sent parcels to staff on the ravaged European continent accompanied by this wish (middle).

In countries where advertising continued during the war, advertisements of course took up topical and notably patriotic themes to attract customers' attention (right). Shell Australia's Christmas cards (far left) echoed similar themes with references to military formations and war theatres.

slowed down to a walk because the managing directors were no longer on speaking terms. Board meetings were strictly limited to the business at hand; in addition Kessler and De Booy had four angry meetings with Bearsted, Waley Cohen, Agnew, Godber, and Legh-Jones to try and solve the dispute. Though we only have the motives of the British directors as filtered through the partisan ears and eyes of Kessler and De Booy, there can be no doubt about their ultimate intentions. The appointment of Godber and Legh-Jones as managing directors of Bataafsche had removed one of the iniquities of the Group's original structure; the suspension of the Dutch non-executive directors left the British directors in a de facto majority. With Europe on its knees, the Group's future and best interests lay firmly in Britain and consolidating the status quo would in the longer term surely lead to an adjustment of the 60:40 ratio.

Shell
the way to Victory!
BY INVESTING IN THE WAR LOAN
and
economise by using
Drag Free
SHELL MOTOR OIL
THE SHELL COMPANY OF AUSTRALIA LIMITED

The motives can be deduced from the kind of objections raised by the British directors. They did not question Royal Dutch's fundamental right to appoint new directors at that point in time, but opposed the candidates, and questioned the urgency of appointments, which they generally declared to be against Group interest. The objections against candidates and timing mattered little to the main argument, in which they contrasted the Group interest that Kessler and De Booy, like the other directors, ought to have in mind, with what was termed the parochial interest of Royal Dutch that the two Dutch directors allegedly had advocated. In an atmosphere rife with chauvinism, the British directors argued that the Group's interest lay in cultivating a British image. Swamping the Anglo-Saxon and Asiatic boards now with foreigners would cause irreparable damage to the Group's standing with the British government and the public at a time when by far the greatest part of sales were in the British Empire and to the British government. This latter part of the argument betrayed the British directors' intentions, for it was patently untrue. After the fall of Europe, the Group's sales in the British Empire still amounted to only a quarter of total sales.[168] The directors really meant to say, therefore, that the Group ought to remain British now that circumstances had at last made it so. That is also how Kessler and De Booy took the argument, pointing out that chauvinism would likely remain widespread after the war and could thus not be a reason for Royal Dutch to forego its appointment rights now, for that would set a dangerous precedent.

In January Kessler and De Booy launched an ingenious proposal which, though probably intended as a solution, only served to worsen the crisis. In both Anglo-Saxon and Asiatic, the company's articles of association had a clause allowing the boards of those companies to delegate their powers to a committee of managing directors. If both companies were now to use that clause to appoint a five-man team of managing directors, Royal Dutch would not need to make the appointments required to sustain its proper representation on the full boards. With this proposal Kessler and De Booy gave away their original concern as feeding on a suspicion against the British majority on the operating companies' boards, because what it boiled down to was that Royal Dutch and Bataafsche's two-tier board model of corporate government should be applied to the London operating companies. On their single-tier boards the managing directors and the non-executives had a joint responsibility for running the business, and they met once a week or fortnightly. By contrast, the Royal Dutch and Bataafsche boards met only once a month; the managing directors were responsible for conducting the business, while the function of non-executives was limited to exercising supervision. Applying that model to Anglo-Saxon and Asiatic thus meant reducing the influence of the British non-executive directors. However, Kessler and De Booy had an additional intention with their concept of a committee of managing directors. They considered the existing arrangement of managerial tasks in the London office unsatisfactory and wanted a fundamental reform. Neither of them had a clear idea of what Godber and Legh-Jones actually did. Now that De Booy worked in London, from the Dutch perspective it would make sense to integrate him into the management of Anglo-Saxon and Asiatic and, for instance, have him take over Venezuela from Godber, who exercised only nominal supervision over the area manager concerned. Some area managers reported to two or even three managing directors so a more rational organization was highly desirable.[169]

Unsurprisingly, the proposal for a committee of managing directors to take over at Anglo-Saxon and Asiatic encountered fierce opposition from the British directors. First Waley Cohen exploded, correctly sensing it to be a threat from Kessler and De Booy to force the board appointments through. Shell Transport would never agree to it, he thundered, would defend itself, take legal action, call a shareholders' meeting, expose the matter in public. When Waley Cohen's anger had spent itself and the meeting calmed down, Bearsted took a principled stand and said he could not delegate responsibilities which, in the eyes of the outside world, a company's chairman ought to exercise. Godber threatened to resign if the idea of a committee of managing directors were accepted, and declared that he did not want to consider a reordering of managerial tasks because he would feel sincerely handicapped under any arrangement other than the existing one. Faced with this opposition Kessler and De Booy withdrew their proposal and then presented a compromise: Royal Dutch would not appoint four new directors, but only three. However, this also proved unacceptable; and in the end, they settled for the appointment, as of June 1941, of four new non-executive directors to Bataafsche only, with a restricted mandate for the duration of the war. Kessler and De Booy were happy about their limited

‘Surely the thing that must be uppermost in our minds is first, to win this war, and second, to make some money for the Group.’ Van Eck

achievement, believing that their firm stance had created a renewed respect for Royal Dutch.[170]

From his New York distance, Van Eck warned them not to let their suspicions dictate their relations with the British colleagues. He wrote to De Booy saying that he considered the row very unfortunate, in being wrongly driven by the motive ‘that we must have a majority on the Board because we don’t trust our friends. Now to my recollection, we have never had any meeting in which the interests of either the Royal Dutch or Shell was discussed or at which meeting we have taken any decisions that affect any of these companies without previous informal discussions. Surely we have never tried or dreamt of putting one over on our colleagues just because we had a majority; that they would do so is unthinkable.

The suspicion that they might do something like it is equally disturbing. You will probably say that this was not your argument of reason, and that you only want to fill the vacancies because of your statutory rights. If this were the only thought in your and Kessler’s mind, I would of course, have no objection, but apparently it is not the only reason, and because it is not, I think the forcing through will do more harm than good.

Surely the thing that must be uppermost in our minds is first, to win this war, and second, to make some money for the Group. To create suspicion in the minds of our colleagues can only be harmful. We can at any time assert our rights or our interests whenever we really think these are in jeopardy, regardless of the fact that we temporarily have a minority on the Board. But I know of no more disrupting element for the happy association of our group than to sow the seeds of distrust.’[171]

Despite Van Eck’s wise words, Kessler and De Booy continued to harbour some suspicion as to the motives of the Shell Transport directors. As a consequence, they set a target for Royal Dutch to have cash holdings 50 per cent higher than those of Shell Transport under all circumstances, so that it could not be outbid.[172] The Group’s odd structure helped them to effect this precaution without the British directors noticing it, for with his seat on the Shell Transport board Kessler could monitor that company’s cash position, but none of his British colleagues was able to do likewise for Royal Dutch. Moreover, their suspicion was not entirely unfounded, for Shell Transport did strive to obtain a majority in the Group. In December 1942, Bearsted, Godber, and Legh-Jones surreptitiously approached the Bank of England with proposals to have the British government secure a change of the ownership ratio from 60:40 to 49:51, either by pushing the Dutch government into arranging a sale, or by making provisions to that effect in a future peace treaty. The details of the plan and the extent to which British government officials discussed it remain a mystery; in the end the Treasury appears to have counselled the initiators to await a more opportune moment, because the Dutch government would never agree to such a momentous transaction during the war.[173]

Apart from sowing mutual suspicion, the row over the board appointments had the unfortunate effect of stalling action over management reorganizations. It was not for the want of ideas. The upset to the business caused by the war was a powerful impetus to reconsider existing arrangements at the top or further down the line. The committee of managing directors idea was certainly the most innovative and far-reaching concept, but in addition the managing directors discussed reorganizing the reporting lines of the area coordinators, ending the anomalous stand-alone position of the Handelszaken marketing operations in the Dutch East Indies by merging it into the main organization there, improving the coordination of the Group’s research, and raising the priority of product development in the research organization.[174] Towards the

[113]

At the end of 1948 Kessler retired and was succeeded by Van Hasselt. Left, the two men are seen (with Kessler on the left) walking in the grounds of Te Werve, Royal Dutch's country club.

end of the war the relationship between the British and the Dutch managing directors appears to have improved, probably as a result of changes in the team. In June 1944, De Booy resigned to become Minister of Shipping and Fisheries in the Dutch cabinet; Van Hasselt succeeded him as managing director of Royal Dutch and the main operating companies. Later that year or possibly early in 1945 Van Eck returned from New York and Oppenheim exchanged Curaçao for London as well to assist in the preparations for the return of peace.

The improvement in relations between the managing directors enabled them finally to take far-reaching decisions. In April 1945, less than two weeks before Germany's capitulation, Kessler and Oppenheim presented the Dutch government with a memo outlining the Group's plan for a much firmer integration of the business. Put briefly, the scheme envisaged transferring the holding of all the Group's assets with the exception of the Dutch East Indies operations to Britain. Bataafsche would sell its shares in European operating companies to Anglo-Saxon and its shares in overseas operating companies, such as Shell Union, Diadema Argentina, Colon Development Company, and Caribbean Petroleum Company, were to be put into a British trust jointly managed by Royal Dutch and Shell Transport. Innocuously and somewhat disingenuously, the memo gave the motives for this transfer as being the desire for British Foreign Office support in rebuilding the continental operations and the need for better protection of the Group's assets overall.[175]

However, from the sequel it became perfectly clear that the managing directors really intended to fulfil the British directors' long-standing wish and anchor the Group more firmly in Britain. The plan appears to have been developed as a straight trade-off with the Dutch directors' desire to have a greater grip on the business by making a committee of managing directors responsible for running the three main operating companies. The documentary record is unfortunately patchy, but the pattern looks clear enough. The Dutch managing directors claimed to have initiated the transfer and trust scheme.[176] They also wholeheartedly supported subsequent plans for giving Shell Transport more influence over the Group. In return, the Dutch managing directors reinforced their grip on the business by the application of the two-tier board system to the British operating companies. By

A bilingual announcement of a measure taken by the German authorities informs Royal Dutch shareholders that they must validate their shares by bringing the certificates or warrants in for stamping.

BEKANNTMACHUNG.

Der Herr Generalkommissar für Finanz und Wirtschaft hat auf Grund der Verordnung über besondere wirtschaftliche Massnahmen vom 28.8.1940 (Verordnungsblatt für die besetzten niederländischen Gebiete, Stuck 23) angeordnet, dass die Anteile der unterzeichneten Aktiengesellschaft zur Abstempelung vorzulegen sind.

Zur Abstempelung verpflichtet sind folgende Personen:

1. die Inhaber (Berechtigten) der Wertpapiere;
2. diejenigen, die die betreffenden Anteile verwalten oder besitzen, in Gewahrsam haben, beaufsichtigen oder bewachen, unter Benennung der Eigentümer;
3. für Anteile, die als Pfand hinterlegt sind, der letzte Pfandinhaber unter besonderer Angabe seiner Eigenschaft als Pfandinhaber.

Massgebend für das in den vorstehenden Ziffern 1—3 aufgeführte Besitz- oder Verwaltungsverhältnis ist der Tag der ersten Veröffentlichung dieser Bekanttmachung im niederländischen Staatscourant (Stichtag). Für die an diesem Tage getätigten Verkäufe besteht für den Erwerber der Aktien die Anmeldepflicht.

Die Abstempelung kann erfolgen bei:

1. De Nederlandsche Bank N.V.;
2. De Vereeniging voor den Effectenhandel;
3. N.V. Nederlandsche Handel-Maatschappij, Amsterdam, und ihre Filialen;
4. Kasvereeniging N.V., Amsterdam;;
5. Schill & Capadose, Den Haag;
6. Van der Hoop, Offers & Zn., Rotterdam.

Bei Anteilen, die nachweislich in den besetzten niederländischen Gebieten nicht zur Abstempelung vorgelegt werden können, genügt die Abgabe einer besonderen Erklärung, die ebenfalls bei den vorstehenden Stellen einzureichen ist.

Die erforderlichen Vordrucke sind von den genannten Stellen zu beziehen.

Die Vorlage zur Abstempelung sowie die Abgabe der besonderen Erklärung haben in der Zeit vom 1. Dezember 1940 bis 31. Dezember 1940 zu erfolgen.

Haag, den 22. November 1940.

N.V. KONINKLIJKE NEDERLANDSCHE MAATSCHAPPIJ TOT EXPLOITATIE VAN PETROLEUMBRONNEN IN NEDERLANDSCH-INDIË.

BEKENDMAKING.

De Commissaris-Generaal voor Financiën en Economische Zaken heeft op grond van de Verordening betreffende bijzondere maatregelen op economisch gebied van 28/8/1940 (Verordeningenblad voor het bezette Nederlandsche gebied, Stuk 23) bepaald, dat de aandeelen der ondergeteekende Naamlooze Vennootschap ter afstempeling moeten worden aangeboden.

Tot afstempeling zijn de volgende personen verplicht:

1. de eigenaren (rechthebbenden) der waardepapieren;
2. zij die de bedoelde aandeelen beheeren of bezitten, in bewaring hebben, controleeren of bewaken, onder opgave van de eigenaren.
3. met betrekking tot aandeelen welke in pand zijn gegeven, de laatste pandnemer, onder uitdrukkelijke vermelding van zijn hoedanigheid van pandnemer.

Beslissend voor de onder eerdergenoemde nummers 1—3 bedoelde bezits- of beheersverhouding is de dag der eerste openbaarmaking van deze bekendmaking in de Nederlandsche Staatscourant. Voor de op dezen dag plaatsgevonden hebbende verkoopen bestaat voor den verkrijger der aandeelen de verplichting tot aanbieding.

De afstempeling kan geschieden bij:

1. De Nederlandsche Bank N.V.;
2. De Vereeniging voor den Effectenhandel;
3. N.V. Nederlandsche Handel-Maatschappij, Amsterdam, en hare Agentschappen;
4. Kasvereeniging N.V., Amsterdam;
5. Schill & Capadose, Den Haag;
6. Van der Hoop, Offers & Zn., Rotterdam.

Met betrekking tot aandeelen waarvan aangetoond kan worden, dat zij in het bezette Nederlandsche gebied niet ter afstempeling kunnen worden aangeboden, is de afgifte van een bijzondere verklaring, welke eveneens bij de eerdergenoemde instellingen moet geschieden, voldoende.

De benoodigde formulieren zijn bij de genoemde instellingen te verkrijgen.

De indiening ter afstempeling, alsmede de afgifte der bijzondere verklaring, moeten geschieden in de tijdsruimte van 1 December 1940 tot 31 December 1940.

Den Haag, 22 November 1940.

N.V. KONINKLIJKE NEDERLANDSCHE MAATSCHAPPIJ TOT EXPLOITATIE VAN PETROLEUMBRONNEN IN NEDERLANDSCH-INDIË.

[114]

the summer of 1946 and possibly earlier, the five managing directors acted as a committee at Anglo-Saxon and at Asiatic, which latter company had just been renamed The Shell Petroleum Company Ltd.[177] As a consequence, the board meetings were now usually held on the same day in succession to each other, perhaps even simultaneously, with separate secretaries following each company's agenda and taking the relevant minutes. This momentous change in the Group's management structure coincided with the resignation of the two original opponents to it. On 12 July 1946 Bearsted retired from all his functions because of ill health; he died on 8 November 1948. Godber succeeded him as chairman of Shell Transport and of the two London operating companies; Frank Hopwood replaced Godber in what now functioned as a de facto committee of managing directors. Godber's position gave him the chairmanship of London meetings, whereas Kessler presided over Bataafsche boards; but in a significant difference, Godber – unlike either of his two predecessors in office – held a non-executive role, while Kessler's was executive.

The stated intention to vest Bataafsche more firmly in Britain also helped to secure, in May 1946, a formalization of the arrangement concluded during the war with the Treasury acknowledging the company as a British resident for currency purposes. This arrangement had given rise to repeated conflicts with officials over the benefits or otherwise to Britain, the Netherlands, Curaçao, and the Group. The Treasury considered itself hard done by and complained about the supposed greater benefits to the other parties.[178] The amounts at stake were considerable. During the war, the Group had obtained an estimated $160 million from the Treasury and thereafter it stood to need $30-50 million a year for reconstruction purposes alone. Despite substantial dollar revenues, the Group ran an overall dollar deficit and the Netherlands looked unlikely to be able to supply sufficient dollars for the Group's planned worldwide investment.[179] With the coming of peace, the Treasury showed itself unwilling to continue facilitating a company with majority Dutch ownership, which in effect would allow the Netherlands to benefit from British access to scarce dollars.[180]

This argument looked fair enough, but it failed to take into account the overall benefits to Britain. The Group's oil sales outside the US remained in sterling, those from the Dutch East Indies and Curaçao included, and the loan account between Anglo-Saxon and Bataafsche had the effect of suspending the need to convert £50-60 million a year into guilders. Taking care of the Group's dollar requirements enabled Britain to support the position of sterling and to obtain a range of goods and services, such as oil from Venezuela shipped by Anglo-Saxon, in sterling rather than in dollars or in other currency. These very substantial invisible benefits meant that Britain profited from the Treasury agreement overall. Dutch officials calculated that during its first seven years, the agreement had cost Britain the equivalent of £487 million in dollars, but had generated £510 million in various currencies, including dollars.[181] The treasury clearly acknowledged the benefits and agreed to formalize the agreement in May 1946. For the Group this agreement had the obvious advantage of enabling it to draw on the British dollar resources, much greater than the Dutch ones, and of not having to operate under two different exchange regimes. In

Koninklijke keert achterstallig dividend uit

Na aftrek van 15% dividendbelasting zal 6.35% in contanten worden betaald

IN AANDEELEN 20%

De directie van de Koninklijke Nederlandsche Mij tot Expl. v. Petroleumbronnen in Ned. Indië, gevestigd te Curaçao, deelt mede dat zij aan de vermoedelijk medio Juli, met toestemming van het Nederlandsch Beheersinstituut te Amsterdam te houden algemeene vergadering van aandeelhouders zal voorstellen over te gaan tot uitkeering van dividenden en wel over de nog niet afgesloten boekjaren 1944 en 1945.

Na de reeds afgesloten en dividendlooze oorlogsjaren 1940 t/m 1943 kan thans worden voorgesteld, met verkregen toestemming der betrokken autoriteiten over 1944 een dividend van 25% uit te keeren, waarvan 5% in contanten en 20% in bonusaandeelen en over 1945 een dividend van 6% in contanten.

De bonusaandeelen, welke van 1 Januari 1946 af in de winst zullen deelen, zullen op grond van de bepalingen van het besluit herstel rechtsverkeer geblokkeerd zijn. Nadere bijzonderheden hieromtrent, speciaal ook met betrekking tot de positie van aandeelhouders in het buitenland, zullen nog worden bekend gemaakt. Voor zoover deze bonusaandeelen met een t.z.t. te verleenen algemeene of bijzondere vergunning van rechtsherstel in Nederland mochten worden vervreemd, zal de opbrengst bij eerste overdracht monetair geheel geblokkeerd zijn.

De verschuldigde 15% dividendbelasting over het volle dividend van 25% en van 6% zal van de uitkeeringen in contanten worden afgehouden, zoodat aandeelhouders netto 6.35% in geld ontvangen.

Hoewel het de directie verheugt, dat de positie der maatschappij zoodanig is, dat zij tot bovenstaand voorstel kan overgaan, acht zij zich verplicht den aandeelhouders er op te wijzen, dat de vennootschap met haar groote belangen in het buitenland, door het ontwerp vermogensaanwasbelasting — zou het wet worden in den huidigen vorm — zeer zwaar zal worden getroffen.

[115]

Above, in 1945 Royal Dutch gave its shareholder a dividend of 25% over 1944 to compensate them for the years 1939-43 when nothing was paid.

Right, Van Hasselt's telegram of 3 January 1947 gave momentous news in the fewest possible words: 'Seat Royal Dutch transferred back to Hague'.

addition, managing directors received assurances as to diplomatic support from the British government.[182]

The Dutch government demurred, however. In November 1945, ministers had objected to the proposed transfer of Bataafsche shares to Anglo-Saxon and to a British trust, causing Bataafsche directors to drop the latter idea. Bataafsche did sell some of its shares in European operating companies to Anglo-Saxon, notably those in Rhenania-Ossag, thus facilitating a smooth liaison with the authorities in what became the British zone in Germany; the transaction may have carried the proviso that it could be annulled should the Dutch government request it at any time in the future.[183] Dutch officials felt that the Treasury Agreement took the Group one step further into Britain and away from the Netherlands; the ministers for Foreign Affairs and for Overseas Territories also considered the arrangement to compromise Dutch sovereignty in bilateral trade relations and in relations with Curaçao. The country's weak currency position clinched the matter, however, and the Dutch government finally assented to the Treasury agreement on condition that the Group fully retain the existing 60:40 ownership ratio.[184]

Thus the planned trade-off for the changes to the board structure of Anglo-Saxon and Shell Petroleum failed to materialize, but, as we shall shortly see, the desire which had motivated it did not disappear.

Back together again On 13 December 1945, Bataafsche held its first board meeting after the war at the Carel van Bylandtlaan. The building was still dilapidated and unsuitable for use as offices, so Royal Dutch and Bataafsche remained based in London for the time being, but one room had been made ready for this celebration of the end of the war and the liberation of the Group's Dutch base. Van Eck gave a festive speech referring to the Group growing into a true partnership during the war, praising the loyalty of the British directors in using their majority on the boards. In his turn, Agnew praised the Dutch directors for contributing so much to building the Group into a superior enterprise. Godber joined in to extol the Group's excellent standing and goodwill, achieved through the indispensable support of the Dutch directors. Just at the moment when Bearsted had claimed to be, with Waley Cohen, the oldest board member present, August Philips entered. Now eighty-one, he had only recently resigned his directorship and clearly attended the meeting to highlight the occasion as a return to normal circumstances.[185] On 13 February 1946 the London operating companies reciprocated when their boards met again at St Helen's Court, Bearsted warmly welcoming the Dutch non-executive directors after six years of enforced absence.[186] In another gesture of celebration, Royal Dutch voted a 25 per cent dividend over 1944, rewarding its shareholders for their patience during the lean years since 1939, when no dividend had been paid. Royal Dutch and Bataafsche formally returned their seat to The Hague central offices

[116]

N.V. De Bataafsche Petr. Mij.
's-Gravenhage.

COPIE VAN TELEGRAM 4A

AFGEZONDEN DOOR: v. Hasselt Judex The Hague 64

GEADRESSEERD AAN: van Diermen Judex Batavia 3-1-47

L C

390 Seat Royal Dutch transferred back to Hague 2/1

= VAN HASSELT =

E54
E56

RA DAB DK

[117]

Left, rejoicing crowds greet Canadian tanks at the liberation of Amsterdam; right, Van Hasselt's memorandum to all departments of BPM.

Netherlands members of our staff may be interested in the following message from Mr. B. Th. W. van Hasselt :

Have just returned from 2 weeks visit to Hollad principally the Hague where I found that fortunately great majority of staff have come through war in fairly good condition although there have been some sad losses and all their experiences specially during last winter simply have been terrible.

Thanks to efficient allied help food situation is rapidly improving and there is no real starvation now although many old cases still require treatment.

[118]

in August 1946; and although Kessler had to wait until January 1947 to obtain the formal recognition as Director-General of Royal Dutch, with that appointment he achieved his life's ambition by becoming the Group's de facto CEO.

Even though the leaders were not yet termed a committee of managing directors, that is what they were, and the unity and efficiency established at the top through the committee's formation gave the Group a new sense of purpose, powering a comprehensive programme of reconstruction, expansion, and revitalization. Now the fateful checks on organizational evolution imposed by Deterding in the 1920s had been eliminated, and at last the mature corporation which had emerged during the inter-war period had a management structure to match. The Group's renaissance manifested itself throughout the business. Four aspects of business policy in the immediate post-war era stand out as showing the shape of things to come. First, moving quickly to recover lost ground in production, the Group closed a long-term supply agreement with Gulf Oil in 1947, which helped to compensate for the lack of its own concessions in the Middle East by ensuring a growing supply of Kuwait crude. Second, in manufacturing, the construction of cat crackers at Houston, Stanlow, and Pernis, coupled with the building of various plant for the production of chemicals, signalled the intention to raise chemicals from an adjunct of oil to a business function in its own right. Thirdly, during April and May 1946, Kessler chaired the first Group research meeting to improve coordination and raise standards. Finally, less visible from the outside, the Group showed a greater sensitivity to the importance of human resource management. In the Dutch East Indies, the question of back payments to staff for time spent without salary under occupation was settled quickly and generously, setting a standard which government departments and other companies found hard to match.[187]

Moreover, the arrangement made no distinction between European and Asian employees, setting aside the rigid divide which had often disfigured pre-war attitudes. In 1947, the Group started experimenting with what became known as regionalization, a determined drive to recruit and train local people for all functions in operating companies around the world. In March 1944 the Group set up a public relations committee, chaired by Kessler, which as 'its first and most important matter to consider' addressed ways to further the training of local staff in Brazil and Argentina. A memo summarized the intentions of this policy by stating that 'In overseas areas more and more supervisory duties entailing greater responsibilities are being allocated to local nationals whereas in the past such work was kept to a very great extent in the hands of the British, Dutch or American employees. This is part of the general scheme of things and cannot be avoided. In fact it must be encouraged if the Group is to become firmly established in a foreign country notwithstanding that there are difficulties and disadvantages to be faced and overcome in following such a policy.'[188]

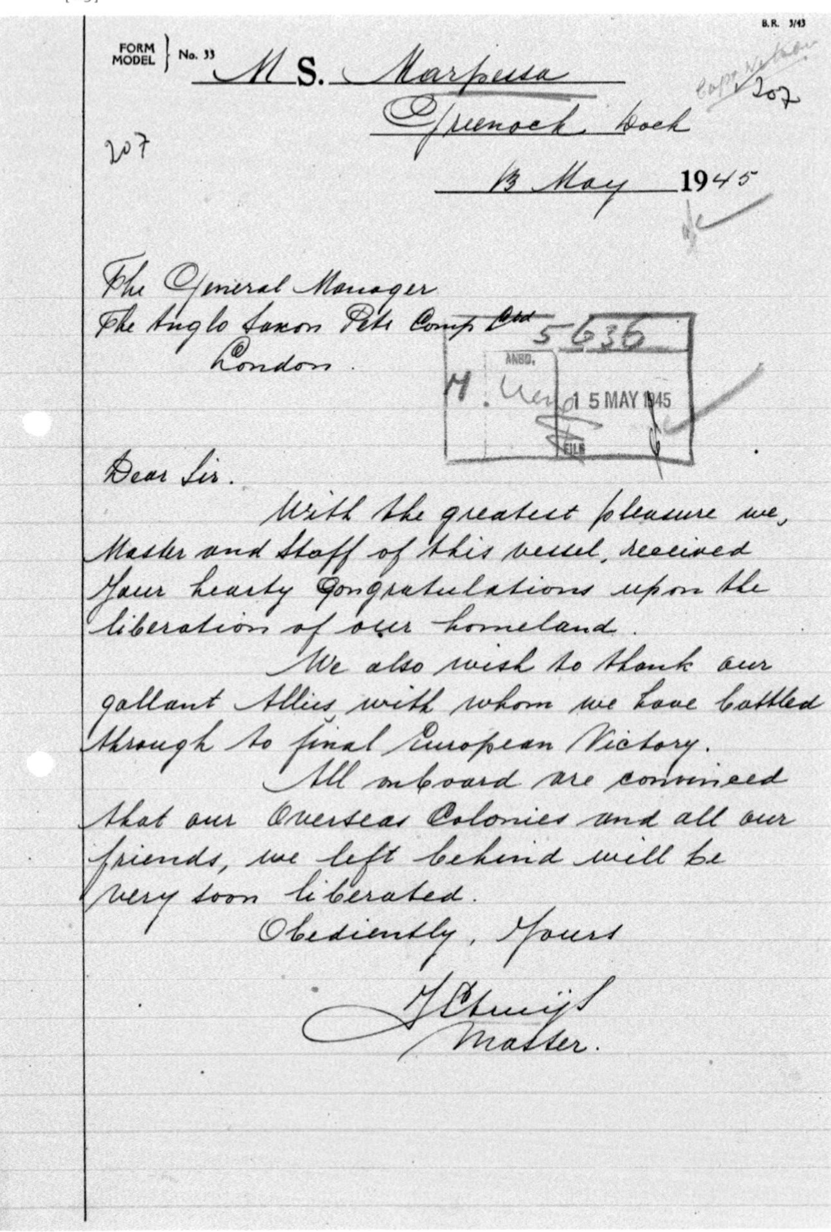

B.R. 3/43

FORM MODEL } No. 33 M.S. Marpessa

Greenock Dock

207

13 May 1945

The General Manager
The Anglo Saxon Petr. Compy Ltd
London.

5636
ANSD.
15 MAY 1945
FILE

Dear Sir.

With the greatest pleasure we, Master and Staff of this vessel, received Your hearty Congratulations upon the liberation of our homeland.

We also wish to thank our gallant Allies with whom we have battled through to final European Victory.

All onboard are convinced that our Overseas Colonies and all our friends, we left behind will be very soon liberated.

Obediently, Yours

[illegible]
Master.

The Master of the Dutch-flag *Marpessa* returns his thanks to the general manager of Anglo-Saxon for the news of the liberation of the Netherlands, May 1945.

This memo marked an important shift in attitudes. Despite the evident success of staff regionalization in Romania and Mexico following the introduction of legislation requiring the employment of local people at all levels, the Group had accepted it only because it was forced to do so. Now managers embraced this policy wholeheartedly as a way to improve performance by anchoring the operating companies more firmly in their environment.

Another aspect of the changed attitudes to staff management was the adoption of a mandatory retirement age of sixty for Group managing directors. The advocate of this proposal, Van Eck, retired in December 1946 and was succeeded as managing director by J. H. Loudon. When Loudon took his seat on the Shell Petroleum board, Godber welcomed him expressing 'the Board's pleasure in the fact that the son of a father so distinguished in his services to the Group had attained to the office of managing director of the company by sheer personal ability'.[189]

Because oil demand accelerated after the war and did not slow down as anticipated, the Group had to raise its expansion and investment plans continually. A projection drawn up in May 1944 forecast a need for £20 million (220 million guilders) of extra capital expenditure in 1946, resulting in a funding shortfall of £11 million (120 million guilders).[190] By 1946 the Group expected to need a total of 210 million pounds, or (2.3 billion guilders) during the years 1947-52 alone. At that time the combined stock exchange capitalization of Royal Dutch and Shell Transport was the equivalent of £350 million. The funding schedule required £105 million to be supplied by April 1948, necessitating Royal Dutch to raise a total of £63 million, almost 700 million guilders.[191] In March 1947, the Nederlandsche Bank told Royal Dutch that the Amsterdam market could not absorb more than 400 million guilders which, even combined with the potential amounts raised in other European countries where Royal Dutch had traditionally sold securities, left a large shortfall. The committee of managing directors decided to cover this deficit by having Royal Dutch sell one-sixth of its share in the Group to Shell Transport, reducing the 60:40 split to a neat 50:50.

This was too neat to be believed. The financial councillor of the Dutch embassy in London, Dr. H. Riemens, who was closely involved with the various negotiations concerning Group finance, immediately suspected that the capital requirement and the timetable for meeting it had been engineered, perhaps with British Treasury connivance, as a pretext for the deal between the parent companies.[192] Circumstantial evidence bears out his interpretation. There was no reason why Shell Transport could not draw on the London market for a higher amount and lend this to Royal Dutch, as had regularly happened between the parent companies in the past, unless the Royal Dutch directors particularly wanted to sell. Indeed, Shell Transport had received such support from Royal Dutch as recently as September 1939, and could thus be said to have a moral obligation to reciprocate.[193] The Dutch managing directors sincerely believed in the desirability of achieving 50:50 and regretted that they failed to carry it through.[194] In May and September 1947, Kessler, Van Hasselt, and Loudon tried to persuade the Dutch Minister of Finance Piet Lieftinck to approve the deal with Shell Transport, using very familiar arguments to emphasize the advantages of going 50:50 – the diplomatic support from the British government, the huge dollar resources supplied by the Treasury, and greater administrative simplicity. They were the very same reasons which had inspired the trust idea.[195]

Lieftinck flatly rejected the Royal Dutch request, but he could only sustain his position if he found a way to let the company raise its share of the projected investment. Therefore the minister asked the banker J. W. Beyen, the Dutch government's special financial adviser during and immediately after the war, to explore the available options. With his customary ingenuity Beyen produced various more or less complex alternatives. One scheme using the Treasury Agreement for a special loan facility foundered on the British Treasury refusing to consider money raised by Royal Dutch for the operating companies resident in Britain as coming under the agreement, fanning Dutch suspicion about connivance. Another plan centring on raising money in New York showed the possibilities there to be by no means exhausted. In a masterstroke, Beyen fed this piece of information back to the British Treasury, which had been led to believe by Shell Transport directors that the company absolutely depended on the London market for issuing £40 million of shares. In November, the Chancellor of the Exchequer Hugh Dalton gave permission for an issue of only £29 million and spaced over four months, effectively denying Shell Transport the funds for completing the transaction with Royal Dutch because Britain could not, or did not want to, afford it. As it happened, the careful husbanding of Dutch currency resources, combined with the hefty premium offered and Royal Dutch's ability to tap into the French and Swiss markets, enabled Royal Dutch to obtain its 60 per cent share of the planned investment by the agreed date, with the bond loan part of the issue being heavily oversubscribed, whereas Shell Transport narrowly missed its 40 per cent target.[196]

With the failure of the investment plan stratagem, the Group managing directors gave up their efforts to change the ownership structure. Moreover, during 1948 the Group moved ahead and raised, through Shell Caribbean, its first big issue in the United States, a twenty-year loan of $250 million from a consortium of insurance companies and secured on Bataafsche's shares in Shell Union. At the end of 1948, Kessler retired as Director-General of Royal Dutch and Van Hasselt succeeded him. Kessler became chairman of the company's supervisory board and retained directorships in the main operating companies, thus continuing his close involvement with the business, but his retirement as CEO closed a tumultuous era in which the Group had moved through war, destruction, and chauvinist confrontation, to achieve a new phase of expansion under a harmonious and far more efficient management.

[120]

Winston Churchill at the opening of the reconstructed Harburg refinery, 4 November 1949.

Conclusion During the Second World War, the Group achieved a fourfold transformation. First, the original imbalance between Dutch and British representation on the boards of the main operating companies was abolished. One vestige remained, the Royal Dutch seat on the Shell Transport board not being mirrored by a Shell Transport seat on the Royal Dutch board, but that was no issue. Second, the top management structure was brought into line with the rest of the business and made more transparent, coherent, and efficient. Many changes and adjustments would still be needed in the future, but these no longer required prior difficult negotiations over the relative positions of British and Dutch managers. Third, the Group's centre of gravity now became firmly lodged in the western hemisphere. The Dutch East Indies never really regained their former importance in production and in manufacturing; Venezuela, the US, and Curaçao were the Group's core and would remain so for a long time to come. Of these three areas the United States may be singled out as the most important single operating area on the basis of Shell Union's great contribution in all business functions. Without Shell Union, the Group could never have emerged so strongly from the war, nor have recovered quite so quickly from the damages inflicted. Fourth, the Group's attitude to staff management changed completely. The old colonial discrimination between people from various national and ethnic backgrounds disappeared, to be replaced by a modern conception based on ability and proven merit supported by in-house selection and training. These transformations put the Group on an entirely new footing, which enabled it to face the challenges of the post-war world.

[1]

Chapter 2

Remodelling the Corporation

In the two decades after the war, with new leadership, new money and a new top management organization, the Group went through a process of reformation that was so far-reaching it could almost be called a renaissance. Its still expanding financial needs were met; the parent companies were listed on the New York Stock Exchange; the lower management structure was thoroughly overhauled; training of and communications with personnel were enhanced; regionalization and internationalization became a deliberate Group culture; and 'The Conference', a regular consultative meeting of the CMD with all non-executive Group directors, was established. Under the persuasive guidance of John Loudon, consensus and coordination became the watch-words of management.

The manual sorting of share certificates in the Netherlands, 1958.

Loudon takes the lead When Walter Samuel died in 1948, his colleagues felt 'it might be said that [he] had been part of the Group structure'.[1] As the second Lord Bearsted, second chairman of Shell Transport and son of its co-founder, his death was indeed symbolic of the changes just beginning within the Group. As the 1940s gave way to the 1950s, the current of organizational change and development continued to flow, and by the 1960s the Group was bonded into a far more harmonious whole than it had ever been. Under the new rules Kessler retired as Director General of Royal Dutch on 1 January 1949, the first New Year after his 60th birthday. He had been only two years in post, but reaching the age of 60 did not mean that senior executives had to stop working for the Group; they could continue in various non-executive capacities, with a much reduced workload.[2] One possibility for retiring Royal Dutch executives was to be appointed to its *Raad van Commissarissen* or supervisory board, and (as noted in the previous chapter) on his retirement from the leadership of Royal Dutch, Kessler immediately joined the *Raad*.[3] Moreover, he became its chairman at once – an influential role, whose incumbent was allowed to stay on until the age of 72.[4]

Kessler's successor Barthold van Hasselt (1896-1960), a Doctor of Law, had joined the Group in 1920. His first ten years of service were spent in Venezuela, after which he was appointed in rapid succession to the posts of general representative in Indonesia, general manager of Mexican Eagle in Mexico City, and general manager of the Group's Venezuelan enterprises. His appointment as a managing director of Royal Dutch came at the age of 48, and as Director General at 52. Under normal circumstances therefore he would have continued in post for eight years, until the end of 1956; but his health was not up to the strains of the job. Only four months into his appointment he was suffering from very high blood pressure, finding sustained concentration difficult, and becoming absent-minded in meetings. His doctor ordered him to have two months' rest, but his physical powers were never great, and he resigned on 31 December 1951, five years early.[5]

This brought about the unexpectedly early promotion of John Loudon (1905-96), by far the most influential of the Group's leaders in the two decades following the war. Taking office at the age of only 46, he brought a young man's energy and vision to the job, and had the time to make the vision real. In this he was aided by the formation of the de facto Committee of managing directors (CMD) immediately after the war. Gone were the chauvinist tensions so deleterious before; the parent companies assigned board seats on merit. Nationality no longer counted.

Consequently Loudon was in a position of extraordinary power for so young a man; yet though his power was directly comparable to Deterding's in the pre-war era, he used it quite differently and it was tempered by his tactful nature. Loudon became a legendary figure both in Shell and the extended oil industry. He was the son of one of the early notables of Royal Dutch, Hugo Loudon, but it was not until after his graduation as a Doctor of Law from Utrecht (1929) and a subsequent visit to the Dutch East Indies that he began to consider joining the business. His father, who by then was chairman of its supervisory board, opposed the idea, saying he felt that John was more suited to the diplomatic service. John in turn refused this, was taken into Bataafsche as a general trainee on 1 April 1930, and was soon sent to the Americas with two main hopes in his mind – that he would work with people who did not know that his father had been an executive of the company, and that he would not have to work as a lawyer, which he thought too dull. Both hopes were fulfilled. A hands-on year in Venezuela was followed by six in the US,

[2]

[3]

Above, during a dance in the ballroom on 7 May 1949 at the Club Miramar in Cardón, Venezuela, Loudon (aged only 43) addresses employees; right, he speaks to members of the press in Curaçao in 1962.

[4]

Honours flowed to Loudon from around the world. On 9 May 1963 he was appointed a Commander in the Order of Orange Nassau; above, on 15 June 1965 Prince Bernhard celebrated Royal Dutch's 75th anniversary and took the opportunity to promote Loudon to the rank of Grand Officer in the Order. Loudon's wife, *née* Baroness Marie Cornelie van Tuyll van Serooskerken, is with them. Below, his colleague Van den Wall Bake wrote congratulating him on an honour from the President of Ecuador.

JHR. MR. J.H. LOUDON

Koninklijke onderscheiding voor Jhr.Mr.J.H.Loudon

H.M. de Koningin heeft ter gelegenheid van Haar verjaardag benoemd tot Commandeur in de Orde van Oranje-Nassau, Jhr. Mr. J.H. Loudon, president-directeur van de N.V. Koninklijke Nederlandsche Petroleum Maatschappij.
Verder werd de heer Loudon door de President van de Franse Republiek benoemd tot Officier in de Orde van het Legioen van Eer. Op 1 mei overhandigde de Franse Ambassadeur hem de bijbehorende versierselen.

[5]

[6]

W.H. van den Wall Bake 13 Januari 1

J.H. LOUDON, Esq.,
St Helen's Court,
Great St Helen's,
LONDON - E.C.3.

Waarde Loudon,

Ik heb vernomen, dat je van de President van Ecuador een hoge onderscheiding hebt ontvangen, voor welk heuglijk feit ik je hierbij mijn welgemeende gelukwensen aanbied.

Met vriendelijke groeten,

culminating as assistant to the vice president of Shell Oil in San Francisco. Returning to Venezuela in 1938, he flourished as assistant manager in Caracas. His first boss there, an American called William Doyle, reported that he was giving 'excellent service', which Godber found 'particularly gratifying (...) we trust that he will continue to justify our selection of him'.[6]

The skills that had caused his father to recommend a diplomatic career were already well developed. On 10 May 1940, when news came through that the Netherlands had been invaded, Frederick J. Stephens (1903-78; chairman of Shell Transport, 1961-67) went to express his sympathy to Loudon, who 'thanked me very wholeheartedly and explained that he would not be in the office for perhaps a couple of days because this situation had created very many problems that he would have to discuss with the Netherlands and British Ministers (...) With his international and diplomatic disposition he could go from Legation to Legation, to the President of Venezuela and to any of the Ministries there and talk profoundly in the language he thought necessary'.[7]

Stephens was visiting Venezuela as the area manager responsible for Group companies there and in Colombia, and his point about languages was well made: John Loudon spoke five languages fluently. Loudon and Van Hasselt managed the momentous introduction of 50/50 profit-sharing with the Venezuelan

[7]

Time magazine front cover, 9 May 1960

government (see Chapter 1), and from the time of Van Hasselt's promotion to the board in London in 1944, Loudon was general manager in Venezuela, tasked with overseeing Shell's entire production and refining there. His success in that role, despite being substantially separated by the war from London helped to make him the 'natural choice' as a Group managing director when Van Eck resigned at the end of 1946.[8]

Virtually everyone who worked with Loudon when he was Royal Dutch's Director General (or president, as the post was renamed in 1956) had the same reaction: there was never any doubt about his authority, but he was never overbearing. When he died at the age of ninety, all recalled his disarming informality in a hierarchical age; it was said to be impossible not to like him. He was a completely international character, charming, persuasive, urbane, clear-minded, decisive, and determined.

He himself remarked that 'every decade has its type of management and its type of people' and he worked in closest accord with others of his generation, amongst them Frank Hopwood (1897-1982) who had joined Shell Transport's board in 1947.[9] Both men had reacted with dismay to the parent companies' arguments that summer on the question of 50:50 Group ownership, but Godber had praised the young directors (and Loudon in particular) saying, 'John and Frank have settled down into their jobs extremely well and we are very pleased with both of them. The atmosphere is good, and John has been extremely helpful during the somewhat difficult discussions in keeping an even balance.'[10] The same could have been said of him at any time of his working life. Rather ironically for one who had refused to enter the diplomatic service, he became regarded as Shell's supreme diplomat.[11] His elevation as senior executive of Royal Dutch and de facto CEO of the Group inaugurated a lengthy period in which Royal Dutch exercised leadership without dominance. Loudon was in his post for fourteen full years, from the beginning of 1952 until the end of 1965, and in the judgement of Sir Peter Holmes (chairman of Shell Transport 1985-93 and chairman of the CMD 1992-93), he was 'certainly the father of the Group in the modern sense of the word'.[12] In an extensive and well-informed article, *Time* magazine reported on the concept of collegiality with observations that still rang true over forty years later: members of the CMD were 'subdued, cautious, and vastly competent in the modern committee manner. All had to pass one prime admission test: they must have compatibility as well as ability. The man who raises his voice or loses his temper is frowned on, the lone wolf considered a troublemaker. This collective leadership, says one manager, "works like a dream. But to be brutally frank about it, if someone ever came here who wouldn't fit in, we'd have our ways of taking care of him" (...) Since the Group has an unwritten rule that decisions are never forced to a vote, Loudon, as the *primus inter pares*, tactfully arbitrates differences, suggests lines of agreement, sounds out his fellow directors (...). Shellmen agree that their Britons incline more to flair and intuition, their Dutchmen to patience and stolidity. Loudon presides over the mix. If he favors a project, it is likely to go through; if he frowns on it, its chances are poor'.[13]

It could be said that that is part of the definition of a chief executive officer, and Loudon himself later acknowledged that the chairman of the CMD 'comes down to the same thing' as a recognized chief executive, with the extent of the CMD's collegiality being largely dependent on its chairman's character.[14] As second senior, Shell Transport's managing director Fred Stephens served as chairman when Loudon was absent. There was no more board-level talk about altering the Group's 60:40 ratio of ownership, and the Group's renewed post-war confidence communicated itself to shareholders: when asked by an American analyst in 1957 if there was any reason to maintain 60:40, a senior Group spokesman said simply, 'I don't believe the Shell Transport holders would want to give up any, and neither do I think the Royal Dutch shareholders want to give up any either.'[15]

The listing of Royal Dutch on the New York Stock Exchange, 1954

After the immediate post-war strictures, the parent companies had no further problems in attracting capital to finance the Group's rapid expansion. When Loudon came to power in 1952, the Group's collective market capitalization had risen in sterling terms from the £336 million or so of 1946 to a total average sterling value throughout 1952 of about £520 million; and in 1953, after converting some debentures into shares, Royal Dutch became the first Netherlands firm to see its issued ordinary share capital exceed the nominal amount of one billion guilders.[16]

However, if an Old World oil enterprise with worldwide operations were to maintain critical mass, access to New World money was becoming vital, and Royal Dutch was eager to gain a listing on the New York Stock Exchange (NYSE). Shell Transport was more cautious, but Royal Dutch's launch on the NYSE in July 1954 was a financial breakthrough for the Group. Before the launch Royal Dutch's market capitalization was hovering around the 3 billion guilder level. Thereafter it rose steadily and quickly until by the end of the year it was over 6.5 billion guilders, and added together, the parents' market capitalizations broke the billion-pound mark.

Limited numbers of shares in the parent companies had been available in the US for many years – Royal Dutch since 1916 and Shell Transport since 1919, in both instances being in the form of bearer shares. The concept has become less familiar in modern days, but unlike centrally registered shares identified by a certificate, bearer shares were identified by a 'Warrant to Bearer'. Warrants were not legal tender and could not be exchanged quite so easily as a bank note, but the principle was similar: without the need for a broker, a warrant could be bought, sold, or exchanged like any other commodity, and its physical possession was proof of possession of the number of shares it represented. To claim dividends the owner detached a serially numbered coupon from the warrant and sent it in when advised to do so by the company.[17] The system made a company's stock more liquid, but also meant that the price of a warrant could be determined by local bargaining. Warrants could therefore sometimes be traded at a discount to the share price on a stock exchange; and such was the case in the 1950s when first Royal

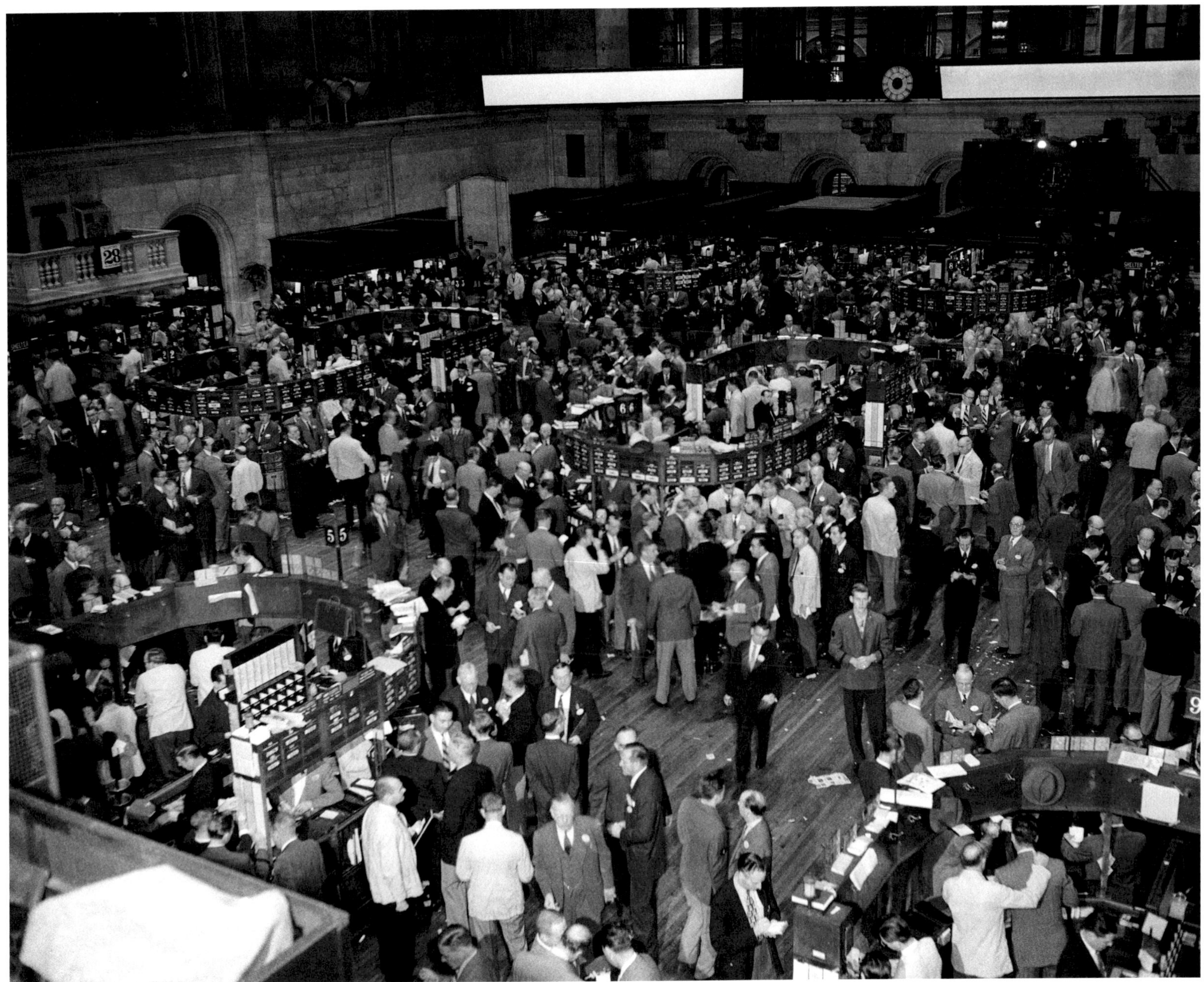

[8]

Floor of the New York Stock Exchange, 1951.

Dutch and then Shell Transport sought their American listings. In addition to the parent companies, Bataafsche had sold bond loans in New York during the 1920s.

The process of listing revealed a strong cultural difference between the European financial system and the American one, with the Americans insisting on much greater transparency. This was nothing new: Royal Dutch had tried to gain NYSE listing nearly twenty years earlier, but after the 1934 Securities Exchange Act, the US Securities Exchange Commission (SEC) was established which required full disclosure to investors of company information. Natural and reasonable as it may seem today, this was a problem for Royal Dutch then; its Articles of Association did not conform to the legislation, and it was unwilling to go the extra mile in order to achieve a listing. In 1936 it abandoned the attempt.[18] After the war, with the pressing need to increase the Group's capital, it was natural for Shell Transport to turn to the London money market; and so as not to compete there, New York was the obvious best source for Royal Dutch outside its traditional markets in the Netherlands, France, and Switzerland.

In 1950 confidential discussions on the timing of a new NYSE approach began between Van Hasselt, Legh-Jones, and Alexander Fraser, the chairman of Shell Oil's executive committee. Van Hasselt passed on a comment from G. B. Huiskamp, Treasurer of Finance Administration in London, who had learned from earlier conversations that in the view of Carl W. Painter, Shell's senior lawyer in the US, the timing was good for a Royal Dutch listing.[19] Van Hasselt suggested that Shell Transport should seek a listing at the same time, but Legh-Jones was doubtful, because large numbers of its shares were already available 'over the counter' in New York at a substantial discount, and he did not feel it would be wise to consider listing Shell Transport while the situation lasted. The men decided that Painter, Fraser, and Harold Wilkinson, the CEO of Asiatic Corporation, would compose a memorandum of the arguments for and against a listing, for the consideration of the parent companies' boards.[20]

The conclusion was that Royal Dutch should seek a listing, but Shell Transport should not do so yet, and Royal Dutch approached the NYSE authorities in 1951. In the ensuing routine investigation it emerged that, as with Shell Transport, there were many Royal Dutch shares already in circulation, including considerable numbers of bearer shares sent over by European investors for safekeeping during the war, which had sometimes been sold for hard cash. However, this did not deter Royal Dutch which, in order to comply with NYSE standards and gain that listing, was now prepared to make whatever changes were necessary in its customary methods. These changes included the granting, to any holders of shares aggregating at least 1 per cent of the issued authorized capital, of rights regarding the nomination of directors and managing directors, more frequent publication of financial data, and the provision of new registered shares with a small nominal value (of which more below).

Simultaneously, a separate matter was approaching completion. For several years the Group managing directors had been conscious of the desirability of publishing consolidated accounts showing the position of their principal operating companies. The British Companies Act 1948 expressly demanded corporations to publish such accounts, but, given that the Group contained over 400 companies of various nationalities working in almost every country and currency, this was an extremely complex operation. In 1949 the first indicative figures were produced. The parent companies' Annual Reports for 1950 contained (in addition to their respective results in their national currencies) aggregated

'When does the Company plan to follow the American corporation practice of issuing quarterly statements to stockholders?'

financial data for the Group as a whole for the first time, expressed in pounds sterling in both reports. The practice continued and the method of consolidation was refined, in order to provide more direct comparison with the figures of other concerns.[21] This followed the appointment on 11 November 1953 of E. Chester Peet, an American, to the boards of Anglo-Saxon and Shell Petroleum in London as head of Group Finance Administration.[22] Having been vice president, Finance, in charge of all Shell's financial and accounting operations throughout the United States from 1949 to 1952, Peet brought excellent qualifications to his task as finance director for the Group, and he was especially responsible for speeding up the production and publication of the Group's results.[23] From September 1954 consolidated Group accounts were presented to the parent boards every quarter, and also served to gauge how well the Group was doing in comparison to itself and to the competition.[24] Although the consolidation exercise was not directly related to the NYSE launch, it can only have helped to make the stock more understandable in the US market, and therefore more attractive to American investors.

In March 1954 Royal Dutch announced that agreement with the NYSE authorities had been reached in principle concerning the conditions of listing. When the news reached the Amsterdam Stock Exchange, the company's quotation there immediately rose.[25] A further change made just before the launch was that Royal Dutch began to publish figures half-yearly, instead of annually, as had been its habit;[26] but the NYSE authorities wanted them quarterly and made such publication a condition of listing, allowing the company a year to implement the change.[27] Now the only

[9]

Left, E. Chester Peet, the Group's head of Finance Administration at the time of the Royal Dutch launch on the New York Stock Exchange. Right, Loudon and others await the launch on 20 July 1954.

[10]

remaining barriers to the listing were alterations to the Articles of Association. Proposed to shareholders by letter on 27 April, accepted at the AGM on 16 June and coming into legal effect on 28 June, these increased the company's authorized capital to three billion guilders divided into four categories: 1,500 preference shares of 1,000 guilders each; 1,498,500 ordinary shares of 1,000 guilders each; 7,500,000 ordinary shares of 100 guilders each; and, as another condition of listing, 15,000,000 ordinary shares of 50 guilders each.[28] New shares would be issued in New York, Amsterdam, and London, the American side being handled by the Chase Manhattan Bank and the others by the Nederlandsche Handel-Maatschappij. The launch was set for 20 July 1954, when Royal Dutch became the first European company to have its shares listed on the NYSE since the end of the Second World War; and it seemed suitable for an event of such importance in the company's history to be conducted with some ceremony. As Director General, Van Hasselt had set the process in motion; it fell to his successor Loudon to travel to New York for the occasion, and when the Exchange's ticker-tape produced Royal Dutch's opening quotation he bought at once, paying $4,059.25 for 100 of the 50-guilder shares and becoming owner of the company's first American-issued share certificate.[29] As he soon discovered, however, it was one thing to overcome the problems leading to a launch; it was quite another to cope with the problems after a launch, when it took place in a different financial culture.

After its launch on the NYSE it took some time for Royal Dutch to adjust to US expectations. Below, a typical letter in; right, a typical letter out.

[11]

TE BEH. DOOR HR Wieringa
COP. GEZ. AAN

H Polish for N Mall

May 25 1955
Banning California
391 Barbour Ave.

Royal Dutch Petroleum
The Hague

Dear Sirs:

I wish to make a suggestion about the management of Royal Dutch Petroleum, I think the financial reports that are sent to the United States should be translated into dollars, Instead of Dutch Guildars. People in this country know nothing about Dutch Guilders therefore the reports have very little meaning. I think the stocks would anjoy a much wider exceptance in this country if the change was made.

I am a stock holder in the company.

Yours Truly

William R. Jones

[12]

JOHN S. HEROLD, Inc.
Petroleum Geologists &
Consultants,
250 Park Avenue,
NEW YORK 17, N.Y.

RA The Hague, 26th November, 1954.

Gentlemen:

We have your letter of November 16th to Mr. Bloemgarten for reply.

We would like to inform you that in accordance with Netherlands custom we do not supply any information which is not furnished generally to shareholders. We therefore regret that we cannot comply with your request for additional information that is not contained in the listing application.

Yours truly,

N.V. KONINKLIJKE NEDERLANDSCHE PETROLEUM MAATSCHAPPIJ

Wg MG. Wieringa

Another irritated letter from a US shareholder. However senior the addressee, the writers expected a personal reply – and usually got it, once the company began to understand the American way of doing things.

Royal Dutch Petroleum Company
30 Carel van Bylandtlaan
The Hague
Holland

Attn: J. H. Loudon, Managing Director

Gentlemen:

For a company attempting to regain its former stature in the eyes of the American Investor, it seems to me--as a stockholder--that without trying too hard, you have made a rather poor beginning insofar as accomplishing your aim.

It certainly seems to me, and I might add to countless other stockholders in Royal Dutch Petroleum, that a company the size of yours, (ours if you will) could assume or absorb the fees/cost for exchanging one security for another and allow a stockholder to own the certificate he prefers without having to pay for the "privilege". I realize of course the main interest was/is to have a security of Royal Dutch listed on the New York Stock Exchange. That goal has been attained, but why not leave a good taste in all our mouths and permit investors to exchange their securities within your company at no cost to them. For the comparatively small amount of money involved, much good will could be created among our investors instead of the attitude now prevailing as a result of the handling and the limitations of the recent stock dividend.

Needless to say, I am intensely interested in whatever reply you choose to make.

Very truly yours,

HHA:rg

[13]

Learning to live with the shareholders Royal Dutch had never before issued shares with a nominal value as low as fifty guilders. The motive was not a sudden outbreak of democratic thinking, but simply because that was the way of the American market. No less significantly, although it was customary in Europe to issue bearer shares, it was not so in the US and for reasons of cost and American custom the company's NYSE issue was of registered shares. To adopt two such important changes on top of the alterations to its reporting system, its Articles of Association, and its accounts shows how eagerly the company desired dollars and at the same time how influential Wall Street was, as far as its new finance was concerned.

The issue's success meant that Royal Dutch suddenly began being bombarded with US shareholders' questions. Both parent companies understood marketing well and, even before the war, managing directors were also fully aware of the importance of good public relations, as their concern over the use of films to project a favourable image of the Group had shown (see Volume 1, Chapter 6). However, investor relations were another matter. Towards the end of 1945 Shell Transport had engaged a consultant from the London Press Exchange to give 'general advice in connection with any questions of public opinion', adding 'it is not at present possible to define very clearly either the scope or the volume of the work. It remains to be seen to what extent we may wish to call on your services'.[30] Royal Dutch had never held a press conference until May 1947, an event that was attended by at least as many curious managers as by members of the press.[31] In 1954, despite receiving last-minute advice from Shell Oil on dealing with American shareholders, it was woefully under-prepared for managing shareholder relations in America.

Written only seven weeks before the launch, Shell Oil's letter of advice appears to have been a Shell Oil initiative rather than a response to a Royal Dutch query. It began by explaining the purpose of a stockholder relations programme, and contained no fewer than twelve separate matters for thought, outlined below. Their common thread was the vital need for organized, regular, full, clear communication with the market.[32]

The first point was a question: would someone be employed at Group level to handle shareholder relations, or would that responsibility be given to an employee of Shell Oil, Canadian Shell

1

ONTVANGEN
18 MEI 1955
W. G. Wieringa

TEGEN... DOOR: Dr. Wieringa
COP. GEZ. AAN:
Copie: Dr. Treep, Fredenburg

Monte Carlo HOTEL
OCEAN FRONT AT 65TH STREET
MIAMI BEACH, FLORIDA

MAY

ROYAL DUTCH PETROLEUM Co.
THE HAGUE THE NETHERLANDS.

GENTLEMEN:

I AM A STOCKHOLDER IN
AND WISH TO BRING TO YOUR
THAT IS AFFECTING THE
AMERICAN PUBLIC IN YOUR

DURING THE PAST FEW
YOUR DECLARATION OF EAR
PAYMENT OF FINAL DIVIDEND
STOCK HAS MET WITH CONS
THE MANY STOCK ANAYLIST
WHO'S OPINION IS HIGHLY
FORMERALLY RECOMMENDED YO
ARE NOW REFUSING TO RECOM
ADVISE AGAINST ITS PURC

THIS SHOULD BE CORRE
BE DONE.

PAY A FAIR EXTRA D
WITH PRESENT EARNING

PUBLISH A CLEAR, CON
STATEMENT SO IT CAN B
PUBLIC.

[14]

Royal
30 Car

The

Gentle

stock
now
of the
with
compa
admin
am
of th
with

abili
fact
quite shrewd. However, there is one
thing I wish to suggest and that is
some program of informing shareholders of
the progress of the company employing
United States dollars. This is not intended
as a selfish request, but rather it
involves promotion of good relations for

[15]

Mr. Bloemgarten
Would you deal with this
I am not sending a reply

76 Glendale Ave.
Hartford 6, Conn.
U. S. A.

November 10, 1954

Mr Wieringa
Deze was bekend in een wedding (?) is

Dear Mr. Kessler:

I am interested in buying considerable stock in your company.

Please tell me about your company in regard to reserves, plans for expansion, future earnings, possible mergers, etc.

Please, if possible, send me a detailed map of your drilling program, and the latest stockholders report.

Any additional information will certainly be appreciated.

Thank you.

Sincerely,

Stanley Zaimor Jr.

P.S. When is your next stockholders meeting? Do you think the dividend will be increased?

[16]

Recu 15/8 23/8

July 11, 1955

TE BEH. DOOR	N. Wieringa
COP. GEZ. AAN	

Gentlemen:

I recently purchased your 50 guilder shares through the New York Stock Exchange, for Investment purposes. In connection with this, will you please send me whatever information about the company which you think I might find of interest.

In particular, will you please let me know what the status of American shareholders was during the last war. Did they receive dividends – were they held in Trust? Could they sell their shares on any market here in the U.S.?

Are earnings reported to shareholders quarterly? When are dividends declared + paid? What is the record of earnings + dividends over the last ten years?

If you have an American office through which I can obtain further information, please advise me.

Thank You

Norman Hoffman

[17]

The summer of 1955 brought Royal Dutch a flood of letters of inquiry or complaint from new and would-be American shareholders – and not just letters. Right, one particularly frustrated shareholder scrawled on his Notice of the AGM, 'Why in hell don't you just send the proxy forms – like all good American companies do – instead of making us write for them. Your practice stinks.'

[18]

ROYAL DUTCH PETROLEUM COMPANY
(N.V. KONINKLIJKE NEDERLANDSCHE PETROLEUM MAATSCHAPPIJ)
established at The Hague

MR LEO PETERS
c/o CARLETON M T
105 W A[...]
CHICA[...]

ANNUAL GENERAL MEETING OF SHAREHOLDERS

May 8, 1956.

[...]d Holders of 50 Guilder
[...]f New York Registry:

Why in hell don't you just send the proxy forms — like all good American Companies do instead of making us write for them. Your practice stinks.

[...] hereby given that the Annual General Meeting of Shareholders of the Company will be held on [...]0th May, 1956, at 11:30 a.m., in the large hall of the Carlton Hotel at Amsterdam, The Netherlands.

AGENDA

1. Appointment of a Managing Director.
2. Appointment of a Director.
3. Appointment of a Director owing to retirement by rotation.
4. Annual Report for 1955.
5. Approval of the Balance Sheet and the Profit & Loss Account for 1955.
6. Distribution of profit, declaration of dividend for 1955.

[...]da, the nominations relating to items 1, 2 and 3 of the agenda, the documents named under items [...]genda with the relevant notes, in accordance with Article 42 of the Netherlands Commercial Code, [...]roposals submitted by the Board of Directors, together with the proposals submitted under item 6 will be available for inspection from today until after the Meeting, and may be obtained by share[...] charge at the Company's office, 30, Carel van Bylandtlaan, The Hague, and at the office of [...]hattan Bank, 11 Broad Street, New York 15, New York.

[...]s of Association provide that anyone who, on the day of a general meeting of shareholders, is a [...] attend and address the meeting and vote thereat, subject to the further provisions of the Articles [...]elow.

[...]re of 50 Guilders one vote may be cast; for each share of 100 Guilders two votes may be cast; [...]re of 1,000 Guilders 20 votes may be cast; but no person, whether voting in person or by proxy, [...]an 120 votes on account of shares held by him, and no proxy for one or more shareholders, as [...]re than 120 votes in all.

[...]**olders of registered shares,** in order to attend and address the Meeting and to exercise such [...]erson or by proxy, must make known to the Company in writing their desire to do so not later [...]day before the date of the Meeting, viz., by **Wednesday, 23rd May, 1956. In the case of regis[...] [...]0 Guilder shares of New York registry, such notice must be deposited on or before the last mentioned date with The Chase Manhattan Bank, 11 Broad Street, New York 15, New York, the Transfer Agent for such shares.**

On behalf of any registered holder of such shares who gives notice of his desire to exercise his rights at the Meeting but who cannot attend the Meeting in person or arrange for a proxy of his own choice, it has been arranged that the Transfer Agent will forward any signed proxy, in which the name of the proxy is left blank and which is deposited with it not later than Wednesday, 23rd May, 1956, to Nederlandsche Handel-Maatschappij, N.V., Amsterdam, and that the latter will endeavour to supply a suitable individual to act as proxy for such shareholder.

Blank forms of letters of notification and of proxies are available on request at said office of the Transfer Agent.

These instructions, which are applicable to registered holders of registered shares do not apply to **holders of bearer shares** in denominations of 100 Guilders or 1,000 Guilders who, in order to attend and address the Meeting and to exercise their voting rights in person or by proxy, are required to deposit their shares against receipt with specified banking institutions in advance of the Meeting in accordance with the instructions contained in the published notice of convocation of the Meeting.

N.V. KONINKLIJKE NEDERLANDSCHE PETROLEUM MAATSCHAPPIJ.

FLB
FNMA
FIC
TREAS
COUPON
ISSUES
TREAS. BILLS
TREAS. BILLS
ISSUE
MARKET
YIELD
BID
ASKED
FHLB
BC
S.B.& H. GOVT. AVGS
1-2
4-5
8-10
OVER 20
OPEN
CLOSE
SVG BK TR
ATS
HARV
RCA CR
LLC
ISC
MACK
USR
SWIFT

[19]

Salomon Brothers & Hutzler, the New York investment firm, handled much of the buying and selling of securities for Royal Dutch. Left, Salomon employees discuss short-term trends, and right, after a visit to Royal Dutch, an employee of another investment firm writes an enthusiastic letter of thanks to Royal Dutch's General Attorney W. G. Wieringa.

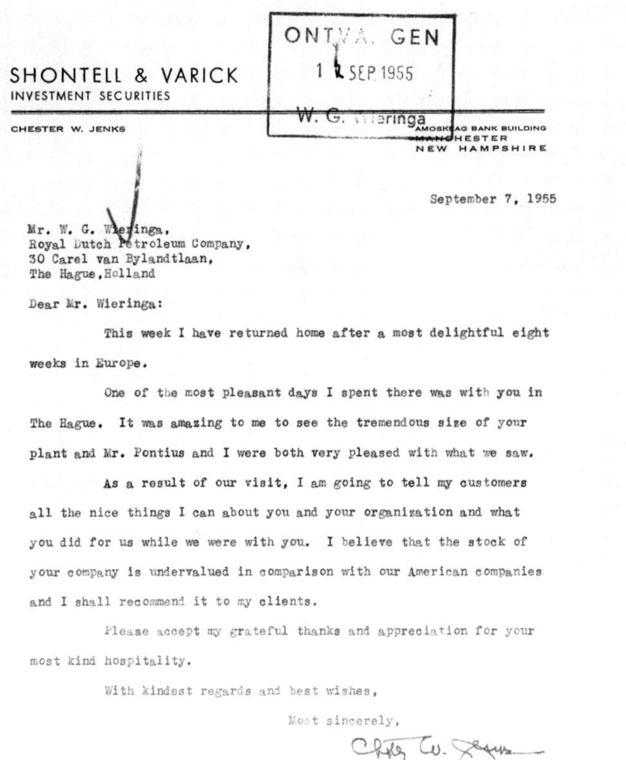

SHONTELL & VARICK
INVESTMENT SECURITIES

CHESTER W. JENKS

AMOSKEAG BANK BUILDING
MANCHESTER
NEW HAMPSHIRE

September 7, 1955

Mr. W. G. Wieringa,
Royal Dutch Petroleum Company,
30 Carel van Bylandtlaan,
The Hague, Holland

Dear Mr. Wieringa:

This week I have returned home after a most delightful eight weeks in Europe.

One of the most pleasant days I spent there was with you in The Hague. It was amazing to me to see the tremendous size of your plant and Mr. Pontius and I were both very pleased with what we saw.

As a result of our visit, I am going to tell my customers all the nice things I can about you and your organization and what you did for us while we were with you. I believe that the stock of your company is undervalued in comparison with our American companies and I shall recommend it to my clients.

Please accept my grateful thanks and appreciation for your most kind hospitality.

With kindest regards and best wishes,

Most sincerely,

Chester W. Jenks

CWJ/D

[20]

or Shell Caribbean? Notably, the American company was not asking *if* shareholder relations would be a dedicated task; that was taken for granted. Similarly, the second point was a statement, or possibly a reminder, of a necessary matter of fact: 'News release to be prepared for release at a meeting with the press when listing approved.' The most striking aspect of that is that it needed to be said at all; and the other ten points for consideration amounted to a total programme for building good shareholder relations. A letter of welcome could be sent to every new shareholder; there could be a meeting with statistical services before the listing; the company's listing application could be widely distributed. A summary of its Annual Report and that of Shell Transport could be advertised in American financial publications; the full reports could also be broadly distributed; current financial information could be released to the press. Meetings should be considered with security analyst groups in New York and other large cities, and with banks, investment trusts, and other large investors, to inform them about the Group; the Annual Reports could be simplified, especially in their financial statements; and lastly, consideration should be given to preparing a comprehensive memorandum on the Group's operations.

The fact that most of those points now seem obvious emphasizes again the difference between the contemporary business cultures of Europe and America. Most of the recommendations were followed as quickly as possible, but Group personnel found that the privilege of American listing brought seemingly extraordinary obligations of transparency. Any American shareholder with any query whatsoever felt fully entitled to write to a senior member of the company, and to expect a personal reply.[33] Some of these questions seemed completely impertinent, such as a request to Royal Dutch's General Attorney W. G. Wieringa for information on a transaction between two Group companies.[34] Never having received such a query before, Wieringa cabled his colleagues in the London legal department for advice. Knowing no better than he, they felt the shareholder was not entitled to such information and said it should not be given.[35]

But other requests poured in, for stockholders' reports, interim earnings reports, future earnings estimates, expansion plans, possible merger options, estimated dividends per share and capital expenditure. One of the large 'stockholders of record', Francis I. Dupont & Co, requested copies of press releases and additional information, and later sent their own lengthy

'...The results of good stockholder relations show up promptly in a more favourable appraisal of its shares in the market.'

questionnaire for completion and return. Harassed by this, an American Group employee told Royal Dutch that even in the US there was no uniform practice of responding to such inquiries, but the company was beginning to realize that its preparations had not included nearly enough cultural intelligence-gathering, and decided to action Dupont's request and any similar ones that might arrive.[36]

There were hundreds, and this one is typical: 'One important brokerage house in New York answered my question in relation to Royal Dutch Petroleum with the short answer: sell – inadequate information ...'[37] Perhaps the most damning criticism came from a member of the exchange: 'It is my impression, and this impression is shared by virtually everybody in the financial community who is interested in your shares, that the company has done nothing to facilitate the understanding of its affairs. On the contrary, it seems that much was done that had the effect of antagonizing the American investor (...) We in America have learned to cultivate the relationship between the management of a corporation and its stockholders who, after all, are the owners of the business. The results of good stockholder relations show up promptly in a more favourable appraisal of its shares in the market. Such a relationship is non-existent in the case of the Royal Dutch Petroleum.'[38]

This was something of an overstatement, since in fact the Royal Dutch had done rather well on the New York market. After an initial slump, the share price rapidly soared and within six months Royal Dutch had become one of the most heavily traded companies on the exchange.[39] As a result the Royal Dutch board had little incentive to respond to urgent requests from some American shareholders to have the Royal Dutch Annual Reports give dollar equivalents for the guilder figures in the financial statements. This very minor change took seven years, until 1962, to accomplish, perhaps because directors feared offending the important and much older contingent of foreign shareholders elsewhere by giving financial summaries in dollars rather than, say, Swiss or French francs.[40] Unfortunately, small Dutch shareholders were suspicious of the unfamiliar registered shares and bought far fewer than had been hoped, even at the 50-guilder level; and as the price rose many scented an immediate profit and sold their shares to willing Americans.[41] Loudon deplored this and sought to encourage a longer-term view, but the lack of widespread understanding of registered shares in the Netherlands militated against him.

It soon became apparent that an important shift was occurring in the distribution of Royal Dutch share ownership. By 1956 a majority of the company's share capital was actually in foreign hands – roughly 4.5 per cent in Great Britain and Belgium, 5.5 per cent in Switzerland, 22 per cent in France as a consequence

Royal Dutch Splits 2½-1

Directors and managing directors of Royal Dutch Petroleum Co. announced they will call a meeting of shareholders in the middle of October to consider a proposal to amend the articles of association.

The amendment will result in a 2½ for 1 split-up of the existing 50 guilder shares.

The split-up will not change the aggregate par value of the authorized share capital or the aggregate par value of issued share capital.

It will change all shares of Royal Dutch Petroleum into shares of 20 guilders par value so that existing shares of 1,000 guilders par value will then be 50 shares of 20 guilders par value; 100 guilders par value will then be 5 shares of 20 guilders par value and 50 guilders par value will then be 2½ shares of 20 guilders par value.

Royal Dutch-Shell Profits Larger

Royal Dutch-Shell Group of companies today reported net income of £134,474,218 for 1954 compared with £130,413,373 for the year before.

Sales totaled £1,851,039,043 against £1,700,714,450.

Royal Dutch Petroluem Co., which has a 60 percent interest in Royal Dutch-Shell Group, reported net income of 195,237,673 Netherlands guilders for 1954 against 161,574,940 for 1953.

of the Group's earlier close relations with the Rothschilds, and 24 per cent in the United States.[42] Would the company remain, as a worried shareholder had asked before the launch, not only 'Royal' but also Dutch?[43] It was starting to look a little doubtful but, over time, Royal Dutch solidly remained both, despite an increasing proportion of foreign share ownership. By 1970, only a third of the shares were held in the Netherlands, one quarter in the US, 18 per cent in Switzerland, and 15 per cent in France. And Royal Dutch shares were far more international than those of Shell Transport which, though also launched on the NYSE in 1957, remained overwhelmingly British-held: 95 per cent in 1970.[44] As a result of that concentration, Britain had the largest total of parent company shares with 40 per cent, followed by the Netherlands with 20 per cent, the United States with 15.6 per cent, Switzerland with 11 per cent, and France with 10.1 per cent.[45]

Royal Dutch-Shell Reports Net Rose in '54; Oil Output Up 5½%

Crude Oil Processed Increased 9%; Continued High Capital Outlays Forecast for This Year

By a WALL STREET JOURNAL *Staff Reporter*

NEW YORK—The Royal Dutch-Shell Group of companies reported net income for 1954 equivalent to $376,527,200 compared with $365,-156,400 in 1953. The figures in the report are shown in pounds. The dollar conversion figure was based on $2.80 to the pound. Gross income of the group approximated $5,183,000,000 com-

Encouraging news clippings appeared in (left to right) *The American Journal* on 20 September 1956, *The Sun* on 10 May 1955, and the *Wall Street Journal* on 10 May 1955.

Towards a closer unity: simplifying and harmonizing

At the end of the 1950s the Group underwent a thorough restructuring in an exercise known as the McKinsey review. This process was so prominent an event that it tends to be remembered as the sole creative programme of the time. That it was the *major* programme cannot be in doubt, yet it was not so much an initiating action as the culmination and reinvigoration of a process that was already in train.

To call this process a plan would suggest a more formal structure than it actually possessed, but its purpose and effect was to simplify and homogenize aspects of the Group's life in the interests of greater commercial efficiency and Group-wide understanding. Two of the earliest post-war steps towards Group homogeneity were in the parent companies' AGMs and Annual Reports. The parents had traditionally chosen the dates for their AGMs without reference to one another, but the practice had occasionally led to confusion, either with different topics being covered in the meetings, or the same topics being covered differently, or different answers being given to the same question. This was overcome by the simple expedient, introduced in 1949, of holding both AGMs on the same day and preparing joint answers for anticipated questions.

As for the Annual Reports, up to and including the ones for 1947 these were genuinely separate reports on the parent companies and were totally different from one another. From 1948, they started to become much more alike, with large blocks of text being identical or very similar, but with many parts that were not. Royal Dutch followed Shell Transport's former ordering of subjects, but the reports' format and presentation remained completely different. A curiosity is that in 1948-57 inclusive the Shell Transport reports consistently referred to 'the Shell and Royal Dutch Group', which they had not done before. Why this was so is unknown – was it perhaps some lingering regret about the failure of the 50:50 bid in 1947, or merely a rather chauvinistic thoughtlessness? Whatever the reason, from 1958 the phrase reverted to 'the Royal Dutch/Shell Group', and the reports themselves became virtually identical throughout in text, format and presentation: they were in fact

In the two decades following the Second World War, the Group renamed many of its companies to reflect their shared Shell identity. The pecten logo too was redesigned, and in some parts of the world lasted an unexpectedly long time. Here, photographed in 2003, an overgrown and ruinous little outlet in Liberia shows the fifth logo and fourth pecten, introduced in 1948.

Group reports, with information about the individual parent company relegated to a position of secondary importance. A third rationalization in the immediate post-war period concerned the assigning of clearly defined tasks to each Group managing director, replacing the somewhat haphazard arrangements current until then.[46] Finally, immediately after the war managing directors took steps towards a closer integration and coordination of research by organizing, in 1946, the first annual meeting to discuss research strategy and budgets (see Chapter 5).

During this period numerous companies were created for special purposes, such as O/Y Kamex A/B, set up to market bitumen and bottled gas in Finland, and NV Rotterdam-Rijn Pijpleiding Maatschappij, for the construction of a pipeline from Rotterdam to the Ruhrgebiet. They were somewhat exceptional, because new companies generally included 'Shell' in their often self-explanatory titles: for example, among those formed in 1957-58 were Deutsche Shell Tankers GmbH, Shell Aircraft Ltd., Shell Production of Argentina Ltd., Shell Australia Securities Ltd., Libya Shell NV, Turkse Shell NV, The Shell Company of Ghana Ltd., The Shell Company of the Bahamas, and British Honduras Shell Petroleum & Development Company Ltd.

In a policy that began between the two World Wars and continued long after, the names of existing Group companies were also being simplified to include the word 'Shell' (see Volume 1, Chapter 6). Many examples could be given, but one of the earliest to be changed post-1945 was Asiatic Petroleum Ltd., the joint marketing company formed in 1903, before the Group was born. On 8 January 1946 this became The Shell Petroleum Company Limited,[47] and in May 1948 the Group's French operating company SA des Pétroles Jupiter was renamed Shell Française.[48] Similarly, on November 1949 Lumina SA and Nafta Societa Italiana became, respectively, Shell Switzerland and Shell Societa Italiana.[49] All were thus able to benefit from the strong 'Shell' identity, compared to which their original names were almost meaningless. Shell Transport and Trading, the eponymous author of the Group identity, never changed its name (except to alter its 'Limited' suffix to 'plc', public limited company, in response to a change in UK law in 1980), but in 1948 Royal Dutch did rename itself, abandoning its original title – 'The Royal Dutch Company for the Working of Petroleum Wells in the Netherlands Indies' – in favour of 'Royal Dutch Petroleum Company'.[50] In the US, Shell Union Oil Corporation became Shell Oil Company in September 1949,[51] and in November 1955 the Anglo-Saxon Petroleum Company, formed in 1907 as the Group's first British operating company, lost its identity (no more meaningful outside the Group than the Asiatic and Jupiter had been) when in the Group's first step towards structural simplification, it was amalgamated with Shell Petroleum.[52] Anglo-Saxon, Shell Petroleum and Bataafsche Petroleum had already been redefined not as 'operating companies' but as 'parent operating companies' and later as 'holding companies', to reflect more accurately their evolving functions, but after the amalgamation of Anglo-Saxon and Shell Petroleum, only Shell Petroleum and

SHELL
ELCOME to
Mr.
promoter
Barber
SHOP

"SHELL"
Shortens Every Road
"SHELL"
Shortens Every Road
PREMIUM
SHELL ECONOMY
OPEN
Silver

'Shell shortens every road' – photographed in 1985 in Caernarvon, Wales, the message was still clear, even if pumps for different grades were empty and the business had closed.

Bataafsche Petroleum remained; then on 21 April 1959 Bataafsche's name was given a more modern spelling, Bataafse, and in 1967 it too was renamed, becoming Shell Petroleum NV. During the late 1950s the management of the tanker fleets was also centralized in two companies, Shell Tankers Ltd. and Shell Tankers NV.

Rounding out the circle, these name-changes were designed to 'complete the revision of the Group companies' nomenclature which had been set in train after the war, to incorporate the word "Shell" as far as possible in the names of Group companies', and in this instance a pragmatic commercial reason was recorded too: experience showed that with a Shell name a company would find it much easier to borrow money.[53]

All these new names were designed to tell the world that the company in question was a Shell company, and within the Group, from the end of the war and continuing through the McKinsey debate, there was a deliberate stimulation of staff solidarity coupled with a more sensitive approach to management. In 1946, Bataafsche inaugurated *contact commissies* or elected staff councils for its establishments in the Netherlands, which anticipated the introduction of a law concerning staff councils. Such councils had existed before, but now received a more formal position. London central office appears to have had a similar institution only in 1951 when a Staff Advisory Committee was set up, initially restricted to advise on catering matters. Until 1962, this was an appointed, and not an elected body, because senior managers felt strongly 'that the correct channel of communication to the staff was via the line Management and that the use of elected Committees, as practised in some other countries, was inappropriate and would tend to result in a by-passing of the Division.'[54]

In 1948, a new pension scheme, designed to bring more uniformity in old-age provisions for all Group staff, replaced the Provident Fund and the pension fund set up in 1938. The scheme adopted uniform retirement ages for men at 60 and women at 55, with reductions for tropical service and fleet service.[55] Shell Oil remained an exception, setting the ages at 65 and 60 respectively. Non-executive directors could even stay into their eighties, and some did.[56] Pensions for directors and managing directors had been introduced immediately after the war. Most employees were already provided for in this regard, but, in an indicator of societal norms, it had long been assumed (not only within Shell, but generally) that directors did not need pensions, either because they were independently wealthy or because their salaries would enable them to provide for their retirement. However, 'taxation over the last few years has been such as to make it impossible' and directors' pensions had become 'a desirable and essential arrangement.'[57]

At about the same time salaries were harmonized by the adoption of a Group Basic Salary (GBS) and, for expats, a standard Local Currency Salary (LCS), an amended GBS to take account of exchange rate variations and salary surcharges depending on local circumstances.[58] A sense of unity manifested itself, through having survived together the suffering of war, and a strong family feeling was maintained by the Group's provision for social and sporting activities and the employees' creation of many active and successful social clubs for motoring, dancing, archery, shooting and so forth, all testifying to a feeling of friendly pride in working for the Group. Because of his own character and his treatment in Deterding's last years, Kessler had long understood the importance of managers getting on with each other and with staff, and in 1957 he wrote to the whole Group, 'Human relationships are *so* important. The Management does not beg for loyalty, but it does try to create a climate in which loyalty is able to flourish.'[59] In this it certainly succeeded: personnel felt valued, and Loudon remarked later that 'the community feeling in Shell is one of its greatest assets'.[60]

In an annual festival of nostalgia and admiration for past technology, the 1962 Hill Climb of the Model T Ford Club takes place on Shell Hill, Long Beach, California. May the best fuel win!

STOP
31

[25]

In 1951, as part of Shell's growing desire to promote an international culture, Henk Bloemgarten (third from below) visited Curaçao to open a cultural centre, reported in the press (below).

[26]

Ir. Bloemgarten opent fraai Cultuurcentrum

Verruiming van de menselijke blik

ag j.l. mocht het Suffisantdorp hoog bezoek ontvangen. H. Bloemgarten, Directeur van de Koninlijke Shell, wilde is maken met de verschillende dorpscommissies en zo et, dat de dorpsbewoners deze directeur van de Konink- n Haag, al was het maar voor kort, in hun midden ben.

illende commissies
n de hoge gast in
gebouwde Cultuur
e bij deze gelegen-
heer Bloemgarten
geopend.
Bloemgarten om-
r n.m., vergezeld
eur van de CPIM/
van Drimmelen,
teur de heer A.
heer H. C. Nieu-
Afd. Arbeidsza-
werd het gezel-
gang van het ge-
loor de dorpsstaf
oofd de heer H.
vaarna de bezoe-
en van de ver-
commissies werd

garten het podium van de zaal en sprak de aanwezigen in een korte toespraak toe, waarin spr. verzekerde, dat zowel de directie van de CPIM als de directie te den Haag een gewillig oor heeft, voor de wensen en verlangens van de commissies, die de spreekbuizen zijn van hun dorpsgemeenschap.

• „Het verheugt mij te kunnen constateren dat er naast Uw belangstelling voor de lichamelijke ontspanning tevens vraag is naar de mogelijkheden voor geestelijke ontspanning en — a' is het dan ook op zeer bescheiden wijze — dit gebouwtje, hetwelk ik zo direct zal openen biedt

CSM en haar werknemers.

Spr. besloot zijn toespraak met de heer Bloemgarten een voorspoedig reis en een tot weerziens toe te wensen. Na deze toespraak was er gelegenheid het gebouw te bezichtigen. De heer Bloemgarten werd rond geleid door de heer van Drimmelen, Nieuwenhoff en Oetelmans, de twee laatstgenoemden hebben in het restaureren van het Centrum een groot aandeel gehad.

Na de rondgang door het gebouw werd er op het terras gezellige zitjes gevormd en onder een koele dronk onderhield zich de heer Bloemgarten met de voorzitters van de verschillende commissies, van wie hij het een en ander wilde vernemen over de gang van zaken in het dorp. Het was wel jammer, dat het uur van scheiden was geslagen, de heer Bloemgarten nam van het gezelschap afscheid, maar niet alvorens hij zijn handtekening als eerste gast had geplaatst in het gastenboek van het Cultuur Centrum, dit voorbeeld werd

deerlampen,
dern tekenbord,
een leeszaal met
op de meest gelezen
ten en kranten
leestafel biedt
lezers.
Het ruime terras
kelijke stoelen.
een groot prachtig
gesteld, die de
aantrekt, een piano
is voor de musicerenden
Voor de verfraaiing
geheel, in elke
met vergulde randen
rijen, die het geheel
aanblik geven.
Voorlopig zijn de
gen van de bibliotheek
diezaal en leeszaal
dagelijks (behalve
van 9 tot 12 en van
n.m..
Mogen wij dit
gen met eveneens
Bloemgarten de wens
ken, dat van
de bewoners van
veelvuldig en
zal worden

In addition to these relatively unplanned steps towards simplification and harmonization, the Group undertook a series of really major harmonizing developments that adhered much more closely to the nature of a conscious plan. One of them, perhaps the most important one, concerned staff regionalization and internationalization. Indeed, regionalization was firmly described as 'our avowed policy'.[61] In later years outside commentators tended to use the two terms interchangeably, and it is true that they both contributed to the same outcome: 'Shell in its staffing is probably the most international firm in the world.'[62] However, within Shell the terms had different meanings, and of the two, internationalization was at first the less important. This was because initially it was used with only one very specific meaning, namely the introduction of expatriate managers of various nationalities into a team that previously had consisted exclusively of managers of one nationality. The best example of this was in Indonesia. There, the expatriate management team was traditionally composed exclusively or almost exclusively of Dutch nationals, but during the turbulent years following the Second World War the Group sought to de-

The launch of the Ultra-Large Crude Carrier *Metula* in 1968 attracted a staff audience epitomizing Shell's international character, with many different nationalities and races being represented.

[27]

[28]

[29]

In fostering an international attitude and the transferability of key staff, the Group took good care of its expatriates. Above left, the clubhouse at Balik Papan in the 1950s, and above right, an elegant dance evening there. Loudon was visiting at the time, and is on the right talking to Jonkheer J. M. van Rijckevorsel.

emphasize its Dutchness: 'As you know we have been studying for some time the pros and cons of putting the expatriate staff in Indonesia on a more international basis than is the case at present. We have approximately 780 expatriates working with B. P. M. 'Works' and Handelszaken and the vast majority of these are of Dutch nationality. Whatever the reasons have been for maintaining this situation up till now, since the recent developments culminating in the unilateral abrogation of the Dutch Round Table Agreements we have received various indications which makes it desirable to bring home to the Indonesians that our interests in their country are not only Dutch but that they are part and parcel of a truly worldwide enterprise. In this connection it is not unlikely that the internationalisation of our expatriate staff would do at least some good.'[63]

Compared to that very limited meaning of internationalization, regionalization (sometimes known in the early days as 'local for local') was far more important: it meant enabling a steady reduction of the ratio of expatriate employees to people from a given region. Before the war, the Group had trained locals for managerial positions in the United States, but otherwise only if forced to do so by legislation, as in Romania, Mexico, and Venezuela.[64] As we have seen in Chapter 1, the Group embraced the principle of regionalization as a guideline for staff policy in 1944. The crumbling of colonialism in Asia immediately after the war reinforced the awareness that operating companies needed strong local roots if they were to survive and prosper.[65] The overall success of the regionalization policy eventually made possible both the Group's fast expansion after the war, and a much greater internationalization than the mere leavening of expatriate management teams; the Group was now able to switch key members of staff from one country to another, in the interests of wider work experience, the transfer of knowledge, and the fostering of a homogenous Shell culture. This did involve some adapting of work styles. In Indonesia, for example, the Indonesian employees' desire

[30]

The young ones were not forgotten, and during summer holiday visits from Europe – for example, here at Suffisantdorp, Curaçao, in 1951 – games were organized for the children of expat staff together with the workers' children.

to do the job as well as possible initially took up a great deal of extra time, so The Hague suggested that 'in future the organization should adapt to the disposition and pace of the Indonesians, because the Dutch striving for perfection often has a slowing effect'.[66] One year later, the organization appears to have settled in. When a London personnel manager on an inspection visit in Djakarta asked whether the Bataafsche management there did not expect work standards to fall as a consequence of staff regionalization, 'the answer was very firmly "no". Methods of approach would be *different* and the total result would not necessarily be lower efficiency'.[67]

As stepping stones towards that, very many examples could be cited of the regionalization policy in practice. Cuba and Argentina were the first two countries, in 1951 and 1953 respectively, in which non-European employees headed the local Group operating companies.[68] In 1957 the Shell Company of Ghana was established; part of its basic company policy was to create openings for locals, and by the end of its first year fourteen Ghanaians held responsible executive positions in the company's business in the Accra district, while the Trans-Volta-Togoland region was administered completely by Ghanaians. Similarly, in Rhodesia (now Zimbabwe) the Group was the first to employ indigenous black people as truck drivers. Although this was obviously more modest progress than in Ghana, it was nonetheless both socially and racially significant: other employers thought it would be impossible, because drivers also had to collect money, sign receipts and so forth. By 1959 there were nine different nationalities in the management of the Shell Company of Thailand; a group of Filipino investors bought a 25 per cent interest in the Shell Company of the Philippines, while its board had five British members and four Filipinos; and in the Federation of Malaysia, the Group employed over 1,000 people, of whom 98 per cent were Malaysian.[69]

As the 'avowed policy' of regionalization spread around the world, the concept of internationalization grew from its first very limited version into the wider and more sophisticated style that became characteristic of Shell. In that phase, the Group developed two further related definitions to describe its employees: 'an "International" is any man who is engaged essentially for long term service outside his home country. Thus far these men have been mainly recruited by central offices; it is, however, the intention that the Operating Companies will increasingly provide their contribution to this category in accordance with central offices' recruitment programme. The criterion for "International" is transferability. (...) A "Regional" is any man who is engaged by Operating Companies (and also by central offices for Operating Companies in Holland and the United Kingdom) to make his career

[31]

[32]

essentially in his own country. For those with Managerial potential this will imply short term broadening assignments to other countries. Apart from such short period, a "Regional" is characterised as being not transferable'.[70]

The key distinction between an 'International' and a 'Regional' thus became the question of whether or not an employee was willing and able to work long term outside his (it was still overwhelmingly 'his') home country. In that definition it is important to emphasize that a 'regional' could just as well be a Dutchman who was only willing to work in the Netherlands, or a Briton who was only willing to work in the UK, as (say) a Filipino who was only willing to work in the Philippines. However, at the same time as it created that definition, the Group was also using the term a 'regional' in its more obvious sense of a person from a particular region, omitting the refinement of whether or not that person was willing to move. So in *that* sense it was perfectly correct for a report covering 1962-64 to say that the Group world-wide employed totals of about 4,000 expatriates and 232,000 regionals of all nations and categories; and in the other sense, with 'Internationals' being mobile and 'Regionals' not, it was equally correct for another report in 1963 to include the gloomy conclusion that 'The unwillingness of regionals to serve abroad is likely to limit progress in internationalisation.'[71] People in Shell understood these different definitions, but it is no wonder if outsiders became confused.

The 1963 report was the Group's first attempt in forecasting its long-range need for internationally mobile managers. In the report, Arab nations and Japan were identified as being particularly difficult to plan for, because of 'the present objection of some Arab countries to the employment of "foreign" Arabs' and 'the special difficulty of such appraisal [for forecasting purposes] in Japan', but despite this it was possible to forecast (even if only tentatively) the shifts in requirements expected at general manager level between 1963 and 1978. In 1963, covering all functions – E&P, Manufacturing, Finance, Personnel, Chemicals and 'Others' – there was already a slight balance in favour of regional rather than expatriate general managers: a total of 314, with 163 regionals (52 per cent) and 151 expatriates (48 per cent). By 1978 it was expected that the total would have increased to 512, and with a significant alteration in the actual and proportional figures, to 371 regionals (72 per cent) and only 141 expatriates (28 per cent).[72]

As may be seen from those figures, expatriate employees tended to be the losers as regionalization progressed. This was exacerbated by 'reduced activities in long-established production areas' such as Indonesia. Total expatriate numbers fell from about 5,600 in 1958 to 4,800 in 1961, 4,000 on average (as noted above)

[33]

[34]

in 1962-64, and about 3,000 in 1965. For the first few years of this process, about two-thirds of the expatriates who became 'surplus to the requirements of operating companies' were found new Shell jobs either abroad or in the home countries, but about 1,500 a year – a number that was expected to rise – had to leave Group service.[73] Redundancy payments were good and the more old-fashioned expatriates did not view regionalization in a kindly light, so quite probably a fair proportion of those who found themselves 'surplus' left more or less willingly. But there was a lifeline for those who wished to remain: 'It will be of the utmost importance to reserve in operating companies abroad suitable career development posts for expatriates, as without doing so the career development of our "international" staff would be seriously handicapped. The Managers of these operating companies should therefore, in consultation with Group Personnel, reserve suitable posts which are intended for the development of promising staff.'[74] Like Royal Dutch's acceptance of the NYSE's demand for shares of low nominal value, regionalization or 'local for local' was not motivated the goodness of the Group's heart but arose for sound business reasons, to accommodate rising local desires for participation. Consequently, the CMD held out against such pressures until managers felt that they had to accept in order to protect the business.[75] Even so regionalization became part of a

Expatriates enjoyed an enviable lifestyle. From left to right, the pool at Shell's sports and entertainments centre in Suffisantdorp, Curaçao, 1954; the swimming pool in the Panaga club, Brunei, 1960s; a domestic interior in the residential area of Cardón, 1948; expat housing at Seria, Brunei, 1960s. Lastly, workers were adequately cared for, but not quite so well, as this 1950s dormitory shows.

[35]

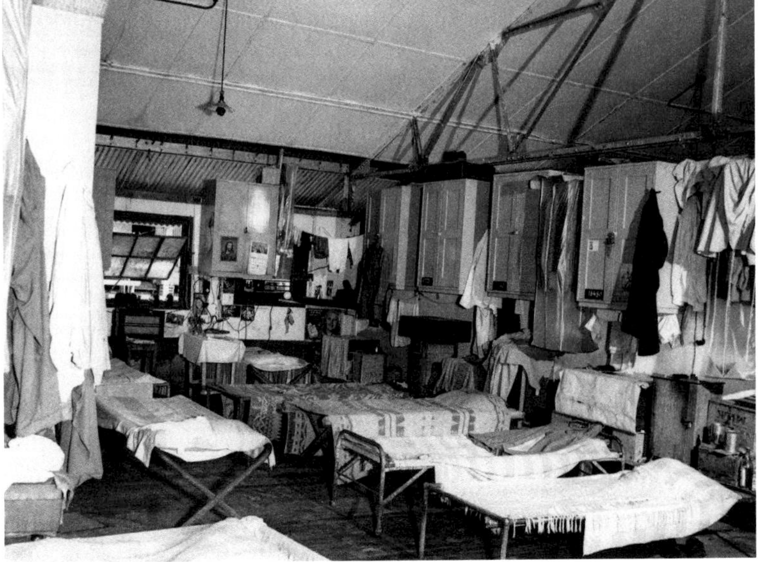

Before the policy of integrating with the local people was established, Shell's expat offices and houses were often fenced off in compounds, as in Venezuela (above right). The houses visible behind the caravan (main picture) are those of Shell's Mera base camp in Ecuador. After exploration there was abandoned in 1948, the site was quickly taken over by Ecuadorians – even though its fences remained.

very large psychological shift within the Group. Its personnel in the various countries of the world should no longer be treated as separate entities; instead, national barriers should be broken down as much as possible. The first implication was that all personnel should become part of the life of the country in which they worked. Outside Europe and North America, this could include the removal of physical barriers: company camps were often surrounded by high fences. Taking them down was a calculated risk, setting reduced security against the hope of a more positive public perception of the company, and did not always succeed. Nevertheless, for the same reason the earlier policy of having one contracted food supplier was abandoned and local ones were encouraged, expatriate staff were encouraged to take an interest in local matters; and the surprisingly liberal idea caught hold that company clubs should no longer be only for the staff but should be available, at least experimentally, to the population as a whole.

Blending in with the locals was further promoted by a gradual change in accommodation: in the early days company buildings abroad had often been built on British or Dutch patterns, but during the 1950s the Group started to employ local architects to produce designs in line with local architecture and in keeping with the country. Having mixed staff could be a mixed blessing: religious facilities had to be provided for people of different faiths, and in British Borneo, for example, foreign labour included Chinese, Indians, and Malaysians. The easiest solution to the question of providing club facilities for the different nationalities would have been to build a different clubhouse for each, but this was rejected specifically because it 'would not have brought about a mixing of the staff'.[76] The preferred answer there was to build a single clubhouse with several wings, so that the nations could mix or not, as they chose.

In breaking down national barriers the second implication was, as we have seen, that staff should become extremely mobile. As was said at the parent companies' AGMs in 1957, 'The practice of drawing personnel from every nation in which we operate and training each one individually in operations most appropriate to his ability will continue. The Group's training and career development need not be bounded by any one country.'[77] This was confirmed the following year: 'The world-wide nature of our business demands a staff which is truly international in outlook, composition and practical experience. Selected personnel of all nationalities are increasingly given the opportunity to widen their horizon and their

[38]

In the 1950s, as part of its new policy of fitting in with local surroundings, Shell began to build houses for its staff in styles that were more in keeping with local architecture, as here in Rio Canario, Curaçao.

experience of the industry by working in other countries.'[78] In these we see the Group's early declarations of intent. By the mid-1960s this wider meaning of internationalization had become one of Shell's outstanding features. In 1969, the general manager in the Philippines was a Venezuelan, in Japan an Australian, in Greece a Frenchman; in Gabon the Group employed an operations manager from Puerto Rico, in Tanzania a personnel manager from Trinidad, in Ethiopia a French finance manager. Shell employed a total of sixty nationalities in that year, forty of which were represented at central offices; two years later the number of nationalities had risen to a hundred, of which sixty in The Hague and London.[79] One acute and usually critical observer in Britain commented about the Group that 'They pushed through – *ahead* of politics – the quick recruitment and promotion of Asians, Africans or South Americans, giving them as much independence as they dared. They tried to avoid choosing local managers by Western or "old boy" standards, and to accept the values of local communities. For many of the old-style administrators the change was appalling. (...) But it was carried through, helped by Shell's hard international experience'[80]

The simple original version of internationalization and the simultaneous ambitious ideal of regionalization had bred a true internationalization. And it was an enduring characteristic. Thirty years later, in the mid-1990s, Sir Peter Holmes, the then recently retired chairman of Shell Transport and of the CMD, commented: 'We really are an international family. In that sense, certainly in the commercial world, we're unique. We're the only company that I've come across in any field which has – today I think the figure is over 65 nationalities working outside their country of origin. Not training, *working*. That's quite a remarkable attainment.'[81]

In this highly progressive period of the Group's history, the old imperialist attitudes that had shaped it from its birth were rapidly replaced, and there was no wish to go back.

Inventing the matrix organization

During the 1950s the Group also thoroughly overhauled its internal organization to eradicate the problems which had remained unsolved since Deterding blocked the management reform proposed by Colijn in the early 1920s (see Volume 1, Chapter 5). Like all big internationally operating companies, Shell faced two fundamental coordination issues, the balance between central offices and local operating companies, and the balance between aggregated business functions as managed from the centre and the infinite variety of local circumstances. In searching for a solution, managing directors adhered to two important guiding principles, first, the need to keep Group operating companies as close as possible to their markets, and second, 'maximum self-reliance', i.e. the delegation of responsibility and authority to the lowest level practicable, backed up by regular reports on progress and results.[82]

As for the balance between central offices and local operating companies, the Group had always attached great importance to respecting the autonomy of operating companies, for fiscal reasons and because managers believed that such autonomy stimulated entrepreneurship. For that reason, central offices always wrote to operating companies in polite terms of 'suggestions' rather than instructions (see Volume 1, Chapters 3, 5, and 7). Even so London and The Hague followed the affairs of the operating companies closely, ensuring coordination through Group managing directors on their boards. During the First World War the first regional or area departments appeared which administered the affairs of local operating companies. Over time, the managing directors came to delegate some of their authority to the heads of regional departments. A. de Jongh, for instance, started as a clerk in Asiatic's German sales department and ended his career in 1938 as area manager Germany with a seat on the board of Rhenania-Ossag, and thus able to handle most issues himself without having to bother Kessler, the managing director concerned.[83]

This evolution had two serious drawbacks. There was no check on the growth of area departments. As Group operations expanded, some marketing areas acquired manufacturing and sometimes E&P operations, and production areas also began marketing products. Moreover, the number of coordinating tasks at regional levels increased. In 1944, for instance, the Group appointed a Regional Adviser Labour Relations at Singapore, and six years later a first regional meeting of industrial relations coordinators was held there.[84] As a result the London area departments gradually assumed more comprehensive managerial responsibilities and spawned an array of functional sub-departments for their areas. London office staff rose from 3,000 on the eve of the Second World War to 5,500 by 1956, making the central office, which housed most of the area departments since they had originated in Asiatic, unmanageable.[85] The fragmentation of staff over more than thirty addresses in the City added further inefficiencies.[86] Despite the increase of the CMD from five to seven members in 1952, the managing directors struggled with a huge workload; in 1951, they had forty-six departmental managers reporting to them at London central office alone.[87] Moreover, individual product departments, ranging from wax and candles to liquid fuel and light oils, dominated the operational side of the London office organization, rendering functional coordination and the balancing between functional demands and regional requirements very difficult.[88] By contrast, Bataafsche had been organized largely along functional lines since the reforms introduced by Colijn and Pleyte during 1916-17, which had also introduced managerial assistants entrusted with keeping the balance between regional and functional interests (see Volume 1, Chapter 3). However, London had a Production Department which overlapped with The Hague, a remnant of the Second World War, and the office in The Hague still had a few area departments overlapping with London, largely but not exclusively the result of letting fiscal or legal reasons, and not managerial rationality, prevail in choosing the administrative domicile of operating companies. Functional coordination and organizational balance entirely depended on the managing directors, an unnecessary burden which prevented them from devoting time to more important matters such as general business strategy.

In October 1955, the Committee of managing directors took a momentous step in rationalizing the central offices organization by appointing eight functional coordinators, respectively for supplies, two for production and exploration (one in The Hague and one in London), marketing, marine, manufacturing, chemicals, and finance. Headed by a managing director, this coordination committee immediately set about drawing the business closer together along functional lines.[89] Group-wide meetings devoted to particular functions and gathering together representatives from all operating companies proliferated during the second half of the 1950s, alongside smaller area meetings for specific subjects in, say, marketing or manufacturing. In addition, the coordinators would have regular meetings with representatives from individual operating companies to discuss the plans and performance of particular business functions. Closer functional integration succeeded in eradicating some surprising anomalies in the organization; it was only in 1958 that Bataafsche started introducing a uniform costing system for drilling and production for all operating companies.[90] The coordinators were really executive vice presidents in all but name; their number rose gradually to reach twelve in 1972.[91]

Once the coordination committee had set to work, the number of departmental managers reporting to the CMD fell from forty-six to fifteen. Possibly at the same time the CMD also delegated the annual appraisal cycles of operating companies, i.e. the discussion of budgets and results, to the area coordinators. In 1966, these appraisal cycles were subsumed into the Unified Planning and Control Machinery or UPM, designed to streamline the existing procedures.[92]

These very important reforms had the dual effect of increasing the autonomy of operating companies while at the same time strengthening the integration of the whole business. For the historian, the far-reaching delegation of authority had an unfortunate side effect. Following the principle of maximum self-reliance,

[39]

[40]

the area and functional coordinators dealt with general business policy and strove to straighten out any problems between themselves. Indeed, the essence of the matrix organization lay in the careful balance between area and functional coordinators which reduced the volume of paper going to the CMD by relying on the ability of the organization to run itself.[93] Issues were taken to the CMD only if they concerned a principle of importance which the coordinators could not decide. The CMD archive has survived largely intact from 1957, but the Group had no policy concerning the preservation of coordinators' archives, of which very little remains for the 1950s and 1960s. Consequently, we can reconstruct the Group's policy for this period only in as far as matters went up to the CMD, which leaves a very large part out. Accident statistics, for instance, or environmental concerns, only became a matter for the CMD during the later 1960s, though coordinators undoubtedly confronted them long before. Moreover, the actual process of policy-making remains hidden from view, so we rarely know the motives, arguments, or data backing particular decisions.

Thus the Group's corporate organization differed completely from the multi-divisional American model propagated as ideal by the business historian Alfred Chandler.[94] It was a unique type of its own, supremely adapted to the fragmented markets of the pre-globalization era, and served Shell well for some forty years. Though in Group tradition McKinsey & Co. is credited with inventing the matrix organization, the consultants found it already functioning and their contribution lay in shaping it to perfection. For the model still offered considerable room for improvement.

[41]

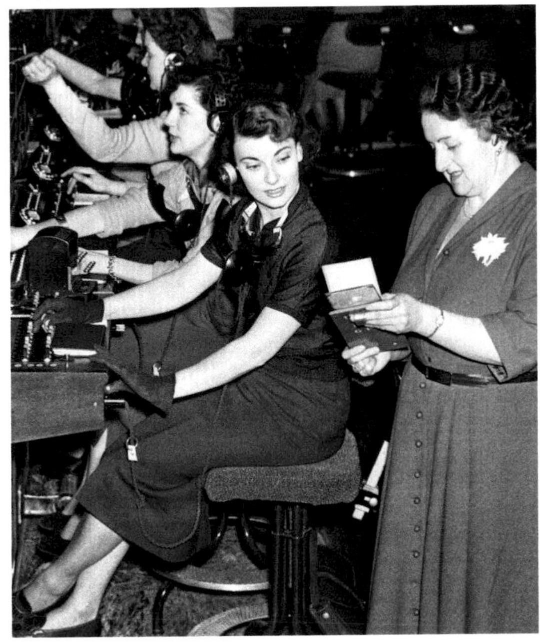

In London, the Group's many scattered office buildings did not help its efficiency. From left, the telephone switchboard at St Helen's Court, 1952; women at work in a typing pool in Shell-Mex House, 1955; and the newly installed telephone switchboard in the Bishopsgate office, 4 February 1953. A staff of 50 tended the network day and night, serving 5,000 people through 80 exchange lines and 1,300 extensions.

Perfecting the matrix: the McKinsey review, 1957-1960

The functional coordinators' committee had hardly been set up when the CMD began considering further steps to improving the efficiency of the central offices. In May 1956 Wilkinson sent a detailed report from New York about the organization of various American oil companies.[95] Probably at about the same time Loudon approached McKinsey & Co. The firm had been recommended to him by a Texaco director, whose company was undergoing a thorough restructuring conducted by the consultancy.[96]

However, all McKinsey's reviews hitherto had been carried out in the United States; so, before deciding whether or not to commission them for a full Group review and let them loose in the Group's Anglo-Dutch heartland, the CMD decided to commission a trial review. Compañia Shell de Venezuela (CSV) was chosen as the testing-ground, presumably because of Loudon's intimate knowledge of the business there, and the company's inefficiency had long rankled with him.[97] Before his appointment as a Group managing director he had never worked in either London or The Hague, so when he reached Group level he brought an unusually fresh eye to the deficiencies of its organization, which had grown in an unplanned, organic way. Some of this he had already seen in Venezuela. When he was assistant manager there, there had been two general managers, one in Caracas in charge of government relations and marketing reporting to London, and another in Maracaibo in charge of exploration and production reporting to The Hague. Loudon was the first person to combine both roles, though the reporting lines remained unchanged. At that time other country managers in the area also reported to different Group managing directors: Peru and Argentina reported to a Dutchman and Brazil and Mexico to a Briton, with little coordination between the two.

The McKinsey trial review in Venezuela got underway during the second half of 1956 and by January 1957 the firm had produced four reports.[98] Amongst other problems, CSV was having difficulty in estimating construction costs and controlling expenditure. Afterwards, noting that it was 'cold comfort that other major oil companies do not appear to be doing any better than we are in this respect', one CSV manager wrote at length about the experience. CSV personnel, he said, were originally enthusiastic about the McKinsey visitation, but it was much more painful and confusing than they had anticipated, and its immediate outcome was a strong feeling that the McKinsey team 'never perhaps properly understood the requirements of a business like ours'. The first reaction therefore was 'a deep depression in company morale', which had not been high anyway. However, CSV staff did grasp the basic principles that McKinsey brought to the fore, and, deciding they must develop a CSV philosophy of management by themselves, they conceded that 'without the "shock" treatment, organized approach and thorough indoctrination, little would have been accomplished'.[99] In central offices this outcome was deemed a success, and shortly afterwards another CSV manager wrote:

'We are well under way to ensuring that complete understanding, acceptance and active support is developed and maintained (...) the recommendations which they made were thoroughly discussed, thrashed out with us and *accepted* by us before finalization. They have, therefore, become in effect, a definite enunciation of CSV Management thinking quite as much as "McKinsey recommendations" and we are actively implementing them throughout the company as CSV's objectives (...) As to the "Authorities", as you know, our draft recommendations in this regard are circulating in the London office.'[100]

By early 1958 the extra cost of the new measures was seen to be justified by the resultant improved coordination and control, and positive reports of progress in Venezuela had already persuaded London central office to extend to McKinsey an invitation for a Group-wide review.[101] This was not unexpected: draft terms of reference had been drawn up in July 1957.[102]

Following those terms, the review's underlying purpose was to gain 'an independent viewpoint in making certain that the enterprise is organized (1) to deal most effectively with the great growth in the volume and complexity of the Group's activities that has taken place in recent years and (2) to provide for the further growth in volume and profits [planned] in both oil and chemicals'.[103]

In particular, the offices of Shell Petroleum and Bataafsche were to be reorganized to improve their coordination; the principle of decentralization was to be preserved and extended; and consideration was to be given to the desirability of establishing a chemicals organization separate from but closely linked with the oil side of the business.[104] Identifying ways of reducing the numbers of employees was one of the McKinsey objectives from the very start: 'The purpose of the McKinsey Study is to make recommendations for improving the efficiency of the human element in our business (...) and the benefits of the McKinsey survey will be apparent only in the indirect form of being able to handle the increasing business and produce the increasing revenue with less of an increase in our human element than would otherwise be necessary; the calculation of direct savings on this basis is very difficult. That the McKinsey survey will result in very considerable savings we feel is beyond question, otherwise we would not be putting ourselves to the very onerous task of converting our organization to fit the new concepts.'[105]

The consultants were reassuring: 'the problems you face are neither alarming nor unusual. However, since the Group is unique its management problems require specialized analysis and solution.'[106] But before the CMD's decision to hire McKinsey could be put into effect, in a Britain still half-crippled by the costs of the war, conducting such an exercise required permission from the Ministry of Labour and National Service: 'The purpose of their visit is to make an impartial and independent study of our Organization with particular emphasis on the role to be filled by London and The Hague. It is felt that the Consultants brought in on this important work should be neither British nor Dutch and as a language difficulty could arise, we have chosen this specialist group in the U.S.A. We visualise a stay in this country of about fifteen months, though during this period there would be visits by odd members. No salaries will be paid, but adequate allowances to cover expenses will be made. To us this is a very important and urgent matter and we submit our application with every hope of your sympathetic consideration.'[107]

Permission was granted and the Group-wide review proceeded. Only a very small team was involved: at its core, just 'two promising men, age between 30/35' from Shell, and four consultants, H. Parker, F. P. Benson, C. L. Walton, and I. S. Wishart, with one or two additional consultants during the field phase of the study and occasional brief visits from specialist consultants.[108] With Loudon, John Berkin, newly appointed to the CMD, acted as the team's sponsor within the Group.[109]

It was later rumoured that many of the ideas in the eventual report were Loudon's own, given timely corroboration by McKinsey.[110] Some initial emphasis on Dutch leadership is noticeable: the two Shell men were J. E. G. Smit (borrowed from Venezuela) and Dr. H. J. Kruisinga, who in 1956 had obtained a doctorate in economics with a thesis on issues of centralization and decentralization in large corporations.[111] Moreover, the agenda-setting meetings were conducted in McKinsey's own offices and The Hague rather than London. The CMD did indeed have a very good idea of what it wanted McKinsey to do before the consultancy had been hired. The terms of reference formulated in July 1957 identified four objectives. First, 'the two head office problem', i.e.

the question of whether the company's line management, now divided, should be concentrated in either London or The Hague and, if not, how to make The Hague's functional authority fit better with London's administrative role. Second, the structure of line management needed examining, notably the position of functional marketing departments within area departments exercising line management. The existing custom of managing directors having primary and alternative responsibilities for the line management of all operations in a given area also required adjustment, as did the relationship between the London Production Administration and The Hague's E&P department. Third, the existing organization left unclear to whom the operating companies reported, and to whom they were accountable, so the respective authority of the functional departments and of the line departments should be better defined. Fourth, the position of the petrochemical operations within the Group, which already been the subject of an internal review, needed to be scrutinized. As a final point, the document stated that 'In considering all the above problems, the possibilities of further decentralisation of responsibility from Head Office to the field should be kept in mind.'[112]

After only three months in the job, the consultants concluded that the dual head office structure simply had to be lived with, because, 'Although this meant inherent difficulties and additional costs, there was a tendency to exaggerate these and to lay all shortcomings at this door. Economies could nevertheless be achieved and friction minimised by "programmed management".' The initial findings also emphasized that 'the control and development of the Group had been by "personal statemanship" and that this should now be superseded by "programmed management" based on a philosophy well understood by the staff. Until now the tendency had been for the organization to be fitted to the individual rather than the contrary, and deference to individual feelings had been expensive both in organization and in cost.'[113] One further subject remained under wraps, McKinsey's conviction that the Group needed to replace its collegial management style enshrined in the CMD with an American-style board headed by a chairman, an executive committee chaired by a president and led by a chief operating officer. In the end Loudon appears to have told the consultancy to drop the idea.[114]

The Shell-McKinsey team proceeded by studying the Group's activities, organizational relationships and decision-making arrangements, working both in London and The Hague to gain an overall understanding. Dividing into three pairs, they then travelled the world to analyse the relationships between operating companies and the head offices. In the first six weeks of 1958 they visited Buenos Aires, Singapore, Paris, Rio de Janeiro, Seria in Brunei, Port Harcourt, Lagos, Melbourne, Sydney, Curaçao, New York, and Nairobi.[115] Quite independently from the consultancy exercise, the managers responsible for the Group's chemicals operations formulated a concept for putting chemicals into a separate organization in The Hague. Responding to this, the CMD decided it would need a detailed balance sheet of the projected advantages and economies versus disadvantages and extra expenses. These were the usual kind of caveat, of course, and more importantly, the CMD not averse in principle: 'The development in view seems to be well in line with the thinking of the McKinsey team.'[116]

In August 1958, the consultants presented a preliminary study of the Group's top management with recommendation for changes.[117] Four months later the CMD received a draft report. The main recommendations concerned a reinforcement of the matrix organization in four ways. First, the matrix model needed to be more rigorously applied to central offices and preferably enshrined

St Helen's Court was the Group's main London office from 1913 to 1962. Here, in 1949, in the entrance hall, the Group's Roll of Honour is partly visible on the right, recording all those employees who were killed in the two World Wars.

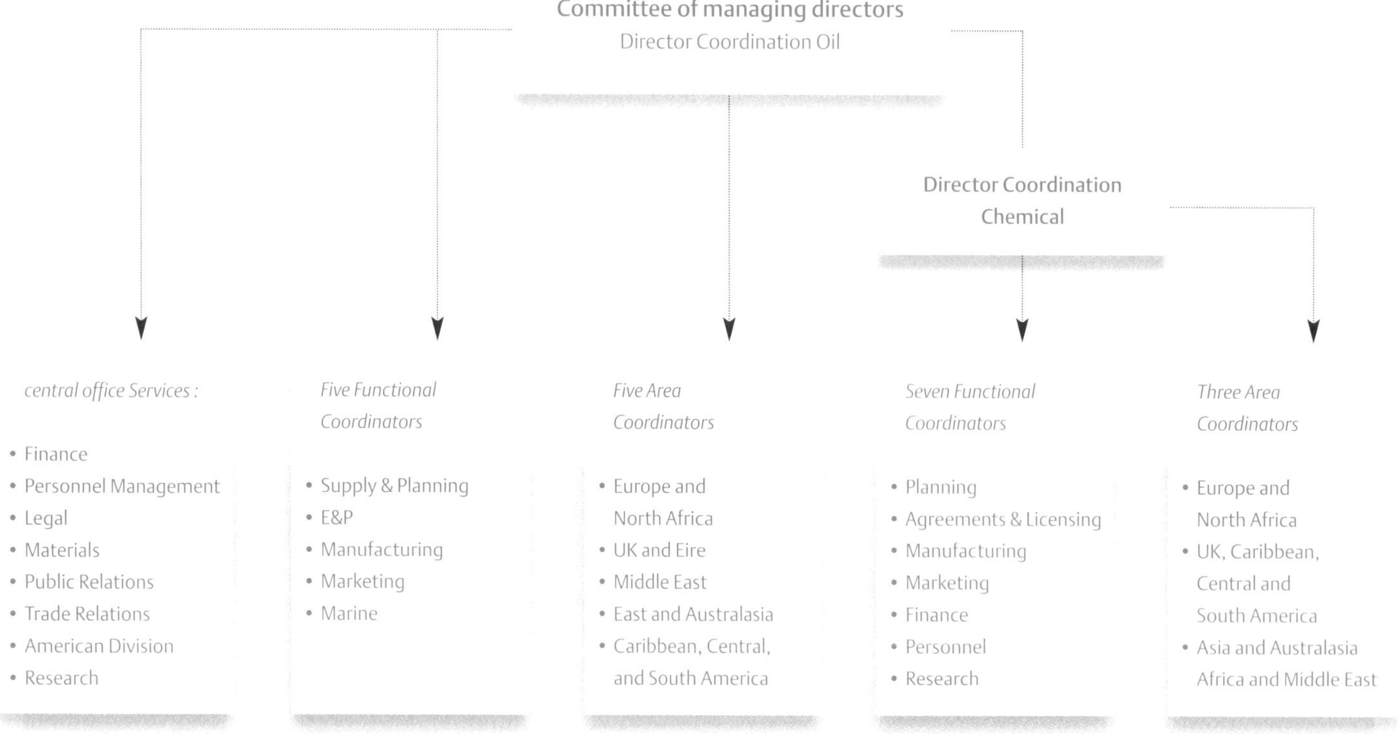

in a separate organization. Second, the position of the four area coordinators was to be strengthened by making the operating companies accountable to them and by defining their position as the vertical link between managing directors and general managers of operating companies, and the horizontal link between functional coordinators and general managers. Third, the checks and balances of the matrix model had to be applied to the CMD too, by making one of them Director of Coordination Oil or DCO, and by having all coordinators report in the first instance to him. His all-powerful position would have to be balanced by giving the other managing directors clearly defined alternative area responsibilities, so coordinators had an alternative director to talk to if they felt unhappy about decisions taken by the DCO. After some adjustments, an arrangement emerged under which the managing directors all had one area of primary concern, a functional concern, and an alternative area concern. Fourth, the chemical operations ought to be entirely separated from the oil side, but remain subordinated to the oil business. The Director of Coordination Chemical (DCC) would not be a managing director, though one member of the CMD would have chemicals as his functional concern.[118] Figure 2.1 shows the Group's top managerial organization after the McKinsey reorganization, with the number of area coordinators increased to five for oil and reduced to three for chemicals.[119]

The CMD read the report over the Christmas break. On 5 January, they accepted the recommendations and notified McKinsey that there was no need for a final report. Taking the McKinsey recommendations one step further, the CMD decided to give the reorganized central offices the form of service companies, with the committee itself becoming an organ of the service companies. This simple device ensured the accountability of the area and functional organizations enshrined in the service companies, and exemplified a fundamental shift in attitude concerning the position of central offices. As Fred Stephens put it, 'There is nothing really new about this, for what we now call the central offices has been slowly changing its role ever since our

Left:
Figure 2.1 shows the Group's top managerial organization after the McKinsey reorganization, with the number of area coordinators increased to five for oil and reduced to three for chemicals.

Right:
Figure 2.2
With the central offices enshrined in the service companies, Bataafse and Shell Petroleum were transformed into holding companies.

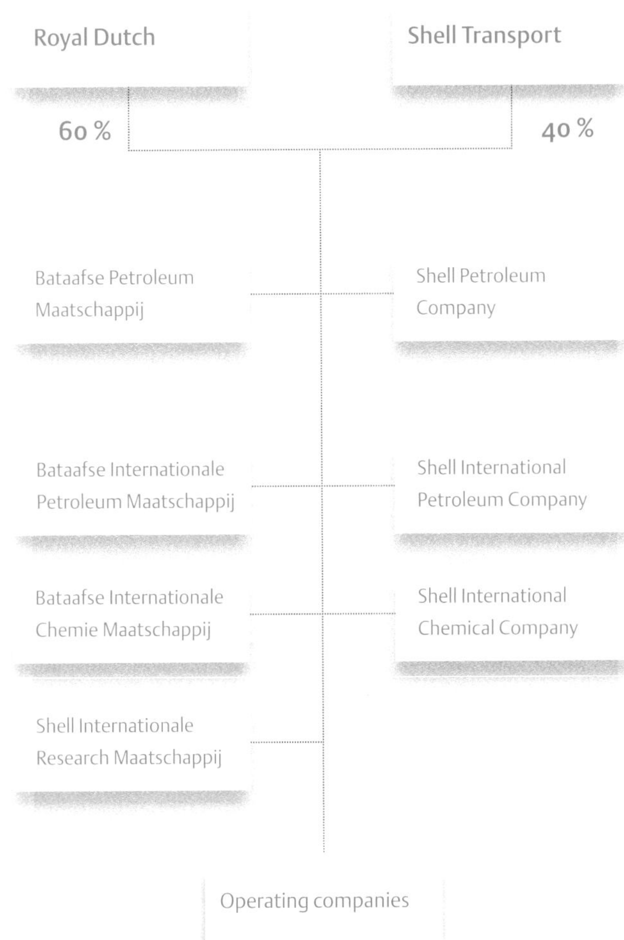

Group was formed [and] the recent reorganization was a deliberate step finally to abolish the old "Head Office" concept and to concentrate on co-ordinating and advising our Group companies rather than managing them.'[120] Initially there were four service companies, one for oil and one for chemicals in both London and in The Hague, but the CMD later set up a fifth one to run the research coordination. With the central offices enshrined in the service companies, Bataafse and Shell Petroleum were transformed into holding companies (Figure 2.2).

On 15 January 1959, the CMD had drafted the outline of the impending reorganization, which was announced to the staff on 16 January to take effect by 1 March, later postponed to 6 April. However, another eighteen months of staff meetings, adjustments, clarifications, and a progress report from McKinsey were needed before the new organizations worked smoothly. The total bill of McKinsey, including the time of the Shell employees involved, came to £267,000, or 2.7 million guilders. By February 1962 the reorganization was estimated to have yielded a reduction of staff at central offices of 18 per cent for an annual cost saving of about £4 million (40.6 million guilders).[121] In 2006 figures, the McKinsey exercise cost £5.6 million for savings of £80 million. The competition was impressed. In July 1961, Jersey Standard's board secretary J. O. Larson paid a visit to London central office to find out more about the Group's recent reorganization. Jersey wrestled with very much the same problems and had just started its own overhaul, but the board was apparently still casting around for directions. Larson particularly admired the fact that Shell managed its business, more widespread than Jersey's, with only seven managing directors against fifteen for his company.[122] After the McKinsey review the Group's management organization continued to evolve, minor changes including the admission of the DCO to the CMD in 1970,

[43]

On the south bank of the River Thames, Shell's new central office in London was built on the site of the former Festival of Britain Exhibition from 1951. Left, the 'skylon' rises above domed exhibition halls; right, seen from the north bank and towering above County Hall, Shell Centre was fully opened for business on 12 June 1963.

but the matrix structure essentially remained in place until the 1990s.

Although, as we have seen, McKinsey & Co. did not invent the matrix organization, the consultancy contributed materially to perfecting it by giving a firm push in the right direction. The review gave a greater sense of urgency to the reorganization, which thus achieved its targets more quickly. Moreover, McKinsey's emphasis on a more rigorous application of the matrix, combined with its propagation of the then-current fad of programmed management, helped to bring greater procedural discipline, and provided the impetus both for extending the model to the very top, and for clearing away old, illogical, and inefficient arrangements, such as Bataafse's Handelszaken marketing arm in Indonesia, run from The Hague but really belonging to London's remit. With hindsight, one of McKinsey's recommendations may be termed a historical error, and that is the separation between oil and chemicals. Chemicals never fulfilled its early promise as a business function in its own right (see Chapter 5 and Volume 3, Chapter 2); separating it from oil gave chemicals a constituency of its own, raising the barriers to exit from the Group. Then again, during the 1950s, managers became more and more convinced that oil and chemicals were fundamentally different businesses, so this particular mistake cannot be laid at McKinsey & Co.'s door. Nor can the consultancy be blamed for the continuing difficulty of controlling the service companies' costs, or for the fact that, over time, the distance between central offices and operating companies tended to grow. By 1970 at least one disillusioned general manager of an operating company interpreted the delegation of authority to mean, doing as you like as long as you produce profits and keep the Group's nose clean.[123] Management at arm's length may indeed have had hidden costs such as demotivation created by distance, or an apparent failure to reap the full synergies of functional integration (see Chapter 4). Combined with the company's policy of offering a job for life to those who deserved it, the consensus-oriented business culture may also have contributed to Shell people becoming, in Spaght's words, 'kind to a fault' rather than hard-nosed.[124] However, Group managers passionately believed in the benefits of the decentralization model; it would fall to a later generation to focus on the hidden costs.

While independent from McKinsey, the CMD devised another managerial innovation known as 'the Conference', intended as a further superior level of Group governance. Once again it was one of Loudon's initiatives.[125] In bringing together the CMD with the non-executive directors of the four main companies, the Conference aimed to streamline the Group's consultative process by eliminating the need for debate in the separate boards. Once the

[45]

[46]

With Shell Centre fully open, colleagues rediscovered the pleasure of lunching together. From left above, office workers in the restaurant of the Downstream building, 1966; senior staff in the 'Sergeants' Mess' on the 24th floor of the Upstream building in 1964; and the serving area in the Upstream building's Main Restaurant, 1964.

Conference had reviewed a particular matter, the boards of the companies concerned would meet to ratify the decisions taken, the host boards on the same day in adjoining rooms, the guest boards the following day in their own country for tax reasons.[126] It was thus really a logical corollary to the formation of the CMD in the aftermath of the war, and the date of its introduction, coming as soon as possible after Godber and Kessler had retired, could not have been accidental. Obviously time was needed to prepare for such a move, yet there is no record either of its preparation, or of the outgoing elderly gentlemen being consulted on the matter. It may be negative evidence but it is persuasive: the old guard were not wanted as part of the Conference.

The first Conference meeting, on Wednesday 20 September 1961, took place not in The Hague, whose influence had so dominated post-war matters, but in London at Shell-Mex House on the north side of the Thames. This was a gesture to the future, because Royal Dutch already had a fine spacious headquarters at Carel van Bylandtlaan in The Hague, whereas Shell-Mex House was the Group's only really good building in London, the rest of the British side of the business being scattered all over London. But on the south side of the Thames, just across the river from Shell-Mex House, the Group's new twenty-six-floor British head office was already taking shape. In 1963, when Shell Centre was fully opened for business, *Time* magazine described it as 'unspectacular-looking' – a fair enough comment by Manhattan standards, even though the building was the highest office in all Europe. The building offered a great range of facilities, including a licensed restaurant on the twenty-fourth floor, nicknamed the Sergeants' Mess, which quickly became the favourite venue where coordinators hammered out their deals over lunch.[127] More importantly, with its opening the Group's theory of a single head office separated by the North Sea became as real as it could possibly be. Even the CMD doubted whether its own theory was true, however. In 1967 the committee seriously reconsidered the dual office structure because, as Royal Dutch president L. E. J. Brouwer put it succinctly, 'we are too conspicuously inefficient in this respect and, what is worse, we are not seen to be doing anything about it', only to find that there were no viable alternatives.[128]

Shell Oil and the matrix model Though no two operating companies were alike, Shell Oil is a good case to demonstrate the effects of the matrix model on the Group's organization. Often considered as a special case, more autonomous and less integrated, Shell Oil's relation to central offices in fact shows an evolution quite similar to that of other operating companies, until litigation brought about an entirely unexpected reverse.[129]

Following the consolidation of the US companies into Shell Oil in 1949, the company also underwent a profound reorganization of its top management structure, involving a clear shift of executive powers to New York. Until 1945, Group managers residing in, or seconded from Europe dominated the board, with six members out of a twelve-man team, including the chairman and the president. Four years later, the Shell Oil board counted seven Dutch or British directors out of a total of twenty members. The chairman, George Legh-Jones, belonged to the Group managing directors in London, but the president, Max Burns, and the chairman of the new Executive Committee, Alexander Fraser, were both originally Scotsmen and now naturalized Americans based in New York. Moreover, whereas before the war the various Shell companies had always had Dutch or British managers in key positions, there were now very few left amongst the vice-presidents of Shell Oil or in the top management of subsidiaries such as Shell Chemical Corporation. Only Shell Development still had a Dutch president, Jan Oostermeyer, but then the company was a 50:50 joint venture with Bataafsche until 1955, when Shell Oil bought out its share.

The effect of this ostensible Americanization should not be exaggerated.[130] True, the number of European directors on the board dropped, but the advent of transatlantic air travel, and more specifically the introduction of jet services in 1958, made control from a distance much easier than before. The presence of Deterding and Godber on the Shell Union board had been fairly nominal. Most issues were discussed and decided by correspondence, with board meetings only formalizing what had already been settled with London. Consequently, the overseas directors needed to attend only once in a while and the large contingent of directors representing the Group's point of view served to secure a majority under most circumstances. From the 1940s trips from London or Amsterdam took a day rather than at least a week, so this need disappeared; the post-war generation of European Shell Oil directors could, and did, attend meetings much more frequently than before.

Moreover, with this managerial emancipation Shell Oil followed behind developments in key Group operating companies such as Deutsche Shell or Shell Française, whose forerunners had always been managed by locals. From the beginning, the Group had kept a close rein on its American operations, mindful of the perceived different style of business practices there (see Volume 1, Chapter 4). In addition, the need to build an organization which would fit in the Group mould necessitated sending over senior staff to effect that integration, and middle managers such as Godber to learn the ropes of the dynamic US oil industry and transfer their experience back to Group operating companies elsewhere. By 1940 this need had virtually disappeared. The US companies themselves had trained a generation of American managers in the Group way of doing things, sometimes reinforced by extensive tours of duty abroad. Burns, for instance, had started his career with Shell in California, then served Group companies in Colombia, Venezuela, and London, before returning to the US in 1946 to head Shell Oil. Consequently, the Group now possessed a pool of American managers sufficiently dyed in the red-and-yellow wool to hold senior positions. London continued to hold a firm grip on all senior appointments.[131]

Even so Shell Oil's scale and scope of operations put it in a class of its own. Throughout the period under consideration, Shell Oil was the Group's single most important operating company by far. The company typically accounted for 20-25 per cent of assets, generated 25-30 per cent of net income, and absorbed around 40 per cent of all investment. Shell Oil also produced some 20 per cent of the Group's total crude and about 30 per cent of its oil products, while manufacturing 40-50 per cent of the chemicals sold by the Group. Finally, Shell Oil was an integrated oil company, and a largely self-sufficient one, neither reliant on supplies from elsewhere, nor exporting much in the way of products. Most other companies were either upstream or downstream, and thus tied to the demands of longer supply chains beyond their control. Shell Oil also had the largest minority shareholding, around 35 per cent of its shares being held by private investors in the US. There were also several pressing reasons for emphasizing Shell Oil's autonomy. First of all, legal reasons: throughout the 1950s and 1960s the threat of anti-trust proceedings against the oil companies remained very much alive, with the US Department of Justice investigating the US majors for restraints on trade both domestically and internationally.[132] One such round of investigations forced Loudon to avoid visiting the US for fear of being subpoenaed, so he met Shell Oil executives in Toronto while the case lasted.[133] The ever-present risk of similar legal action put a premium on keeping Group and Shell Oil operations distinctly separate, so as not to implicate each other in any proceedings.

On another level, the laws protecting shareholder rights also required clear demarcations, to avoid litigation from the minority for neglect and violation of duties. Then there were prohibitions on the export of technology to Communist countries, which needed to be taken into account in the research exchanges between Shell Development and Bataafse. Finally, the US fiscal regime demanded very close attention to the transparency of the relationship between the Group and Shell Oil, from the seconding of staff to sharing the costs and benefits of research between Group companies. These legal and fiscal concerns did not suddenly arise after the Second World War. They had been part and parcel of Group relations with the US almost from the start, and had provided the rationale, in 1920, for setting up a separate service company in New York, Asiatic Corporation, to act as an independent contractor for Group oil and materials supplies, rather than letting the local operating company perform such functions. They had also inspired the Simplex Agreement of 1929, which had laid down a basic framework for the exchange of research results and costs between Bataafsche and Shell Development, and was later continuously adapted to new circumstances by successive Research Agreements.[134] In addition, there was a regular exchange of information through staff secondments. During the mid-1950s, some fifty people a year came over from Shell Oil to Europe, for training or for specific assignments, and a similar number travelled in the opposite direction.[135]

However, Shell Oil's managerial leeway and funding arrangements were different in degree, and not in kind. The matrix structure gave all companies considerable room for manoeuvre in determining policy. Companies had to finance their own operations as far as possible with retained earnings, topped up with bonds or bank loans; raising equity from the Group remained very much an exception. Budgets were, of course, submitted to central offices for approval, but local managers could determine their own business strategy, provided that agreed targets were met. Managers also

In 1953 the 'Central College of Shellmanship' issued 'STATO, or The Shellman's Snakes and Ladders'. The positive and negative points on the board still ring true in many businesses today.

[48]

OR THE SHELLMAN'S SNAKES AND LADDERS

RULES:

1. Any number of players can take part. To obtain the best results the game should be played with loaded dice.
2. Any player landing on a square already occupied by another aspiring Shellman is allowed to retain his Status.
3. Any player landing at the bottom of a ladder immediately moves to the top rung.
4. Any player landing on a snake's head immediately loses Status to the position of the tail.

100 You qualify as a 100% Shellman	99 You fail to recognise a Managing Director	98	97	96 You admit to never having gone abroad on a business trip	95	94	93	92	91
81	82	83	84	85	86	87	88 You are forced to admit you have no secretary	89	90
80	79	78	77	76	75 Your best contact is fired under a mysterious cloud	74 You are invited to a Directors' cocktail party	73	72	71
61	62 You are asked to co-ordinate something	63	64	65	66	67	68 You cannot avoid taking the minutes of a meeting	69	70
60	59 You are seen arriving at 0900 two days running	58	57	56	55	54	53 You are invited to lecture to a course	52	51
41	42	43	44 You are heard making a phone-call to New York	45	46	47	48	49	50 Your bulging briefcase disgorges kippers in the lift
40 You fail to take the opportunity to have coffee with your chief. Miss one turn	39	38	37	36 You get your tea on a tray	35	34	33		31
21	22	23	24		26 You let it be known casually that you've been talking to the Ruritanian Ambassador	27	28 You get a conference table	29	30
20 You get a carpet	19	18	17	16	15	14 You admit to never having been on a course	13	12	11
1	2	3	4	5 You pass off someone else's day's work as something you ran up in the train	6	7	8	9	10 You get an office to yourself

(Issued by the Central College of Shellmanship)

had considerable discretion in allocating any surplus funds; during the late 1960s, for instance, Shell Oil diversified into real estate, including shopping centres. Companies discussed their budgets, results, and problems with the functional and regional organizations in the service companies, but did not formally report to them. Shell Oil went through similar cyclical appraisal discussions in London and The Hague as did all operating companies, though until 1960 the functional reviews were restricted to E&P, chemicals, and research; it was different only in having a London department to itself, in having two Group managing directors on its board, and, during the 1960s at least, in its president being allowed to attend the CMD on a regular basis. Thus Shell Oil was fully inside the service company structure.[136]

Like all Group operating companies, Shell Oil also carefully cultivated a corporate image to suit its local environment and emphasized its identity as a fully American company, owned by Americans, managed by Americans for Americans, producing and selling American products. From the early days, Group marketing strategies had bolstered this chameleonesque attitude by positioning Shell as a global brand with distinct regional shades of colour (see Volume 1, Chapter 6). From the 1960s, Shell Oil sent out a similar message of local roots with global ties by sponsoring a television series called *The Wonderful World of Golf*, in which top golf players competed head-to-head on courses in the US and around the world. The very popular series portrayed the company as American but part of a big international family. It ran from 1961 to 1970 and was resurrected in 1994.[137] In one crucial aspect of marketing, however, Shell Oil remained a case unto itself. No other operating company would have been allowed to use its own, slightly different variety of the pecten brand image. Until 1998, Shell Oil did, and the company also had different shades of red and yellow.[138]

[49]

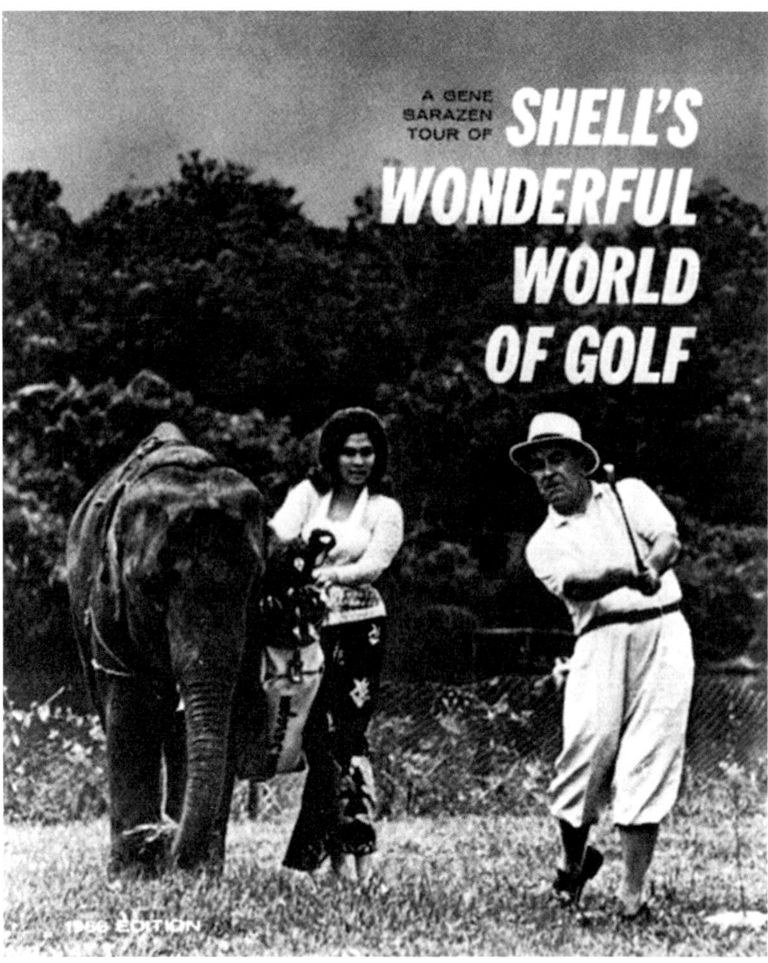

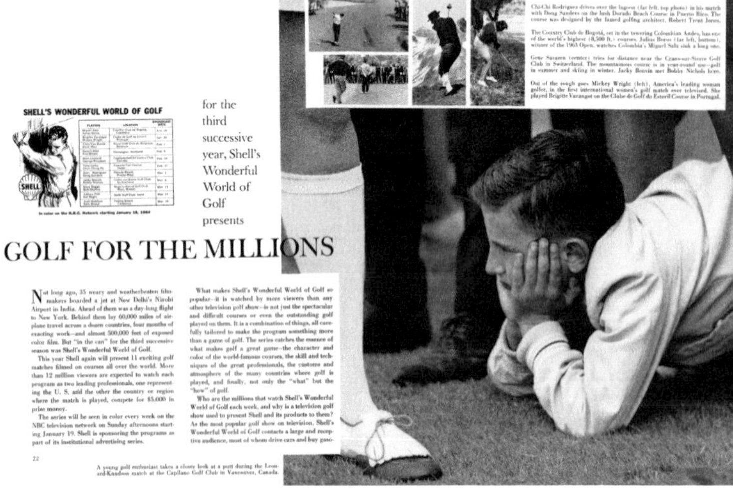

[50]

In cities like London and New York, television had just begun to spread before the Second World War interrupted its development. Shell quickly perceived its value as an advertising medium and in 1949 sponsored the first full demonstration of television in Australia (above). Left, above and below, shows the fascination of its 1960s sponsored series *Shell's Wonderful World of Golf*, revived in 1994.

Close personal ties also cemented Shell Oil and the Group more firmly together. In 1961, the Group's New York lawyer C. W. Painter was appointed a non-executive director of Royal Dutch. He remained on the board for only two years and was succeeded by Eugene Black, former president of the World Bank. As Shell Oil president from 1960, Monty Spaght regularly attended CMD meetings and in 1965 he became managing director of Royal Dutch, the first foreigner in that position.[139] This was, in a way, a reverse

[52] Monty Spaght

[53] David Barran

[54] Gerrit Wagner

[55] Dr. S. A. Ballard

integration, balancing Spaght's succession in the same year to Loudon as Shell Oil chairman. He thus became the first American in that position, which he himself regarded as proof that Shell Oil had finally become fully emancipated.[140] He remained an exception as such, however. The Shell Transport and CMD chairman David Barran succeeded him as Shell Oil chairman in 1970, to be followed one year later by the incoming Royal Dutch president, Gerrit Wagner. Two more Shell Oil men took up senior positions at Group level. In 1963, Harold Gershinowitz became the first chairman of the Group Research Council and six years later Dr. S. A. Ballard was appointed to head the Amsterdam laboratory. With so many Americans coming to Europe whilst Shell Oil built closer ties to the Group, one might say that the Group, rather than Shell Oil, became Americanized.

By the mid-1960s Shell Oil was fully integrated with the Group at all levels and in all business functions. Such was the level of trust between the CMD and the Shell Oil management that an important deal such as the El Paso takeover in 1964, which gave the American company a badly wanted expansion into the southwest, could be cleared in two transatlantic telephone calls between Spaght and Loudon.[141] The CMD considered the links between central offices and Shell Oil 'quite good, probably adequate for management's needs'.[142] Indeed, the ties had grown sufficiently close for the CMD to consider winding up Asiatic Corporation and transferring its functions to Shell Oil, but when an efficiency operation had eliminated the need for this step, the service company was allowed to continue.[143] Events then took an unexpected turn. In June 1969 a group of disgruntled Shell Oil shareholders started litigation in New York, a case that became known as Halpern vs Barran. The plaintiffs accused Barran and fourteen other Group and Shell Oil directors of neglecting their duties towards the minority shareholders and subordinating the company's interests to those of its dominant shareholder, the Group, by having Shell Oil buy oil and other goods and services from Group companies at prices substantially above market prices. In short, the Group had prevented Shell Oil from making more profits than it had, to the detriment of the minority shareholders.[144]

Based as it largely was on circumstantial evidence, Halpern vs Barran was not a very strong case, at least initially.[145] The case dragged on for twelve long years, however; the two sides only reached an out of court settlement in 1981.[146] The experience of complex litigation with discovery proceedings involving requests

Harold Gershinowitz

for comprehensive information and documents made central offices acutely aware of the need to ensure that commercial relations between Shell Oil and other Group companies were always fully competitive and transparent. In July 1970 Dennis Kemball-Cook, Shell Oil's Chief Operating Officer, reported to the CMD about an E&P joint venture with Shell Canada, commenting that 'the local directors of the companies concerned had noted with appreciation the fact that the holding companies and their officers had scrupulously refrained from taking part or even advising upon the negotiation of the deal in view of their different interests in the two companies. The result had been a genuine arms-length, and at times hard-fought, negotiation'.[147] This example showed a hidden cost of the extreme decentralization practised by Shell Oil. Simulating market conditions between Group companies for legal or fiscal reasons only served to create transaction costs between firms without reaping the rewards. After all, although negotiations between Group companies might be hard fought, conducted at arm's length, and creating competitive conditions, but such exercises were designed to keep operating companies on their toes and not to open business to outside contenders.

In September 1972, Royal Dutch president and Shell Oil chairman Gerrit Wagner spelled out the risks involved and reiterated guidelines for proper conduct. Unless Shell Oil could convince American courts of its managerial independence and the fairness of its transactions with other Group companies, the litigation with minority shareholders would never end. Indeed, those other companies might even be implicated in lawsuits and exposed to investigations by the American tax authorities as well. By extension the same held true for dealings with Asiatic Corporation. The heightened awareness of legal and fiscal risks in the US had clearly forced managers to change their minds about winding down this company and transferring its functions to Shell Oil.[148] For his part, Wagner and other Shell Oil directors with close connections to other Group companies somewhat artificially distanced themselves from board discussions if they felt that their impartiality might be in doubt.[149] It is a moot point, however, whether the dangers spelled out by Wagner were as real as he made them out to be. European Group managers tended to panic at the prospect of litigation in the US, whereas Shell Oil managers treated it casually as a fact of life.[150]

Then again, if the Group's model of corporate organization at times strained the credulity of minority shareholders, judges, or fiscal officials, the resulting problems originated from the fact that the model differed so much from the then common conception of corporations as seamless, multidivisional, integrated, and hierarchical businesses. Given the Group's passionate belief in the benefits of the decentralization model, the CMD probably did not even consider taking Shell Oil private there and then. Reversing the integration achieved during the 1960s was the only option available, which over time led to relations between the Group and Shell Oil being ring-fenced.

Staff matters As one of the smaller consequences of the central offices reorganization, the Group adopted a different term to refer to staff following the proposal from an internal report that Shell should start 'to use the word "personnel" instead of "staff", being more in line with the McKinsey view'.[151] A simple but valid psychology said this change would make employees sound and feel more like human beings and less like units, but in the event no linguistic diktat on the subject was issued, and instead both words tended to be used interchangeably. Nevertheless the proposal was part of the general trend of the period, away from patterns of the past. The Group took the care of its personnel very seriously. A statement of business aims drafted by the CMD in 1962 put the creation of an environment to keep the staff motivated and enthusiastic at the top of the list, before the generation of shareholder returns.[152]

Quite apart from the central offices reorganization in 1959-60, the number of people employed by the Group, and staff costs as a percentage of sales, continued to fall throughout the 1950s and 1960s (Figure 2.3). By contrast, taxes paid rose from a third of sales in 1961 to nearly 40 per cent in 1972. Goods and services supplied remained steady at around 40 per cent, while dividends absorbed only about 3 per cent throughout the period. For the people who had to go, redundancy terms included enhanced pension rights if aged over 50, or lump sum payments if younger.[153] Standard retirement terms were liberal by comparison with any large employer, and usually included both a lump sum payment and a pension. Some UK and Netherlands companies gave bigger lump sums but smaller pensions, and American companies gave less on both counts. In the UK, for tax reasons each lump sum case had to be treated individually on an ex-gratia basis, without terms being openly declared.[154] Sometimes the CMD created special provisions for individuals.[155] A 'retirement bonus' would be up to one year's salary. G. C. K. Dunsterville, a CSV manager who had handled much of the McKinsey trial in Venezuela, received $30,000 on retiring, and in response wrote a very warm personal letter to Loudon.[156]

There was of course another side to the coin: loyalty and discretion were highly prized, and following a spate of

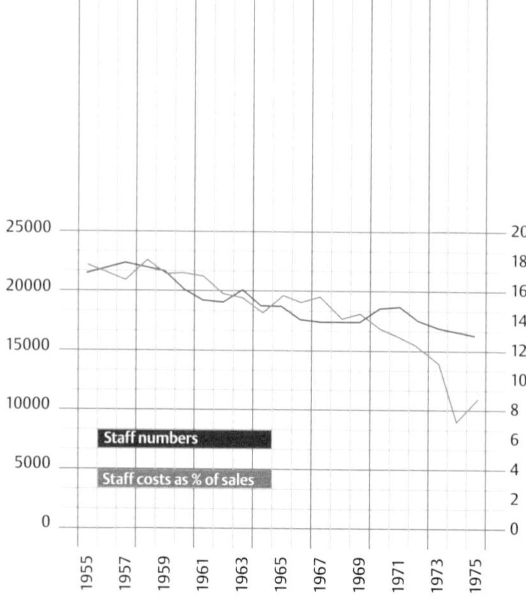

Figure 2.3
The Group's staff numbers (left scale) and staff costs as a percentage of sales (right scale), 1955-1975.

embezzlement in Belgian Shell, an implicated board member was required to stand down. Nevertheless, he was told that after 'our ultimate date of separation' and under certain conditions unspecified in the record, he might be offered a position as an adviser. Wishing to avoid the negative publicity of a prosecution, the Group's main desire in this instance was 'not to cause a rumpus and to disassociate the separation as much as possible from the recent defalcations', i.e. just the 'kind to a fault' attitude which Spaght diagnosed.[157] But if fiddling the books was only a venial sin, disloyalty was practically a mortal one, as evinced when a Danish colleague – recently retired after forty years' service, but still a non-executive director of Dansk Shell – decided that he was bored and that his pension was inadequate, and took up a position with the Danish subsidiary of Getty Oil. As soon as this became known in the Group, he was asked to resign from the board of Dansk Shell. A senior Shell colleague went so far as to describe his move as a 'defection', and, using the man's nickname, Loudon personally wrote to him expressing his shock: 'Dear Jørgy (...) I am sure you will agree with me that after such long service one never quite loses one's Shell identity, even on retirement. While you are of course a free agent and I would be the last to try to tell you what you should or should not do, I would have thought that when you were thinking of accepting a responsible position with a Shell competitor, and in your own country at that, you might at least have contacted us before actually committing yourself (...) We certainly would not have stood in your way, but we do feel that it would have created a better atmosphere all round if you had told us about it beforehand.'[158]

Despite their dismay, Shell executives agreed that it would be advisable to maintain friendly business contacts with their former colleague, since he knew so much about their Danish operations; and in a further development, a review of the salaries and retirement benefits of top-level operating company executives in Europe was proposed 'as protection from recurrence of this type of event'.[159]

Salaries and benefits while in work were naturally a matter of perennial interest for paymasters and recipients alike, and the Group's basic policy was to try and offer conditions that were at least as good as those of its competitors, and preferably a bit better. This sometimes required some creative thinking, as in late 1957 when the British Chancellor of the Exchequer announced an anti-inflationary wage freeze. Whilst understanding his reasons, Shell's privately proposed interpretation was that 'it was felt that freezing should apply only to salary scales and not to merit increases or normal annual increments'.[160] Such 'normal annual increments' posed their own problem, however. A so-called 'funnel system' for control of salary progress was introduced in Venezuela in 1950 and in London in 1952, and at first proved valuable in bringing an orderly pattern out of an uncoordinated picture. However, McKinsey had criticized the 'wage for age' principle including the funnel as holding back the better performers while over-paying the pedestrian, with age being too great a factor in salary progress.

The firm advocated introducing a merit-based system based on clearly defined job classifications.[161] Shell Oil had such a system, which reduced the emphasis on age and instead decided individual salaries through other criteria – relative job weight, performance, potential, and likely career opportunities – with staggered merit increases introduced throughout the year, rather than annual increases. This system was commended by London Personnel to Group managing directors, in order to 'reduce the "Christmas box" mentality and get men used to the idea that they work for a *salary* not a raise; that the reward for a good year's work is a good salary, not the size of the increase'.[162] However, personnel managers in The Hague had counter-arguments, the main ones being that everybody understood the year-end system, and that spacing increases unpredictably throughout the year would keep employees on tenterhooks. They also protested that The Hague had many adjustments to make before its higher-paid people were even on an equal scale with London. But further evidence came in from Shell Refining, Shell Chemicals, and Shell Oil, all of which showed that compared to competitors, the year-end handout was contributing to a levelling of pay between the Group's best personnel and the ordinary. The arguments from The Hague were dismissed and a plan similar to Shell Oil's was proposed, to start in 1960.[163] Adoption of another Shell Oil remuneration scheme, stock options for senior executives, was postponed until 1966.[164]

It was probably as a consequence of the debate about job weight and salaries that the Group introduced its job classification system with numbered categories from fifteen to one, then lettered categories from A to D, and finally the unclassified group for the top executives. We have been unable to establish the precise date of introduction, but the system was in place by March 1960.[165] Having clear job classifications also enabled Group personnel managers to start planning recruitment for particular functions, allied to systematic assessments of individual performance and future potential. In 1962, Shell started long-range surveys to establish future needs in particular job categories. Four years later, the Group sought support from the social sciences to solve some of its problems in personnel management. An American behavioural scientist did a survey on senior supervisory personnel, and a Dutch university was commissioned to research methods of forecasting managerial performance of individual employees. After a pilot study of two years, the CMD tested the initial findings on a group of high executives. The directors were sufficiently impressed to commission a sequel and to incorporate the methodology into a new staff appraisal system from September 1968. One year later, the CMD had instituted a monthly session to discuss a list of potential high-fliers identified with the categories defined by the university research. The university survey results were published in 1970, after which the method became a standard part of Group management assessment exercises (see Volume 3, Chapter 4).[166]

Despite clear classifications the difficulties of establishing fair rates of pay and benefits remained, and these came in two forms, each affecting the other. On the one hand was the need to safeguard the Group's competitive position and reputation as a leading employer in any given country; on the other hand, there was also the need to ensure that Group employees doing similar jobs or at similar grades in different countries received levels that ensured them comparable standards relative to the cost of living in the country they were working in. Thus, for example, in 1959 ICI introduced a fifth week of holiday for staff above a certain age and a certain salary, forcing a review of possible repercussions of this on the Group's policy regarding holidays, and in 1960 a general salary increase in the UK became essential.[167] There had been none since 1957, and in the interim Unilever, Fisons, IPC and the major clearing banks had all made adjustments of around 5 per cent, and Courtaulds 10 per cent. Esso and ICI were known to be likely to fix on 5-7.5 per cent, and BP had made increases totalling 15 per cent in the previous two years. It was pointed out to the CMD that

'The fact that others have made adjustments since 1957 would not necessarily substantiate a recommendation that that we should do likewise since our whole salary structure might have been superior. However it appears that, on the contrary, our competitive position has slipped [and] there is a strong case for the recommendation that an increase of 7½% be applied generally.'[168]

This was accepted, as was a simultaneous recommendation that staff in the Netherlands should be given a general increase of 6 per cent on top of their normal merit increase, 'to keep them in line with the movement of wage and salary levels throughout the country'. In the mid-1960s a CBI survey confirmed the Group's belief that 'Our graduate salary scales are generally comparable with those of other industrial firms and the academic world', but at the same time recent British university expansion meant that 'young scientists are often able to command salaries in academic posts which industry does not feel justified in paying'.[169] It was hard enough merely to maintain comparability between the UK and the more highly-taxed Netherlands. In those home nations the question of the expatriate status and benefits of a Dutch employee living and working in Britain, compared to a British employee living and working in the Netherlands, was often revisited, with housing loans being a particular bugbear, along with differences in taxation. Beyond the home nations, the problem was made much greater by the Group's increasing internationalization, and disturbance allowances and bonuses on reassignment from one area to another often featured as topics of debate. Discretionary allowances sometimes facilitated attractive salaries, but there were serious difficulties in recruiting American and Canadian nationals for Group service outside North America, particularly because of pay, until Shell Oil showed itself willing to cooperate in finding American nationals firstly from outside the Group and secondly from its own personnel. At the same time the parent companies decided that steps should be taken to increase the flow of non-American Group employees to taking up positions with Shell Oil.[170]

From 1951 until 1965, Shell Lodge provided a comfortable country house base on the banks of the River Thames near London for training the Group's senior executives.

[57]

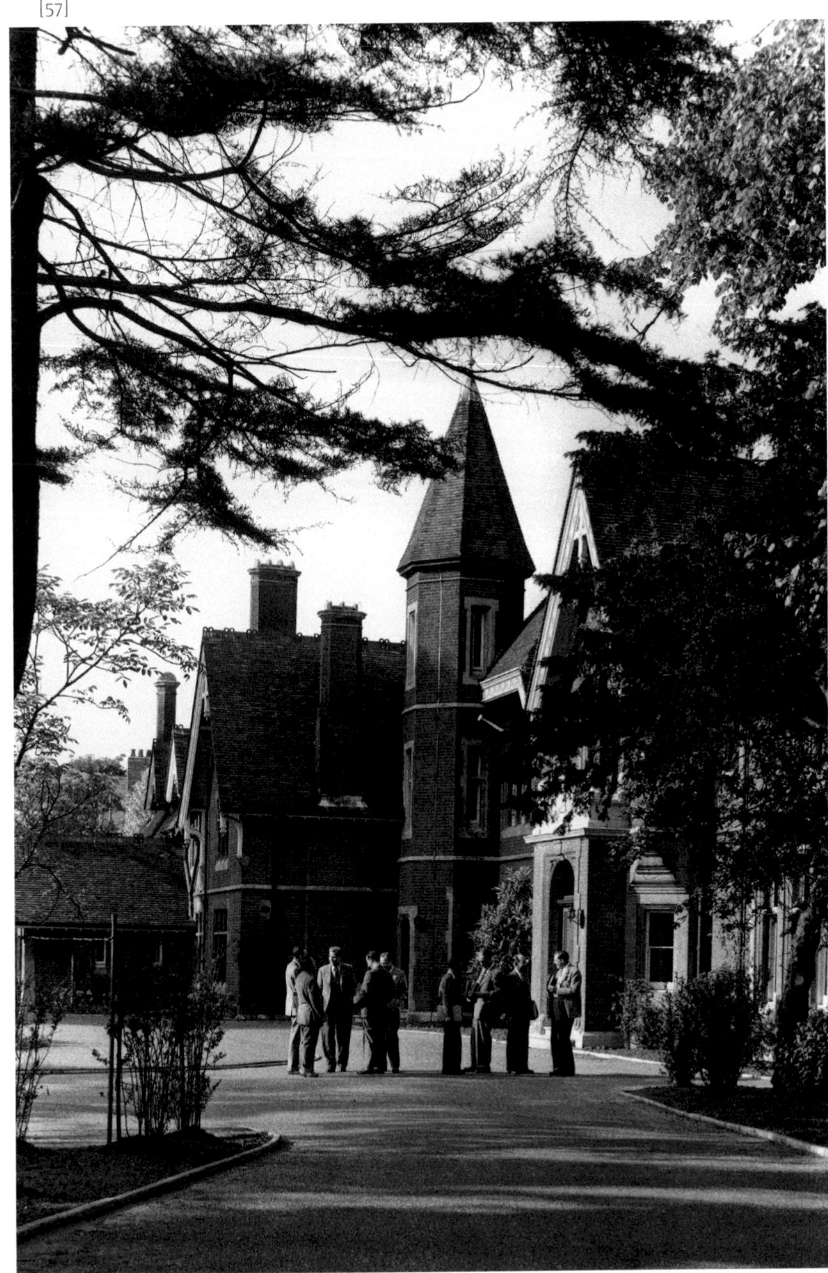

The evolution of training

Organizing programmes training employees for middle and senior executive positions became a prime topic for managerial attention during the war, partly in connection with the emerging staff regionalization policy.[171] The Group's rapid post-war expansion turned training into a key concern, occupying much of management thought, and staff time. The earlier informal arrangements for selection and training on the job needed to be formalized and extended. In October 1947, Bataafsche decided to set up three training centres at Schoonebeek, Amsterdam laboratory, and at Pernis, with accommodation for a total of 200 trainees, replacing the earlier informal training schemes at those locations. To support the new training programmes, the company's staff department organized a special sub-department which in 1952 opened splendid country house facilities for courses.[172] The Group also needed to rethink its training policy and ensure that staff from outside Shell's traditional employment pools in Britain and the Netherlands would get an equal chance to get into programmes and succeed. At the same time, the CMD raised its commitment to higher education in the Group's two home countries, endowing various universities there with chairs and scholarships.[173]

Britain was chosen as the centre for higher management training. Close by the Lensbury Club, at Teddington near London, the two houses known as Shell Lodge occupied a nine and a half acre riverside site. From 1951 this enviable location served as the centre for the Group's executive courses. Numbers attending were limited to 100 a year, with long-term residential courses taught by Shell people for Shell people in a country house atmosphere. The country house element, so evocative in the austere 1950s of more expansive pre-war days, was a deliberate factor in the choice of venue, and with its aura of privilege and great comfort Shell Lodge

[58]

soon 'acquired a reputation and personality of its own in the eyes of large numbers of our staff all over the world.'[174] Initially the Executive Course was intended for marketing staff only. Members at first received four weeks of instruction, after which they went to France and to the Netherlands to undergo practical training. In 1957, the course was extended to cover staff from all Group functions, with an unchanged annual intake of 100. This resulted in greatly increased competition for entry and an increase in the calibre of successful nominees. About the same time a decision was also taken to institute 'junior' courses, mainly for regional staff aged between 30 and 37 who it was hoped would eventually reach senior executive positions. London was the venue for the Marketing Staff Development Course, and The Hague for the Production Staff Development Course, with both being strongly supported by management in the respective areas. Again the annual intake for each was about 100, and although much larger numbers could theoretically be catered for, the managing directors considered that the junior courses should be available only to a maximum of 150 of the best.[175] From 1956, Shell Oil ran a management course of its own, which was open to Group staff from other operating companies in the western hemisphere.[176]

With courses like these, the Group finally achieved a greater homogeneity in its managerial training, which helped to efface the differences between Dutch and British recruits so evident before the war (see Volume 1, Chapter 5). We have not found the detailed content of the General Executive Course, but it was focused on the oil industry and its quality was evidently high, because in 1958-59 various non-Shell members of the Iraq Petroleum Company, in

Diverse moods: training could be both light-hearted and very serious, as these 28 Shell men discovered when attending the Houston Regional Management Course in 1969.

[59]

Chapter 2

which the Group held a 14 per cent share, sought entry to it. At this time BP, the 40 per cent owner of IPC, actually shared the Lensbury Club with Shell, and did so until 1964. Moreover, Shell was already accepting marketers from Shell-Mex and BP Ltd. onto the course, and in 1956 an IPC man had been accepted. He had been excluded from one confidential session, but with all these shared elements there was clearly a case (which was pressed hard by IPC) for including more of their people. From Shell's viewpoint there was an equally clearly a case against it: 'Anyone attending this course hears a lot about management thinking and policies in the Group and the question is whether the indirect benefit we shall gain from the attendance of senior members of IPC staff on these courses outweighs the risks inherent in passing on confidential material to executives who have no direct Group allegiance.'[177] The SM & BP people were acceptable because they were always permanent staff who had grown up in that organization. Even that was not entirely satisfactory to Shell, but there was no way of getting out of the arrangement. It seemed it might be possible to accept IPC people for only part of the course, and IPC was willing for its executives to be excluded from sessions if necessary; but the whole issue was important enough to be referred to the CMD, and their reaction was that 'considerable embarrassment would result both to IPC and Group personnel in any Course that the former attended, and that as far as the Group were concerned, there would be no commensurate compensating advantage'.[178] The matter was then referred to Loudon, who decided that the IPC request should be turned down. When it was suggested to him that this might drive IPC men into the arms of BP, he asked if BP ran any similar course. Investigations

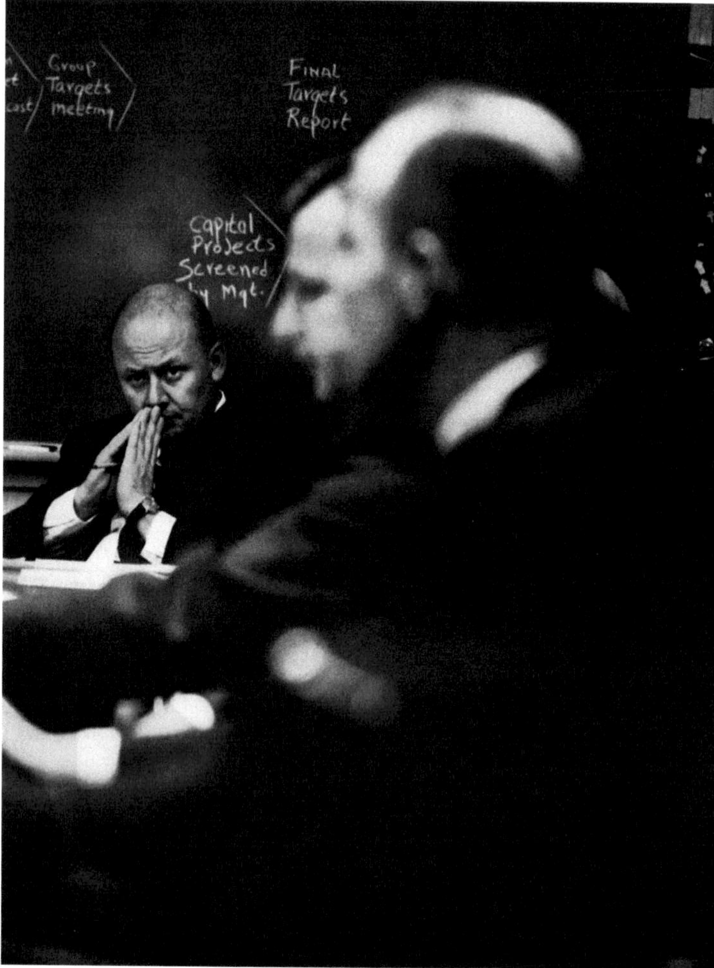

[60]

revealed that they did not, but only one-month courses in business administration for lower to middle management. IPC already sent men to these, but 'they are not oil industry courses'.[179] Thus IPC remained excluded from the privileged atmosphere of Shell Lodge and the commercial advantage of its senior course.

The Group liked to keep control not only of those who were permitted to attend but also of the content, even when input was offered by government. Late in 1959 the British ambassador in Venezuela suggested that CSV Industrial Relations personnel might attend a two-week course run by the Foreign Office that was 'designed to throw light on the philosophies and techniques of communist-inspired elements in labour unions and similar bodies, and enabled policies and tactics that had been employed in some areas to be recognised at their true value as soon as they appeared in others'.[180] This was declined as being inappropriate for personnel employed abroad, but the CMD did think that senior Industrial Relations personnel from central offices could benefit, and could thereafter disseminate their acquired knowledge. A few weeks later, the Foreign Office expressed its willingness to run special courses whenever necessary, with follow-up courses whenever Communist techniques changed. The CMD reaction remained the same – but this time for an interestingly different reason, namely that they felt that non-British employees would be excluded.[181]

Promoted as 'an opportunity to grow' (left), the 1964 Shell Management Study Course in the US was taken very seriously, whereas executives on a senior course in 1961 played 'Simplex'.

By 1960 the Group had already been aware for some years that European management ideas were lagging behind those of the US.[182] Game-playing as a means of learning – an American innovation – was introduced into management training days in 1959, in the form of 'Simplex' or Simulation Planning by Executives. Described as 'a dress rehearsal for decision-making', Simplex was based on the concept of business games, an idea derived from military staff colleges' war games. Shell was among the pioneers of the method outside America – an unexpected spin-off benefit of the McKinsey review, because G. R. Andlinger, the inventor of the concept, was a McKinsey employee. Sixteen or more individual participants would be divided into four competitive 'companies' for an 'exhausting and exhilarating' game. Other companies including Unilever, J. Lyons & Co, the British Motor Corporation, and Kodak showed great interest, and it was adopted by organizations as different as the pharmaceutical firm of Boots and Birmingham

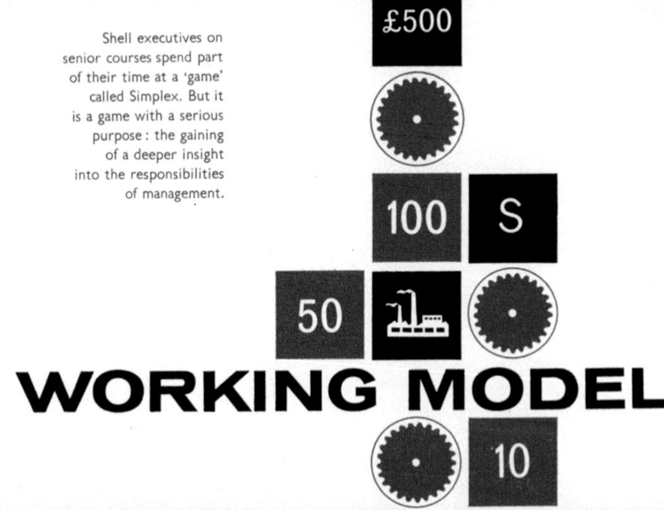

Shell executives on senior courses spend part of their time at a 'game' called Simplex. But it is a game with a serious purpose: the gaining of a deeper insight into the responsibilities of management.
£500
100
S
50
WORKING MODEL
10

'BUSINESS GAMES' is one of several names given to an interesting field of activity that has been developing in many places in the last 15 years or so.

During the second world war, classroom exercises in which strategic and tactical situations were simulated realistically proved to be very helpful in developing skill and judgment in officers under training, and the possibility suggested itself that similar methods might have important applications in time of peace.

At one time, when industrial concerns were smaller than they are now, companies tended to evolve their business policies over the years by trial and error, but in today's competitive conditions, where the importance of the decisions facing senior executives and the amounts of capital involved are often much bigger, such methods can lead to catastrophe. Today's businessman is faced with highly complicated situations in which there are large numbers of variable factors, including some whose effects may well be unknown, and it is in training executives to take a large number of variables into account simultaneously that the business game promises some of its most notable results.

The thing that really distinguishes business games from other educational techniques, however, is their dynamic nature—the fact that they can be so constructed that the decisions taken by the participants can be quickly translated into terms of their effect on the market, so that the business environment is constantly changing, as it is in real life. An executive who takes part in a business game, in fact, must face the consequences of his mistakes, in all their cold reality, almost as soon as they are made.

Training Division of Shell International Petroleum Company Ltd. has worked out its own version of a business game, largely through the enthusiasm of Mr. Bruno Eldon, who began his Group career in Roumania some 30 years ago, and is now responsible for developing management training. It is known as Simplex, a code name standing for 'Simulation Planning by Executives'. This name was deliberately chosen in order to emphasise that the 'game' is not a frivolous pastime, but a scientifically planned exercise, based on the hard facts of competitive business, and intended to achieve clearly defined and limited ends. It consists, to quote Mr. Eldon, in "a controlled business situation in which a team competes against intelligent adversaries and/or environment to attain predetermined objectives."

Fair play

Shell International Petroleum's version of the business game—another type, employing an electronic computer to evaluate the results, has been employed by Bataafse Internationale Petroleum Maatschappij N.V. in The Hague—goes back to an article by G. R. Andlinger, published in the *Harvard Business Review* in 1958. This article aroused considerable interest among the staff of Training Division, who decided to investigate the possibilities of a business game becoming a useful addition to their armoury. After a series of pilot runs, in which a number of senior members of the Company co-operated, various modifications were made to the Andlinger version. These were designed to increase the realism of the project and to fulfil three essential elements that make for its proper functioning: acceptance of the model as a fair simulation of a real business situation, easy assimilation by the participants, and avoidance of all unnecessary complication in the work of the umpires.

A session of Simplex will normally take up a day and a half. An introductory brief is given by the Project Leader, who explains the origin of Simplex and sketches in the background. He then explains, in some detail, the ground rules and the general working of the project.

The course members are split up, as a rule, into groups of four, each group representing a separate 'company'.

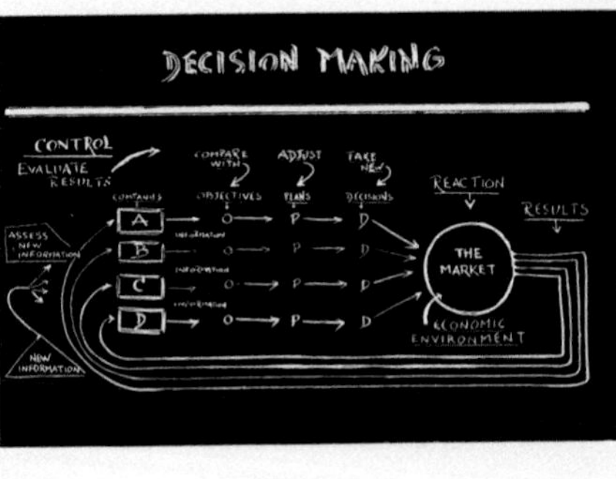

(Above) A diagram illustrating the flow of decision-making. Companies absorb information, formulate objectives, make plans, take decisions. The market, affected by the general economic environment, reacts. Companies evaluate the results of decisions in the light of new information, compare results with new objectives, adjust their plans, act by a further set of decisions. Then on to the next . . .

(Left) Deadline 10.20: by the time shown on the model clock the Company must have made its decisions. The planning-board and counters lie on the table: 12 salesmen are in the field, one is still in training; production seems to be going well. But they must meet that deadline. This photograph was taken during the second 1961 Senior Chemical Course.

240

241

[61]

University to suit their own needs. By 1961 over 250 Shell men in London and The Hague had taken part in Simplex,[183] but in that year a brief backward swing of the training pendulum came when the regular meetings of another Shell body, the Management Study Group, were suspended because of 'a widespread feeling that management training had been somewhat overdone in the past when viewed against the need for maximum economy in operational costs'.[184] D. W. Mullock, the Group Personnel Coordinator, argued (and a somewhat doubtful CMD agreed) that it should be reinstated at least for 1962. One spin-off from the management training programmes was the formulation, during 1962, of the first of the Group's business aims and principles which has already been mentioned. The CMD did this in response to complaints from senior managers and management trainees that they did not really know what the business was trying to do.[185] The aims ranged from the duties of management, the scope of the business, the need for productivity and innovation, staff conditions and participation, to the value of competition and private enterprise. They also emphasized the need to operating at the highest standards of business integrity and social responsibility.[186]

The Group's thinking about senior management training continued to evolve, away from sole provision in luxurious surroundings. It was also in 1961, when the Management Study Group was under question, that 49-year-old David Barran was appointed to the board of Shell Transport. Barran was a Cambridge history graduate whose potential had been spotted soon after he joined Asiatic in 1934. As a new Group managing director it is unclear exactly how much influence he had over the evolution of training and to what extent it was coincidental with his presence on the board, but in 1962 a three-week Business Introduction course was introduced in cooperation with Regent Street Polytechnic in London, to inculcate 'business-mindedness' in staff joining the marketing and financial sides. Its success encouraged the running of more courses, including for graduates going into the technical functions.[187] Whether or not Barran had much directly to do with this, at the end of October 1963 he had a meeting with the moral philosopher, former ambassador and Oxford provost, Lord Franks. Their discussion showed that their views were in substantial agreement, and later that same day Barran addressed the CMD on the subject of 'Education for Management'. Referring to growing British national interest in this with the creation of bodies such as the Foundation for Management Education, he suggested six guiding points for the future.[188] With the caveat that education in business (accountancy, finance, business practices, and so on) should come before education in management, the CMD agreed to all six points; and these meetings were among the steps that led, following the Franks Report of 1963, to the establishment of the London Business School and the Manchester Business School.

Training spread worldwide as an important part of the Group's regionalization. Below, from left, in 1969 all Shell retail site dealers in Australia went through a six-week course at Shell training centres, with this group receiving instruction on car engines; in Atlanta in 1968 trainees learn about greases; in Nigeria in the 1960s others are instructed in palynology; and at the joint Shell-BP Trade School in 1960, a technician explains the working of a diesel engine.

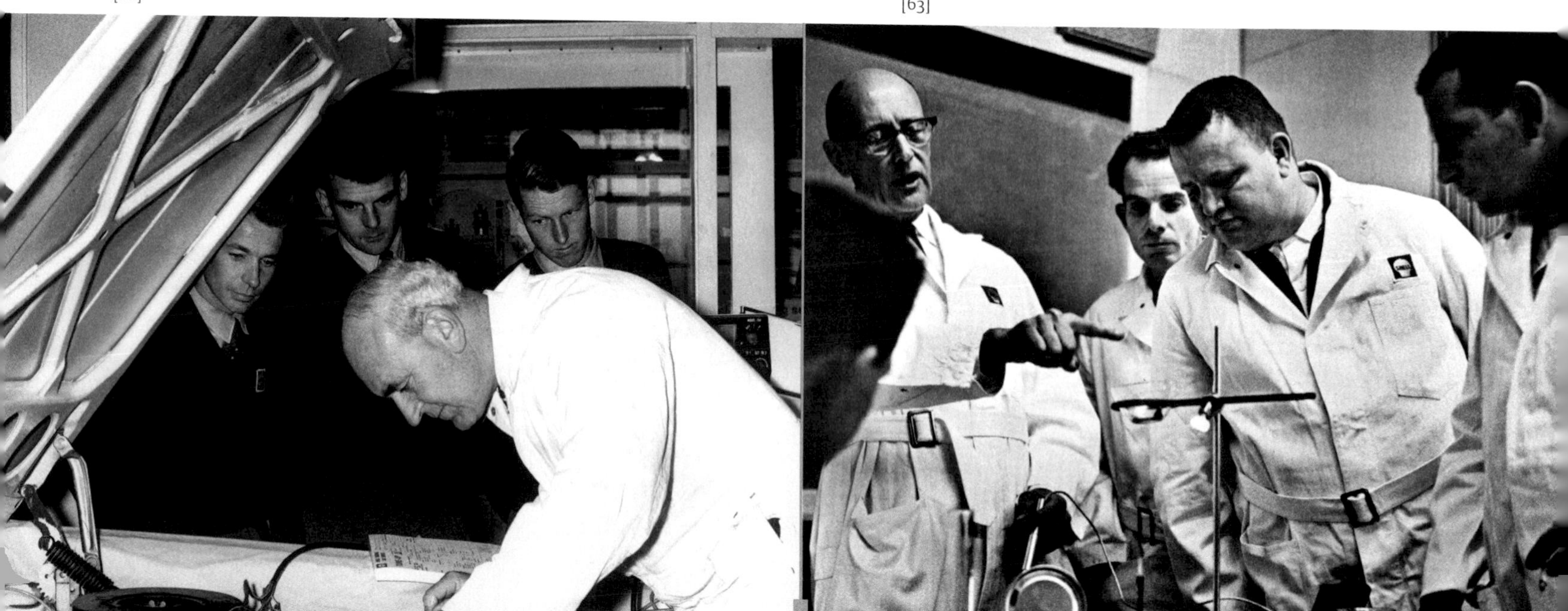

In the face of that strong new breeze, the days of cosseted training at Shell Lodge were numbered. From the beginning of 1963 the Lensbury Club was increasingly used for accommodation of trainees and sometimes for actual training. In 1965 the club was due a £100,000 (1 million guilders) upgrade of its facilities, and by a conservative estimate it was calculated that running costs of at least £24,000 (240,000 guilders) a year could be saved if all training was moved there. Accepting that some of the aura of Shell Lodge would be lost with a move to a larger training centre, it was nonetheless felt that the maintenance of Shell Lodge purely for prestige or sentimental reasons could hardly be justified. Worse still, for those who had loved it, 'we now have reason to believe the country house atmosphere is not only not necessarily conductive to the learning process, but may even be a hindrance'.[189]

Early in 1967 a comprehensive paper was prepared on the future of management education courses. Three levels of internal course were to be run by the service companies, including a study group for senior executives – a new course for lettered-category staff aged 40-45, and therefore designed on a participative rather than direct-instruction basis. There would be four five to six week courses a year totalling sixty-four participants, with analysis of the

political, economic and social environment in which the Group operated, reassessment of modern business techniques, and discussions led by outside speakers on the Group as an integrated system. Theses on various Group policies would also be presented by visiting professors, a novel aspect of the curriculum which the CMD welcomed; but they were more wary about the integrated system discussions, because 'This part of the course also included (...) a concept of group learning in the context of a team approach to the solution of problems. The mutual criticism inherent in this method of instruction was viewed with some caution, but it was agreed that it was worth trying, subject to review after the first course. It was stressed that it would have to be made very clear to participants that the object was to teach, not to test.'[190]

A second tier of internal training, the 'Group Development Course', provided four-week courses and was aimed at middle-level managers with prospects of promotion, in all functions and nationalities, with seven courses a year totalling 126 participants aged 38 on average, and covering such topics as business finance, planning, decision making and control, research computer applications, personnel and man management, and effective group work. The third tier, the two-week 'Graduate Introduction Course',

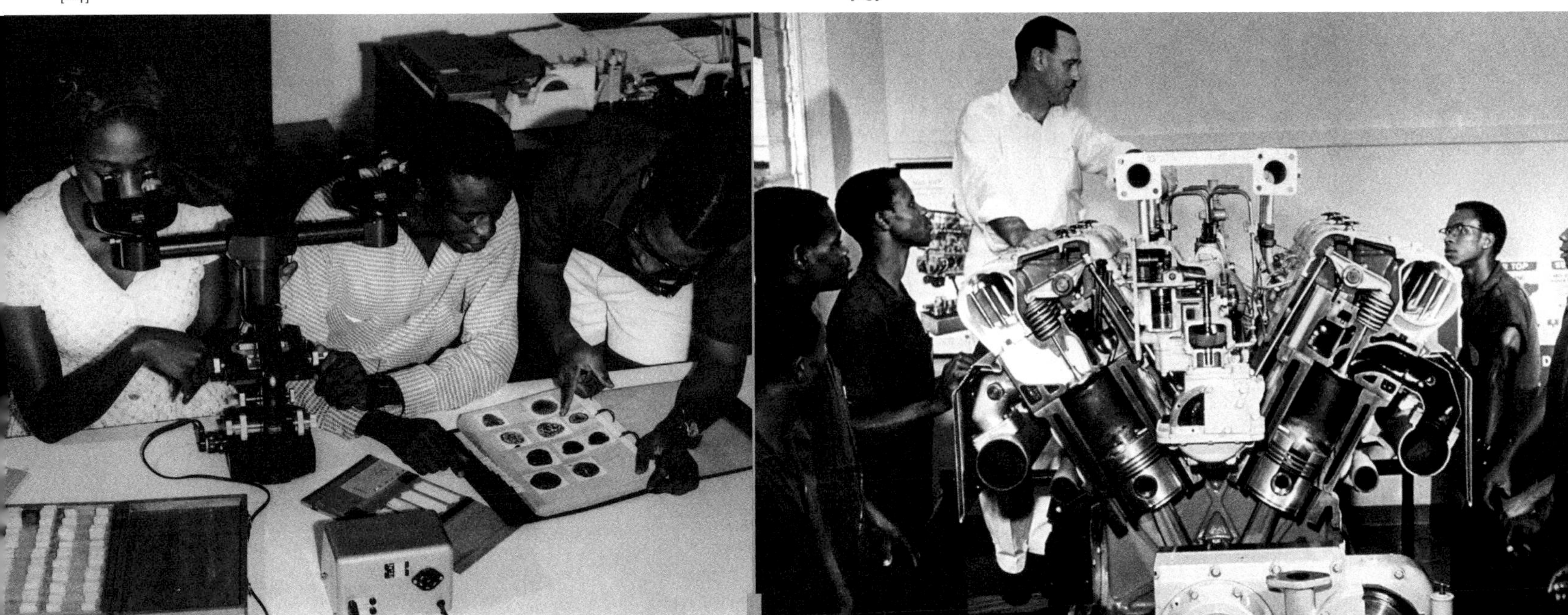

ran 14 times a year with a total of 336 students drawn from all graduates recruited through BPM and SIPC in their first year of service, covering the history and organization of the Group, the background to the oil industry, and an introduction to basic management practices. In addition to those three internal tiers, external management courses were to be arranged for senior executives at Harvard, MIT, Stanford, Columbia, London, and Manchester, and for middle-level managers at numerous universities and similar establishments in North America, the UK, and continental Europe.

By 1970 considerable progress had been made on this basis. At senior level, thirteen study groups had been run with a total of 188 Shell personnel. One of the 188 – his identity is unrecorded – had also attended a senior level outside management course at Columbia, and a further ten (not among the 188) had attended such courses either at Columbia, Stanford, MIT or Harvard. A notable detail in this record is that in addition to the 188 Shell employees, these senior level Group courses were made available to seventeen individuals from outside Shell, including members of the Post Office, IBM, Algemene Bank Nederland, Unilever, and many other such organizations – quite a contrast with the late 1950s, when IPC and everyone other than employees of Shell-Mex and BP Ltd. was rigorously barred from participation.[191]

At middle-level, 612 Group personnel had attended the Group Development Course introduced in 1967, for personnel of all regions and all functions. Other selected staff had completed management courses organized by Shell Oil and Shell Canada, and as a notable indication of regionalization in action, fifty-seven non-English-speaking personnel had completed middle-level outside courses in the English language in North America, Europe and the UK. Plans for 1971-75 were based on the results of a questionnaire completed by a representative selection of operating companies and all coordinators in the London office, but before long it was recognized that the conflicting ideals of regionalization and central organization created problems and cross-currents. From these in 1972 there was 'a general feeling that the regionalization of jobs, particularly in developing countries and the allied running down of expatriate staff, have caused a certain lack of cohesion and some fall in standards in training activities in the Group as a whole. The problem has been insufficiently investigated and more work needs to be done, especially to determine which role the Centre should play in the initiatives. In addition to the programme run at the Centre, it will probably be necessary for the Centre to do more in the way of assisting in surveys of training needs, in regional and local courses.'[192]

It was recognized that training should not be undertaken centrally if it could be done as well or better locally, and training divisions in London and The Hague by now were closely integrated, with technical training being largely concentrated in the Netherlands and marketing, finance and other non-technical functional training in the UK. Nevertheless, by deliberate policy, much more time was spent on lending support to regional training programmes, such as in marketing companies in Africa and the Far East, while the 'home countries' were experiencing a much increased demand for manufacturing and E&P training in laboratories and plants. Individually tailored programmes, though sometimes inevitable, were avoided as far as possible because of their expense. Marketing training was led by the plans of the companies concerned – merchandizing in Europe, LPG marketing in the Far East, and automotive dealer training in Africa.

Many learning packages were translated from English into a variety of other tongues, but as regionalization advanced and 'although English is the universal Group language', fluency in at least one language other than one's mother tongue was seen to be of increasing value and a necessity for everybody of ability and ambition: 'with increasing regionalization a minimum proficiency in other languages becomes more and more important for the effective functioning of service company personnel; there should for instance be more staff speaking French and possibly German.'[193]

Consequently, Shell's General Executive Course was not its only training innovation to excite outside interest. In the late 1950s Shell Venezuela became the Group pioneer for intensive language training, an experiment which was so successful that in 1960 it was adopted by Shell International in London, because 'For anyone in the multi-national Shell world, languages play a vitally important part (...) Methods which can step up this process of language learning are therefore of considerable interest'.[194] Known then as a 'language switchboard' and later as a language laboratory or language resource centre, the ground-breaking system was widely reported in the press and sparked great curiosity outside Shell: the London office received intrigued visitors from the War Office, the City of London College, Bristol University, Birmingham University, the University Institute of Education in Oxford, and HM Inspectorate for Schools.[195]

Regionalization was also thought to be affecting the training effort in manufacturing: 'It may well be that one of the immediate effects of regionalization is a decline in standards in overseas plants (...) In order to be successful, standards of teaching and facilities need to be high.' Start-up projects such as the Brunei LNG scheme and a refinery in Cyprus certainly required much attention from the centre, but against these demands it was found that the Group's E&P training arrangements, focused on its laboratories and its training centre at Wassenaar in the Netherlands, were a source of relative strength: 'apart from Exxon, most other oil companies lack such a concentration of in-house expertise as Shell has, both in Europe and the US. (...) There is probably no better example of a close integration between work in the Function, the operating companies and training as in the E&P training centre at Wassenaar.'[196]

By 1972 Shell had forty-two training centres in thirty countries, and about seventy-five expatriates working as training specialists. The Group also had a clear recognition of four 'problems and projects for development' in its training programmes. First, the planning of any training needed to include its cost and cost recovery. Second, training needs had to be identified in liaison with line management, and criteria established to evaluate the effect of training. Third, it was important to keep a keen eye on the outside world, especially universities, business schools and other companies, and to make effective use of outside resources. And fourth, special attention had to be given to the problem of early motivation and induction of young graduates, and the training needs of older employees whose skills might otherwise be obsolescent. In short, after fifteen years of the evolution of training, the Group had learned some of the eternal truths of effective training – not least that its evolution would be a never-ending process.

A chemistry class in the 1950s, teaching young people about gas at the Group's Amsterdam Laboratory.

Conclusion

During the 1950s and 1960s, the Group underwent a profound reformation, which included financial, cultural, psychological, and generational changes. There is a strong impression of a younger generation discreetly taking charge with a combination of foresight and determination, and that Royal Dutch's more youthful board was the driving force, assisted by the less senior board members of Shell Transport. How much this was a nationality matter and how much a generational one is difficult to know, but between the parents at this time there is more than one echo of the time when the Group was formed. Dutch commercial ambition and ability in 1947-52 was undeniably greater than British, as had been the case in 1906-07; and one cannot help remembering that just as Marcus Samuel had been thirteen years older than Henri Deterding, Godber was almost old enough to be Loudon's father. In both eras it looks as if, for any Dutchman or Briton with ideas about seeking a position of influence in the oil industry, it was better to be young and Dutch rather than old and British. Thus the Group continued to benefit from very dynamic leaders.[197]

The reformation affected all the Group's main stakeholders. The New York listing taught new and vital lessons in shareholder expectation and in the art of communicating with the public, lessons that the Group in future would forget or ignore only at its peril. The reorganization of central offices during the second half of the 1950s profoundly changed both the position of the top managers and the relations between Shell and its operating companies. For these relations, decentralization and delegation became the watchwords, but at the same time, central offices launched comprehensive programmes for recruitment, training, and remuneration designed to provide the Group with highly motivated personnel, characterized by a recognizable common outlook, able and willing to work wherever the business might need them. Before the war, most employees felt an allegiance to the operating company for which they happened to work; during the 1950s, the Group's employees became Shell people.

Thus the managerial decentralization was balanced by a distinct centralization in the form of a more uniform corporate culture. The 1950s also witnessed the emergence of a Shell style, perhaps best characterized by the notions of balance, a penchant for open and informed debate, and consensus. The CMD epitomized this style, its chairman acting very much as first amongst equals. His authority rested on merit and ability, no longer on nationality and Royal Dutch's majority vote. This momentous change was buttressed by Barran, on succeeding Brouwer as chairman of the CMD in 1970, becoming the first Briton to lead the Group, and by the absence of any scheme in the later alternation between Dutchmen and Britons in that position. Anglo-Dutch confrontations did not disappear entirely, but the occasional flare-ups were more in the nature of ritual jousts than the open fights of the late 1930s and 1940s.[198] Moreover, balance, debate, and consensus were the norm throughout the organization, stimulated by decentralization, delegation, and the need for dialogue between functional and regional interests. These attributes served Shell very well during the 1960s and 1970s, though by then decentralization, notably in the extreme form applied to Shell Oil, began to show some hidden costs.

[1]

Chapter 3

Into ever deeper waters

Following the Second World War, the major oil companies divided the newly found riches of the Middle East between them, creating a new and stable order for the oil industry which, though rocked from time to time, lasted for twenty-five years. The Group took full part in the alliances between the majors and over time came to derive a growing share of its production from the Middle East. It succeeded quite quickly in recouping the large losses of production during the years 1938-48, but despite intensive E&P efforts, it remained crude short. Throughout the period Shell strove for an even geographical spread of supplies, but during the 1960s economic nationalism and the rise of OPEC made the established pattern of supplies unattractive and managers consequently shifted resources to offshore operations and exploration outside the OPEC area. Following a huge discovery in the Netherlands, natural gas became increasingly important.

After the Second World War, car ownership rocketed and traffic congestion became common, as here on 45th Street, New York, in 1954.

A new order for the international oil industry With the discovery of large crude reserves in the Middle East during and immediately after the Second World War, the oil industry's operating environment changed dramatically. Kuwait and Saudi Arabia, countries which many experts, including the Group's top geologists, had believed to offer no profitable opportunities, turned out to harbour immense fields equivalent to the greatest found until then. Nor did the pace of discovery slacken after the initial successes; throughout the period under consideration, the volume of estimated world oil reserves continued to rise rapidly year after year.

Simultaneously the industry balance shifted. In 1947 the United States, the world's biggest producer and a net oil exporter, became a net importer. This intensified concerns about the security of crude supplies to the western world, both in official circles and amongst the major oil companies. The advance of communist regimes in Europe under the tutelage of the Soviet Union had led to the emergence of two opposing political and military blocs separated by an Iron Curtain across that continent. This so-called Cold War had made the British and American governments particularly concerned about threats to the stability of political regimes in the Middle East and for the need to provide them with a regular income in the form of oil revenues from long-term supply contracts. For their part, the oil majors sought to prevent the sort of uncontrolled crude glut which had devastated the world market during the late 1920s and 1930s. To this end the companies entered into a set of long-term contracts which effectively tied up most of the new fields to joint ventures between them. The Kuwait supply contract which the Group concluded with Gulf in 1947 was one such agreement. Also in 1947 Anglo-Iranian Oil Company (AIOC, later BP), joint owner with Gulf of the Kuwait Oil Company (KOC) and like Gulf badly in need of a marketing network to offload its surging crude supplies, agreed to long-term supply contracts with Standard Oil New York (Socony, later Mobil) for Kuwaiti oil and with Jersey Standard and Socony for oil from Iran. In November 1948, the shareholders in the Iraq Petroleum Company (IPC), that is to say the Group, BP, Jersey Standard, Socony, the Compagnie Française des Pétroles (CFP), and Gulbenkian, revoked the old Red Line Agreement, which had bound them for twenty years to undertake operations within the area only through IPC. Jersey Standard and Socony were now free to take part in Arabian American Oil Company. Originally a joint venture between Texaco and Standard Oil California (Socal, later Chevron), Aramco faced rapidly rising production from its vast fields in Saudi Arabia without a marketing network to take it up, which the company acquired by letting Jersey Standard and Socony in. As supply contracts, the deals with which the majors carved up the oil fields in Iran, Kuwait, and Saudi Arabia between them were fundamentally different from the 1928 Achnacarry agreement, which had attempted to freeze market shares in the face of rising overproduction. Through ingenious clauses penalizing overlifting by any one of the partners, they were also far more effective in controlling the majors' output than the As-Is agreement had ever been. Their collective strength in controlling output and markets led to Jersey Standard, Socony, Texaco, Gulf, Socal, Royal Dutch/Shell, and BP becoming known as the Seven Sisters, with CFP not counting officially as a sister but, on the strength of its share in IPC, being treated as a cousin.

These contracts received the full blessing of the British and American governments, and also of the host governments, which were entirely reliant on the majors for marketing their oil. In effect, the contracts established a new order for the world oil industry, based on an informal coalition of oil companies, western

One of the refineries belonging to the Arabian-American Oil Company (Aramco, later Saudi Aramco) in 1947. The revocation of the Red Line agreement enabled Jersey and Socony to join Texaco and Socal in Aramco's operation.

[2]

[3]

In Iran in 1953 the 'oil question' generated rising social unrest (left), and (right) crowds opposing the Shah placed posters of Mosaddiq on the gate of the Parliament building in Tehran.

[4]

governments, and host governments working towards a stable development of the Middle Eastern resources. It was a resilient coalition, because the parties to it were each other's economic hostage.[1] The oil companies wanted control over resources, the western governments political stability, security of supply, and cheap oil, the host governments rising revenues; by splitting the rents from oil exploitation amongst them, the parties all got what they wanted, at least initially. Tested by three crises, the coalition proved its mettle. The first of these crises began in 1949 when host governments, inspired by the Venezuelan example, started pressing to change the concession conditions into 50:50 deals. The American and British governments pressed the oil companies into accepting it and at the same time eased the pain by introducing regulations under which the companies could offset expenditure on royalties and taxes abroad against corporate income tax at home. In 1950, Aramco agreed to 50:50 with the Saudi government, Kuwait following in 1951 and Iran in 1952.

Meanwhile the second crisis had erupted in 1951, when the Iranian prime minister Mosaddiq nationalized the country's oil industry. After two years a tight embargo on Iranian oil, aided by Anglo-American covert intelligence action, brought about Mosaddiq's fall. When the new regime proved willing to let western companies return to operate the country's oil industry, the majors formed a consortium to take over 60 per cent of BP's position in Iran. BP kept 40 per cent overall.[2] The Group fought hard to get a large share and it finally obtained 14 per cent. 'It was difficult to agree as to the percentages', Loudon later recalled. 'We at Shell were anxious, not being in the Middle East, to get a large percentage.'[3] The five American companies got 7 per cent each, CFP 6 per cent, and a joint venture of nine US independents 5 per cent. This last partner was included at the insistence of the US government which, on special order claiming an overriding national interest from US President Harry Truman, had also halted anti-trust proceedings against the American companies so they could join the Iranian consortium. Bataafsche took up the Group's stake. Because under the terms of the deal the operator could be neither British nor American, the two companies formed to carry out the new agreement – Iranian Oil Exploration and Producing Company, and Iranian Oil Refining Company – were both incorporated under Netherlands law and registered in Iran with their management and operational headquarters there. In addition, the consortium's first two managing directors were both Shell men.[4]

The third crisis broke out when the Egyptian President Gamel Abdel Nasser nationalized the Suez Canal in 1956. Wanting to make a stand against what they perceived to be the threat of Arabian nationalism, Britain and France secretly agreed with Israel to launch a joint attack on Egypt and occupy the canal zone, but when the

Nationalization of the Suez Canal in 1956 provoked one of the worst crises in Britain since the Second World War. Below, after making his announcement in Alexandria, Egyptian President Gamel Abdel Nasser returns in triumph to Cairo, and (bottom) shocked Londoners read the news.

[5]

[6]

invasion got under way the waterway had already been blocked by ships scuttled in its approaches. As a consequence the main flow of oil to Western Europe came to a standstill while tankers were being rerouted past the Cape of Good Hope. Acts of sabotage in Syria and Kuwait reduced supplies still further; Saudi Arabia proclaimed an oil embargo on Britain and France. In the end the much-feared interruption of oil supplies failed to materialize. Swift action by governments in Europe and the United States collaborating closely with the majors in an operation known as the Oil Lift ensured that sufficient oil remained available to meet a demand reduced by economy measures which various countries had adopted. Meanwhile political pressure from the US forced Britain and France to accept a humiliating defeat and withdraw their troops from Egypt. In April 1957, the Suez Canal reopened for traffic and the third crisis was over.[5]

Powered by Middle Eastern crude, the world entered the hydrocarbon age, oil replacing coal as the main source of energy for all uses. As the real price of crude gently declined, the number of motor cars skyrocketed; industry, power stations, and households switched to oil; and oil found applications in a rapidly widening range of products, from pesticides to plastics. Between 1948 and 1972, oil consumption in the United States tripled, in western Europe it rose by fifteen times, and in Japan by 137 times.[6] However, the Seven Sisters' cartel proved unable to prevent the independent oil companies from spoiling some of the game. The US tax incentives that had been created for the majors also stimulated American independents to acquire concessions in the Middle East, and host governments there, keen to increase their revenues, welcomed them. Headed by the flamboyant Enrico Mattei, the Italian company ENI/AGIP also entered the Middle East, offering the

Seen from a bullet-riddled Port Said lighthouse, a Norwegian tanker lies idle in the Suez Canal – one of many ships trapped for months on end when the Canal was blocked.

[7]

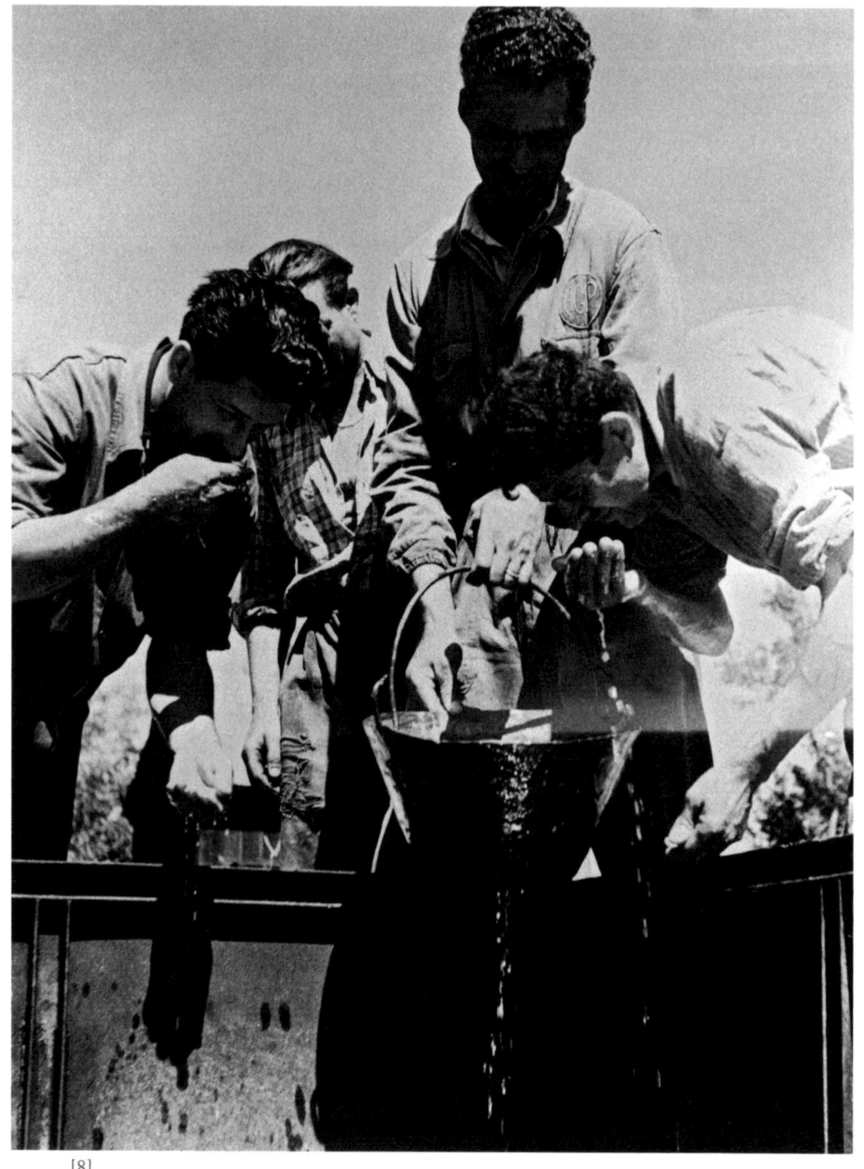

[8]

Shah of Iran a 25:75 profit split and government participation in return for a concession. A Japanese consortium set up a joint venture with Saudi Arabia based on a 46:54 split.

This competition pushed concession prices up, and crude prices down. In 1959 the resumption of oil exports by the Soviet Union, the discovery of large fields in Libya, and the imposition of import controls by the United States to protect small Texas producers from the competition of low-cost Middle Eastern oil, increased the pressure on crude prices still further, exposing a fundamental weakness in the established industry order. The agreements between the majors themselves and between the companies and the host governments were built around 'posted prices', i.e. official crude prices used by the companies for internal accounting purposes and based not on the actual cost of production in the Middle East, which was very low, but on the very much higher cost of production in the United States. The rapid growth of crude supplies outside the majors' control opened a

[9]

Following a discovery of oil in the Po valley in northern Italy in 1949, Italian experts smell and test the crude (above), but the supply was not enough to satisfy the head of ENI/AGIP, Enrico Mattei (right, in 1962), whose negotiations with oil-rich countries threatened to destabilize the international industry.

widening gap between market prices and posted prices, which became more and more fictional, a phenomenon well captured by the industry saying that only fools and affiliates pay posted prices.

The divergence benefitted the host governments, which now effectively received more than 50:50, to the detriment of the oil companies. At first the majors would not consider cutting the posted prices to market level, because they were reluctant to upset their cherished order and because the tax benefits compensated their transfers to the host governments anyway, but by the late 1950s the pressure had become too great. In 1958, 1959, and 1960 the majors cut their posted prices, raising howls of indignation from the host countries, which understandably saw these cuts not as a necessary adjustment to market conditions, but as an unfair and unilaterally imposed reduction of their revenues.

This indignation inspired an initiative from Nasser, by which seven Arab nations agreed to unify their policies towards oil companies operating in their territories. Delegates discussed ways of doing this at the first Arab Oil Congress at Cairo in April 1959. When the conference was announced, 'we [in Shell] were sceptical, and doubted whether the Arabs could organize an event on such a scale. However, the Congress (...) turned out to be a success for the Arab League, the Arab countries and many of their leaders.'[7] By cutting the posted prices on the eve of the conference, the major oil companies had contributed materially to that success. The unilateral action had convinced the representatives of five major producing countries, Iran, Saudi Arabia, Venezuela, Kuwait, and Iraq, of the need to take action and ensure at least consultation over such cuts in their revenues. They spelled out their intentions in a secret and unofficial 'Gentlemen's Agreement' which recommended that their governments build a united front to countermand the power of the oil companies.[8]

Though the oil majors held regular and close consultations with each other, they do not appear to have discussed cuts in posted prices between them. When on 9 August 1960 Jersey Standard reduced its posted price again, BP 'heard the news with regret'.[9] John Loudon told Jersey's chairman he had 'put the cat among the pigeons'.[10] Though the Group, in line with the other companies, followed Jersey's lead, it followed unwillingly and made a much smaller cut (four cents a barrel, compared to Jersey's fourteen cents) in its posted price.[11]

The cuts in posted prices precipitated a last-straw reaction from the five signatories to the 1959 'Gentlemen's Agreement', who now accepted its recommendations led by Venezuela and Saudi Arabia, the countries responded by forming, on 14 September 1960, the Organization of Petroleum Exporting Countries or OPEC, intended as a producers' cartel to counterbalance the power of the majors. The creation of OPEC in itself did not change the established order of the oil industry, but during the 1960s the organization gradually acquired a position which finally enabled it to overturn the old conditions from 1973.

Expanding operations in a competitive environment

Buoyed up by abundant supplies and burgeoning demand, the Group expanded at a stupendous rate. Crude production rose by a factor of ten, from about 630,000 b/d in 1946 to 6.4 million in 1972, or from 32.8 million tons a year to 332.8 million. In other respects the company did very well, too (Table 3.1).[12] Between 1951, the first year of fully reliable consolidated figures, and 1970, assets grew 170 per cent in dollar terms, sales and profits almost 180 per cent. In terms of assets and refinery throughput, Shell remained level with Jersey Standard, and by 1956 the two companies ranked respectively third and second in the league table of the world's biggest companies by sales, after General Motors.[13] However, the sharp decline in the return on assets after 1955 shows the Group struggling with rising competition. Some companies performed far better. Between 1950 and 1970, Texaco sales grew fastest, quadrupling from $1.5 billion to 6.3 billion, followed by Jersey Standard, which saw its sales almost quadrupling from $4.4 billion to 16.5 billion.[14] The other majors maintained a tempo similar to that of the Group: Mobil sales went up by over 160 per cent, Gulf 130 per cent, Standard California almost 190 per cent. Only BP lagged behind with a mere doubling of sales.[15]

Jersey's faster growth originated first of all in the much greater weight of the company's American operations in the total. The imposition of import quotas insulated the US market from the increasingly heavy competition elsewhere in the world. With only 19 per cent of refinery runs, Jersey's American operations produced nearly half of the company's profits. By contrast, Shell Oil refinery runs and net profits averaged about a quarter of the Group's total and could thus not boost its profits by the same degree. Outside the US, Shell did as well as Jersey Standard.[16] Until the mid-1960s, the rest of the Group grew slightly faster measured by assets than Shell Oil, but the American company was rather more profitable. Indeed, Shell Oil appears to have been the exception to the rule that subsidiaries of non-American multinationals tend to perform relatively poorly there.[17]

Pounds	Net assets	Sales	Net profits	Return on assets
1951	1,183	1,436	121.3	
1955	1,239	1,488	168.1	
1960	2,151	2,004	178.9	
1965	2,771	2,636	226.1	
1970	3,730	4,667	394	
Dollars	**Net assets**	**Sales**	**Net profits**	
1951	$3,285	3,988	337	10.3%
1955	3,441	4,132	467	13.6%
1960	5,976	5,568	497	8.3%
1965	7,696	7,322	628	8.2%
1970	8,882	11,113	938	10.6%
Guilders	**Net assets**	**Sales**	**Net profits**	
1951	12,516	15,194	1,284	
1955	13,110	15,743	1,779	
1960	22,530	20,991	1,874	
1965	27,706	26,359	2,261	
1970	32,153	40,229	3,396	

Table 3.1
The Group's main financial data, 1951-1970 (figures in millions).

Almost crowded out by oil derricks in 1947, a ribbon of highway runs through Signal Hill, California.

[10]

[11]

Second, according to its own analysis the Group had a heavier tax burden.[18] Third, partly as a consequence of having a smaller production in the Middle East than most US majors, the Group had a cost disadvantage of up to 20-25 US cents per barrel of crude landed in Europe. This was largely a consequence of distance to source, that is to say, the difference between Venezuela and the Middle East. In 1961, for example, the Group drew about 700,000 barrels of crude a day, or almost 40 per cent of its total production, from Venezuela at a cost of $1.40 a barrel, whereas the 539,000 barrels of Kuwait crude received under the Gulf contract cost only $1.13 per barrel. Rising supplies from the Middle East, Libya, and Nigeria were expected to halve this amount by 1970. Jersey Standard had a similar disadvantage, however.[19]

Fourth, for a considerable time Shell maintained outside the US a barrel cut favouring the low-margin middle distillates over the higher distillates such as gasoline with its better margins (see Chapter 4). Marketing reports from the early 1960s blamed the drop in unit proceeds about half-and-half on price competition and the adverse product mix.[20] From the 1920s until the early 1950s the Group produced mostly fuel oil, with gasoline coming second and kerosene fractions making up the rest (see Figure 4.3, p. 298). The rise of chemicals, the appearance of new products such as jet fuel, and changes in processing technology boosted the demand for middle distillates, which passed first gasoline and then fuel oil to become the main product group by the early 1970s. By 1959, supply planning surveys were based on meeting minimum requirements for gasoline and maximum requirements of middle distillates, with fuel oil becoming the residual category.[21] Chapter 5 will discuss whether the overall contribution of chemicals to the Group's business warranted the new departure taken in the 1950s; it suffices to point here to the competitive disadvantage of a less profitable barrel cut.

Fifth, Shell's attitude to downstream differed markedly from that of its main rivals. The American companies thrived on their protected home market, generating funds which fuelled their overseas expansion. Moreover, they could offset their overseas investments, including service stations, against US income tax, which enabled them to expand their marketing networks and increase market share at marginal cost. This created a fiercely competitive environment. The Group, having the most widely spread operations of the majors, also suffered most from this competition and consequently its market share outside the US steadily eroded, from around 24 per cent in the mid-1950s to 17 per cent in 1966, and 15.8 per cent two years later.[22] Group managers faced this steady erosion with equanimity. Shell could not compete on the same terms, which meant that stable profitability was preferable over growth. As Gerrit Wagner put it in 1967, 'We do not believe in percentage and volume *per se*. It has to be profitable. There are some people who go for volume because they believe that profit will automatically follow sooner or later. I do not unqualifiedly subscribe to that'.[23]

Near Odessa, Texas, in 1947, an oil derrick on Shell Oil's TXL field nears completion. Its scale is given by the human figure half-way up.

[12]

Men rigging an oil well somewhere in the US. People suffering from vertigo were not wanted.

This focus translated into the conservatism of managers preferring a secure trade over sharp fluctuations, prudent objectives over ambitious growth targets. Throughout the 1960s the Marketing Oil function consistently underestimated industry growth and thus sales opportunities. A more aggressive policy was only adopted during the late 1960s.[24] However, the emphasis on profitability also favoured maintaining brand strength over keen pricing, promotional campaigns over price cuts, and large filling stations at prime sites over a dense network, building a consumer loyalty which enabled the Group to charge two to three cents per gallon more at the gasoline pump than the competition.[25] As a result Shell's profits grew as fast as, or faster than, Jersey Standard's profits.

Finally, Jersey's market capitalization was noticeably different: in 1946-49 it was fairly stable at about 40 per cent higher than the Group's, but then, with share issues in 1949 and 1950 and a stock split in 1952, it began a steep and continuing rise until it reached a high of $4.9 billion, or £1.77 billion – well over three times the Group's. This rise was mainly a consequence of Jersey's position as a favourite on the New York Stock Exchange, at the heart of the world's largest economy, which lifted Jersey's shares to a level far higher than the shares of either of Shell's parent companies could achieve.

The Group's supply pattern Group crude production grew slightly faster than world output, notably during the 1950s when managers strove to rebuild operations to their former level and achieved, by 1952, approximate parity of production with Jersey Standard, which had surged ahead in wartime. At the end of the war Shell produced barely 8 per cent of the world total, down more than 3 per cent on the Group's 1938 figure. It took the company five years to regain its pre-war share of world production and from the early 1950s its share was back at 11 to 12 per cent. Most of the huge leap in world output came from the new fields discovered in the Middle East, but advances in technology contributed a very substantial share as well, as we will see below. The rise of natural gas production was even more spectacular; in 1972 the Group sold 6 billion cubic feet, up from just over 400 million cubic feet in 1950.

Two features of the Group's supply pattern stand out (Figure 3.1). First, the centre of gravity in production gradually shifted back to the Eastern hemisphere. The share of the Americas dropped from about two-thirds in 1950 to one-third twenty years later. The Middle East steadily rose in prominence, most of the Group's oil

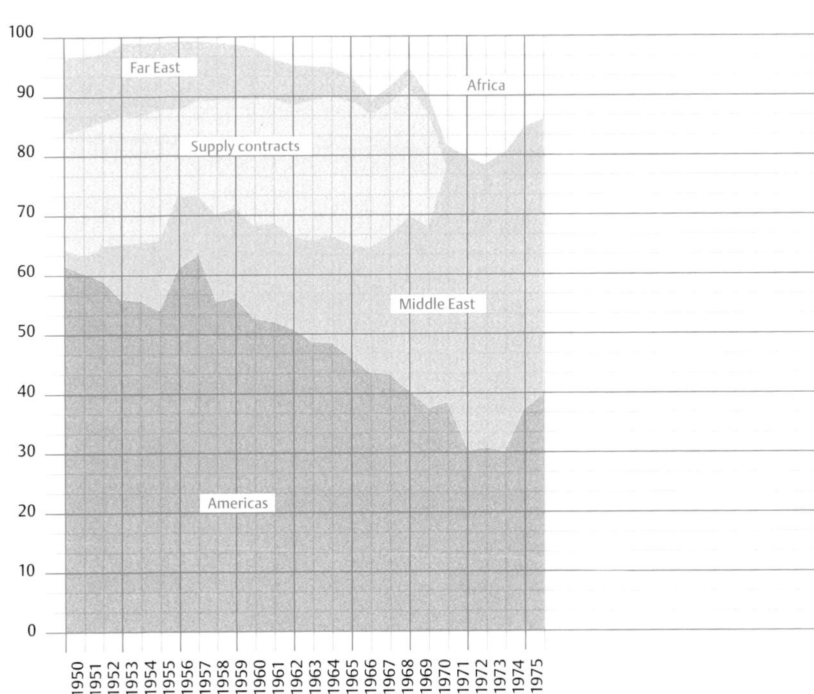

Figure 3.1
Group crude supplies by region as a percentage of the total, 1950-1975.

Chapter 3

derived from supply contracts also coming from that area. In 1967 the Middle East supplanted the western hemisphere as the Group's most important crude supplier, and Shell took up some 20 per cent of the region's output.[26] Seven years later Iran became Shell's largest supplier of crude, pushing Venezuela into second place, Nigeria coming third and the US fourth. The importance of supplies from the Far East, i.e. Brunei and Malaysia, remained steady after the departure from Indonesia in 1965, about which more below. After the loss of production from Romania through the nationalization of the oil industry there in 1948, Europe's contribution dropped to a steady 1 per cent, which gently started rising when the North Sea delivered its first oil in the early 1970s. Thus, contrary to common perception, the Group derived a considerable and rising proportion of its oil from the Middle East, though, unlike some of the American majors, its sources were spread over a large number of countries and not concentrated in one or two. Moreover, its volume of supplies from main sources such as Venezuela and the Middle East did not differ much from that of Jersey Standard.[27] The Group sought to achieve a judicious equilibrium between security of supply and crude oil cost, that is to

[13]

say, managers tolerated instability if the cost was low enough, or a relatively high cost if the source was very stable.[28]

Second, Shell was a classic crude-short business, heavily reliant on buying crude. In 1966, the Group had a market share outside North America of 16 per cent, but resources, i.e. reserves and long-term supply contracts, for less than 13 per cent of the market. Jersey Standard had a similar market share with resources for more than 15 per cent; all other majors were crude long, BP spectacularly so, having 8 per cent of the market, but enough crude for 20 per cent.[29] On average, the Group depended for 20-25 per cent of its supplies on long-term supply contracts, of which the contract with Gulf was by far the most important.[30] The terms of the Kuwait contract made Shell almost the third concessionaire in that country, and from 1965 the company reported its supplies from Gulf as if deriving from its own production, which as a matter of fact they did not.[31] In addition Shell Oil never had a sufficiency ratio, i.e. the percentage of refinery runs covered by its own production, of more than 60-65 per cent, which was comparatively low for an American company.[32] Thus the Group depended for a total of 25-30 per cent on purchases.

[14]

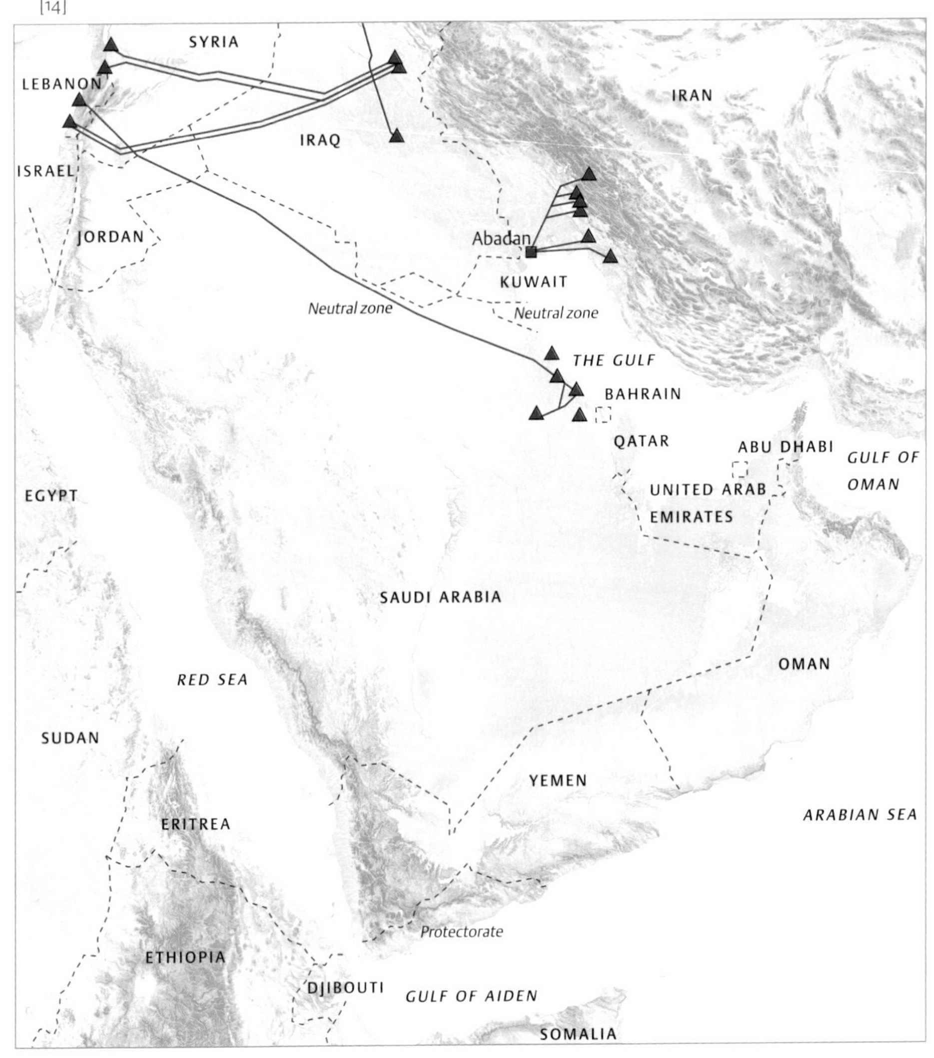

On 31 July 1954 the 40th anniversary of the completion of the first well in Mene Grande, Venezuela, was celebrated (left). The well had been brought in by the Caribbean Oil Company, forerunner of Compañia Shell de Venezuela. On the right, the major oilfields and pipelines in the Middle East.

— Pipelines
▲ Oil producing area's and/or refineries

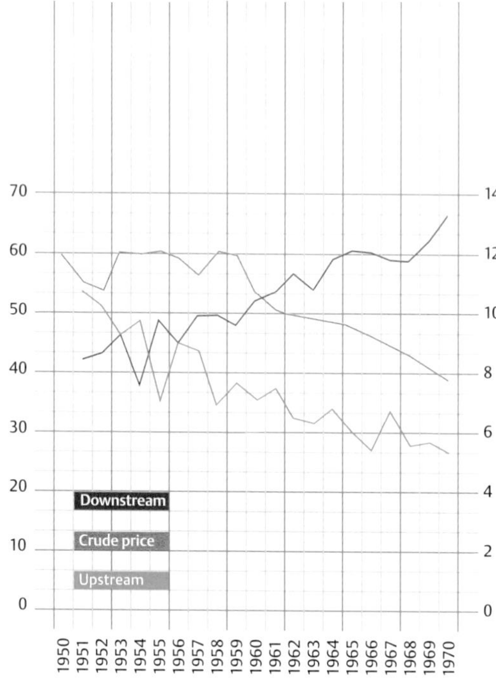

Figure 3.2
The Group's upstream and downstream investments as a percentage of the total (left scale), and real crude prices in 1999 dollars (right scale), 1950-1970.

Under the circumstances this made sense (Figure 3.2). From the late 1950s, real crude prices steadily declined, eroding the profitability of E&P operations. At the same time competition flared up in manufacturing and marketing, as one oil company after another sought outlets for the plentiful Middle Eastern supplies; and as OPEC countries began moving towards setting up their own downstream operations, Shell decided to reinforce that function so as to entrench itself against the expected competition. Consequently, the Group shifted its overall investment from upstream to downstream. The United States formed the exception here. Because the import restrictions kept cheap crude out, Shell Oil had to maintain its E&P budget, with the effect that its share of Group upstream investments rose from just over half to two-thirds. By 1970, however, the Conference started to have some doubts about the then current balance between upstream and downstream investments, asking 'might not the Group find itself at a disadvantage as the result of a disproportionate concern with increasing market share, without the backing of adequate crude resources – an aspect generally commented on in investment circles, where the Group's relative lack of own crude availability tends to be characterised as a weak point in relation to other international majors'.[33]

This fear was assuaged by pointing to the general crude surplus, which was expected to last for the foreseeable future.

The Group's crude shortage also meant that it spent more than any of the other majors on E&P. According to its own figures, the company spent a total of US$1.1 billion during the years 1951-66 on exploration outside North America, against US$750 million for Jersey Standard, with the other majors all remaining below the 450 million mark.[34] This divergence widened during the later 1960s. BP estimated that, over the years 1963 to 1970, Shell spent $900.5 million and Jersey 652.6 million, towering over the others with less than $390 million.[35] At first sight the Group's amounts appear misspent because other companies found more hydrocarbons for less money. During 1951-66 Jersey, for instance, discovered oil and gas reserves of 23 billion barrels for a finding cost of 3.4 US cents a barrel; during the later period the company found 15.4 billion barrels at 4.2 cents a barrel. BP topped the league with a splendid figure of 30.9 billion barrels for only 1.1 cent per barrel during 1963-70. Compared to that, the Group's performance appears very poor indeed, only 20 billion barrels at 5.7 cents a barrel during 1951-66, and 13.4 billion barrels for 6.7 cents a barrel during 1963-70. Not only were Shell's finding costs consistently above average for the majors; after 1966 they exploded. During 1968-70, the average nearly quadrupled to 24 cents, leading the Conference to question the sustainability of that level with some insistence.[36]

For all the appearance of inefficiency, the high E&P spending and finding costs made sense given the Group's position and strategy. First of all, the United States absorbed 50-60 per cent, and in some years as much as 75 per cent, of E&P investment. Because of the American import restrictions, Shell Oil had to keep up a large effort if it wanted to sustain its growing market share, at very much higher costs than those in North Africa and the Middle East, where the other majors concentrated their E&P. Second, from the mid-1960s the Group deliberately switched its efforts away from OPEC countries, discouraged by the political climate there and the gradual upward pressure on crude cost as a consequence of tax increases.[37] As a result the Group had, by 1971, the third lowest reserves of all majors, but by far the lowest proportion of reserves in the Middle East. Jersey Standard, BP, Texaco, and Socal had 70-80 per cent of their reserves in the Middle East, whereas for the Group this was about 55 per cent. Conversely, Shell's reserves outside the Middle East were bigger than any of the majors and the company's exploration teams were active in more countries as well.[38] Table 3.2 shows how, in 1966, the Group planned its E&P spending to support a long-term shift to supplies from West Africa and Australasia.[39] Moreover, in its global searches Shell put its money increasingly on offshore, more so than the other majors, as we will see shortly.

As for the contributions of individual countries to the crude supply, Venezuela continued as the Group's mainstay, production there quadrupling between 1946 and 1965 to just over one million b/d, one and a half times as much as that of the US, in second place. That was Venezuela's peak, however; from the mid-1960s supplies levelled out with a downward trend, and in 1972 the US had caught up again, both countries producing around 900,000 b/d. Moreover, by the mid-1960s the Group considered the conditions in Venezuela insufficiently attractive for further exploration activities, which were accordingly no longer actively pursued.[40]

From 1953, Shell also received supplies from Canada, which after a hesitant start rapidly rose to a steady level of 200,000 b/d during the 1960s. The large volumes of natural gas that Shell Canada obtained from the Jumping Pound field generated an interesting diversification. The gas was 'wet and sour' and had to be turned into 'dry-sweet' gas by the removal of hydrogen sulphide, carbon dioxide, water vapour, and natural gasoline before it could be used as either domestic or industrial fuel. At the very time when the Jumping Pound sulphur extraction plant came into operation in February 1952, there was a world-wide shortage of sulphur, providing a double market. Suddenly an optimistic piece of planning could bear fruit. The gas processing plant had been designed so that extensions could be added with the minimum of interruption to its normal functioning. Willing buyers for both gas and sulphur were now easy to find, and by 1959 the plant, fed by eleven producing wells, had already been expanded three times.[41] In the Middle East, Qatar production rose gradually during the 1950s, but very rapidly in the 1960s following the expansion of the offshore operations, peaking at 311,000 b/d in 1973. As for Oman, in 1954 IPC had formed Petroleum Development (Oman) Ltd. or PD(O) to start prospecting in Oman, but it gave up two years later. The Group, convinced that there was oil to be found in the country, had then taken over PD(O), together with CFP and Gulbenkian, and struck oil in 1962. Omani production started in 1967 and increased very fast to 2-300,000 b/d, later even reaching 800,000 b/d.

Table 3.2
The Group's reserves, expected discoveries, and projected E&P spending, 1966 and 1967-1971.

	Reserves 1966 billion barrels	Expected discoveries 1967-1987 billion barrels	Projected E&P expenditure 1967-1971 $ mln
Central/South-America	8.2	1	22
Europe/North-Africa	1.4	1	26.3
Middle East	2.5	6	43
West Africa	1.9	7.3	129.6
Australasia	0.9	3.5	90.3
Total	14.9	18.8	311.2

[15]

Operations in the Niger delta.

Another new production country for the Group was Nigeria. Shell and Anglo-Iranian began joint exploration there during 1936-37, with Shell as the operator.[42] After interruption by the Second World War, exploration resumed in 1948, but it took Shell-BP Petroleum Development Company of Nigeria Limited (Shell-BP Nigeria) until 1953 before it discovered oil onshore. In Nigeria as in Louisiana, the problem of traversing swampland had to be overcome, and there the solution was the Albion-Cuthbertson Water Buffalo, a giant amphibious crawler tractor. Built by Albion Motors Ltd. in the UK, this invention was able to pull twenty-eight tons of machinery while spreading the load so widely and delicately that the pressure exerted on the boggy ground was less than that of a man. In 1956 Shell-BP also found oil offshore, in the Niger delta and the coastal area east of it. 'We are sitting on the outside of a safe, very hopeful,' said one senior Shell executive, 'but we have not yet got the key to see what is inside.'[43] Nigerian onshore production commenced in 1958.[44] In 1959, when a fifty-fifty profit-sharing agreement was enacted in the Nigerian parliament, 6,000 b/d were produced, accelerating in a rising curve to 178,000 b/d in 1966.[45] Had it not been for the Biafran War – Nigeria's brutal and disastrous civil war of 1967-70 – production would undoubtedly have reached new records. Instead, there was a decline as dramatic as the former increase. Exports ceased in July 1967, and in 1968 production crumbled to 22,000 b/d. This revived very rapidly once the war was over, production reaching 620,000 b/d in 1972, the Group honouring its contractual obligations, undertaken in a democracy, in a completely different political climate: the war marked the end of Nigerian civil government until 1979.

Two further outstanding features of this period cannot be read from the data presented above, and those are the increasing use of secondary recovery techniques and the rapidly expanding offshore operations. Taking the recovery techniques first, Shell Oil appears to have been a pioneer amongst Group companies when, in 1949, it started experimenting with various methods such as injecting hot liquids and later steam into the wells so as to increase the fluidity of viscous oil and coax it out, the main object being to raise crude yields from maturing wells and prolong their economic life. A pilot project at Benton (Illinois) proved a great success, resulting in 1 million barrels (nearly 140,000 tons) being recovered within two years. By 1960 secondary recovery had become increasingly important for Shell Oil, and the company had even diversified into building water supply systems for such projects for other oil companies. In 1966 Shell Oil obtained 15 per cent of its crude from secondary recovery, which was expected to rise to 22 per cent by 1973. Meanwhile the Group had also introduced the technique at its Schoonebeek operations in the Netherlands, and in Venezuela.[46]

[16]

[17]

Operations in the Niger delta in the early to middle 1960s were made difficult by the natural terrain, consisting mainly of waterways separated by mangrove swamps. Clockwise from left, a seismic shot on dry land; other members of a seismic party; a drilling crew in the swamp; and a seismic party moving off to a new location.

[18]

The offshore challenge Offshore held great attractions for the Group. There were huge prospects in areas around the world with little risk of political interference; the higher costs could be partly compensated by getting better concession terms; and it was a new frontier offering great opportunities for experimenting with new technology.

Shell Oil drilled its first successful offshore well, in a joint venture with Pure Oil, as early as 1937 on a site one mile off the Louisiana coast, in the shallow waters south of the Mississippi delta in the Gulf of Mexico.[47] After the war Shell Oil resumed operations there in 1949, two years after Shell Venezuela had drilled its first wells in Lake Maracaibo. In 1950 Shell Oil brought in the South Pass Block 24 field, judged to be of similar importance to the legendary Coalinga and Signal Hill fields.[48] The early offshore operations were conducted either in marshland from moveable drilling rigs mounted on pontoons resting on the ground, or in shallow waters from rigs on platforms raised above sea level from a submersible section or from stilts resting on the seabed. In addition, there were fixed platforms resting directly on the seabed and anchored with piles driven through the legs into the seabed. Meanwhile Shell Oil had also joined, in 1949, a study group for offshore geological and geophysical study formed by Continental Oil (later Conoco), Union Oil, and Superior Oil. This CUSS consortium, after the initials of the participants, concentrated on research along the Californian coast, for which it developed innovative technology, particularly in ship design.[49]

These methods worked well in water up to 200 feet or about 70 metres deep. In 1962, Shell Oil introduced a deep floating platform capable of working in waters of up to 1,000 feet (300 metres) and able to drill and complete wells on the ocean floor at that depth using a remote-controlled robot with a television camera. The platform rested on 100-feet (30 metres high)

[21]

Some of Shell's earliest offshore exploration was undertaken not at sea but in the waters of Lake Maracaibo, Venezuela. In 1948, at Laguinillas on the western shore of the lake, Compañia Shell de Venezuela operated a harbour for launches and other small vessels. Behind its jetty (left) can be seen some of the hundreds of rigs that dotted the lake, one of which is in the process of being erected (right) by a giant crane perched on a barge, about 1950. The building on the barge shows the scale.

[22]

[23]

stabilizing columns attached to a submersible structure with ballast tanks. Once in place, the ballast tanks would be filled with water and sunk to a depth of about 40 feet or 13 metres, where wave action is very low. Eight mooring lines kept the platform in place, with special instruments ensuring a minimum of movement away from the wellhead.[50] Shell Oil had earlier, in 1961, launched the *Eureka*, an exploration ship with an innovative automatic positioning system through controlled propulsion which kept the vessel in place during exploratory drilling.[51] Shell Oil also funded an entirely new type of ship from CUSS, the first purpose-built exploration vessel, which was launched in 1962.[52] Shell Oil pursued offshore E&P with vigour. By the mid-1950s, the company already drew 52,500 b/d, 12 per cent of total production, from operations in the Gulf, nearly a third of the industry total; this had risen to 274,000 b/d or 43 per cent of supplies in 1972.[53] Following the exploration with CUSS, Shell Oil also undertook offshore operations on the coast of California, but strenuous efforts undertaken over a very long period and at great cost failed to yield commercial results in Alaska. BP's discovery, in 1969, of the large Prudhoe Bay field caused considerable consternation at Shell Oil, which had no leases in that particular area because geologists had considered it insufficiently attractive.[54]

Shell Oil's overall offshore successes stimulated the Group to undertake similar ventures around the world. In 1954 a 1,400-ton moveable drilling platform set to work five miles off the coast of Qatar; and in 1961 the Dutch operating company NAM used another, named *Triton*, to start drilling in the southern North Sea off the Dutch coast. The Qatari platform came to grief during the night of 27/28 December 1957 when, having just completed a 12,000-foot well (its second, which like its first was dry), it was wrecked in a storm. Twenty-two men lost their lives and the platform and ancillary equipment, together worth £900,000, were a total loss, forcing a review of the Group's entire offshore Qatari programme. Despite the tragedy the decision was made that it should proceed,

[24]

To create the necessary sound-waves, dynamite was often used in offshore seismic exploration, as in this sequence from 1953. From left to right, crewmen show considerable familiarity with the explosives; ready to fire; the resultant explosion; and studying the data for a preliminary interpretation.

[25]

and in October 1959 a new platform, named *Seashell* and built by NV Gusto Shipyard at Schiedam, began its 6,400-mile journey from the Netherlands, towed by two Dutch tugs. Here we see one of the forerunners of the great fixed platforms that later sprouted in the North Sea, for although Seashell was still limited to only 90 feet of operational water-depth, it was designed to drill several wells from one site to a depth of 15,000 feet, and to withstand gales of up to 100 miles an hour and waves up to 30 feet high. Also in 1961, offshore exploration began on the coast of Sarawak, while the Group obtained a concession for offshore exploration in Kuwait, where drilling commenced in 1962. Two years later, a long-term E&P review concluded that Shell urgently needed to devote more funds to offshore exploration. Because it would become increasingly hard to find sufficient reserves to match the pace of growing demand, the higher offshore cost would be likely to prove economical; they were very necessary, too, for without significant new discoveries the Group faced a serious shortfall in supplies during the 1970s. The review included sketches for the equipment required to perform E&P operations in water up to 1,600 feet deep. The innovative solutions including a 'flip-top floating storage', later to become known as a spar.[55]

Consequently, during the 1960s Shell's offshore operations proliferated around the world, boosted by new technology enabling drilling in ever greater depths of water, by the scarcity and rising cost of on-shore concessions, and by the attractions of an environment 'where the opportunities, challenges and risks are great'.[56] Following Shell Oil's successful use of geophysical survey ships, the Group chartered two more ships to carry out worldwide comprehensive geophysical surveys in 1968.[57] Three years later, the Group launched a new type of drilling ship, the *SEDCO 445*, capable of drilling exploration wells in up to 2,000 feet (600 metres) of water to a depth of 20,000 feet or 6,060 metres beneath the seabed.[58] The offshore operations received another boost when Shell decided to direct its E&P investment away from OPEC

03|198

[27]

Under construction in Schiedam in 1959, the *Seashell* mobile drilling platform (main picture) replaced an earlier unnamed platform that had been wrecked off the coast of Qatar. Left, after its commissioning ceremony, the Ruler of Qatar Sheikh Ali bin Abdulla bin Jassim Al Thani leaves the platform, followed by his son Sheikh Ahmed bin Aku Thani, his representative with the oil companies.

countries. From the early 1960s, the Group began directing its E&P research on the one hand towards operations in ever greater depths, and on the other, towards lowering offshore costs by developing the seabed completion of wells, floating production and storage facilities, high-sea terminals, and ways to perform human activities in deep water. By 1968 Shell ran deep-diving training programmes for production staff and divers and had developed a special rapid decompression technique. Trials were under way in Qatar and Sarawak with the first seabed completed wells. Combined with specially designed equipment for deep water operations these preparations gradually opened opportunities for E&P in waters beyond the Continental Shelf of 600 feet (180 metres) or more. In 1971, the E&P laboratory started research into what it called a spar, a vertical floating container for up to 300,000 barrels to facilitate production operations, crude storage, and tanker loading on the high seas. With the discovery of the Brent field the following year the expectations for the North Sea were amply fulfilled, and the technology developed came just in time.[59]

In 1973, the Group estimated that offshore wells supplied around 20 per cent of the world's crude and 5 per cent of the natural gas. We have not found such figures for Shell's own production, but given its long-standing commitment to this type of E&P, this figure was probably higher, and its concession acreage suggests the same. In 1972, the company held concessions covering 2.2 million square kilometres, of which 46 per cent was on land, and 54 per cent off shore. Two years later, when the first oil shock triggered a scramble for concessions, the offshore share rose further to 43:57. By that time Shell had by far the biggest offshore E&P acreage of the majors; half of E&P drilling and 70 per cent of E&P expenditure went to offshore operations.[60]

[28]

[29]

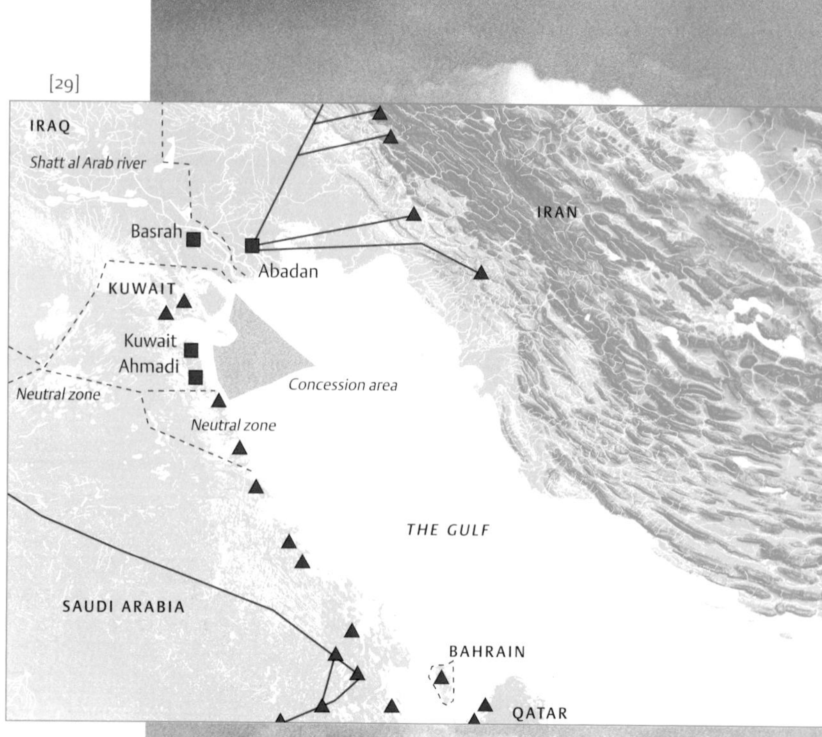

The agreement signed (above) in January 1961 by Fred Stephens and His Highness Sheik Sir Abdullah As-Salim As-Subah, the Ruler of Kuwait, gave the Group a very valuable further access to Middle Eastern oil (concession area mapped, inset).

Offshore concession Kuweit, 1961

— Pipelines
▲ Oil producing area's and/or refineries

Orient Explorer, the first mobile platform constructed in the UK, was hired by the Group for exploration in the high seas off Borneo, and was the first such platform to operate there. Commencing its long journey east in February 1959, it reached its destination in 1960, and oil was found in 1963. Here the platform is seen being towed through the Suez Canal.

Shell Oil's overseas ambitions

The Group's E&P successes in the Middle East and Africa, and notably the low production prices in the Middle East, did not have uniformly beneficial effects, for these tended to highlight the isolation of Shell Oil, locked into the protected American market. The growing divergence between American and world crude prices put the company at a disadvantage, both in respect to other Group companies, and in respect to the American majors, which used their Middle Eastern supplies for a rapid overseas expansion. Moreover, the low upstream costs in the Middle East radically cut their average exploration and production cost per barrel, which gave them more to spend on E&P in America. Shell Oil executives regarded this as a penalty for their membership of the Group, for which they wanted compensation.

In December 1957, Shell Oil president Max Burns wrote to Loudon arguing that, in order to remain competitive, Shell Oil needed access to cheap oil and, with costs in the US rising, would have to start looking for oil abroad. Since he expected that it would take some ten years to find a suitable supply, he wanted to start planning now. Among the areas of specific interest, Burns mentioned southern Nigeria. According to him, the area closely resembled the Louisiana Gulf coast and offshore waters with which Shell Oil was well acquainted. The joint venture between Shell and BP held leases there which would have to be surrendered in 1959, so exploration had to start soon. Shell Oil had experienced staff available; in collaboration with Bataafsche operations could quickly get under way.[61]

This letter was part of a campaign which, by the time it arrived in London, had already been going on for some time.[62] As early as 1948, Shell Union had participated in a bid for a concession in Kuwait's Neutral Zone which failed for reasons unspecified.[63] The company had then dropped the matter, since any overseas venture would have diverted resources from domestic opportunities.[64] Group managing directors were opposed in principle, wanting Shell Oil to continue drawing any supplies needed from Group companies rather than from its own E&P. Allowing Shell Oil to obtain crude from third parties would of course have run counter to the principle of integration and might also have inspired operating companies elsewhere to claim the same right to buy crude from the cheapest supplier, rather than rely on the Group E&P which, with the rising flood of Middle Eastern crude, was increasingly perceived as expensive. In any case in January 1958 Shell Oil withdrew its request, considering exploration outside the US now 'unattractive', presumably because by then the conclusions had emerged from a study into how, without direct access to cheap overseas crude, Shell Oil could still be put in a sufficiently competitive position. Focusing on finding financial compensation of some $20 million, the report of the investigation had presented an exercise in fiscal engineering. Under American law, taxes paid to governments abroad over crude purchased there could be offset against taxes due in the US. To obtain maximum advantage, Shell Oil was to purchase foreign oil in such a way that as much as possible of the total cost consisted of tax payments. This was achieved by having Shell Oil set up a trading company in Kuwait, Shell Enterprises, to buy crude at cost price from Gulf under the Group's agreement with that company, and shipped also at cost price. The Kuwait oil did not go to the US, however, for Shell Oil did not at all want to take the high-sulphur crude into its system. Half the oil was sold back to the Group in exchange for sweeter, i.e. less sulphurous, Venezuelan crude; the other half was sold, at posted prices and full freight rates, to Shell Canada, which then passed it on to one of the European companies. It took a great effort to get

Printed in Shell's internal *Long-term Exploration and Production Development Review, 1964*, the diagrams below were created at a time when it was only possible to drill in relatively shallow waters. As preliminary plans for working in much deeper waters, they are a striking demonstration of the Group's habit of thinking beyond the technical limitations of the day.

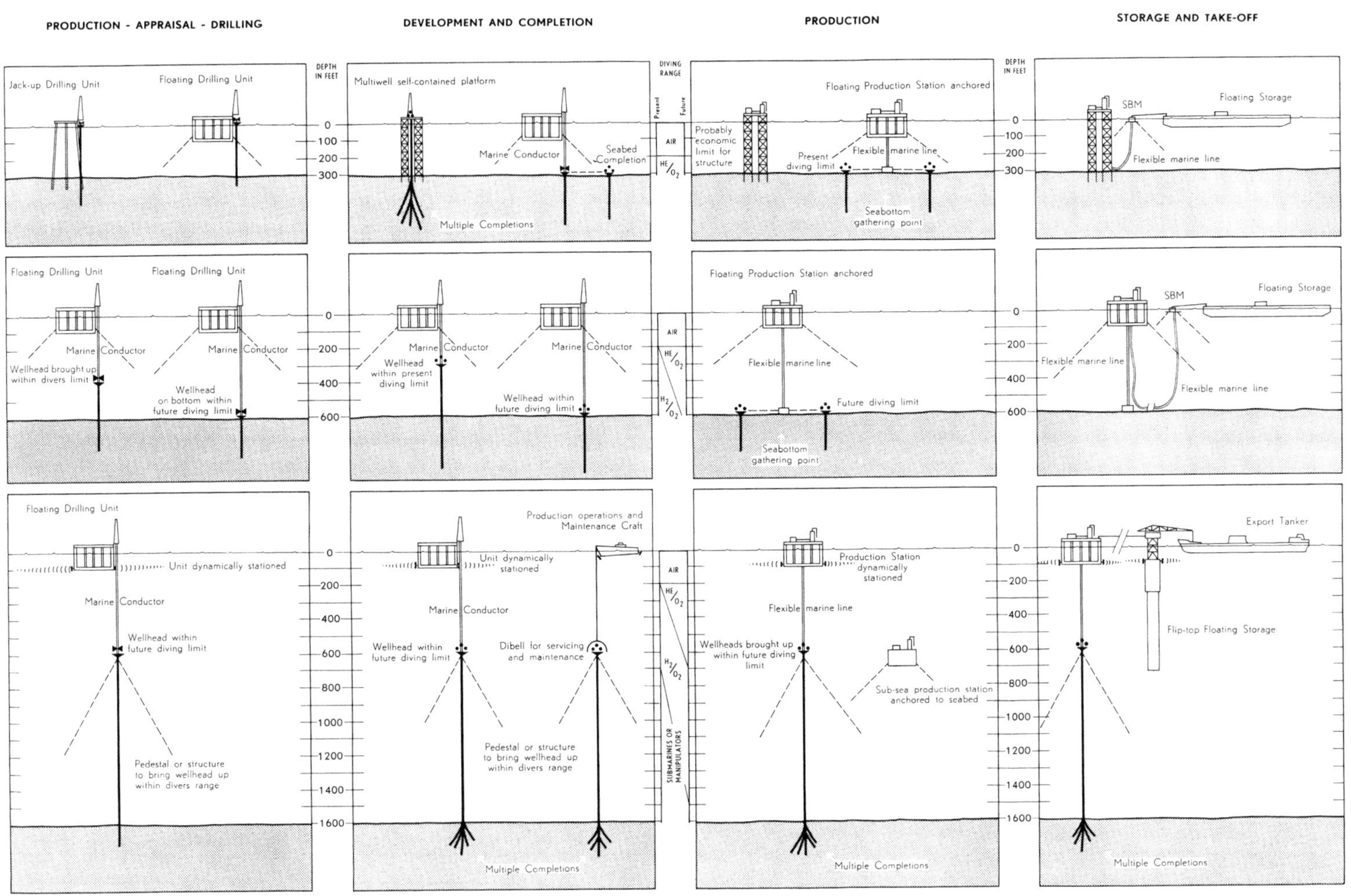

Gulf's agreement, the company harbouring deep suspicions about the motives behind the construction, but in July 1958 Shell Enterprises could finally start trading.[65]

This elaborate charade would have delighted Deterding who, in an earlier phase of the Group's history, had excelled in devising convoluted schemes to avoid tax, manage profit flows, or compensate lost earnings. One would not have expected to find it in a more mature corporation, however. The whole episode serves to demonstrate how shallow Shell Oil's ostensible stand-alone position within the Group really was. Within two years the arrangement ran into difficulties. Shell Canada no longer wanted the Kuwait crude from Shell Oil, since independent traders offered shipments at lower prices; Shell Oil itself came to doubt whether tax inspectors would accept its position, trading crude in a carousel with three Group companies. The scheme was then amended to have Shell Oil buy directly from BP or from Gulf in Kuwait. By 1966, the CMD considered that the whole arrangement had lost its purpose.[66] Three years later Shell Oil's supplies became subject to legal scrutiny in the Halpern vs Barran case (see Chapter 2). One of the complaints by the plaintiffs alleged that Shell Oil had had to buy oil substantially above market prices from Group companies, rather ironical considering that the Shell Enterprises scheme had been designed to boost Shell Oil's profits. However, the pricing of supplies to Shell Oil was not always and under all circumstances fully competitive; as we have seen above, Group managers were equally concerned about that.[67]

With the end of the Kuwait arrangement Shell Oil once again raised the question of expanding its upstream activities outside the US. Shell Oil's total E&P spending, i.e. capital expenditure plus running costs, had nudged steadily upwards, from around three dollars a barrel during the late 1950s to about 3.80 in the mid-1960s.[68] This cost was clearly considered an upper limit given the prevailing market circumstances. A memo from April 1966 argued that it had become very difficult to find viable propositions for that price within the US, so Shell Oil had cash to spend but lacked suitable E&P opportunities.[69] Moreover, during the same decade Shell Oil's share of Group capital expenditure on E&P had risen sharply, from some 60 per cent to over 80 per cent, whereas its share of Group crude production had dropped from more than 20 to less than 16 per cent. Shell Oil's E&P performance thus had wider relevance, because the Group as a whole was spending more but getting less.[70] If this trend continued, Shell Oil expected to face a crude deficit by 1975.[71] The CMD rejected Shell Oil's request in June 1966. Foreign ventures no longer offered any real tax benefits. The company's crude position did not really need propositions abroad either, they would simply be alternatives to domestic ones. Moreover, the wider Group interest would lose any E&P advantage that Shell Oil might gain outside of North America. The question would be reconsidered as and when the company began to run out of viable E&P prospects in its home territory. Monty Spaght (Shell Oil president 1960-65 and thereafter the company's chairman until 1970) accepted this conclusion, but sounded a warning by pointing out that it would only be palatable to the outside shareholders 'as long as Shell Oil's import requirements of some 50,000 b/d was being acquired at competitive prices, and some concern had therefore been expressed at the forecast deterioration in the economics of the present supply arrangements'. The meeting agreed with him that Group supplies to Shell Oil had to be 'fully competitive', evidently not something which was a matter of course.[72]

One may doubt, however, whether Shell Oil really wanted to set up E&P operations outside of the US. In a 1975 interview, Spaght

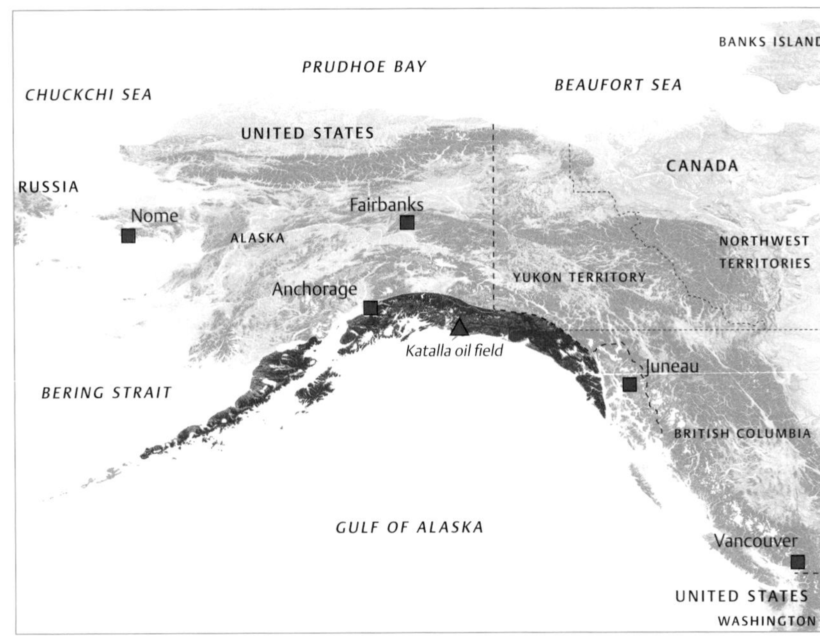

[32]

General area being explored by Shell

Southern Alaska experienced a classic oil rush in the late 1950s. In the more arduous north, despite three years of effort and expense, Shell failed to find oil and suspended exploration there in 1959. ARCO and Humble together made the first discovery in Prudhoe Bay on 26 December 1967, confirmed in 1968. The main competitors thereafter were ARCO, Jersey and BP, with Shell and others also present. Prudhoe Bay proved to be the largest field ever found in North America, the 18th largest in the world, and possibly the most environmentally contentious of all.

emphasized that it had not. According to him the Group had offered it twice, and both times Shell Oil had turned it down wanting to concentrate resources on domestic opportunities.[73] Spaght was quite outspoken about the matter, commenting that 'In my time, it was quite clear it was too late; there was nothing I ever saw that I wanted to go overseas for. And, the fact that Shell Oil is overseas now, is probably awfully late in the day. What one finds overseas today is of very questionable value.'[74]

This suggests that Shell Oil's effort to get access to overseas oil was part of a game between the company's executives and the CMD about Shell Oil's overall position and financial relationship with the Group, rather than a serious objective on its own.

But the matter now had a momentum of its own, to which outside events added weight. The gap between US oil consumption and its own domestic production continued to widen, rendering the lifting of the existing import restrictions more and more likely at some date in the future despite new and promising oil discoveries in Alaska. Shell Oil's own gap between sales and production also continued to widen. In 1969, managers accepted that there would at some time in the future be a need for Shell Oil to undertake E&P operations outside America, and decided to review the question every year.[75] A year and a half later, the CMD reviewed a long-term US energy forecast and concluded that 'it is more than ever necessary to plan the Group's future requirements on a worldwide basis, including Shell Oil Company; and that Shell Oil Company cannot continue to operate as a self-sufficient integrated domestic company'.[76]

From a Shell Oil problem, the E&P situation in the US had finally become a Group concern, and Shell Oil at last received permission to start operating outside North America. In March 1971 the company signed the contract for its first E&P project abroad, a joint venture with Continental Oil Company and the Colombian state oil company Ecopetrol. Propositions in Ecuador, Peru, Senegal, Somalia, Papua New Guinea, South Korea, and Gabon followed.[77] As Spaght had foreseen, none of these ventures had more than marginal success. Talks with National Iranian Oil Company about acquiring crude and product supplies in return for downstream facilities in the US did not yield results either. Given Shell Oil's lacklustre enthusiasm for overseas exploration, this is hardly surprising.[78]

[34]

[35]

More familiar in Texas than in Europe, 'nodding donkeys' and other paraphernalia of the oil industry (main picture and above) became a common sight in Schoonebeek, the Netherlands.

Strategic partnerships: NAM and Gulf Before the Second World War, the Group conducted most E&P operations on its own, but after the war strategic partnerships assumed ever greater importance. Indeed, Shell's main Middle Eastern operations, in Kuwait, Iran, Iraq, Qatar, Abu Dhabi, and in Libya, were all strategic partnerships; and in addition, the E&P in the Netherlands, the overture to the North Sea oil boom of the 1970s, was also a joint venture. We will look in some detail at two of them. Whereas the Group took part in the Middle Eastern partnerships with enthusiasm, it entered the Dutch one most reluctantly.

The Netherlands partnership came about with the formation of NV Nederlandsche Aardolie Maatschappij (NAM), incorporated in September 1947 as a 50:50 joint venture between the Group and Jersey Standard, with Bataafsche as the operator. Behind this lay a history that, as Jersey acknowledged, reflected well on the Group. In 1933 the two enterprises had discussed the possibility of a Netherlands joint venture and the Group had offered its agreement in principle, but Jersey had 'inadvertently' (its own word) failed to accept until 1938.[79] The agreement made then was that Bataafsche would explore in the Netherlands and Jersey in Cuba, and each would share whatever was found equally with the other. Jersey found nothing in Cuba and the agreement lapsed from memory, but with Bataafsche's discovery of the commercially viable Schoonebeek field near Coevorden during the German occupation, one of Jersey's employees remembered the 1938 agreement. In the intervening years Jersey had cancelled various agreements, which Shell understood to include the Netherlands' joint venture proposal. Jersey's understanding was different. The letter describing the 1938 agreement was unearthed and presented to the dismayed Group, and a confirmatory copy was found in

[37]

13th March, 1946

My dear Wilkinson,

Further to my cable Close 216, I would inform you that the Hague file containing the 50/50 agreement with Jersey regarding oil development in Holland has now been found.

As you will see from the attached copies of the most important documents in this file, it is quite certain that such an agreement was made at the time. There is no actual letter from Jersey New York in this file, Bolton's letter being the last, but this is clear enough.

Moreover, shortly before his recent return to the States, Bolton had a meeting with Leyds, the actual subject being Germany, but on that occasion he made it perfectly clear that he remembered the agreement regarding Holland.

Yours sincerely,

(Sgd.) D. TH. W. VAN HASSELT

[38]

Chapter 3

Bataafsche's files.[80] Shell also found 'considerable confusion in the minds of the Jersey Production Department as to their policy'.[81] In New York, Wilkinson 'emphasized (...) with the utmost forcefulness' to Orville Harden, one of Jersey's vice presidents, 'that this matter had been badly handled by their people, and (...) had added considerable confusion and hard feeling, and of this Harden was very appreciative'.[82] The Group managing directors eventually concluded that 'Jersey's argument boils down to their coming in where they think we have got something and staying out where this seems to suit them better. However, all this is now water under the bridge and I still think we must abide by our decision to accept Jersey as partner with the best possible grace.'[83]

Viewing NAM as its 'most promising venture' in Europe,[84] Jersey was impressed by the Group's honesty. Shell could easily have said that the second copy of the 1938 letter of agreement had been lost or destroyed; its files had been moved several times

Jersey Standard was impressed when Shell allowed Schoonebeek to be a joint venture, honouring an old and half-forgotten agreement. Above and right, production and drilling; top right, Van Hasselt's letter to Wilkinson in New York, confirming the existence of the agreement.

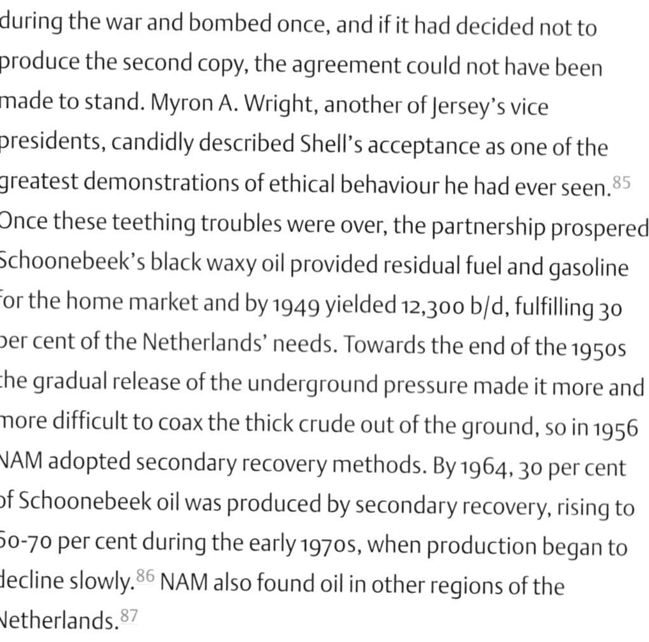

during the war and bombed once, and if it had decided not to produce the second copy, the agreement could not have been made to stand. Myron A. Wright, another of Jersey's vice presidents, candidly described Shell's acceptance as one of the greatest demonstrations of ethical behaviour he had ever seen.[85] Once these teething troubles were over, the partnership prospered. Schoonebeek's black waxy oil provided residual fuel and gasoline for the home market and by 1949 yielded 12,300 b/d, fulfilling 30 per cent of the Netherlands' needs. Towards the end of the 1950s the gradual release of the underground pressure made it more and more difficult to coax the thick crude out of the ground, so in 1956 NAM adopted secondary recovery methods. By 1964, 30 per cent of Schoonebeek oil was produced by secondary recovery, rising to 60-70 per cent during the early 1970s, when production began to decline slowly.[86] NAM also found oil in other regions of the Netherlands.[87]

By then oil had long become a secondary concern to NAM. In 1948, the company had found some natural gas near Coevorden, which was then sold to that city's gasworks, but in 1959 NAM had found an enormous natural gas field at Slochteren in the province of Groningen.[88] When the news reached Jersey's head office in New York, a senior executive is reported to have shouted 'I'll buy it all'

Secondary recovery techniques became ever more valuable, in order to maximize output from an oil well or oil field. Left and right, steam produced by these ovens was injected into a well to make thick viscous oil more fluid and easier to extract.

and instructed his assistant to take the necessary steps, only to be told that Jersey already owned half of it.[89] With estimated reserves of 1.6 billion cubic metres, the Slochteren field was claimed to be the world's biggest.[90] After extensive negotiations with the Dutch government, the concession for exploiting the gas was given to a partnership called Maatschap Groningen, in which NAM participated for 60 per cent, the Dutch government for 40 per cent through DSM, a state-owned mining and chemicals company. In combination with the taxes levied on Maatschap, this construction gave the government 70 per cent of the profits. In 1963 the two partners set up a joint venture with the Dutch government, Gasunie, for building a network of pipelines and distributing the gas. Shell and Esso held 25 per cent each, DSM 40 and the state 10 per cent. As the Gasunie network spread over the country, Dutch households, industry, and other intensive energy users such as greenhouse growers switched to gas on a massive scale. By 1974 the Group sold more gas in and from the Netherlands than it did in the United States.[91]

The Slochteren discovery started a flurry of exploration within the Netherlands and on the Continental Shelf in the North Sea. In the summer of 1963 the German joint venture with Jersey Standard, Brigitta (Shell 50 per cent) proposed starting

Drukkerij & Publiciteit
BERNAERTS N.V.
aardgas
voor
warmte, komfort
en gezelligheid.
GEEN BEROEPSLEGER
JONGSOCIALISTEN
GEEN BEROEPSLEGER
ABORTUS VRIJ
JONGSOCIALISTEN
ABORTUS VRIJ
6

[41]

The Netherlands' bonanza – natural gas for everyone. Below, the gas distribution network soon covered most of the country.

— Gas transport network
▲ Gasfields

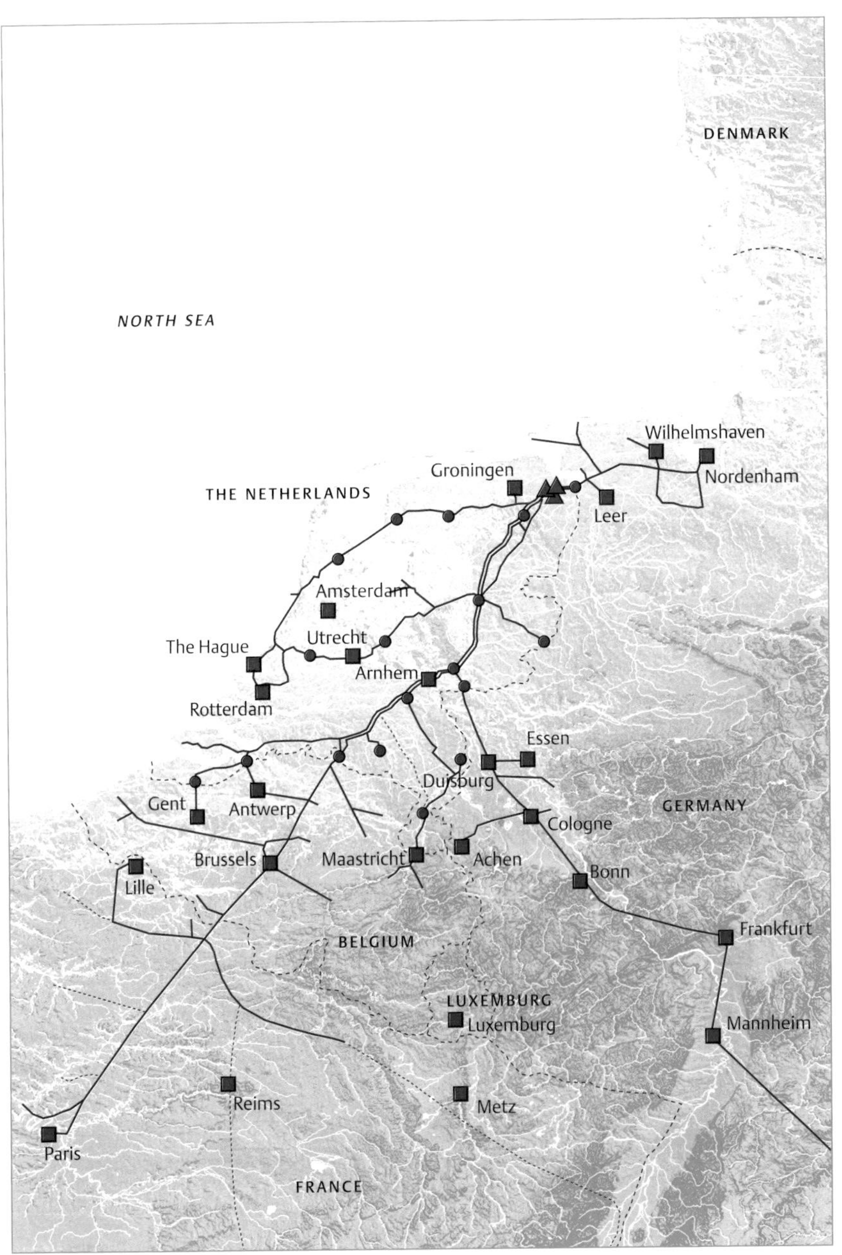

[42]

negotiations to join a ten-company German offshore consortium.[92] In the autumn of that year a partnership between Shell, Gulf, and A. P. Møller gained from the Danish Government rights over its Continental Shelf the first grant of rights over any offshore area of the North Sea.[93] Finally in 1964 the Group formed a fifty-fifty joint venture with Jersey's European affiliate Esso, which was operated by Shell UK Exploration and Production Ltd., or more simply Shell Expro. The agreement that brought it into existence was only sixteen pages long, but the venture's purpose was to explore for oil and gas in the North Sea; and before long it developed into the largest capital investment made by either of the partner companies anywhere in the world.[94]

The venture could have been a tripartite one, because, acting on the premise that the Netherlands' gas-bearing geology could extend beneath the sea between Holland and England, Shell, Esso, and BP together had already undertaken a joint geophysical survey of the southern North Sea basin.[95] However, BP withdrew from the arrangement in December 1963, being unwilling to tolerate Shell's restrictions on the exchange of information with companies outside the survey.[96] Moreover, BP rightly guessed that its record as a direct contributor to Britain's fuel economy would bring it some preferential treatment in the auctions for blocks.[97] However, in the first round of six-year licensing in September 1964, when 380 out of 1,000 blocks, each of 100 square miles, were on offer, the joint venture with Esso operated by Shell Expro received seventy-five of the ninety licence blocks it had applied for, and acquired overall more than twice the area granted to any other applicant. There was an interesting division between the blocks north and

A pipe-laying ship on the North Sea – a small indication of the vast quantities of material needed in building the network.

National sectors of the North Sea

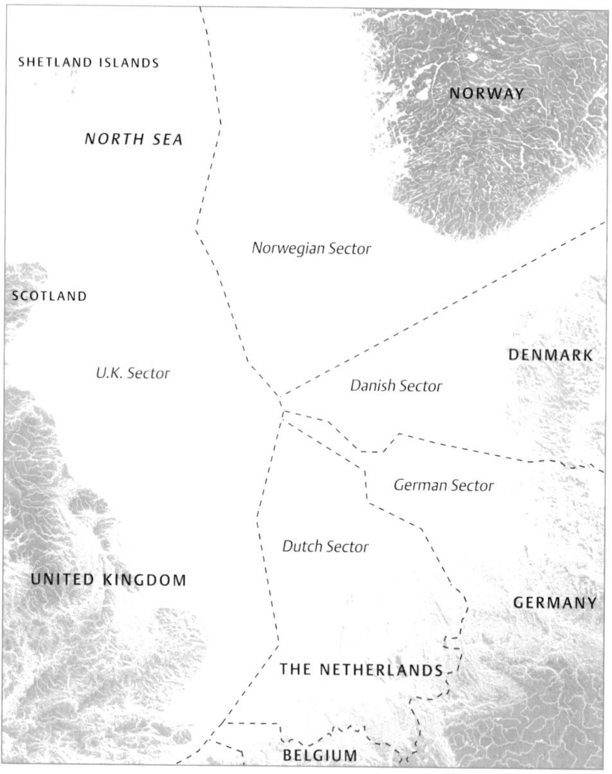

[43]

south of 56 degrees north, the line of latitude running from Edinburgh to Denmark. South of the line the Group applied for forty blocks and received twenty-five; north of the line, which was universally seen to be a more difficult and assumed to be a less promising environment, it applied for fifty and received them all. Even Shell considered the southern North Sea to be 'more interesting' than its northern half, but no one suspected the enormous fields that lay hidden in the north.[98] As those began to unfold later in the decade, Shell Expro's pre-emptive bids proved their worth. At the end of 1965, Shell still did not expect that the North Sea would add much to its crude oil reserves, but by 1970 eleven fields had been found in the North Sea and Shell had ordered three new submersible rigs and one deep-water jack-up rig, all specially designed to cope with the rough conditions, to expand operations there.[99] Due to the difficult natural conditions, the capital investment costs in the North Sea were ten times higher than those in the Middle East, but the lack of political risk alleviated the difference.[100]

During the 1950s the Group's involvement with natural gas deepened in another way as well. In 1959 Constock International Methane Ltd. converted a cargo carrier into the first ship for transporting liquefied natural gas (LNG), the *Methane Pioneer*, which took successful trial cargoes from the US to Britain. Making natural gas liquid requires a temperature of minus 165°C, so transporting LNG meant building a ship and installations able to withstand such deep refrigeration. Understanding the future prospects of LNG shipping, Shell acquired a 40 per cent interest in Constock, which was renamed Conch International Methane.[101] In 1964, Conch accepted two new and purpose-built LNG carriers, *Methane Progress* and *Methane Princess*, which brought LNG from

These evocative images of work on the North Sea rigs show the combination of massive engineering and sheer muscle-power that was essential for the job.

[47]

Shell delivers LNG to the Sodegaura terminal in Japan (left), the world's largest LNG receiving terminal. Right, inside the cathedral-like vault of one of Sodegaura's storage tanks.

[48]

Algeria to the UK and France. The Group then went on to develop its first major LNG project in Brunei for contracted customers in Japan, which turned out to be a technological as well as financial success. In December 1971 Shell delivered its first shipment of LNG to Japan.[102] Capturing and shipping LNG at last offered an alternative to the practice of flaring associated gas in remote areas.

The other partnership of 1947 provided the cornerstone of Shell's post-war revival. Group geologists failed to recognize the potential of the Middle East, turning down concession offers in Saudi Arabia during the mid-1930s because they were convinced that there could be no oil there (see Volume 1, Chapter 4). In 1938 a joint venture between AIOC and Gulf, KOC, struck oil in the Burgan field and began to produce prodigious quantities after the post-war reopening of the shut-in wells. Towards the end of 1945 Gulf approached Shell to discuss ways of marketing its share of oil. After two years of negotiations, the companies signed a complex agreement which effectively gave the Group a half share in Gulf's Kuwait production, initially for ten years, which in 1956 was extended for another seventy years, until 2026.

The Kuwait contract gave Shell an invaluable asset.[103] As early as 1948, a member of staff could hardly believe the Group's 'outstanding' good fortune in being party to 'the prodigious bounds being made here by the KOC. After a brief three years of active life, production is overtaking AIOC after as many decades (...) Thirty-six wells producing 20,000 barrels a day'.[104]

The intricate profit-sharing terms of the contract welded Gulf and the Group into very close partners over Kuwait.[105] The complexity of the contract derived from the need to lay down profit-share arrangements for an enterprise with completely unknown, but potentially huge, ramifications. In addition, the Group wanted to ensure that Kuwait oil would not flood its markets. Gulf's initial proposition involved its oil being actually tracked and identified through Shell's network, which was impossible. The two companies therefore adopted a profit-sharing scheme covering the entire supply chain from wellhead to consumer and based on a 'realization figure' – a theoretical calculation of the profit margin to be shared between the parties, following rules set up in the agreement and based on Shell's actual market proceeds and costs in selected territories and refineries. At the same time, Gulf undertook not to increase its market share in areas where the Group sold Kuwait oil, and the Group agreed to process Kuwait crude for Gulf in certain refineries such as Pernis.[106]

Though from time to time the partners bickered over clauses and interpretations, the overall relationship between the companies remained good, indeed so much so that, in 1954, Gulf suggested that Shell should buy out half Gulf's interest in the concession. Both sides studied the matter carefully, but it failed because BP had the right of first refusal on any such sale.[107] Moreover, the profit-sharing formula used proved sufficiently workable to serve as a model for the contract between the Group and Amerada about Libyan oil. This deal was also a compromise, Shell wanting an outright purchase of the concession concerned from Amerada, and the American company preferring to sell half its concession plus a specified but variable volume of oil from it for a profit share. The preliminary agreement was signed in 1963, but it took another three years of negotiations before the Group began receiving crude.[108] Having the Amerada deal was important for the Group, for Libya Shell was unsuccessful in finding any crude and ceased exploration in 1965, having spent over £22 million without result.[109] Production rose rapidly from 105,000 b/d in 1966 to 372,000 five years later, but then tapered off, the Group receiving only 96,000 b/d in 1974, when Libya nationalized its oil industry.

Confronted by nationalism in Egypt and Iraq The Group also lost access to oil through nationalization in Egypt and Iraq. Having expanded production nearly sixfold during the war, Anglo-Egyptian Oilfields Ltd. enjoyed continued success during the late 1940s, striking oil in 1946 and again the following year, in a joint venture with Socony Vacuum.[110] Output peaked at 34,000 b/d in 1951 and then fell abruptly, reaching a low of 7,000 b/d in 1956, when, following a military coup which ultimately brought Gamel Abdel Nasser to power, the political climate in Egypt changed profoundly. The Egyptian government, formally independent from Britain since 1936, had already shown itself eager to get a grip on the foreign oil companies, which resulted in Jersey Standard giving up its E&P operations there in 1948 because of the deteriorating conditions. For the same reasons, the Group had curtailed its investments and operations. The necessary overhaul and extensions to the Suez refinery were postponed until better times,

[49]

and at the same time Anglo-Egyptian accelerated its programme of staff regionalization. The foreign oil companies had subsequently accepted the government's demand for a price agreement, which included a provision for the companies to transfer their head offices to Egypt, but once Anglo-Egyptian had fulfilled its obligation, the government had cancelled its commitment to the agreed prices in 1951.[111] The military coup failed to quell the political unrest; public protests and outbreaks of violence against the oil companies continued, inspiring the Group, coordinating closely with the British Foreign Office, to prepare plans for the evacuation of expatriate staff and their families.[112] In 1954, the Egyptian government forced the oil companies into accepting the handling and distribution of Russian oil, acquired under a barter agreement, by threatening to expropriate their installations.[113] That same year, Britain withdrew from the canal zone and transferred the management of its oil handling and tanking installations to Consolidated, Shell's marketing joint venture with BP for Africa.[114]

In 1956 the Egyptian government responded to the invasion by Britain and France with the sequestration of all the Group's assets. Anglo-Egyptian Oilfields, Shell Chemicals and the Group marketing company Shell Company of Egypt together constituted the largest single unit of British investment in Egypt, valued at about £40 million. However, the sequestration was not an anti-Shell but an anti-British measure. Through painstaking private diplomacy and talks in neutral Italy, the Group reached agreement with the Egyptian government about a return of its assets in December 1958, a full year before the British and Egyptian governments re-established diplomatic relations. As part of the agreement, the Group accepted working under an official oil import monopoly; in accepting the deal, the CMD considered the return of the properties more important than the likely profitability of the business.[115] Anglo-Egyptian Oilfields received new

[50]

Symbolic of rising anti-western Arab nationalism, President Nasser (left) leads his guest Nikita Khruschev, leader of the Soviet Union, through the Temple of Luxor to see the Aswan High Dam under construction on 21 May 1964. President Abdul Salam Arif of Iraq is present on the left of the picture. Right, workers in the Suez Canal zone are searched by soldiers.

[51]

concessions and all Shell's Egyptian properties were desequestered in stages, the last in July 1959. Production resumed at a modest level of 2-3,000 b/d, but the revival of business proved short lived: in 1964, after a last-ditch attempt offering Nasser a 50-50 joint venture, the Group's Egyptian businesses were nationalized.[116] Again this move was not anti-Shell but anti-British, prompted by Britain's continuing support of Israel, where Shell had ceased operations in 1958. Some other foreign-owned oil interests in Egypt remained intact, and there was 'no doubt that Shell has suffered this discriminatory treatment in Egypt because of its partly-British nationality'.[117] Compensation from the Egyptian government was agreed in the summer of 1966 at £9 million, payable in quarterly installments over eight years, and with effect from 1 July 1966.[118] The first three payments were made smoothly until the explosive outbreak of the Six Day War between Israel and Egypt (5-10 June 1967), when payments were frozen on an account at the Bank of Alexandria pending the reopening of the Suez Canal.[119] Nonetheless, the Group worked patiently to restore Egyptian confidence, successfully developing new business relationships (especially with its old company, renamed Misr Petroleum), and in 1974 was able to recommence exploration there, acquiring offshore and onshore acreage.[120]

During the Arab-Israeli Six-Day War in 1967, Shell suffered on both sides of the divide. Left, Shell's refinery at Port Suez blazes after an Israeli attack, and right, journalists and oil company employees examine damage to pipelines near the city of Irbid in Jordan.

[52]

The Group's output from Iraq stood at more than 21,000 b/d in 1945, but post-war export was hampered by the regional political situation. IPC could neither resume pumping to the Shell-BP refinery at Haifa, nor complete the second pipeline through the land that had become Israel. However, its Iraqi production figure remained fairly constant until 1949, when IPC struck oil at Zubair and Nahr Umr in its Basrah concession, and a large steady rise began, reaching a peak of 161,000 b/d in 1955. Much of this passed by pipeline across Syria, but on 3 November 1956, Syrian engineers acting on Nasser's orders retaliated against the Israeli invasion of Egypt by sabotaging the pipeline. The immediate result for IPC was that Iraqi production fell badly for two years.

With the end of the 1956/57 Suez crisis, Iraqi production levels swiftly returned to their earlier level, but in 1961 a new military regime started flexing its muscles. IPC's concession acreage in Iraq was enormous, covering more than two-and-a-half times the size of the United Kingdom, or nearly ten times that of the Netherlands. The company was active on only a tiny part of its concession, which had led to talks with officials about surrendering surplus acreage. The Iraqi government also wanted a 20 per cent government participation in IPC plus a greater than 50 per cent share of profits, neither of which the IPC shareholder companies wanted to grant. By September 1961 Shell glumly considered that there was little reason for optimism about the outcome of these talks.[121] This proved all too accurate: in December 1961, the government unilaterally reduced the concession's area to 750 square miles – just 0.5 per cent of its original extent.[122]

Losing 99.5 per cent of its acreage was not disastrous for IPC, because the company's entire prolific production came from its 750 remaining square miles, but the expropriated area did include a newly discovered and very promising field, North Rumaila. Nevertheless, the reduction and the continuing political violence in the country dented the IPC shareholders' confidence in Iraq.[123] In 1965, negotiations between the company and the government led to a draft agreement about a partial restitution of the IPC acreage, which included North Rumaila, in return for a lump sum plus an undertaking of the IPC shareholders, minus Jersey Standard, to form a joint venture with the state oil company Iraq National Oil Company (INOC, set up in 1964) for exploring the remainder of the original concession area. The agreement never came into force. First a new dispute arose when the Syrian government demanded a 50 per cent increase in the transit payments for oil passing through its territory, together with £40 million in alleged under-payment over the previous ten years. Talks between the company and the

Post-war nationalism was also rife in the Far East, particularly in the former Dutch East Indies. In 1945 Achmad Sukarno declared independence, with himself as President. Four years of sometimes heavy fighting ensued.

Below, photographed on 20 July 1948, Sukarno receives an enthusiastic welcome from some of his people on Sumatra, and right in 1949, greets others after the Dutch had ceded sovereignty.

[53]

government failed to produce a result. In December 1966 Syria seized IPC's equipment and installations. The Iraqi government ordered IPC to stop pumping oil through the pipeline until the dispute was resolved, though at the same time demanding that the company pay royalties on the normal level of exports.[124] Just when a settlement of the dispute with Syria finally appeared to bring ratification of IPC's agreement with the Iraqi government closer, the 1967 Six Day War between Israel and its Arab neighbours broke out, fanning anti-western feelings throughout the Middle East. As a result the Iraqi government revoked the draft agreement and gave INOC exclusive rights in the entire country, except for IPC's remaining acreage. In November 1967, despite the fact that France's CFP was a member of IPC, and against protests from Britain, the US and the Netherlands, INOC signed a long-term contract with the French state-owned oil exploration company Entreprise de Recherches et d'Activités Pétrolières (ERAP) for extensive joint exploration; and in December the USSR promised to supply INOC with drilling and transport equipment, and to undertake a geological survey of northern Iraq.[125] Thus the awakening nationalism in Iraq left IPC with a tenuous foothold in that country. Despite the uncertain circumstances, the company did manage to raise its output during the later 1960s; in 1970 Shell received a record 384,000 b/d from IPC. The situation remained inherently unstable, however, until Iraq finally nationalized IPC's properties in 1972.

Struggling to keep operations in Indonesia

The Group's longest and ultimately most frustrating struggle with nationalism took place in Indonesia. Using reports from employees who had managed to escape the Japanese occupation, and American intelligence damage assessment reports, London managers had drafted comprehensive plans during the war for the reconstruction of installations to commence as soon as the archipelago was liberated by the Allied forces.[126] However, these plans took the restoration of the Dutch colonial regime for granted and thus overlooked one crucial factor. On 17 August 1945 Indonesian nationalists, led by Achmad Sukarno and Mohammed Hatta, proclaimed the independent Republik Indonesia Serikat (RIS). Stepping into the vacuum between the collapse of the Japanese regime and the arrival of Allied troops and Dutch administrators, military forces of the new republic moved to occupy strategic locations and regions, including oil fields. It took another year before the Group regained control over fields and installations and started rebuilding, but some areas remained out of bounds to employees for several more years, and the RIS never restored North Sumatra, Royal Dutch's cradle, to the company. Reconstruction

[54]

proceeded at speed wherever possible. In 1946, Balik Papan, Bendul, and Wonokromo resumed processing with all or at least part of their installations. Pladju followed in 1947 and five years later the Group considered the rehabilitation of its accessible plants virtually completed.[127] The ambitious expansion plans were partially shelved, however, awaiting the outcome of political developments. Capital expenditure for Indonesia peaked in 1949 and then dropped sharply to around £2 million a year.

Meanwhile Indonesia had gained full independence with the transfer of power and the withdrawal of all troops and government personnel by the Netherlands in 1949. The Group's relations with the RIS government were fraught from the start. Indonesia regarded Bataafsche, and by extension the entire Group, as tainted by its close association with the former colonial regime, a not unreasonable charge given the company's involvement with the strenuous Dutch efforts to retain power through military force in 1947 and 1948. Moreover, Bataafsche continued to dominate the country's oil industry and also acted as the informal leader of a cartel with Stanvac, the former NKPM, and Caltex, which shared production and marketing between them on the basis of a formal 'let alone' agreement with the Dutch government in 1947.[128]

For its part, the Group struggled to adapt to the new situation, and never really succeeded. It tried hard to transform the business in Indonesia, to shed colonialist attitudes and reinforce the ties with the local community, but its actions, whether they concerned the building of schools and housing, the organization of personnel training, the adoption of regionalization, wage policy, or the maintenance of police forces, always appeared to be out of step

[55]

LABOUR TROUBLE AT BOMBAY
Bombay, Aug. 4. – Sailings of motor tanker OVULA and steamer LOMBOK delayed about 24 hours on account of anti-Dutch action of trade unions but ban now lifted - Lloyd's agent.

[56]

Shell struggled to return to Sumatra after the Second World War. In 1947, left, new pipelines are being laid. Right, local workers survey the damage done by republicans to Bataafsche's installations at Pladju.

with developments. Shaped by the experience of the 1930s and 1940s, Group policy towards nationalism and the emergence of OPEC was generally wise, detached, and inspired by a sense for long-term interests. Yet one gets the impression that Royal Dutch/Shell had lost its touch during the first years of dealing with Indonesia, perhaps because that country had been so central in moulding the business that managers could not come to understand, let alone accept, the historical change which had made the company an alien there in the eyes of the new government. Then again, the old condescending attitudes died hard. Preparing imminent talks with officials about a new law governing the petroleum industry in 1951, one central office memo summarized the Group's position as follows: 'In any case it must be recognized that the Indonesian government are very ignorant and a lot of

[57]

...Deze treinreis verliep uitstekend; aan alle stations en haltes waren grote menigten samengestroomd om voor het eerst in 8 jaar weer blanken te zien. Leger en politie kwame er overal aan te pas om de nieuwsgierige menigte op afstand te houden. De houding der bevolking was niet ongunstig; beledigde uitdrukkingen werden in het geheel niet vernomen.

In 1951 a railway excursion was organized in an attempt to assess the possible recovery of territories in Aceh. The report above notes the curiosity of the locals in seeing white people for the first time in eight years.

[58]

[59]

[60]

education will be required before they can be made to understand the sort of principles involved. It would seem most necessary that the first stage should be educative, so that having offered to negotiate, the companies should then present a memorandum setting out in simple terms such basic conceptions as: the need for Indonesian oil to be competitive; the need for capital investment to earn a proper reward; "Profit-sharing" must include some provision for "loss-sharing"; world oil prices – how they are arrived at and how applied; the economic necessity for Indonesia of fully integrated companies'.[129]

The failure to make a leap of the imagination may have been exacerbated by central offices, quite contrary to the general trend towards decentralization, keeping an increasingly close control over the Indonesian operations. Bataafsche's Djakarta office never had the kind of managerial independence enjoyed by local operating companies; indeed, it only became an operating company in its own right in 1960. Consequently, the creative tension between central strategy and local knowledge which had leavened, for instance, the policy debates between London and

[63]

...e oorzaak, dat de noteringen van de Amsterdamse beurs van gisteren niet in ons bezit kwamen.

Hoh... 22.22 Avondgebed ; 22.25 Beursnieuws, Nieuws. Sluiting.

De boorterreinen in Atjeh

Arbeiders niet tot teruggave aan BPM bereid ?

„Ondanks het feit, dat er tekort heerst aan tal van voor de olie-industrie belangrijke machinerieën, leveren de olieterreinen van Noord-Sumatra thans voldoende olieproducten, om in de behoeften van geheel Noord-Sumatra te voorzien", aldus verklaarde de heer Abdoelrachman, hoofd van de „Oliebedrijven Noord-Sumatra", in een onderhoud met Aneta.

De heer Abdoelrachman, die sedert de Japanse bezettingstijd op de boorterreinen werkzaam is, vertoeft momenteel te Djakarta teneinde contact op te nemen met de regering betreffende de status van de olieterreinen in Noord-Sumatra...

hadden herbouwd, werd vernield.

Na de proclamatie van 1945 werd door de aanwezige arbeiders, onder leiding van de heer Abdoelrachman, aan het herstel begonnen.

De boorterreinen liggen alle op Atjeh's grondgebied en wel in Pase (Lhok Seumawe), Djulu Rajeu, Perlak (Peureula), Rantau Tamiang, terwijl de raffinaderijen zijn gelegen in Pangkalan Susu en Pangkalan Brandan.

De maandproductie van olieproducten is ongeveer als volgt: Benzine: 1963 ton; Kerosine: 490 ton; Dieselolie: 360 ton; Solarolie 16 ton; As-olie ruim 3½ ton.

[61]

[62]

Below, a 1948 newspaper report questions whether workers occupying Bataafsche's drilling fields in Aceh will return them to the company. Above, Shell's regionalization is seen in action in the middle 1950s. Left to right, in the past the man under the sunshade would have been a European, but now is an Asian; a worker carries out an oil survey in the Sumatran jungle; a worker operates the master valves controlling pipelines at a Sumatran oil refinery; and other workers manhandle pipes ready for laying.

Mexico during the 1930s, or between London and Venezuela in the 1940s, never developed to the same degree between central offices and Djakarta. Moreover, until 1958 the Indonesian operations suffered from the fact that the organization was split into two quite separate departments, one for E&P and manufacturing, and one for marketing, with separate reporting lines, the former to The Hague, the latter, still called Handelszaken, formally also to The Hague, but really to London. The general representative in Djakarta was the senior Group manager there, yet he was not, in fact, the head of the organization, his main task being the maintenance of the company's relations with the government. The responsibility for running the operations lay with the managers of the two departments who did not necessarily have similar interests.[130]

Nor would it have been easy for central offices to relinquish control at a time when the Djakarta office, while attempting to redefine its position, faced serious operational difficulties as well. The loss of oil fields restricted crude production, which passed the 1938 mark only in 1954 and then stalled. In 1951, the Indonesian parliament froze the issuing of new concessions. As a result the Group had a shortage of oil and needed imports to sustain its market share, which nevertheless slid from 78.5 per cent in 1953 to 68 per cent in 1957.[131] Bataafsche emphasized its urgent need for new concessions at every opportunity, but none were forthcoming because the government could not, or would not, resolve the political stalemate about the new mining law to replace the regulatory framework left by the Dutch regime.

Another bone of contention between the Group and the government was the status of their joint venture, NIAM, which had become a very large part of Shell's operations, and its major source of profits. Following independence, the Dutch government's share in the company had been transferred to the Indonesian government, but the company's head office remained in The Hague, with a Dutch board and staff. After long and difficult negotiations, the Group accepted the government's demand to have NIAM transformed into an Indonesian company with its seat in Djakarta and with Indonesian board members, but only in the face of a veiled threat that it would otherwise be nationalized. Though Bataafsche had realized the desirability of making itself agreeable by giving in to the RIS demands, its stubborn defence failed to inspire much goodwill and did not save NIAM either. When the transformation finally took place in 1958 the company, renamed Permindo in 1959, had become worthless, because its concessions were due to run out two years later and the government did not appear likely to renew them.[132]

Despite a gradually deteriorating business climate due to inflation, exchange controls, high taxation, and political uncertainty, the Group did generally well during the 1950s, net profits from marketing being estimated at £10 million during 1949-56.[133] However, from 1957 the Group's position in Indonesia worsened when, in line with other Dutch companies, Bataafsche suffered from the fall-out of an organized propaganda campaign against the Netherlands over West New Guinea, still in Dutch possession but claimed by Indonesia as part of its territory.[134] This caused the Group to step up the pace of staff regionalization and to replace the Dutch job titles with English ones. To reduce the

[64]

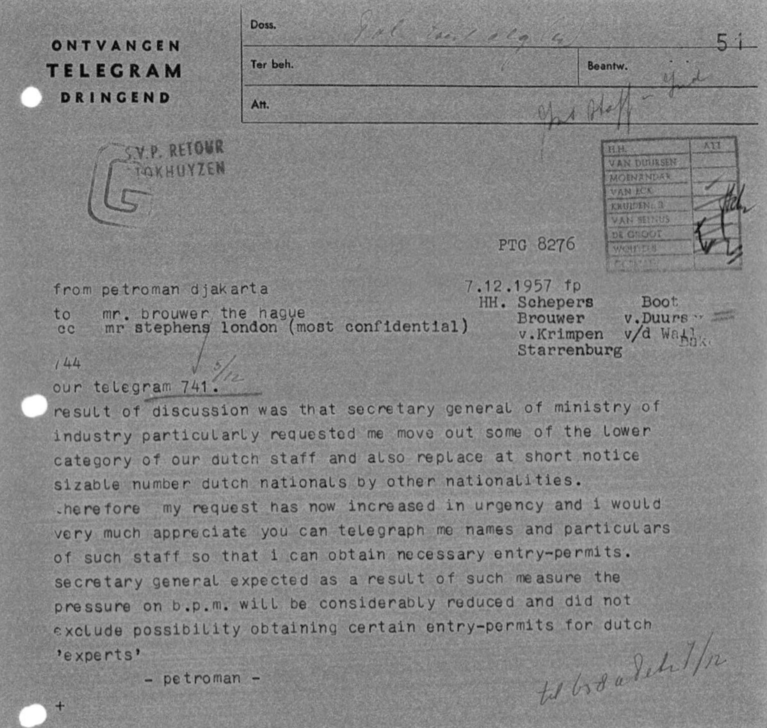

ONTVANGEN
TELEGRAM
DRINGEND

Doss. 51
Ter beh. Beantw.
Att.

S.V.P. RETOUR
TOKHUYZEN

PTG 8276

from petroman djakarta 7.12.1957 fp
to mr. brouwer the hague
cc mr stephens london (most confidential)

HH. Schepers Boot
Brouwer v.Duurs
v.Krimpen v/d Wall
Starrenburg

744

our telegram 741.
result of discussion was that secretary general of ministry of industry particularly requested me move out some of the lower category of our dutch staff and also replace at short notice sizable number dutch nationals by other nationalities.
therefore my request has now increased in urgency and i would very much appreciate you can telegraph me names and particulars of such staff so that i can obtain necessary entry-permits.
secretary general expected as a result of such measure the pressure on b.p.m. will be considerably reduced and did not exclude possibility obtaining certain entry-permits for dutch 'experts'

\- petroman -

\+

As anti-Dutch feeling increased in Sumatra in December 1957, Shell's man in Djakarta sent this urgent telegram to Brouwer and Stephens requesting details of Group non-Dutch employees who could be sent as replacements.

I N T E R N A T I O N A L I S A T I O N

S H E L L   I N D O N E S I A

(Per 1st August 1958)

[65]

	Dutch Expatriates:	Non-Dutch Expatriates:	Total Expatriates:
end 1957	657	30	687
end 1958	300	300	600 (50/50)
1st Aug. 1958	472	120	592 (±80/20)
end 1958 (Est.)	384	216	600 (±65/35)

A. Vacancies to be opened (300-30) 270
Vacancies opened 246
Vacancies unyet opened 24

B. Vacancies opened 246
Candidates nominated 186
Vacancies not yet nominated 60

C. Candidated nominated 186
[illegible]

From our AG-126 (25/2) and the attached up-to-date summary you will observe that the internationalization percentage, even after fullfillment of our present vacancies, falls short of 50% and with 60 vacancies against which no nominations have yet been made. Indications are that our percentages by the end of 1958 will be close to 65:35...

Further extracts from telegrams (above and below) show a small part of the Group's efforts in meeting the challenge of providing non-Dutch staff.

...there are already some shell british staff in indonesia. we are sending some british schoolmistresses soonest but will revert with details schooling facilities...

exposure during the time it took to train Indonesians for middle and higher ranking managerial tasks, Group managers also adopted a policy of internationalization, replacing Dutch expats with other nationalities. At the end of 1957, the staff consisted of 812 expatriates, 784 of whom were Dutch, 1,340 Indonesians, and 20,112 Asian labourers; two years later, the number of Dutch expats had fallen to 240 out of a total of 550.[135]

In December 1957 the pressure increased further, with wholesale confiscations of Dutch companies and real estate, and the suspension of entry permits for Dutch citizens. The Group's business very narrowly escaped harm when the general manager B. Scheffer succeeded in turning away a band of 400 armed men who appeared one night to occupy the Djakarta head office.[136] Bataafsche responded to the violence with steps, as radical as they were belated, to change the structure and appearance of its business. The two separate departments were at long last integrated into a single organization, similar to the one adopted in Venezuela during 1957. Three British employees were appointed to the top managerial positions, and plans were prepared to reorganize the operations into a Djakarta-based company, PT Shell Indonesia. To give this company an entirely different formal nationality, its shares would be held not by Bataafsche, but by Canadian Shell and Shell Overseas Exploration Ltd., with managers reporting to London, not to The Hague. At the same time Loudon urged the Dutch government to give up its resistance to handing over West New Guinea to the RIS if it wanted to avoid damaging the Group's business further, though Shell, with its rearguard action over NIAM so similar in intent and purpose to the Dutch intransigence over New Guinea, had only itself to blame for at least part of that damage.[137]

[66]

We gather from your telegram that information of new staff going your area has not yet reached satisfactory or desired level We asume that this is mainly due to the fact that on this side we are not sufficiantly au courant of all constantly fluctuating conditions stop

The telegrams continued to sketch Shell's difficulties (above), with Dutch staff still in Indonesia became increasingly fearful (below).

Having just again visited all friends feel compelled report that Dutch staff who scheduled to leave this area are increasing-ly worried about their future, especially those with children stop
Of course our policy and underlying reasons are generally appreciated but it is ambarrassing not to be able to tell them a little more (?)

[67]

The new general manager in Indonesia, L. A. Astley-Bell, immediately showed a keen awareness of the need for a fundamental change in attitudes, writing in December 1958 'In the present circumstances BPM, in our opinion, might do well to make clear that we are going to stay in Indonesia for many years unless we are forced out by developments beyond our control, such as adverse safety or economic conditions, crippling legislation and the like. This message should demonstrate our real understanding of, and proper sympathy with, Indonesia's problems and interests, which by the nature of our business we fully share with the country. Our basic attitude is one of considerations wider than those of pure commerce. Our activities in Indonesia, like in other parts of the world, are those of a competent oil industry with international resources, experience, and skill, offering, like we did in the past, beneficial co-operation with the government and people of Indonesia. Although we are international, both in operation and shareholding (some shares must be in Indonesian hands), many of our interests, as operators in Indonesia, are identical with those of the country. So are our problems (...) In line with our thinking new investments like Tandjung and probably others that may follow (with inflow of foreign currency) would well fit in the "we stay with you" philosophy which we prefer to the "we change our face" attitude (...) Emphasis on the benefits of our staying [with expressions such as] "the likely results of the discontinuance of our operations in Indonesia" would make many Indonesians feel rather irritated. Are we vital? Mexico still lives'.[138]

In the summer of 1958 the Committee of managing directors was, despite the recent upheavals, cautiously optimistic about the prospects in Indonesia. They considered that further investments could be made, provided they had pay-out times of two years or less. With a return on assets of around 3 per cent, the business was not very profitable, but the Group still held about 70 per cent of a market growing at an annual rate of 15 per cent. It seemed better to waive the claim on the North Sumatra fields for instance, as a grand goodwill gesture. There appeared to be a good chance of extending the NIAM concessions, because the government was not in a position to take over that company. The Group's concession covering the Tandjung fields in Kalimantan looked an outright winner, its estimated reserves of 200-300 million barrels even opening possibilities for long-term supply contracts with Shell Oil's Pacific Coast refineries. Even so managing directors, aware of the looming dangers, concentrated firmly on the medium term: 'Although the present political situation was not encouraging and no guarantees could be given, it was felt reasonable to assume that the oil industry would not be subject to nationalization in the next five years by which time it was expected that the Tandjung project would have paid off.'[139]

The main threat consisted of the draft petroleum law, the preparation of which held everything in suspense. The law was likely to lay down requirements for companies operating in Indonesia, for instance about board appointments, and as long as

[68]

Sukarno gained powerful anti-western friends: left, he rides through Cairo with Nasser in 1955, and below, in 1960 he and Khruschev were very close.

[69]

these clauses were unknown the Group could not properly proceed with forming Shell Indonesia. More and more draft clauses leaked out which, though unfavourable for the oil companies, still sparked parliamentary debates about the desirability of an immediate nationalization of the industry. By September 1959 the Group's mood had turned more pessimistic. The government had just confiscated 90 per cent of all cash holdings in return for long-term public bonds, proclaimed a Mining Rights Bill voiding concessions if the working of them was deemed insufficient, and also issued a decree providing a framework for the nationalization of mixed Dutch-foreign companies and foreign enterprises domiciled in the Netherlands. This latter action had prompted the Group to enlist American and British diplomatic action in Djakarta to get a promise that the decree would not be applied to the oil companies, all three of which were, at that stage, domiciled in the Netherlands.[140] The CMD considered cancelling the £4 million Tandjung project and decided against it, because 'cancellation of the project would serve clear notice on the Indonesian Government that the Group had lost interest in the country and intended probably to withdraw, with the result that Government action tending to accelerate such withdrawal would probably follow. It was agreed that, although the situation in Indonesia was very difficult for the Group, it had proved good practice elsewhere to stay and ride out local storms and it was agreed that this policy should also be followed in this case. As a secondary consideration, it was regarded [as] unwise to contemplate withdrawal at a time when, with other companies such as Stanvac and Caltex continuing their operations, support for the companies in their difficulties was forthcoming from Governments concerned. It was hoped too that the intention of the Group to continue with the Tandjung project might be turned to good account with the Indonesian authorities when seeking improvements in local operating conditions although, judging from previous experience, no great appreciation from that quarter was to be expected.'[141]

[70]

In Makasar, chief city of Celebes (Sulawesi), the largest island in east Indonesia, a huge portrait of Sukarno overlooks a mass rally in 1962. After addressing the crowd, Sukarno led them in a round of patriotic songs, despite the sudden onset of rain.

In October 1960 Sukarno finally signed the petroleum bill into law, the Indonesian parliament ratifying it in February 1961. The act reserved hydrocarbon exploitation to state companies and required the private oil companies to renegotiate their existing contracts and agreements with the government. The new agreements would in effect turn the oil companies into contractors for the state enterprises and it soon emerged that a 60:40 profit split in favour of the government would be taken as the basis. Coordinating closely with each other, Shell, Stanvac, and Caltex entered into negotiations with the government, which proceeded in a fitful manner. The Group took care to keep Eugene Black, the president of the World Bank, informed of all developments, so as to ensure that the bank would exert subtle pressure on Indonesia to respect foreign investment if the country wanted to qualify for loans from the bank.[142]

Once again rumours about an impending nationalization of Shell Indonesia became loud and persistent, presumably amplified by officials wishing to keep pressure on the company to accept the new terms.[143] Nothing happened, but the government took a series of measures which made life for foreign business, and particularly the oil industry, increasingly difficult: a 25 per cent devaluation of the rupiah, tighter foreign exchange regulations, introduction of special taxes raising the public share of oil revenues to 70 per cent, and, in the case of Shell, a refusal to accept the transfer of the Shell Indonesia shares to the Canadian holding company. Given the Group's prevailing slender margins, these measures combined to render the continuation of operations, including the Tandjung project, entirely uneconomical. In addition, the Group faced the likelihood of blocked cash amounting to a potential £35 million by the end of 1963. Nor had the government

Exemplifying the passions that underlay the 'New Guinea question', this road-side poster praises Sukarno as liberator of the Papuans from the oppressive and ugly white people.

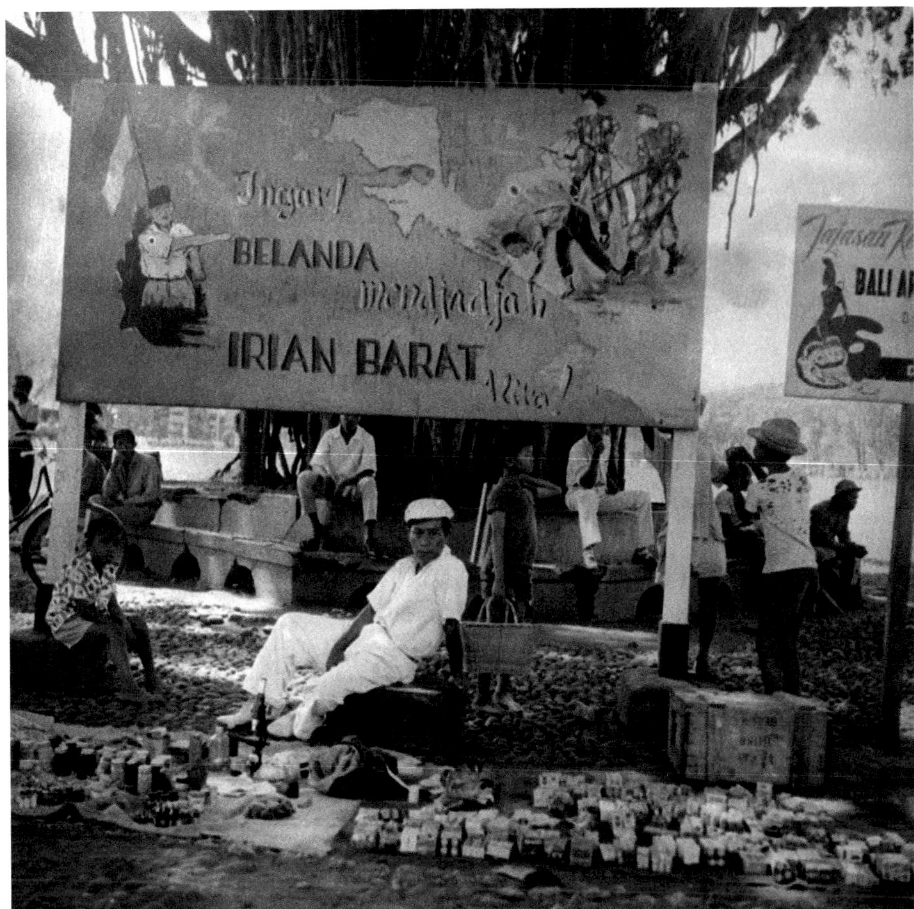

[71]

offered any compensation for the Group's half-share in Permindo, which was nationalized and then wound up following the expiry of its concessions in December 1960.[144]

Though these measures applied to all oil companies, Shell felt singled out because relations between the Netherlands and Indonesia had deteriorated to a point where armed conflict over West New Guinea seemed inevitable. In talks with Group managers, Indonesian officials always linked the government's anger over Dutch political inflexibility to the Dutch 60 per cent share in the Indonesian operations, regardless of Shell Indonesia's status.[145] Group managers were always relieved if, in August, Indonesian Independence Day had passed without Shell Indonesia being nationalized.[146] The danger of a war was finally averted when, during the summer of 1962, the Netherlands at last accepted the transfer of West New Guinea to the RIS. Meanwhile the talks between Shell Indonesia and the government had progressed, very slowly, to a point where a package deal over all outstanding issues appeared possible.[147] The stumbling block remained the 60:40 split, which Shell and Stanvac refused to consider. Caltex, however, wavered, and in November 1962 broke the companies' common front by accepting it. The RIS now posed an ultimatum to the Group and Stanvac; by 1 December they would have to agree to 60:40 and in addition to accept selling their marketing assets on signing the new contracts, with their manufacturing assets following after a period of ten to fifteen years. In return, the terms for exploitation concessions would be extended to twenty years, and the companies would finally obtain concession rights over new areas. If the companies refused, they would suffer unspecified consequences.[148]

Chapter 3

Main picture: On 19 April 1962, with the Netherlands and Indonesia coming close to war, 800 Dutch soldiers embarked in the *Zuiderkruis* (*Southern Cross*) for the long voyage east. Below, at Djakarta airport Indonesian students welcomed volunteers from Singapore under the banner 'Let us drive out the Dutch'. By the diplomatic leadership of U Thant, charismatic Acting (and later permanent) Secretary General of the United Nations, war was avoided and the administration of West New Guinea was temporarily transferred to the UN until 1 May 1963. Netherlands-Indonesian relations further improved when the Netherlands Foreign Minister Dr Joseph Luns visited Indonesia in 1964. Departing from the Netherlands on 24 July (below), his nine-day visit ended with an agreement to restore diplomatic relations.

Showing how close the two nations had come to war, fires rage in the Indonesian port of Probolingo after the arrival of Dutch troops in 1962.

The Group was quite prepared to relinquish the marketing operations, which were by now marginally profitable anyway. After a quick enquiry among country general managers whether this step would not have repercussions elsewhere the CMD authorized Shell Indonesia to sell. Nor did the CMD object to the principle of selling the manufacturing assets as well, but only on the expiry of the new contract, i.e. after twenty years, and as part of a comprehensive package including agreements on pricing and exchange rates.[149] The December ultimatum lapsed and talks continued in their customary dilatory fashion, Indonesian officials not making haste, until the end of February 1963, when the government issued a new ultimatum, and when that had also passed a stiffer one, which threatened the cancellation of all existing agreements and a unilateral imposition of new conditions, including a 80:20 profit split, unless the proposals were accepted. The oil companies understood this to mean a de facto nationalization, and they appealed jointly for diplomatic action. In a striking departure from US policy over the Mexican and Venezuelan negotiations, when the State Department told the oil companies to accept the demands and fend for themselves, the Kennedy administration decided to lend support. The American ambassador in Djakarta warned the government that the demands threatened to upset the international programme for economic stabilization in Indonesia, where economic mismanagement had created rampant inflation and serious social unrest. In addition President Kennedy sent a special mission to Tokyo for a summit with President Sukarno in May. The common front between the oil companies was now restored after Caltex's earlier breaking of ranks; and when the Indonesian delegation attempted to eliminate the Group from the talks on political grounds, the two American companies refused to consider a deal without Shell.[150] Liaising with oil company

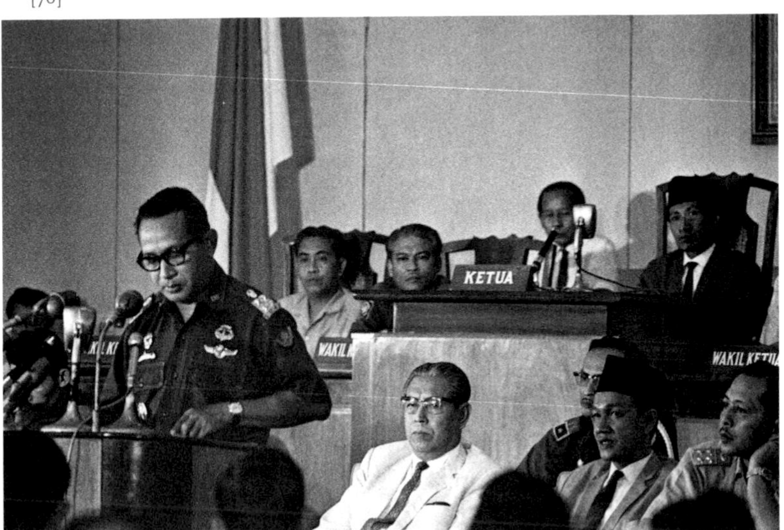

[76]

After years of corruption and chaos in Indonesia, a bloody but abortive attempted Communist coup in 1965 led to an anti-Communist takeover by the military, led by General Thojib Suharto, who assumed full executive powers and here is seen on 4 March 1967, addressing Parliament for the first time. Although a virtual dictator, he stabilized the nation and greatly improved its international relations.

executives flown in for the occasion, the presidential mission brokered a deal between the Indonesian government and the oil companies which was signed by the two sides on 1 June 1963.[151]

Despite having had to concede the 60:40 split and substantial signature bonuses, the CMD thought that the agreement 'covers virtually all the major points at issue and resolves them in favour of the established companies'.[152] The deal took the form of an operating contract between Shell and the state company Permigan to supply the Indonesian market with crude and oil products at cost plus fixed fees. Exploitation rights for existing concessions were set at twenty years, and for 2 million hectares of new exploration areas at thirty years, but the Group would have to relinquish 25 per cent of its areas after five years, and another 25 per cent after ten years. The marketing facilities were sold and a schedule agreed for the sale of the manufacturing installations, both to be valued at 60 per cent of the original cost, and not at book value, which in most cases was only a fraction of cost. The agreement also settled a range of disputed fiscal and monetary questions. Finally, the contract had a clause raising its contents to the status of a national law which, tied to a provision appointing a Swiss judge as umpire in any arbitration case, the Group deemed 'a milestone along the road to giving international status to agreements of this type and providing greater protection against the whims of the local legislature'.[153]

For all its comprehensiveness and sophistication the Work Contract, as it became known, did not provide an immediate solution to all of the Group's difficulties in Indonesia. Arising from public excitement about a political dispute between Indonesia and Britain over Malaysia, the harassment of British staff in Kalimantan continued, forcing the evacuation of their families and the immediate replacement of the employees in October 1963.[154] Moreover, the Group could not have foreseen the rapid worsening of political and economic conditions in Indonesia which followed. An intense power struggle between the Communist Party and the army paralysed the economy and crippled public finance, pushing the rupiah into free fall and causing state companies and government departments to run up large debts with Shell Indonesia for product supplies. Notwithstanding the increasing unit costs of manufacturing, Indonesia remained a competitive source for Group oil for the time being, although in the spring of 1964 this was not expected to last for long.

Barely a year later the situation had worsened. The government exercised further undue pressure on Shell Indonesia for a further regionalization of staff and in March 1965, Shell considered it merely a matter of time before it effectively lost control over the operations in Kalimantan through the forced transfer of the business to Indonesians.[155] The CMD discussed withdrawing from the country but, mainly because of the high cost of finding alternative supplies, decided to try and keep the operations going if possible until the end of 1965.[156] In August the committee considered selling the remainder of the Group's assets in Indonesia to the government, ahead of schedule; Stanvac had already started talks about transferring the manufacturing installations.[157] The following month an escalation of violence confirmed the worst fears about Indonesias future. The army defeated an attempted coup and then went on the rampage; mass killings of suspected political opponents are said to have cost hundreds of thousands of Indonesian lives.[158] Negotiations with the government proceeded as rapidly as could be expected under the circumstances. By mid-December the two sides were close to an agreement, and before the year's end the Group had sold its Indonesian business for $110 million, to be paid with surplus products over a projected period of five years.[159]

Thus, after more than seventy-five years, the Group formally but unceremoniously withdrew from the country which had given birth to Royal Dutch. Not all the ties were cut; Shell continued to sell some products, such as branded lube oil, bitumen, and chemicals in Indonesia and retained good relations with the state oil companies, but the company's own presence disappeared.[160] Since 1946, it had been a very tough struggle to keep viable operations going in Indonesia, and the CMD clearly lost faith when the circumstances deteriorated so badly during 1964-65. The decision to sell the E&P and manufacturing operations was not a precipitate one and the terms of the sale were considered good. When he was asked later if he regretted the sale, Loudon replied firmly: 'No I did not. It was the Group's judgement that it had to be done. (...) Whenever dealing with these producing countries, you always have to know when you should let them cut off a finger and don't wait until they cut off an arm.'[161]

Yet one cannot escape the feeling that, over time, Shell did come to regret it. After the bloodshed of 1965, the political situation gradually stabilized; General Suharto emerged as the strong man of a new government which at last succeeded in halting the crippling inflation, reorganizing public finance, and in creating a firm basis for economic growth.[162] Having stagnated for nearly twenty years, Indonesian oil production picked up again, attracting new producing companies as it did so. By 1975 more than thirty foreign oil companies were active in the archipelago.[163]

The Group clearly also saw new opportunities. As early as September 1966, the CMD was considering new E&P ventures in Indonesia, though nothing appears to have come of them.[164] In January 1969, managing directors discussed an E&P project in Kalimantan and laid down that 'the history of our experience in Indonesia must not be allowed to color our assessment of future opportunities there'.[165] Three months later, the CMD gave the go-ahead for the project commenting that 'the Group has more reason than most to be active in Indonesia, on the understanding of course that any venture is economically attractive, and that the prospects for security and stability are judged to be satisfactory'.[166] During the late 1960s and early 1970s the Group had some small E&P ventures in Indonesia, none of them yielding spectacular or even rewarding results. Other oil companies were seen to be doing nicely out of Indonesian oil while Shell, having given up its place at the table, had to make do with some crumbs. We must thus conclude that, in Indonesia, Shell had shown great tenacity, only to give up just short of reaping the rewards.

Coming to terms with OPEC

As noted above, the Group's attitude to the organized economic nationalism of OPEC was a moderate and wise one, presumably because managers recognized the shape of things to come earlier than other companies and adjusted the company's long-term commitments accordingly.

The formation of OPEC in September 1960 wrong-footed the oil companies. In October, the Arab Petroleum Congress was due to take place in Beirut, so the companies had to decide quickly on the attitude that their delegations would have to take. The American companies wanted to attend as silent observers, as they had in 1959, but the Group preferred its delegation to take part in the proceedings. The CMD accepted that this might put the Shell representatives into an awkward position in debates over posted prices and consultation on reductions, but the committee considered that 'Otherwise, if they remained silent, the Governments concerned might well assume – as had been the tendency in Cairo – that the companies were not willing to co-operate in ensuring the success of the congress and were, in effect, attempting to sabotage the occasion.' Moreover, remaining silent could leave the field to Enrico Mattei, who sought to bolster ENI/AGIP's position by posturing as the white knight amongst oil companies. The CMD feared that he 'might seize upon Beirut as a suitable forum at which to make statements intended to please the producer governments but quite inconsistent with other remarks which he had made from time to time for the benefit of consumer governments; it would be a pity, to say the least, if their self-imposed silence were to deny the companies the opportunity to challenge such statements and to draw attention, by comparison

Chapter 3

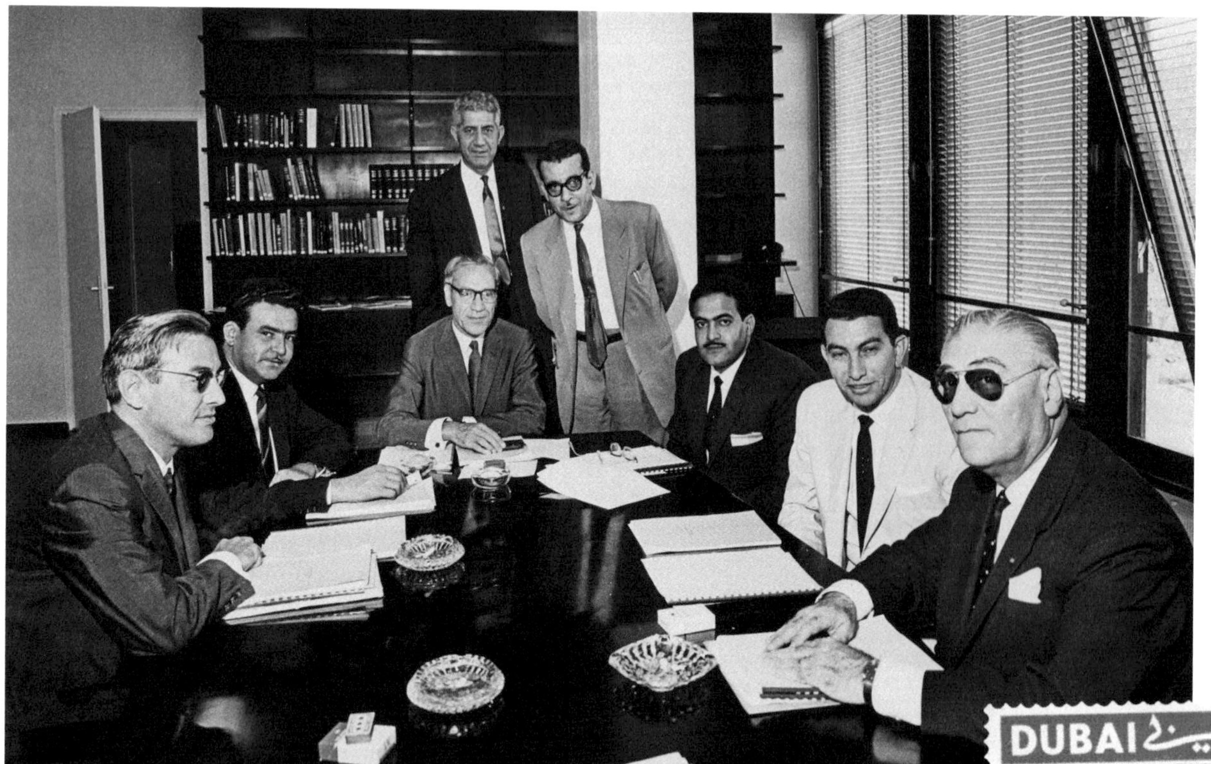

[78]

Main picture, leaders of the first OPEC conference met on 10 April 1960. Proud of their oil industries, OPEC member nations often issued postage stamps showing the source of their wealth. Venezuela and the United Arab Emirates appear on the previous page; below and right, Dubai, Muscat and Oman – and although it was not part of OPEC, Brunei quickly followed suit.

with Mattei's other recent statements, to his contradictory remarks in relation to producing countries and consuming countries respectively.'[167]

In a typical example of the Group's style of delegating authority, the CMD left it to the representatives to decide whether or not to participate. The available records do not indicate to what extent the delegation did take part.

Stung by the second reduction in posted prices, the conference showed militant intentions towards the oil majors. The companies reacted very differently to this new mood. Jersey Standard and apparently also Mobil were not really impressed and wanted to try and break up the producers' cartel by an outright refusal to cooperate.[168] BP dismissed OPEC as an empty threat. The cartel would disintegrate of its own accord; in defending its interests, the company 'would rely chiefly on the superior technical knowledge and experience possessed by the major international companies and on the fact that the producing Governments were at present not in a position to run even their own sectors of the industry as efficiently as we can'.[169]

By contrast, the Group took a wait-and-see attitude, which was based on three broad considerations. First, obstruction by the majors would be counterproductive and only incite the producing countries to be more militant. Second, whereas western nations had been the dominating political power, providing a cushion on which the major oil companies could fall back if necessary, this was no longer the case and it was therefore very much up to the companies themselves to keep their relations with producing countries in good order. Finally, in dealing with the producing countries, the companies' best card would be on the one hand, to

[79]

strengthen their relations with the governments of those countries, and on the other, to reinforce and widen their grip on markets.[170] As always, the majors' best hope lay in a close collaboration between themselves, but uniting them would be difficult given 'their competitive instincts and jealousies'; like it or not, 'The time has come when the industry (or some of us) will have to hammer out some realistic policies and attitudes. Particularly if more ground has to be given (as it will be) (...) Otherwise we may finish up in a stampede'.[171]

To facilitate such collaboration, the CMD freed one of the coordinators to devote more time to OPEC related matters. A Concessions Study Group, set up to analyse terms and conditions, would talk to their opposite numbers in Jersey Standard to search for common ground.[172] By March 1961, Jersey had come around to sharing the Group's wait-and-see attitude, though BP still demurred.[173] During 1961 the majors obtained the blessing of the US State Department and the Department of Justice to establish a closer consultation in the form of the McCloy Group, an informal gathering of the company CEOs to discuss policy chaired by John McCloy, a veteran lawyer in the international oil business.[174]

It must be said that, during OPEC's early years, BP's assessment of the situation held considerable attraction, for the producers' cartel was beset by internal rivalry and, given a continuing world surplus of oil, could not bring much power to bear.[175] The oil companies continued to talk to individual governments of producing countries without the OPEC organization succeeding in imposing its agenda. This tactic of divide and rule appeared to work, mitigating OPEC's original demands and stalling its further evolution.[176] By 1966, however, the OPEC members had gradually understood that the only way to achieve results lay in sacrificing their individual aims for a common policy.[177] The organization had also started to achieve results: the oil companies had not reduced posted prices since 1960, and in calculating what they owed to governments they had accepted that their 1 per cent marketing allowance (a variable figure worth up to two US cents a barrel) should be set at a steady half a cent a barrel. Slowly gathering confidence and cohesion, OPEC was gradually building itself into the equation, and was less and less easy to overlook.

Through those years the Group's policy towards OPEC was consistently cooperative – although not to the extent of self-destruction.[178] In general, 'Shell's philosophy concerning OPEC was that good relations, based on frank exchange of information, would be to our long-term advantage, but this view was not shared by all our competitors'.[179] The Group's conviction about the merits of transparency and a conciliatory approach, which of course held an acknowledged and natural element of enlightened self-interest, failed to convince the other majors active in the Middle East, so Shell remained caught in a fundamental dilemma together with them. They were incapable of agreeing on a policy to counter the growing strength of OPEC; nor could they act individually, bound as they were by the various partnership contracts. And even if the majors could agree, they would have difficulty in convincing the independents to toe the line. In 1969, the Conference noted with regret that 'the tension evident between the producer Governments and the companies could, at least to some extent, be

attributed to the intransigence often displayed in the latter's negotiating tactics, despite the lessons to be learnt from past experience and present realities. Mr. Barran pointed out that the Group, at any rate, has a clear conscience in this matter, since we have acquired a certain reputation for our "conciliatory" approach; but with some of the US companies consistently favouring a hard line, and with the pace necessarily tending to be set by the most obdurate, we are faced with a difficult problem especially having regard to our minority role in the Middle East.'[180]

Unable to take positive action, the majors really played a waiting game and let the initiative slip to OPEC. During the summer of 1970 the Group tried to set an example by holding out on its own against Libyan demands, backed by a forced shutting in of wells, for a 55:45 split and a rise of thirty US cents in posted prices, giving in only long after all other oil companies had accepted.[181]

In December 1970 the oil companies finally appeared to muster the necessary agreement between them. Libya had set various new demands only months after reaching a settlement, including another rise in posted prices. Iran had raised its tax rate to 55 per cent; not to be outdone, the Venezuelan government had increased its rate to 60 per cent and had also declared that it would henceforth set the tax reference values for oil. A group of producing countries on the Gulf had called the oil companies to Tehran to negotiate new demands. Algeria had suspended oil shipments to France. Finally, an OPEC conference in Caracas had declared a 55 per cent tax rate to be the desirable minimum and had also threatened to cut off supplies to companies which refused to pay.[182]

Thus the stampede foreseen in 1960 had finally begun. On the initiative of David Barran, who had just succeeded Jan Brouwer as chairman of the CMD, the majors, supported by CFP plus most of the independents and after obtaining anti-trust clearance from the US Department of Justice, now set up a joint steering group for mutual consultation and support for negotiators, called the London Policy Group.[183] They also issued a joint statement calling for a comprehensive settlement between the companies and OPEC, to be valid for five years. To bolster their unity, the companies further agreed to help each other out if Libyan supplies were cut, in effect a safety net designed to back the companies negotiating there.[184] Even so the common front failed to achieve its objective of halting the stampede. The companies first had to negotiate separately in Tehran with the Gulf states. In February, they accepted the 55 per cent minimum plus a rise in posted prices for a five-year period. Then talks in Tripoli began about prices in the Mediterranean, which ended in April with the oil companies forced to concede a higher rise in posted prices, much to the chagrin of the Gulf states. Meanwhile Algeria had nationalized 51 per cent of CFP's company there, and negotiations with Nigeria, Saudi Arabia, and Iraq continued.[185]

These events led the Group to rethink its position. A long-term planning review presented to the Conference in September 1971 concluded that the industry stood at a crossroads. With the growth of energy demand likely to continue at 5-6 per cent per annum, all proven reserves of oil and gas would run out by 1985, and all expected deep water reserves, oil shale, and tar sands by 2010. Shell would thus need to diversify if it was to survive, but in addition the Group had to influence the likely scenarios through governments. Those in producing countries no longer considered their interests to run parallel with those of the oil companies; 'they are themselves subject to severe nationalistic pressures, and the companies, whose main concession era is approaching an end, are handicapped by partnership arrangements in taking any positive initiatives'. Consequently, the Group should start working on the

[81]

[82]

Even in the latter 20th century, oil did not always travel by pipeline, tanker or railcar. Every morning at Fesclum near Tripoli in Libya, this small donkey (main picture) would head for the warehouse to collect a new load of kerosene. Libya was another member of OPEC to issue postage stamps celebrating its oil industry (above). Syria in contrast had quite a small oil production and was not part of OPEC, but (above left) found a political theme for its stamps: 'Arab's oil for Arabs'.

[84]

Left, more OPEC postage stamps, from Algeria; main picture, the trials of desert exploration – an Algerian seismic camp in a sandstorm.

governments of the consuming countries in order 'to secure or retain their confidence and foster the belief that the oil companies will remain of continuing vital service to them through diversity of resources and technical ability. We should encourage them to stimulate the development of other energy resources and to prevent the wastage of energy. Confirmation of a need for rapid reorientation by the industry from upstream affinities to downstream affinities was one of the major points to come out of the Planning Review'.[186]

Some members of the Conference questioned the assumption of continuing high demand growth, arguing that price rises, environmental concerns, and energy conservation were likely to lower it, but the main outlines of the review were accepted.[187] In the short run, however, the Group could not afford to remain standing. E&P expenditure for 1972 was up 20 per cent on 1971, the total sum spread judiciously over geographical areas so as to minimize likely political interference.[188]

Despite the Tehran and Tripoli agreements, the power struggle between the oil companies and the governments of producing countries continued. When in August 1971 the United States administration suspended the convertibility of the dollar into gold, which amounted to a devaluation of the currency on the open market, the revenues of oil-producing countries suffered a parallel drop, causing them to call for a renegotiation of the agreements. Libya imposed a new exchange rate on payments for crude deliveries and sequestered bank balances of companies which refused to pay the rate. The country also raised the issue of participation, demanding a 51 per cent stake in the ENI company working there.[189] Participation had been mooted before; Saudi Arabia had discussed it, on and off, with Aramco since 1963.[190] Now it instantly became a hot topic. Other OPEC members started making similar claims to Libya, but there was no unanimity within the organization over what exactly members wanted to achieve. A delegation led by Aramco entered into negotiations with a number of Gulf states about participation, while the Iran consortium held talks in Tehran on changing the operating conditions there, which initially centred on the Iranian government wanting to start its own downstream business. In July 1972 the two sides reached broad agreement on this and other points, but when two months later the negotiations with the Gulf states ended with a deal giving these states a phased 51 per cent participation by 1983, Iran posed an additional demand. National Iranian Oil Company would replace the consortium as producer and sell it part of the crude produced under a 20-year contract. In February 1973 the consortium accepted the outline of this arrangement, hoping against hope that it would not lead to repercussions with the other OPEC members. Company lawyers occupied themselves with legal constructions designed to create 'an enhanced legal situation at intergovernmental level in international law' so the agreements would stand.[191] Before the year was out an entirely new situation had emerged, however (see Volume 3, Chapter 1).

In October 1962, Dutch nationals and Eurasians flee from the Indonesian takeover of former Dutch-controlled territories.

Conclusion

The Group's E&P function was in many respects the heart of the business. For much of the time, E&P absorbed a substantial share of investment, and it may have generated most of the revenues. Comprehensive data have not survived because Shell switched to profit and loss accounting by business function only in the 1970s, but in 1961 the CMD reckoned that E&P generated £133.6 million of profits of the Group's total of £162.1 million outside North America.[192] Even if this figure was probably inflated by affiliates paying posted prices, it does suggest that E&P generated Shell's largest revenue flow. In addition, the function also attracted the limelight with tales of daring exploits and discoveries, cutting-edge technologies, and towering hardware. This prominence of E&P was largely a consequence of Shell's position as a crude-short business, with large interests in the United States, and a strategy directed at balancing supply regions so as not to become too dependent on a single one. It was a comparatively expensive policy, notably so during the build-up of offshore investment during the second half of the 1960s, but it paid handsome rewards when, after the first oil shock, balanced supplies and offshore operations became the norm throughout the industry. The same was true for gas, which rapidly developed from a sideline into a main business for Shell long before most other companies recognized its value.

The switch in E&P policy and the accompanying gradual shift in investment from upstream to downstream were both the consequence of the ever stronger economic nationalism of the producing countries. Overall, the Group reacted sensibly to these developments. Managers wanted a firm line with close collaboration between the majors, but at the same time they showed full understanding of the real forces behind the events and advocated dialogue and openness above confrontation. It was only in Indonesia that their nerve failed, after a long and often bitter struggle. However, this disappointment undoubtedly also contributed to the policy changes mentioned above, and thus may have helped to bring about the reorientation from which Shell was later to profit.

302

Chapter 4

Creating competitive advantages downstream

Plentiful and cheap crude supplies created strong competition in downstream operations. Throughout the 1950s and 1960s, the Group's manufacturing, transport, and marketing functions faced a tough, dual challenge, the need for a continuous expansion to meet racing demand, while keeping costs to a minimum in the face of sharp competition. Using favourable tax legislation, the American oil companies could accept downstream as a zero-margin business, which Shell could not. Shell responded with a dual strategy, economies of scale combined with product differentiation, supported by its strong brand, at the same time working hard to achieve better functional and operational integration, and juggling with conflicting policy demands on individual functions.

Pernis by night in 1953

[2]

OFFICIAL FRONT WHEEL & AXLE ALIGNMENT SERVICE
29 PREVENT SHIMMY, HARD STEERING & TIRE WEAR 29
BLUE RIBBON SERVICE STATION, INC.
WHITE HOUSE GARAGE.
DAY AND NIGHT PARKING.
TEXACO
AUTO REPAIRS
29 AUTO RADIO CO. SALES SERVICE EXPERT INSTALLATIONS
SYSTEM
BRAKES
BROWNIE'S
WELDING
PLYMOUTH- SERVICE - DODGE
SHELL MOTOR OILS
EXPERT AUTO REPAIRS
GLASGOW IRON WORKS 28 AND SUPPLY CO.
PALUMBO
The 100% ANTI-KNOCK "regular"
SHELL MOTOR OIL

Figure 4.1
Manufacturing productivity at Shell Oil and the rest of the Group, 1955-1970 (b/d per man).

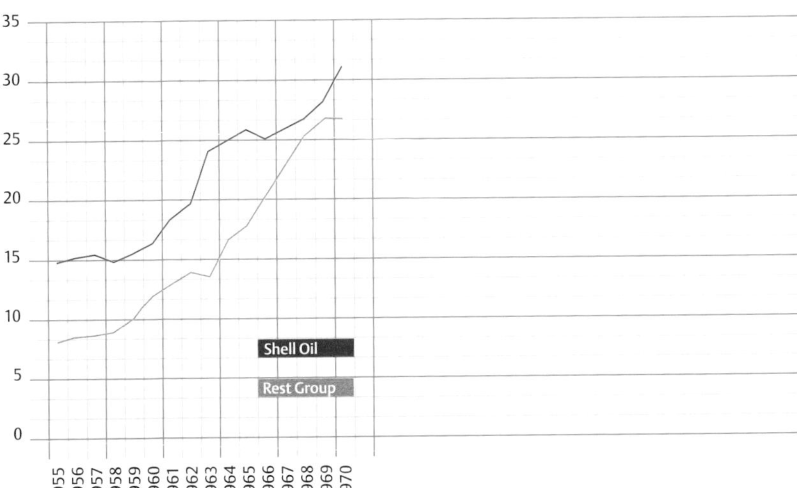

The elusive ideal of Shell Oil As discussed in Chapter 3, Shell Oil was the Group's single most important operating company by far, and it was also a very profitable one. Until the mid-1960s, the rest of the Group grew slightly faster measured by assets than Shell Oil, but the American company was rather more profitable, with a net return on assets of 7 to 8 per cent, whereas the rest of the Group only made 4 to 6 per cent. Indeed, Shell Oil appears to have been the exception to the rule that subsidiaries of non-American multinationals tend to perform relatively poorly in the US.[1]

There were several reasons for Shell Oil's greater profitability, all of them connected to the structure of the North American market. First of all, the import restrictions protected the US companies against the sustained pressure on prices which characterized, for instance, the European market. Shell Oil's overall marketing strategy was similar to the one practised by the Group in Europe, i.e. a focus on profit rather than volume. Consequently, Shell Oil's total number of service stations remained largely the same, varying around 22,000. As marketing managers weeded out smaller and less profitable stations and replaced them by larger stations at better locations, gasoline sales per station tripled between 1950 and 1970.[2] The number of Shell stations owned by independent operators steadily declined, but in 1970 these still made up about half of the total, unlike in Europe, where the operating companies generally strove to own most stations.[3] The competition from established marketers prevented Shell Oil from re-entering in force the mid-continent retail markets given up during the early 1930s, so the company adopted a step-by-step approach, adding districts to its network as and when a profitable opportunity arose.

In the mid-1930s the White House Garage in New York offered everything for the motorist – including Shell Motor Oil.

Second, Shell Oil could afford a more profitable barrel cut. With almost half of total sales during the early 1960s, gasoline dominated marketing in North America. Asiatic Corporation handled the liquid fuel imported from Venezuela, and the chemicals function relied primarily on natural gas, reducing the need to make feedstocks. For the rest of the Group, lower-margin fuel oil and middle distillates generated most sales, leaving only 18 per cent of the total to gasoline. The chemical industry used naphtha for its feedstocks, limiting the volume available for upgrading into gasoline. Third, though for most of the 1960s Shell Oil had growing staffnumbers, these people were very productive. Manufacturing productivity was rather higher than that of the rest of the Group, though the gap narrowed considerably during the 1960s for reasons which we will see below (Figure 4.1). Shell Oil's net revenue per staff member comfortably exceeded the figure for the rest of the Group to start with, and grew faster than that for the rest of the Group (Figure 4.2).[4]

[3]

This particular difference reflects both Shell Oil's better overall marketing position, the import restrictions, and the higher volume of gasoline in total sales, and its lower cost base, largely the result of scale economies at all levels of organization. In 1960, for instance, six refineries served Shell Oil's entire US market of about 180.7 million people. The six countries of the European Economic Community had a population of similar magnitude, 172.2 million people, but served by no fewer than fourteen refineries, some of them, like Pernis, very big and serving several countries, others such as Monheim in Germany much smaller. Europe produced only a fraction of the crude that it consumed, imported most of it by sea, and transported most products by barge or road; the US of course produced most of its own crude and the bulk of the products was transported by pipeline, some of it by barge. Steady improvements to communications and traffic helped Shell Oil to consolidate its marketing efforts into an ever shrinking number of regions and divisions, but Group marketing in Europe remained tied to staunchly independent operating companies which would not or could not collaborate across national boundaries. Borders cut across logistic lines everywhere, preventing the optimization of supply routes. By contrast, Shell Oil could choose its supply routes more or less as managers pleased, though their long-standing wish

Pumps and pumpkins: with their fruit giving a strange if unintended visual echo of the pecten across the road, two men stroll past a competitor's outlet somewhere in America.

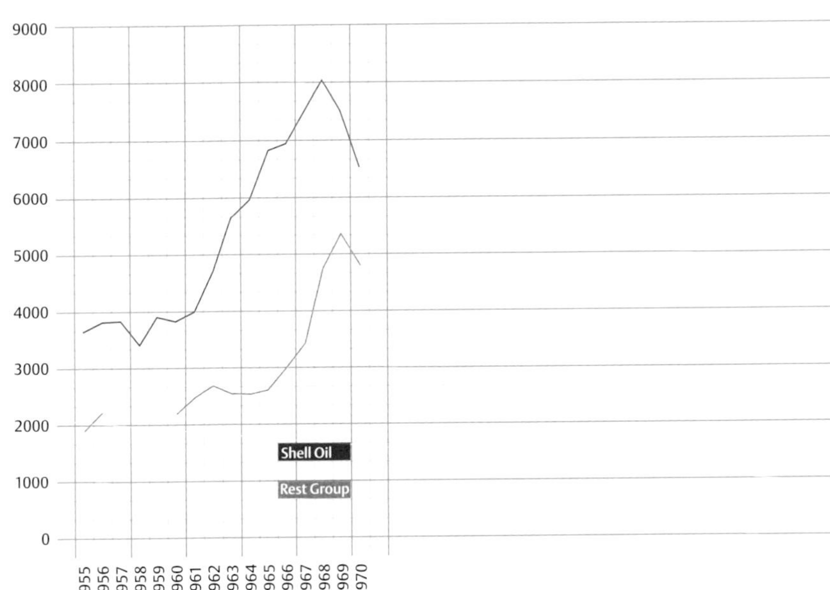

Figure 4.2
Net revenue per staff member for Shell Oil and the rest of the Group, 1955-1970.

for an East Coast refinery was destined to remain unfulfilled.

The US operating company also served as a source of managerial innovation. It was Shell Oil president Max Burns who, in 1947, first suggested that Group managers should use consultants for solving the organizational problems which manifested themselves at central offices during the late 1940s.[5] As related in Chapter 2, Shell Oil also gave advice on investor relations prior to Royal Dutch's New York listing, and Loudon approached McKinsey after hearing from a Texaco executive about the firm. On a practical level, the need to catch up with developments in the US during the Second World War created a flow of secondments and appointments during the immediate post war period, which included Chester Peet's transfer to London to become Group finance director. Lower down the managerial ladder, productivity teams from Shell Oil trained marketing managers for Deutsche Shell and helped to reorganize the marketing organizations in Burma and Indonesia.[6] A refinery efficiency team which toured several installations during 1958-59 was also headed by a man from Shell Oil, though the management advisory committee visiting a number of operating companies to survey managerial efficiency, was not.[7] On a different level, in 1953 Shell Oil also centralized and professionalized its corporate charity by setting up the Shell Companies Foundation to distribute gifts and grants. At first the Foundation received cash gifts from the US companies, but in 1961 Shell Oil started transferring assets such as office buildings and concessions, which were then leased back.[8] The institutionalization of charity donations enabled Shell Oil to give substantially more. In 1969, the Group's charity donations amounted to £3.6 million, of which £2.5 million was in the US, Canada, and Venezuela.[9]

Another key management area which Shell Oil pioneered was the use of computers for administrative data processing. As early as 1961, the accounting work for the entire marketing network of 22,000 service stations was centralized in just two computerized data centres, one in California and one in Oklahoma.[10] In 1966, the then Shell Oil president R. C. McCurdy briefed the CMD about the pitfalls and profits of the company's new information and computer services. Shell Oil had started to use computer systems to deal with the enormous volume of repetitive data processing now done by hand and employing about a third of total man hours. The scheme centred on the setting up of six regional centres to process the data supplied by offices across the United States. Implementation had been slower than anticipated, however, and the volume of information dealt with smaller. Using words which will sound all too familiar to modern ears, Dick McCurdy blamed the problems on the fact that 'Original estimates of the time, money, and technical skills needed to develop the service had been far short of reality, and the intractability of a number of problems had not been fully appreciated.'[11] More specifically, he singled out four key problems in the switch from manual to electronic processing. First of all, the importance of standardizing information, since the unit cost of information depended so much on the quality of the data submitted. Then the training needed to change existing business procedures had been substantially underestimated. Third, the risk of obsolescence; by the time a system had been fully tried and tested, there was a good chance that it no longer suited the environment for which it had been designed. Finally, regional variations in business practices had seriously complicated the attempt to develop the six data centres simultaneously.[12] Shell Oil progressively consolidated its data processing so that by 1972 the number of centres had been reduced to three, and in 1975 only a single one remained.

The example of computerization highlights the fundamental difference between Shell Oil and the rest of the Group. Working in a very large market with marginal cultural differences and few economic barriers, Shell Oil could afford to pioneer economies of scale, because the company could expect to reap the benefits. None of the other operating companies could do that to the same degree, or even hope to get the any benefits at all from imitating the American example. Shell Oil could concentrate most of its head office services first in New York and subsequently move that office with all research laboratories bar one to Houston, as it did in 1971.[13] The Group could not follow suit and merge central offices into one. Shell Oil could also, during 1965-66, forge the value chain from wellhead to consumer closer together by reorganizing its core

[4]

[5]

[6]

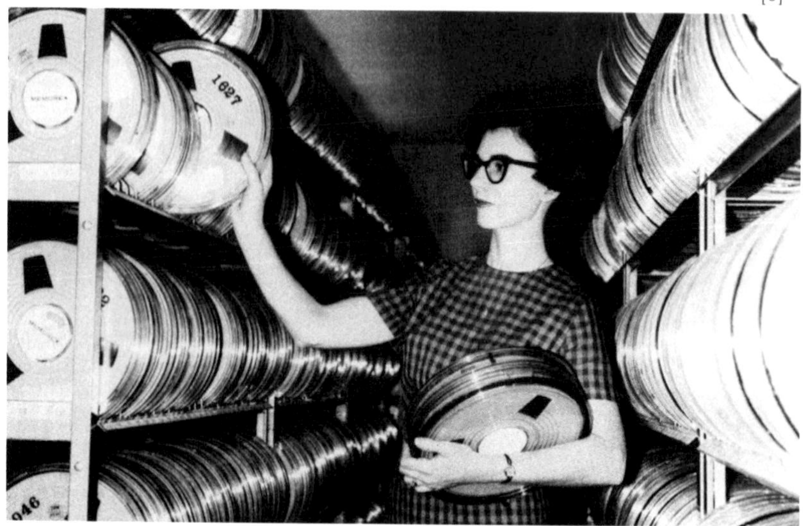

As the pace and complexity of business increased after the Second World War, Shell Oil adopted computers nationwide. Left, from top to bottom, one of its data centres in 1964; electronic data processing in the personnel department, Los Angeles Data Service Center, 1966; and a tape librarian at the Houston Data Service Center. Earlier, in 1949, the London office had chosen a different solution, a mechanized accounting system that used punched cards. By the mid-1950s this was still deemed the best answer to the problem for the time being, although it was also recognized that electronic processing would eventually prevail. In 1958 London office ordered its first 'high-speed electronic computing system' from Ferranti, but punched cards were still in use in Shell Centre (below right) as late as 1969.

operations into a single division, Manufacturing, Transport and Supplies, and Marketing, also known as MTM.[14] For an integrated company working on the scale of Shell Oil this arrangement made sense; for other operating companies or Group central offices it did not, because of the varying importance of individual functions, the overriding importance of national boundaries across business functions, and the carefully maintained balance between functional and regional interests at central offices. It was only during the second half of the 1960s that the Group began to achieve closer integration between the business functions, turning downstream from an abstract concept into an economic reality.

Consequently, during the 1950s and 1960s Shell Oil served as a dynamic, enterprising, and innovative example for other Group companies largely because the American market was considered as showing the shape of things to come. In many respects, however, Shell Oil was, and remained, the odd one out within the Group, simply because companies elsewhere did not develop in the same direction, or with the same speed. Some of Shell Oil's innovations travelled well, others took time to catch on elsewhere, still others, hailed as epochal concepts at the time of their introduction, remained dead in the water or did not take off with the expected speed, even in the US, as we shall see below. We will now turn to look at the development of the Group's individual business functions.

[7]

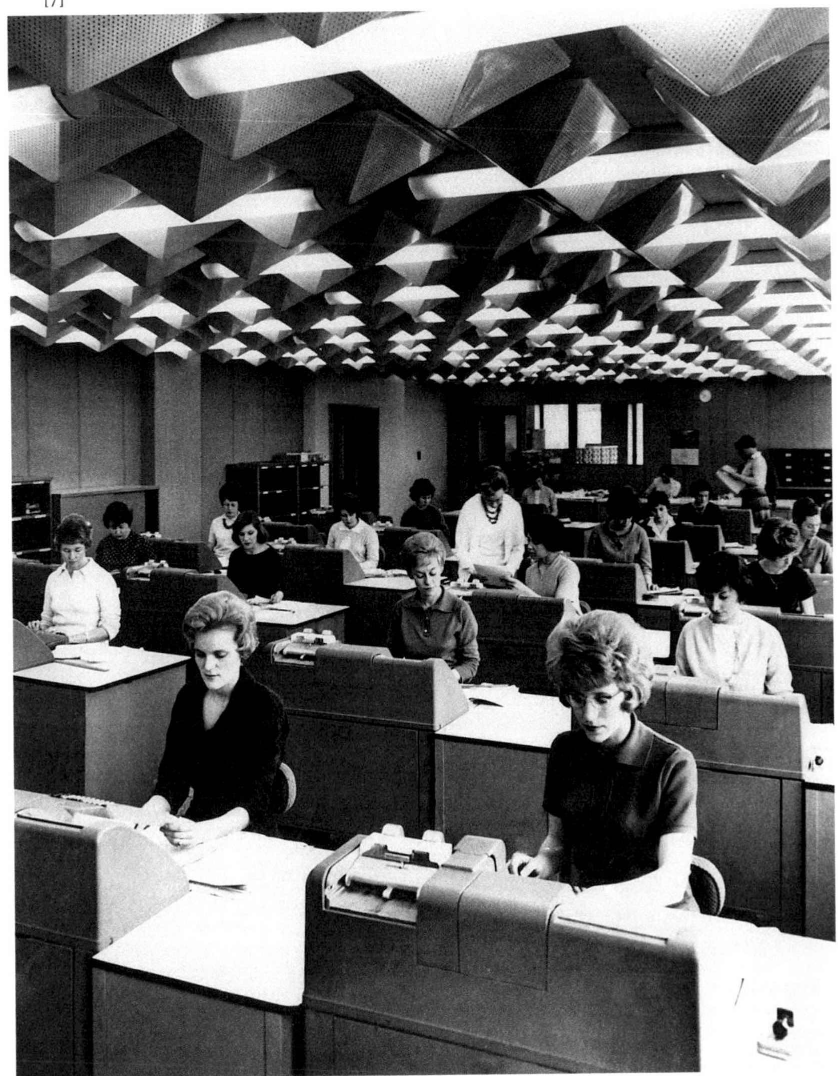

Seeking synergies in manufacturing oil The Group's manufacturing policy initially concentrated on building large manufacturing complexes centred on fluid bed catalytic crackers and capable of turning any particular type of crude into a wide range of oil products, feedstocks, and chemicals. Building on the experience gained during the war with cat crackers at Wood River and Wilmington in the US, three more such installations were constructed at Houston, at Pernis near Rotterdam, and at Stanlow (UK); the fourth, intended for Balik Papan, was indefinitely postponed in 1949 following the political unrest in Indonesia.[15] The Houston unit came on stream first, in 1946, and commercial production of synthetic glycerine, that famous breakthrough, started there two years later.[16]

The Pernis refinery near Rotterdam serves as a good example of Group policy. In May 1947 Bataafsche organized its first public press presentation to announce plans for expanding the crude refinery with a conventional atmospheric distillation unit, a vacuum distiller, two wax crackers, and a de-waxing unit, so the slack wax from the very paraffinic Schoonebeek crude could be used as feedstock for the ester salts needed to produce Teepol.[17] The Group's 1947 agreement with Gulf Oil necessitated further substantial and immediate extensions to process the Kuwait crude obtained by the deal, much lighter but more sulphurous than the Schoonebeek or Venezuelan crude processed until then. Known as the Middle Eastern Crude project (MEC), this bold expansion envisaged a total capital outlay at Pernis and Stanlow of almost 140 million guilders (£13 million) in five years to treat an extra 1.2 million tons of crude in 1949, rising to 2.5 million by 1951. The main items of expenditure were a third crude distiller, two more high vacuum distillers, the first fluid bed cat cracker to be built in Europe, a platformer (about which more below), a lube oil blending and filling

Shell Chemical scored a world first in 1948 when its synthetic glycerine plant (below) came on stream in Houston. At Stanlow in the UK on 28 June 1951 (right) huge gantries raise a steel fractionation column for the refinery's second distillation plant.

[8]

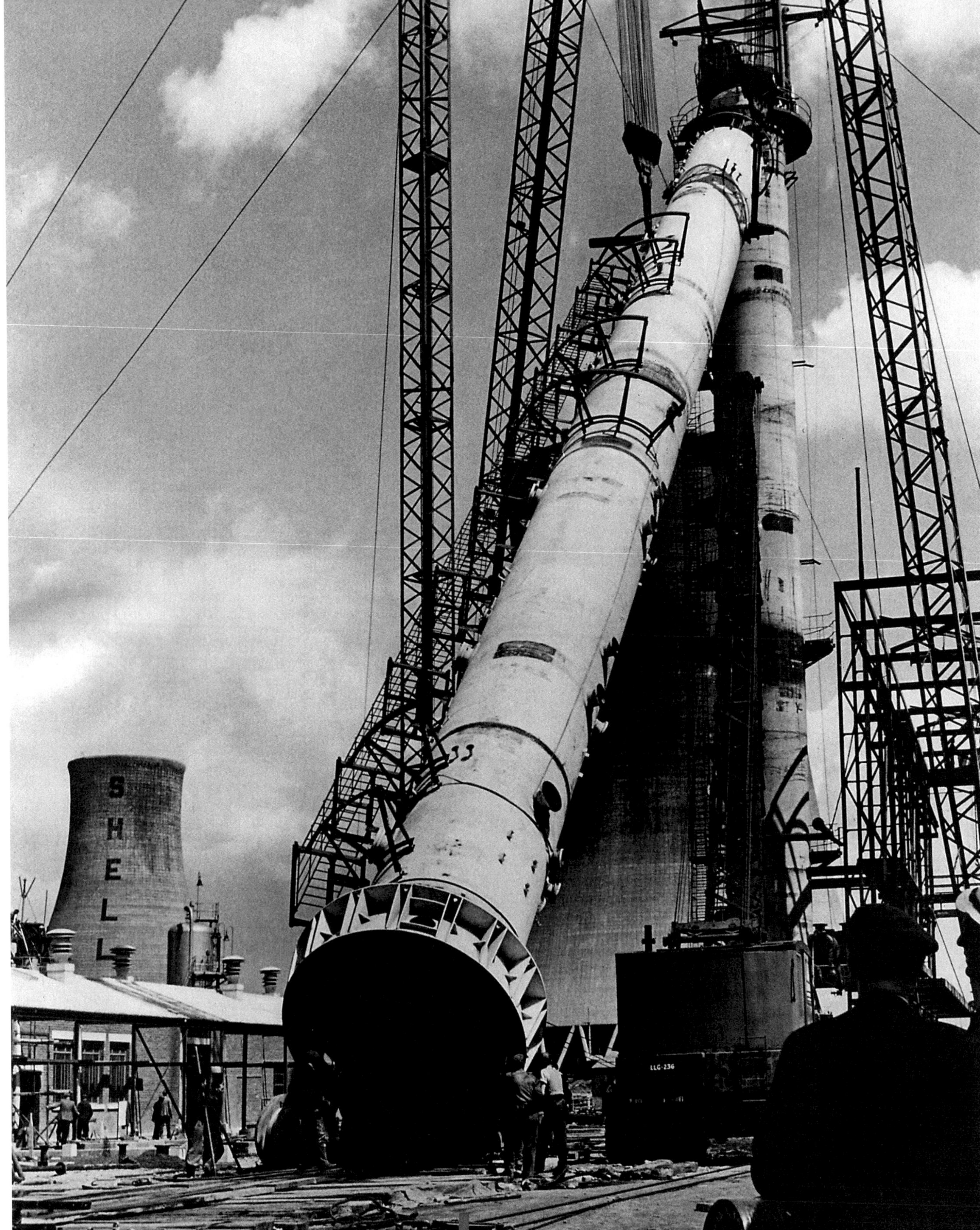

[9]

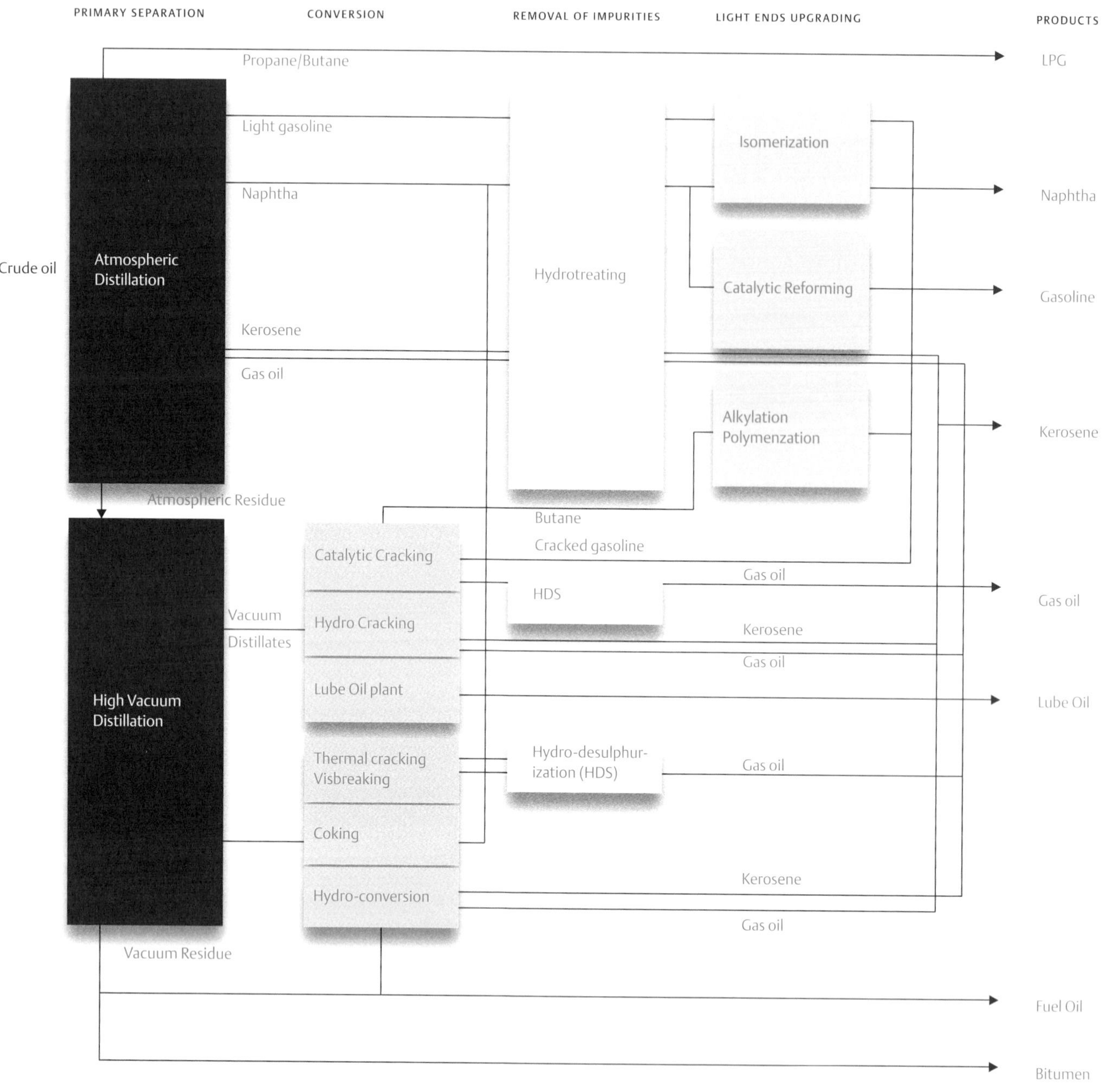

With the introduction of secondary processing, manufacturing oil became an articulated process involving several stages designed to upgrade intermediates to high-quality products.

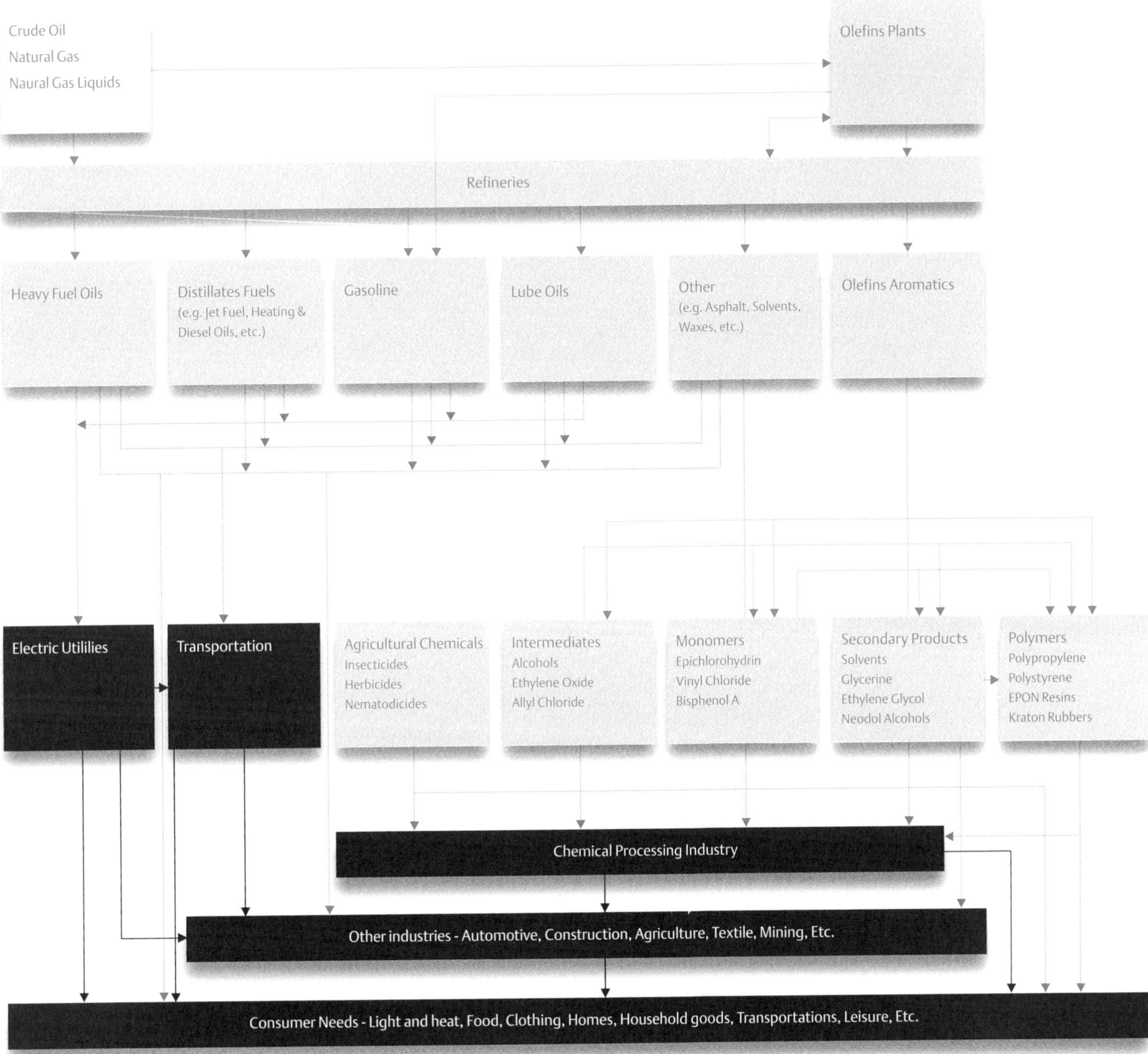

The diversification into petrochemicals during the 1950s and 1960s widened the oil industry's scope enormously, as this Shell Oil scheme from the early 1970s demonstrates.

[10]

As the Pernis refinery began to recover from war damage and came partly back on stream again in 1946 (left), Ir. A. H. S. Welling of BPM shows reporters around the site. The refinery provided employment for many, but to have such a large installation on one's doorstep (right) was a mixed blessing at best.

plant, and a second processing plant for butane and propane gas. A separate programme for building or buying residential accommodation to house the staff and their families cost another 13.5 million guilders or £1.3 million, obtained by temporary loans from the Group pension fund and an insurance company. By December 1951 most of the new plant had come on stream, raising Pernis throughput to 122,000 b/d, or about six times the pre-war figure. Just over half of the crude processed came from the Middle East, some 10 per cent from Schoonebeek, and the rest from Venezuela. About a third of production was sold in the Netherlands and the rest elsewhere in western Europe, with some special products supplied to Group companies further afield.[18]

With this expansion, Shell finally realized its intentions, dating back to the 1927 decision to leave the Charlois location for Pernis, of building a large refinery as the hub of its European operations (see Volume 1, Chapter 7). The complex enabled the Group to integrate its continental manufacturing operations and to develop an overall programme balancing supply and demand, achieving at the same time economies of scale and increased flexibility. In the US, gasoline was the main product of most refineries, but the Group's European refineries produced relatively more gas oil, fuel oil and kerosene, than gasoline, and no lube oil. Reconfiguring each of these refineries to meet the rising demand for gasoline and lube oil would have been uneconomical given the volume of investment required, the size of the individual markets, and the increasing sophistication of the components needed. Instead, concentrating the mass production of intermediates enabled the Group to retain a basic configuration for its other European refineries. Intermediates issuing from Pernis, such as components and dopes for high-octane gasoline or lube oil, were partly blended and finished on site, but mostly shipped for blending or final processing to refineries abroad, in volumes and specifications tailored to local needs. Conversely, these installations sent their residuals, such as heavy fuel oil, to Pernis

[11]

[12]

for further processing. Moreover, most refineries were configured to one or two specific intakes, but, as a consequence of the Group preference for flexibility in supplies, Pernis could process almost any type of crude, so surplus supplies from shutdowns elsewhere could be sent there. Large-scale production yielded impressive returns: Pernis quickly became the Group's cheapest manufacturing unit. In 1956, its operating costs were half of those at Curaçao or Cardón, a third of Berre-l'Étang, and less than a fifth of Stanlow.[19]

The need for a large, centrally located refinery to process Middle Eastern crude, combined with Shell's wish to balance its European manufacturing operations with a large, integrated petrochemical complex, clearly outweighed other considerations, such as the Group's preference for siting refineries near the source rather than in consuming countries. Buttressed by agreements between the majors, this had been the prevailing industry practice before the war. In 1940, some 14.5 per cent of Group refining overall

[13]

was done in consuming rather than producing countries, but in Europe the principled position was already slipping, local processing reaching 60 per cent of sales.[20] As a rule, the Group had only sited installations close to consumers if fiscal discrimination favoured local processing, as in France and Germany, or if practical considerations such as defence requirements did so, as in the British case with Shell Haven and Stanlow.[21] Pernis thus showed a marked change in this particular policy. Otherwise the huge post-war rehabilitation and expansion programme, which in 1947 had reignited the 50:50 Group ownership debate, followed established policy in designating Cardón in Venezuela, Curaçao, Balik Papan, plus a new refinery planned for Sumatra, as the Group's main processing centres outside North America.

As newspapers reported (above right), Shell's Pernis refinery became the site for much of the Group's new manufacture of chemicals, and at the opening of the new cat cracker on 9 November 1949, Van Hasselt (bareheaded, above) led the Minister of Economic Affairs J. R. M. van den Brink on a tour of inspection.

[14]

ataafse chemische fabrieken Pernis geopend

a-oorlogse bijdrage aan industrialisatie van ons land

(Van een onzer verslaggevers)

ER belangstelling van vele honderden genodigden heeft anochtend de Minister van Economische Zaken, prof. dr. M. van den Brink, officieel de chemische fabrieken ge- welke de Bataafse Petroleum Maatschappij (Kon. Shell- annex haar raffinaderij-installaties, te Rotterdam— heeft gebouwd, als een van haar eerste na-oorlogse bij- aan de industrialisatie van Nederland. Gepavoiseerde n in de petroleumhavens, wapperende vlaggen en een nde gasflambouw gaven een feestelijk tintje aan de gheid.

taat van minutieuze studie

ieuw gebouwde schaftlo- e grote, uitgebreide ter- de B.P.M. te Rotterdam- aren omstreeks 11 uur iteiten aanwezig, verte- gers van ambassades, n takken van dienst der Rotterdam, afgevaardig- t bedrijfsleven, van be- lustrieën, aannemers en s. Ir. H. H. Buss, direc- raffinaderij der B.P.M. eerst allen welkom, in er prof. van den Brink. n kort historisch over- et wel en wee van het rende de bezetting, toen ernield werd. Nadat de k hersteld was, ging de uitbreiding over en zó rein te Pernis een che- strie door de bereiding hetisch — beter gezegd — wasmiddel en een en thermoplastiek van ychloride-type. Beide in door Nederlandse et B.P.M.-laboratorium n uitgewerkt, aldus ir. Buss. Met leedwezen constateerde hij, dat ir. S. J. van Peski, die hierin een belangrijk aandeel had, deze dag van opening niet meer heeft mogen beleven. Er kwam, na de bouw, een tweede uitbreiding, ter verwerking van meer ruwe olie met de daaraan verbonden veredelingsinstallatie voor de eindproducten. Beide projecten tezamen noodzaakten tot uitbreiding van de stoom-, water-, electriciteitsvoorziening, verladingsinstallaties, tankenbouw etc. Met enthousiasme hebben allen zich op de uitbreiding geworpen en ir. Buss dankte tenslotte allen die aan de totstandkoming der nieuwe fabrieken hadden medegewerkt. De B.P.M. zal het researchwerk voortzetten om te zorgen dat Nederland de voorste plaats in de productie zal kunnen bezetten.

Rede minister v. d. Brink

In de rede, welke de minister van Economische Zaken, prof. v. d. Brink, ter gelegenheid van de opening van deze nieuwe fabrieken te Pernis hield, gaf hij nogmaals als zijn mening weer, dat toenemende industrialisatie voor ons land een levensvoorwaarde is.

Het eenparige streven van de West-Europese landen naar een hogere export, gaat gepaard met pogingen tot slechting van tariefsmuren en andere belemmeringen, welke op het ogenblik nog het internationale handelsverkeer beperken. De concurrentie zal in de toekomst scherper zijn dan voor de oorlog.

„Slechts landen, die goederen kunnen aanbieden tegen lage prijs en van een uitstekende kwaliteit, zullen in staat zijn hun levensstandaard te handhaven en mogelijk in de toekomst op een hoger niveau brengen".

Hierop begaven allen zich naar de wasmiddelenfabriek, waar minister van den Brink, na zijn rede te hebben uitgesproken, het eerste vat, onder de flitsende lampen van fotografen en filmoperateurs, vulde met het in deze fabriek te vervaardigen wasmiddel. Vervolgens werd met een aantal autobussen een rondgang gemaakt langs de raffinaderij-installaties, zodat men zich overtuigen kon van de uitgebreide nieuwbouw-werkzaamheden, welke in de raffinaderij aan de gang zijn. Deze nieuwbouw beoogt een belangrijke uitbreiding van de raffinaderij-capaciteit en omvat o.m. de bouw van een katalytische kraakinstallatie.

Enige cijfers

Door de nieuwbouw zal het stoomverbruik met circa 300 ton per uur toenemen, hetgeen betekent, dat er per uur 300 ton voedingwater gereinigd en verdampt moet worden. Het koelwaterverbruik zal ongeveer 30.000 m3 per uur bedragen, dwz. dat deze hoeveelheid per uur door de koelwaterleidingen zal moeten worden gepompt naar de verschillende verbruikscentra en vandaar door retourleidingen weer moet worden afgevoerd. Hiertoe moesten vele kilometers leiding van 100—150 cm. diameter dwars door de oude fabriek worden gelegd.

In totaal zullen ca. 1.000 km. productenleidingen aangelegd moeten worden.

Het electriciteitsverbruik zal stijgen tot 180.000.000 kWh per jaar bij een belasting van 20.000 kW, d.i. 2½ x het electriciteitsverbruik van de gehele provincie Friesland over 1948.

These refineries, with Pernis, Stanlow, and Shell Haven, were the main beneficiaries of the £125 million to be spent on new refining capacity.[22] As discussed in Chapter 1, the choice of Cardón as a refinery location was part of the Group's concession agreements with the Venezuelan government, under which the Group had agreed to restrain capacity at Curaçao and to build Cardón into its main Caribbean processing centre. The refinery came on stream in 1949.[23] The Group spent a total of nearly £300 million, or 3.2 billion guilders, during the years 1948-53 on refinery construction. Some 80 per cent of the amount spent outside the US was on plants in Indonesia, Venezuela, Curaçao, the UK, and Pernis.[24] Seventy per cent of processing outside the US was done in these five main centres. During 1945-55, Group refinery intake expanded from 35 million tons a year to more than 100 million tons, or from over 600,000 b/d to almost two million b/d.[25]

However, by the mid-1950s the Group policy on refinery construction had already begun to follow the Pernis example, shifting in favour of building new capacity nearer consumers. There were several different motives for this change of heart, first of all market defence. Since 1949, Burmah-Shell had been in pressured negotiations with the newly independent Indian government about the building of a large, modern refinery, which the government wanted to obtain so as to save on foreign currency by adding the product value locally rather than importing it. Burmah-Shell was at first most reluctant to consider such a project as it was thought uneconomic unless the products could be sold at 10 per cent above Gulf parity prices. However, when its Indian market share appeared threatened, with product supplies from Abadan doubtful after Mosaddiq's nationalization of Anglo-Iranian operations, and with both Caltex and Stanvac planning refineries on the subcontinent, the company hastily agreed in December 1951 to build a refinery on an island near Bombay (Mumbai), which came on stream in 1954. The saving realized by importing crude for local processing rather than finished products was calculated at more than £2 a ton. It was a profitable project, too; by 1971, the Group's equity had been repaid 2.5 times.[26]

Market defence extended as the tide of decolonization gathered momentum across Asia and Africa. The Group started building a range of refineries to protect market positions established under colonial rule but now exposed to new political regimes with different priorities and objectives, such as the need to save foreign currency, to provide jobs, and to boost national prestige by shining new technology. Several Central American countries also began to press the oil companies for local processing

[15]

[16]

[17]

No place for star-gazers here: like any refinery of the period, Pernis by night in 1953 was a hideous source of light pollution and air pollution.

[18]

facilities. The viability of such projects depended on whether the inefficiencies of small-scale refining would be offset by an assured market share or other factors, so wherever possible, the Group collaborated with other similarly-placed oil companies in a policy of holding out to obtain a maximum of fiscal or other incentives, such as tied off-take requirements for other marketers or exclusive exploration rights. The refinery would then be built as a joint venture. Shell usually sought a partnership with members of what board memos typically termed 'the industry', i.e. the recognized majors. The Group usually partnered Caltex, Mobil, and Stanvac, sometimes Esso and BP, and in former French colonies also CFP, invariably with the intention of thwarting newcomers such as the Italian ENI/AGIP, despised and feared because of its readiness to enter into joint ventures with government participation.

Shell's diffidence towards ENI/AGIP was partly inspired by the general principle of never helping newcomers to establish themselves, and partly by a deep mistrust of Mattei's intentions. As discussed in Chapter 3, Mattei, not content with the room for expansion which the widening gap between posted and real crude prices opened for the independents, sought to advance his company by positioning ENI/AGIP as the champion of fair deals. He granted much more favourable terms to host governments, thereby upsetting carefully maintained agreements and price structures. In line with the other majors, the Group treated him with great circumspection because managers feared he could not be trusted as a business partner. Feeling rejected by the established powers from whom he craved acceptance, Mattei widened his offensive, publicly excoriating the majors as 'the seven sisters' working to keep oil prices unduly high and threatening to expose business practices amongst them. Antagonism now turned to open hostility with publicity salvos from both sides and threats of litigation. The Group became utterly reluctant to even quote prices to ENI/AGIP for fear of any business information being used against Shell.[27] Immediately after Mattei's sudden death, the CMD sought 'with the utmost caution and delicacy' to find out whether the Italian company would change its policy.[28] When this turned out to be the case, Shell gave up its antagonistic attitude, the CMD agreeing, after consultation with the other majors, that 'provided ENI were indeed prepared to act in a reasonable manner in their commercial relationships on the African continent (as set out in the agreed note of our discussions with them dated 7th May), then the established marketers' relationships with ENI should be on a normal commercial basis like any other oil company under similar circumstances. It is clearly understood that ENI must first establish themselves on the market in the country concerned, each case thereafter to be dealt with on its merits as it arises.'[29]

The sisterly approach to newcomers encountered increasing opposition from the independents and from emerging national oil companies in Asia. Proposals from 'the industry' for joint venture refineries in Thailand, Korea, and Ceylon were rejected in the face of alternative offers. The Pakistani government skilfully played a consortium of Shell, Burmah, BP, and Stanvac against other contenders into conceding a 40 per cent local participation in the proposed Karachi refinery. Other countries soon followed this precedent and the Pakistani government repeated the procedure in 1962 to get the Group to build a bitumen and lube oil plant.[30] Shell came to own substantial and sometimes majority stakes in thirteen such defensive refineries, often also acting as operator. Most of them had a fairly small capacity of 10-20,000 b/d and possessed only basic processing equipment, but this resulted in the lowest operating costs of all Group refineries outside the US. Moreover, being joint ventures with the main local marketers, the refineries

[19]

[20]

Above, delivery to Curaçao of a section of High Vacuum column number 3A, 5 April 1953. Right, strengthening rings are welded onto the column and the adjacent column 3B.

usually operated at near capacity levels, disproving the original doubts about their economic viability.[31] Though initially built for defensive reasons as well, Mumbai and Karachi did not really belong in this category. Both had state-of-the-art processing units, including cat crackers and catalytic reformers, for a full range of products, even if this had required governmental pressure in the latter case.

Security of supply, a traditional concern for Shell, provided another motive for moving refinery capacity from producing countries to consumers. The Group never regained sufficient production in Indonesia to maintain its export trade there; indeed it had to import a rising volume of crude to satisfy local markets. Coming as it did on top of the other operational handicaps imposed by local conditions, this check effectively defeated the plans for a new petrochemical complex in the archipelago to serve the region in the way that Cardón and Curaçao, the UK installations, and Pernis did in their parts of the world. Because of the continuing uncertain political outlook, the Balik Papan cat cracker project was postponed indefinitely in September 1949. Four months later the Bataafsche board did sanction a budget to reconstruct the reformer at Pladju, but only because of its short pay-out time of one year. From 1950, investment in Indonesia steadily dropped as Bataafsche restricted itself to a care-and-maintenance policy. The equipment ordered for building a new Edeleanu plant at Balik Papan was redirected to Pernis.[32] The volume of oil refining in Indonesia stagnated as the Group built the new required capacity in Brunei, Malaysia, Japan, Australia, and New Zealand, to process crude from the Middle East or from Venezuela. As late as 1955, 60 per cent of the Group's manufacturing in Asia was done in Indonesia, but in 1960 it was already less than 50 per cent and by 1964, a year before the transfer of the installations to the Government, the share had sunk to below 30 per cent. Together the refineries in Malaysia and Singapore now processed almost as much crude as the Indonesian ones.

Meanwhile the Group had adopted an additional policy of spreading capacity rather than concentrating it. The share of the original central locations in processing outside the US steadily dropped: after touching 60 per cent in 1960, it was only 40 per cent ten years later. The Singapore refinery, near the old site of Pulau Bukom where Mark Abrahams had built the Tank Syndicate's first bulk installation in 1890, eventually became the Group's biggest manufacturing complex in Asia; like Pernis it was a balancing refinery and the main regional centre for certain streams, notably lube oil manufacture and blending. By 1971 it processed more crude than the

Below, delivery to Curaçao of part of the fractionation column of the cat cracking unit, 31 July 1956, and right, the installation of the upper part of Stabilizer number 3 on the refinery's polymerization unit, 13 February 1930.

[21]

CPIM 34

During the latter 1960s and 1970s the Group's Singapore refinery (below) on Pulau Bukom, close by the Samuels' original tank storage site, became its largest Asian manufacturing centre.

Expansion at adjacent Pulau Bukom-Ketjil began in 1970 (below right) and continued until virtually the whole island was a tank farm.

[23]

[24]

Group's combined Indonesian installations ever had after the war, and with a capacity of 460,000 b/d it was second only to Pernis.[33]

Even so, the other refineries in the region were all substantial operations in their own right, configured to supply a full range of products to the markets concerned. After the transfer of the Indonesian refineries, Japan became for a while the Group's biggest manufacturing operation in Asia, overtaken by Singapore in 1971; and the regional installations were sufficiently sophisticated to function as test beds for new technology. Coming on stream in 1964, the cat cracker at Clyde in Australia pioneered a new integrated energy recovery system, in which the heat and pressure in the catalyst regenerator gases drove a turbine to power the unit. Since 1959 there had been a string of successful trials with simpler energy saving units at Pernis, Hamburg/Harburg (Germany), Berre-l'Étang (France), and Oakville (Canada), but the ambitious Clyde configuration suffered many initial problems, because if one of its interdependent parts failed, and its waste heat boiler system often did fail, the whole plant had to close down. Several years of costly experiments were needed to correct this.[34]

Once under way the shift of manufacturing capacity to consumer countries was considerably boosted by fundamental changes in transport economics, about which more below. Continuing increases in the deadweight capacity and speed of tankers translated into rapid and continuous decreases in unit costs, rendering the transport of crude rather than products economically viable. The need for flexibility in crude supplies, as exemplified by the expropriations in Iran, the supply constraints in Indonesia, and again by the 1956 Suez crisis, also made carrying crude preferable. Before the war, Shell's processing in consuming countries represented an annual volume of crude shipped of some 5 million tons on total product sales of 37 million.[35] We do not know when managers decided to change overall transport policy from shipping products to shipping crude, but the switch appears to have been largely completed by 1959, since in the Annual Report for that year the section on transport moved up from its traditional slot between manufacturing and marketing to the one before manufacturing and immediately following exploration and production.

Pipeline construction also reinforced the shift, notably in Europe. Transport by pipeline was by far the most economical means of conveying oil. Barges cost almost twice as much as rail and three times more than pipelines. However, the threshold capacity of a pipeline, i.e. the cost at which its operation became profitable, usually exceeded the capacity of a single refinery, so most pipelines were built as joint ventures.[36] The Group had a 40 per cent stake in the Rotterdam-Rhine pipeline, which in 1960 started pumping crude to the new Godorf refinery, centrally located near Cologne in the Ruhr, Germany's industrial heartland at the time. The Godorf refinery was one of four new refineries, with a combined capacity of about 230,000 b/d, built by the oil industry and fed by the pipeline from Rotterdam and another one from Wilhelmshaven.[37] By 1972 Godorf's initial capacity of 80,000 b/d had more than doubled to 180,000 b/d, overtaking the Group's old refineries at Harburg in the north and Monheim further up the Rhine. Though entirely supplied by pipeline, most of Godorf's output was still shipped by barge and thus was subject to the river's seasonal limitations to traffic, although considerably less than Monheim's operations.

As a shareholder in the Société du Pipe-Line Sud-Européen, the Group also participated in the building of a crude pipeline from Marseilles to Strasbourg, Karlsruhe, and Ingolstadt in Bavaria. Launched in direct competition with an ENI project for a pipeline from Genoa to Ingolstadt, the Southern European Pipeline had a

Like cities, refineries are often subject to development and change. The Group's Norco refinery in Louisiana, which began processing heavy Mexican crude in 1919, is seen here undergoing modernization, probably in the 1960s.

[25]

whiff of sisterly action about it. Since the pipeline's crude coming up from Marseilles would undercut by a considerable margin the existing product supply lines to eastern France, southern Germany, and northern Switzerland, the Group had to have new refineries in those areas. Three medium-sized joint venture refineries were built, at Reichstett-Vendenheim near Strasbourg, at Ingolstadt near Munich, and at Cressier near Neuchâtel, with a combined capacity of 164,000 b/d. When the pipeline opened for traffic in 1963, the industry had built refineries with a total capacity of 500,000 b/d in southern Germany to take advantage of the cheaper supplies.[38] Until then the high cost of transport by barge or railroad from the north and west had kept oil prices in the area markedly higher than elsewhere in Germany, but now the difference narrowed considerably for gasoline, and dramatically in the case of fuel oil.[39]

The building of Godorf and Ingolstadt formed part of the continuing rapid expansion of manufacturing. Between 1955 and 1970, Group refinery intake rose from 100 to 250 million tons a year, or from almost two million to over five million b/d.[40] As the demand projections emanating from the Supply and Planning function climbed ever higher, the hectic pace of manufacturing expansion, averaging 6.5 per cent a year, became an end in itself, putting tremendous pressure on the Group's financial and management resources. In the early 1960s it took only about three years to realize a new refinery project, at a cost of $600-800 per daily barrel, but in the face of relentlessly rising demand even that was often too long and too expensive.[41] In 1962 manufacturing worked at full capacity and it was no longer possible to meet the seasonal variations in demand with spare capacity.[42] During the winter of 1962-63, a lack of processing capacity caused Shell to buy 500,000 tons of middle distillates and still to lose 250,000 tons of business.[43]

Refinery efficiency teams had brought some solace in the past, but now the Manufacturing Oil staff had to find ways, somehow or other, of creating capacity quickly. This could sometimes be done simply by modernizing units. During the mid-1960s, Shell Oil embarked on an ambitious expansion and modernization programme at Norco, Martinez, and Wood River. Crude intake capacity at Martinez rose by 75 per cent, but new and more efficient installations raised gasoline output six times.[44] Simpler expedients included raising the capacity of processing units above the design ceilings and by so-called 'debottlenecking' operations, that is, the removal of specific constraints in the overall product flow.[45] The rationalized capital approach, i.e. trimming design specifications to exact requirements, helped to free capital for investments elsewhere. Such programmes could achieve substantial savings. Trimming design cut the capital cost of four refinery projects by a total of £14.6 million in 1962. Three years later, a debottlenecking project at Karachi raised output by 16,000 b/d, or 66 per cent of the original total there. In 1969, similar programmes at five other refineries created another 70,000 b/d of extra capacity.[46] However, design rationalization halted before one of the Group's operational principles, flexibility of supplies. Refineries had to be able to process different crudes. Given the history of recurrent ruptures in supply lines, it was reasonable to insist on this feature, but it added cost and complexity which Esso, for instance, avoided.[47]

Mirroring the shift in marketing discussed below, Group processing in Europe rose from 28 per cent of Shell's worldwide total in 1955 to a peak of 44 per cent in 1970.[48] Most of this growth occurred in the major markets of the UK, France, and Germany. Only Italy remained a persistent problem. From the late 1950s the Group suffered a shortage of Italian refining capacity which seriously handicapped its efforts to compete effectively. Its La Spezia site, near Rome, could not be expanded and the Italian government refused permission to build a new complex, since this would aggravate the existing national overcapacity. Processing deals with other oil companies brought some relief, without however lifting the constraints.[49]

Gradually European manufacturing became decentralized by the building of substantial new capacity in Scandinavia, Eire, and Switzerland, all of which had until then been supplied from the Netherlands and the UK. This growth of demand and processing facilities elsewhere nibbled at the original rationale behind Pernis, and the complex saw its share in European processing drop from 36 to 21 per cent. Its cost leadership also came under threat from new highly efficient Group refineries such as Fredericia in Denmark, on stream in 1967 with processing units neatly arrayed around a single smoke stack and a single control room.[50] Such a lay-out could not be easily replicated on the heavily congested Pernis site, where new installations had to be shoehorned into any, not necessarily the best, space available. Even so, regular efficiency audits and flying teams helped to spread best practices rapidly throughout the Group, for initial manufacturing cost differentials between new and older installations narrowed within a few years.[51]

During the second half of the 1960s Shell clearly achieved a greater degree of operational integration. As noted above in Figure 4.1, the productivity gap between Shell Oil and the rest of the Group narrowed considerably. This new synergy also affected Group accident ratios. Industrial safety was always a prime managerial concern, but evidently one left to local operators with little direction, if any, from the centre. In June 1968, perhaps spurred by two serious accidents at Pernis earlier that year, Manufacturing Oil presented its first Group-wide survey of accident ratios. Shell Oil had a frequency rate (lost-time accidents per million man-hours) of 4.1, markedly below the US oil industry, API, and manufacturing industry averages, and less than a third of all Group companies outside North America with 13.0. Shell Oil's severity rate (seriousness of lost-time injuries) was also significantly lower.[52] To improve the safety record, Manufacturing Oil had installed a Group Industrial Safety Committee, which started out by drumming safety consciousness into all levels of the organization and by organizing plant safety audits. Within two years, the frequency rate of Group companies outside North America was down from 13.0 to 10.0 and Manufacturing Oil confidently expected the figure to be at the Shell Oil level by 1975.[53]

Black gold and precious metal: first developed in America, 'platforming' significantly improved the octane level of gasoline by passing it over a reusable catalyst of platinum and so reforming the molecules. As part of the Group's international transfer of technology, an early platformer (probably the first in Europe) is seen under construction at Pernis on 9 September 1951.

Built in 1961, the chimney stack of the second platformer at Pernis – 115 metres (374 feet) high – was one of the tallest in the world.

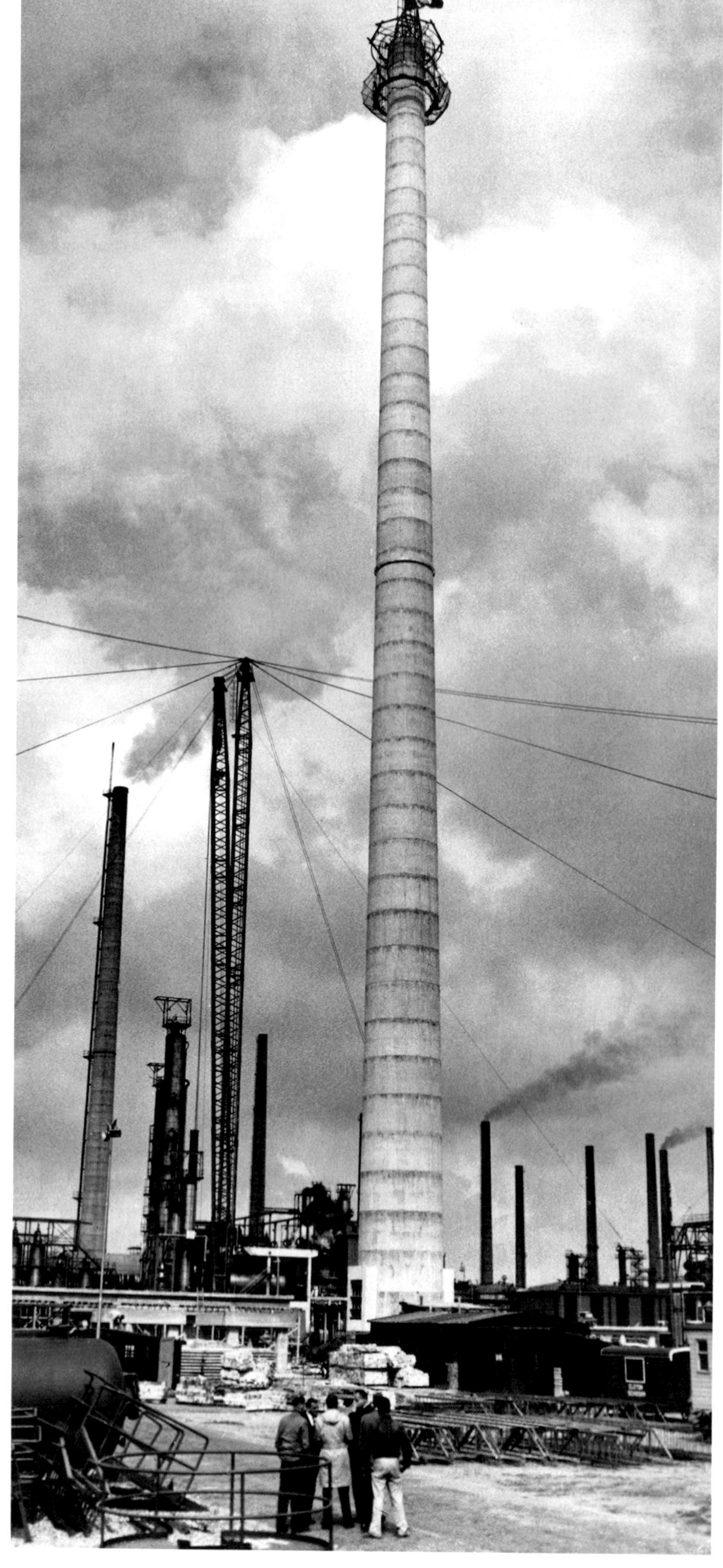

The rapid expansion of primary crude distilling capacity across Europe and elsewhere was paralleled by an equally rapid and important expansion of secondary processing and conversion during the 1950s and 1960s, another development which challenged the pivotal function exercised by Pernis. Secondary processing had its origins in the drive to develop processes, on one hand, for making high-octane components for gasoline, and on the other, for making more gasoline by conversion. Catalytic cracking converts the heavy part of the crude oil by cracking the heavy hydrocarbon molecules into lighter ones, including gasoline components such as naphtha with relatively low octane numbers. Reforming by dehydrogenation rearranges the molecules of light components with a low octane number, such as naphtha or heavy straight-run petrol, that have straight carbon chains, into branched and aromatic hydrocarbons with higher octane numbers. Catalytic crackers only make economic sense on a big scale because of their capital cost, but reformers become commercially viable on a much smaller scale.

The Group adopted secondary processing on a very wide scale, reflecting its long-standing commitment to make high-quality components at the lowest possible cost. As discussed in Volume 1 Chapter 6, the Group had used thermal reforming processes on a considerable scale since the 1930s. However, a more efficient dehydrogenation process had in the meantime been invented by Vladimir Haensel of Universal Oil Products and patented in 1947, with a reusable platinum-based catalyst. Platinum reforming or 'platforming', a word that became a registered trade mark, involved passing low-octane gasoline over the precious metal, with the result that over 90 per cent of the feedstock was reformed to an octane rating of over 80.

Subsequent process improvements raised the efficiency still further. In contrast, the older method of thermal reforming yielded only 70 per cent, with an octane rating in the 70s. Platforming was rapidly adopted as a main source of high-octane gasoline components throughout the oil industry, starting with Shell and Socal.[54] The platformer in the Pernis MEC programme came on stream in 1950, and in 1951-53 Shell Oil converted its two toluene installations at Wood River and Wilmington to the same process. Brand new platformers followed at Wood River in 1952, Houston in 1953, Martinez in 1954, and at the new Anacortes (Washington State) refinery and the rebuilt Harburg installation in 1955.[55] By the early 1960s the Group relied heavily on platforming for its gasoline supplies, with fourteen installations outside the US in 1962. In addition, the platformers were used for manufacturing components such as benzene and toluene.[56] Rising concerns over the health risks of leaded gasoline further boosted the importance of high-octane 'platformate' to avoid the use of lead additives in reducing engine knock. In 1970 the number of installations had risen to forty-seven.[57] Thermal reforming made a modest comeback from about 1963 in the form of a process adapted from it called visbreaking, which raised the yield of middle distillates and also increased refinery flexibility.[58]

Another secondary process, hydrotreating, took off in a similar way. Hydrotreating really continued the Group's long involvement with hydrogenation processes going back to the early days of Bergius's experiments during the 1920s. Reacting oil fractions with hydrogen removes unwanted elements, most importantly sulphur, and stabilizes others, such as diolefins, which may form gum during storage.[59] Since platforming produces hydrogen, hydrotreating became a matter of course in refineries equipped with platformers; Houston and Wood River obtained the first feed hydrotreater for their catalytic reformers in 1956, Norco and Wilmington following two years later. For hydrodesulphurization of kerosene and gas oil, a vital requirement in the processing of sulphurous Middle Eastern crude, the Group developed its own three-stage Shell Trickle Process, though not without a royalty deal with Texaco over a patent infringement claim. The first unit came on stream at Stanlow in 1955.[60] The refineries at Durban and Godorf were the first to be designed as integrated crude distillers with a platformer and hydrodesulphurizer.[61]

The success of hydrotreating convinced Group engineers that the future of the oil industry lay with hydrogenation and they developed a range of selective hydrogenation processes, including applications for lube oils, paraffin waxes, kerosene, and gasoline. From hydrotreating Shell moved on to build its own hydrocracking process, using a catalyst developed at Emeryville. Hydrocracking raised the gasoline yield from 45 to 75 per cent. The first such unit began operating at Martinez in 1965, and Norco, Houston, and Wood River soon followed.[62] Adding hydrogen was also considered the best option for treating heavy oils contaminated with sulphur and metals, such as those derived from Venezuelan crude, to cracker feedstock. This could alternatively be done by coking, i.e. carbon removal, a cheaper method especially suited for obtaining gasoline, but Group engineers preferred hydroconversion, partly because of their partiality to the technology of hydrogen addition, partly because they considered coking wasteful and messy, and partly because the hydro conversion resulted in higher quality products, thus recouping some of the extra cost. As part of Shell Oil's expansion and modernization programme, Wood River received a conversion unit.[63] In 1974, the Group collaborated with the Venezuelan government in a research project to assess the scope for applying this technology in that country.[64] To generate the extra volume of hydrogen required for the hydroconversion of heavy oils, Shell developed a residue gasification process in the 1960s which was widely licensed, yet only adopted by Shell for the Hycon installation at Pernis, built in the late 1980s.[65]

Manufacturing Oil thus became far more sophisticated and articulated during the 1950s and 1960s. Distillation, as the separation of crude into fractions, was relegated to the position of an initial stage of a complex refining process, and the ideal of sharp fractionation to make high-quality, straight-run products gave way to preparing feedstocks for the downstream units with the aim of maximizing the value of hydrocarbon components by fine-tuning successive processing steps. This evolution went hand-in-hand with remarkable productivity gains. In 1961, the Group employed seventy men to manufacture 2,700 b/d. By 1965, the number had fallen to forty-five, and in 1970 it was only twenty-seven.[66] Unit costs did not decline to the same degree, because conversion and hydrotreating were energy-intensive and the cost of platinum made platforming an expensive process. By 1970 the Group had at any one time stocks of platinum worth £15 million, which meant that savings in catalyst use remained amongst the first research priorities of Manufacturing Oil.[67] Even so, unit costs fell steadily from 66 US cents a barrel in 1958 to 40 in 1969, but then the tide turned. Costs first edged up slowly, then exploded during the early 1970s, reaching $1.39 a barrel in 1975.

Wage inflation was the main driver behind these spiralling expenses. Between 1966 and 1971, the costs per member of staff in continental European manufacturing doubled; for the whole Group outside North America, the rise amounted to more than 170 per cent.[68] Environmental regulations also raised costs. Until the mid-1960s the Group viewed pollution primarily from the point of efficiency, only taking action if and when savings could be made (see Chapter 6). Wider concerns did not figure very largely, and environmental regulations imposed by official authorities were resented. In 1965, the Manufacturing Oil report complained bitterly about the expenses incurred at Neuchâtel by the 'excessively stringent' air and water pollution regulations in Switzerland. However, five years later managers had clearly accepted the fact that environmental costs would remain part of the game, estimating that conservation cost the Group 4 to 8 per cent of manufacturing investment, which then was running at just over $500 million a year. Also, lead times for new capacity had risen sharply: developing a green-field refinery now took six to seven years, as opposed to three in the early 1960s.[69] Yet despite these drawbacks, and although the growth of demand had shown distinct signs of slowing down since 1967, Group planners confidently expected to need more capacity, forecasting in 1970 that primary capacity would have to be raised from four to five or even six million barrels a day within five years.[70]

They were wrong. Two years later the mood suddenly changed. The de facto devaluation of the US dollar in 1971 provoked a crisis in the international monetary system and strong exchange rate fluctuations. A global recession set in and demand dropped. For the first time since 1958, the Group sold less oil than the year before. Manufacturing Oil found itself facing rising costs with capacity to spare, commenting ruefully that 'in the latter part of the sixties the Group had faced a situation where capacity coming on stream had quite often already been outpaced by demand. It now seems certain that demand levels foreseen for the objectives period and for which new facilities were planned, will not be as high as forecast a year or so ago (...) The Group's position as regards manufacturing capacity is in general not very different from that of its competitors which limits the extent to which costs could be reduced by hiring out capacity.'[71]

Projects for new refineries in Limburg (the Netherlands) and Graz (Austria) were cancelled and plans for a hydrocracker at Stanlow, selected with Limburg to introduce hydrocracking to the European operations, were shelved. After more than twenty-five years of unremitting expansion, managers suddenly faced overcapacity and a need for contraction.

Transport: economies of scale afloat and ashore The Group's marine operations showed constraints similar to those in manufacturing. In this case, the drive for lower costs was sometimes at cross-purposes with the preference for stable shipping facilities, which favoured a high degree of ownership and long charters over the vagaries and lower costs of the open market.

With the surplus of tankers that became available immediately after the war, the rebuilding of Shell's fleet proceeded very rapidly. The nearly 880,000 deadweight tons lost was almost entirely replaced by the end of 1946, and a year later the fleet was actually larger in tonnage terms than it had been pre-war.[72] A major part of this swift replacement came from the purchase of US-built T2 tankers, the equivalent in the tanker world of the famous

[28]

Built in Oregon in 1944 for the US Maritime Commission, then bought and renamed by Anglo-Saxon in 1947, the 16,600dwt *Tomocyclus* was one of 19 T2 tankers taken into Shell's fleets to replace wartime losses. Her sister *Thelidomus*, whose certificate for transiting the Suez Canal is below, had an identical construction and ownership history. Both ships transferred to the CSM flag in 1954 until their scrapping in 1961.

[29]

KINGDOM OF THE NETHERLANDS

SUEZ CANAL

SPECIAL TONNAGE CERTIFICATE

Name of Ship	Number of National Tonnage-Certificate	Home Port	Tonnage on National Tonnage-Certificate Gross	Net
„THELIDOMUS"	10174	'S-GRAVENHAGE.	10672.76	613680

Details of tonnage for the above-named Ship when passing through the Suez Canal

The spaces measured for Gross Tonnage in this Ship comprise the following and no others, viz.:—

1. Space under the Tonnage Deck included part(s) of double bottom compartment(s) number(s) 27/25 S.B. **available for draintank** — Cubic Metres 26938.19 — Registertons 9509.1

2. Space or Spaces between the Tonnage Deck and the uppermost deck: Lower Tween Deck / Upper Tween Deck

3. Closed-in spaces under or in permanent constructions above the uppermost deck, viz.:—

Space between uppermost deck and shelterdeck with side openings . . . . . 156.97 / 196.97

dry cargo 'Liberty ships'. In 1947 Shell bought nineteen, each of 16,613 dwt, replacing more than one-third of its wartime losses more or less overnight. All at first flew the Anglo-Saxon house flag, and most stayed in that fleet, but two were sold to Eagle Oil in 1949, three were transferred to the Group's Dutch flag in 1954, and one was transferred to Compañia Shell de Venezuela in 1961.[73]

The world tanker fleet nearly quadrupled in tonnage between 1938 and 1960, rising from 17 to 64 million dwt. Its profile of ownership altered as it grew. In 1938, while governments owned 7 per cent, 39 per cent was held by independents and 54 per cent by oil companies. By 1960, reflecting changing economic opportunities, the government-owned element had slipped to 5 per cent and the independents and oil companies had almost exactly changed position, with independents holding 58 per cent and oil companies 37 per cent of the world fleet.[74] Although at any given time Shell's oil would be transported by a mix of owned and chartered vessels, the company's general policy differed from the general trend by a preference for ownership over charters. As a result, the Group fleet was one of the largest privately owned in the world until the 1970s.

The most striking trait of the post war era was the increasingly rapid rise in tanker dimensions, culminating for Shell in the mid-1970s with the Group's L-class, eighteen ships displacing 315,000 dwt each. Any one of them had the carrying capacity of at least twenty-one inter war tankers, which had a typical capacity of 10-15,000 dwt. There were several distinct steps in this process, starting with Shell's order in 1947 for three tankers of 28,000 dwt. The Group's next increase was to 33,000 dwt, the first, *Vexilla*, coming into service under the British flag in August 1955, followed by the 38,000 dwt Z-class, with two British-flag and two Dutch-flag ships, ordered in 1955 for delivery in 1957. The Z-class was of some significance because of the relative dimensions of the Suez Canal. The Canal had been enlarged from its original size, and would be

Completed for Shell (Bermuda) Overseas Ltd. in August 1955 at Birkenhead, UK, the 33,752 dwt *Vexilla* was the first of the eight-strong V-class and served the Group until her scrapping in July 1972.

again, but at the time it could not accommodate vessels greater than 32,000 dwt. The Zs could of course be partly off-loaded on approaching the Canal and replenished after leaving it, but – as shown by the fact that they were ordered before rather than after the Canal's closure in 1956 – their size indicated the beginnings of a shift away from regarding the Canal as a vital waterway.

The technology already existed to build much larger tankers, and the 1956 closure accelerated the move towards such vessels. The year 1959 saw the launch of the world's first oil tanker to exceed 100,000 dwt, the 114,356 dwt *Universe Apollo*. The Shell fleet then contained nothing of remotely similar size, but on 18 October 1960 the 71,250 dwt *Serenia* was launched on the Tyne, followed on 30 June 1961 by *Solen*, both for the British flag. These ships and their sisters (*Sepia* for the Dutch fleet, and *Sitala* and *Sivella* for the French) had originally been ordered at 33,000 dwt, the order being changed in 1958.

Although Japanese shipyards were already emerging as very keen competitors, the orders for most of what became the fifteen-strong 70,000 dwt D-class were purposely placed with British, Dutch, German, and Swedish yards. A French yard, Penhoet, also benefited, mainly because of a French government decree which stipulated that French-flag vessels must be used for trading there. Chartering was not a viable option since French owners were not prepared to build vessels as large as Shell required, but the pill was

[30]

[32]

[31]

Main picture and inset, the launch of *Sepia* on 17 February 1961, blessed by Mrs H. Wilkinson. Built at Birkenhead (UK) and destined for the Dutch flag, *Sepia* marked a quantum leap in the Group's shipping history: originally intended as a standard 33,000 dwt ship, her design was radically altered in 1958, and she became one of the first of the 72,891 dwt S-class. Even such large ships soon seemed too small, and *Marinula* (right, under construction at the Odense yard in Denmark) was completed for Shell Tankers (UK) Ltd. in August 1968 at 198,637 dwt. But the trend had gone too far: in 1974 *Marinula* was sold to Shell Tankers BV and laid up in Brunei Bay.

[33]

considerably sweetened by the fact that substantial subsidies were available for French-built French-flag vessels – but were soon to be reduced. The prices under negotiation were consequently so good that the Group was even considering bringing forward some of its next year's building budget to buy an extra two ships.[75]

The competitive offers of Japanese shipyards could not be ignored indefinitely, however. A study by the Marine Section on the costs of larger crude tankers concluded that for a round voyage from the Gulf to north-west Europe, routed laden via the Cape, a 150,000 dwt vessel would have an advantage of four shillings (20 pence) a ton compared with a 75,000-tonner, the optimum size by then for passages laden through the Suez Canal.[76] The Group's next orders (the 115,000-122,000 dwt N-class, ordered for delivery in 1966-68) not only took Shell's individual vessels through the 100,000 dwt line but also included its first Japanese-built tanker, *Niso*, which became part of its Dutch-flag fleet.[77] By then, in a departure from past Group practice of undertaking its own ship-design work down to the last detail, enquiries were being made on a 'design and quote' basis;[78] and the subsequent M-class showed the irresistible rise of Japanese builders, long schooled by the American company National Bulk Carriers in the construction of very large ships. The class itself was huge, with twenty-two ships; each ship was huge, typically displacing 205,000 dwt (a model that became an industry standard, defining the Very Large Crude Carrier or VLCC); and while eight of the class were built in Japan, only one was built in Britain. Through lack of experience, however, their introduction was marred by tragic accidents. In December 1969 three VLCCs – two belonging to Shell – exploded during tank-cleaning operations at sea. On 14 December *Marpessa*, from the Group's Dutch fleet, blew up and sank on her maiden round voyage. She was the largest vessel, civil or military, ever to sink, and two crewmembers died. Fifteen days later, *Mactra*, from the Group's British fleet, blew up. She remained afloat and mobile under her own steam, but one crewmember was killed and others were seriously burned. Shell had not suffered a tank-cleaning explosion since 1943. For two to occur almost simultaneously demanded the Group's close investigation, and with the third explosion (the Norwegian *Kong Haakon VII*, just one day after *Mactra*) the ensuing two-year enquiry became an industry matter, seeking to establish the most probable cause and best solution (see Chapter 6).[79]

With these large tankers, the oil majors rerouted more and more of their supplies to Europe from the Suez Canal to the Cape. In 1966 BP and Esso each already shipped 65,000 b/d by that route; Shell, 160,000. All the other oil majors' Cape-routed imports into Europe at that time were in the 10-20,000 b/d range.[80] By the spring of 1967 the three main competitors' Cape-routed European imports had risen to 178,000 b/d (BP), 297,000 b/d (Shell) and 309,000 b/d (Esso).[81]

As a consequence, the rising political tensions in the Middle East during 1966-67 caused little concern. Shell wrote a contingency study which concluded that a closure of the Suez Canal would create short-term tanker tonnage problems but not materially affect oil supplies.[82] In the spring of 1967, the Group actually had a surplus of tonnage that had to be 'dissipated' by slow steaming, dry docking, and routeing cargoes around the Cape, with the ships returning in ballast to the Middle East via Suez.[83]

When hostilities broke out in June 1967 and again closed the Canal, however, the available surplus turned into a deficit. To maintain supplies, the Group quickly needed to find an unexpected two million dwt or more of shipping. The completion of one 100,000 dwt tanker was brought forward by six weeks at a cost of £60,000 ('money well spent'), but by the Group's own admission, BP was chartering more quickly and more skilfully, and in a highly competitive rush for extra ships the market rose steeply – freight rates tripled within ten days, from 32 to 95 shillings (in decimal terms, from £1.60 to £4.75) a ton.[84] Ships were chartered for Group loading in the Gulf and in the Caribbean at a total cost of £31 million, 'reflecting the very high market rates now prevailing', and other measures were adopted to maximize the use of existing tonnage, including the selective deferment of dry-docking, faster turn-around in ports and terminals, and the maximum use of options on time-chartered vessels.[85]

Meanwhile, senior managers of the Group sought advice from ex-managing directors who had been involved in the 1956 crisis,[86] calculations were made on how many days' or weeks' stocks would last in different scenarios,[87] shipping tonnage reports were made almost daily, and from time to time the British Foreign Office was provided with confidential reports. But though the supply situation caused many headaches, the archives convey no sense of alarm; and this was confirmed many years later by Shell Transport's chairman at the time, David Barran: 'the closure of the Canal, although it was tiresome, wasn't in any way crippling'.[88]

Redeploying its trade flows, the Group weathered the 1967 supply crisis extremely well; so, indeed, did the industry as a whole. One of the correct forecasts in the Group's pre-crisis contingency study was that any interruption of supply would be 'of relatively short term duration, i.e. assumed to last for the duration of the third quarter of this year'.[89] By drawing down stocks and enhancing production in unaffected parts of the world, supplies in general and to Europe in particular were kept strong overall, with western European demand in the second half of 1967 being fully met and with stocks rebuilt to more than the pre-crisis level, including a reasonable margin for the winter.

[34]

The Group continued to have ships built in European yards, but *Niso* (seen here in January 1967) was built in Japan and marked a step towards new and cheaper shipyards. Clearly visible in the elevation plan (below), the pecten on her funnel is the design in use from 1963 to 1972, when Loewy's new design, commissioned in 1971, was introduced.

[35]

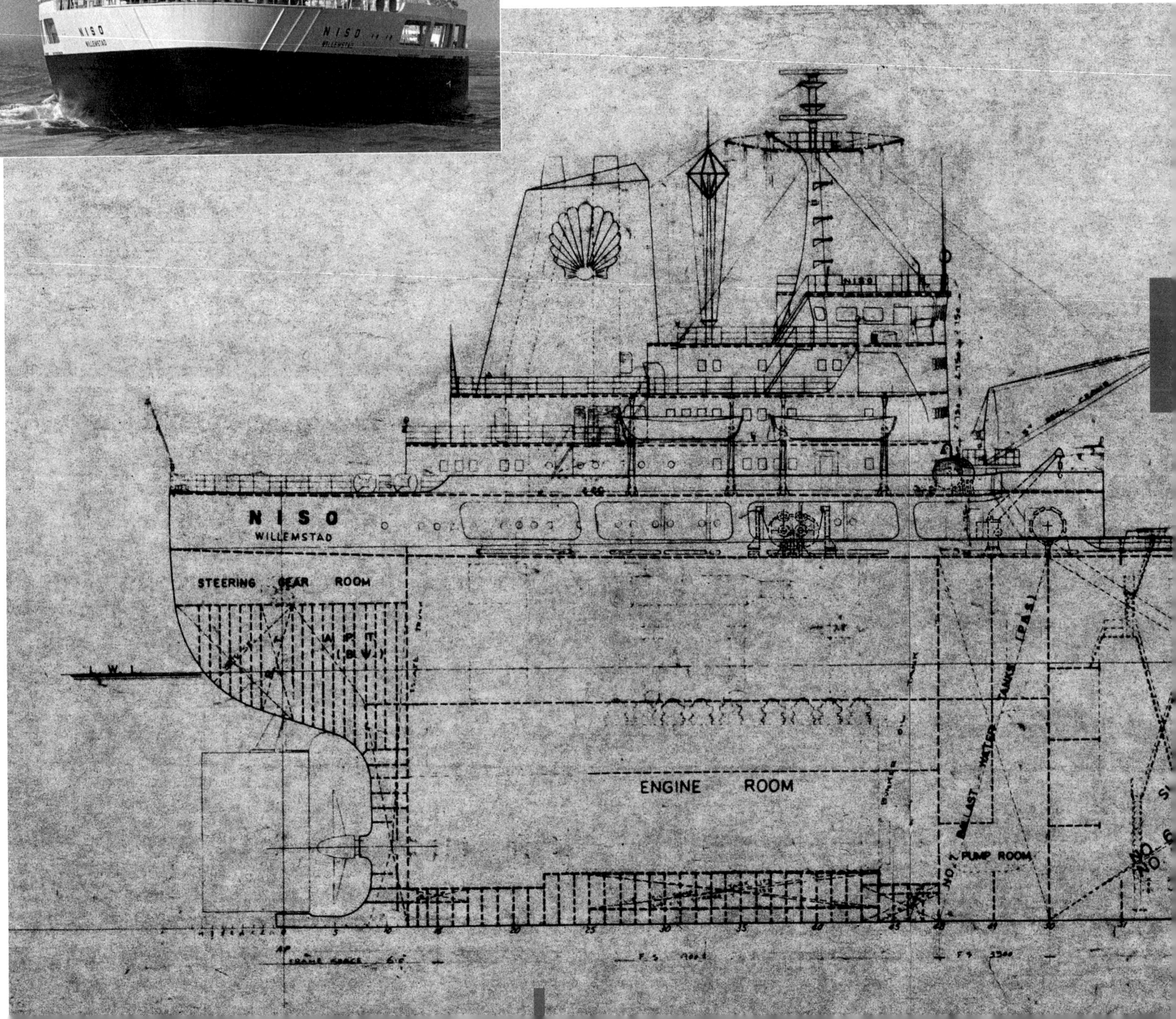

[37]

One of Shell's largest ships, the 210,000 dwt *Metula* of the ill-fated M-class, under construction by Ishikawajima Harima Heavy Industries in Aioi, Japan, in 1968. Taken into the CSM fleet on 25 September 1968, she ran aground in the Magellan Strait on 9 August 1974.

[36]

[38]

From 1968 routeing around the Cape of Good Hope became standard for VLCCs. Here that year, 12 miles off the Cape, *Marisa* is receiving supplies and mail by helicopter.

After the emergency, the trend towards bypassing the Suez Canal accelerated. The Group already had on order twenty-nine VLCCs; the first, launched in 1967, was due for delivery in 1968. Such ships could only transit the Canal in ballast: for carrying oil to Europe, the Cape route would become the standard. As an industry analyst wrote at the time, the closure of the Canal had come too soon, and as a Shell commentator remarked, 'in the light of the closures of 1956 and 1967, the oil industry could quite easily and with very little extra freight cost, dispense with the use of the Suez Canal in the shipment of crude from the Middle East to the West'.[90] The Group's founders would have been astonished.

Simultaneously three other inventive aspects of the Group's shipping operations were proceeding. First, as we have already seen in Chapter 3, Shell entered into the transport of LNG in 1960. Second, loading tankers was considerably aided by single-buoy mooring or SBM – basically the refinement of an ancient maritime technique, in which a vessel is moored to a buoy with sufficient depth of clear water all round, enabling it to swing freely with wind and tide. Its modern version originated with the Danish navy as a means of mooring small warships, and after development by Shell in the Netherlands it was first put to use by tankers in Miri, Sarawak, in August 1961.[91] But the tankers were not simply moored: the SBM was situated at the end of a submarine pipeline, and through its complex engineering was delivering oil to the receiving tanker. Shell also used the system fifty miles off Qatar at Idd el Shargi, the world's first completely offshore oilfield, and gained a great technical advantage over competitors by being able to load tankers direct from the field.

The third of the Group's sea transport developments in this period stemmed directly from the increasing size of tankers. Their growth brought economies of scale, of course, but also certain

At the time of her launch here at the Deutsche Werft yard in Hamburg, Germany, in 1966, the 71,917 dwt *Drupa* was a large ship, but only two years later she began work as a lightening tanker in UK waters, taking part-cargoes off much larger vessels. In 1976, having been converted into a North Sea shuttle tanker, she ran aground and lost 400 tons of oil. She continued to serve the Group until being scrapped in 1993.

[39]

physical difficulties, because the largest new ships had overtaken the ability of ports to accommodate them: indeed the VLCCs were so big that when fully laden they could not transit the Channel, which was a bit of a problem for those destined for Rotterdam. Shell's solution, introduced early in 1968, was 'lightening ship', that is, pumping part of a VLCC's cargo into a smaller vessel, which through the benefit of cheaper VLCC long-haul carriage costs was calculated to save a £ million a month.[92] Like SBMs, lightening ship was essentially a modern version of an ancient maritime practice, and it was a mark of the speed of tankers' growth that the typical lightening ships were members of the 70,000 dwt D-class, and even of the N-class. By any other standard the 115,000 dwt Swedish-built *Naticina* (completed in September 1967) was a big ship, yet from 1972 she was employed in lightening still bigger ones.

Economies of scale applied in Shell's transport operations on shore – specifically pipelines – as much as at sea, and for a similar reason: an increase in size raised the capacity by a disproportionate amount.[93] With these considerations and the world's escalating demand in mind, it is no surprise that the post-war Group chose not only to extend its pipelines in all parts of the world but also to make

[41]

The mammoth *Metula* at work.

[40]

them as large as circumstances permitted. In the first post-war decade and in the United States alone its wholly or partly-owned network of 'trunk pipelines' (as distinct from the smaller and more numerous 'gathering lines') grew from 3,845 miles to 5,331 miles – an increase of 1,486 miles or 35 per cent in simple length.[94] Unfortunately the data have not been located to calculate accurately the volumetric increase, but given that the new lines were generally at least 20 inches in diameter, that figure would undoubtedly be much greater. The three new major US pipelines constructed were all cost-sharing joint ventures, enabling the most efficient supply from oilfields to refineries. Built and operated by the Texas Pipe Line Company, the Basin pipeline system (Texaco 44 per cent, Shell 34 per cent, Sinclair 13 per cent, and Empire 9 per cent) was completed in June 1949, pumping 385,000 barrels a day from New Mexico to refineries in Oklahoma. With a planned maximum capacity of 296,000 barrels a day, the Ozark system from Oklahoma to Illinois and Indiana (Texaco 45 per cent, Shell 55 per cent) came into operation one month later, crossing five rivers – including the Mississippi – beneath the riverbed, and the Rancho system from West Texas to Houston (owned by seven companies including Shell Pipe Line Corporation, 38 per cent, as constructor, operator, and main owner) came on stream in April 1953 with an opening capacity of 345,000 barrels a day. As an indication of the magnitude of these feats of civil engineering, the manufacture of the pipes for the Ozark system alone involved eighty-five days' continuous operation by the fabricating mill; a sixty-wagon freight train carried newly-made pipes, to the field locations once every thirty-six hours; and during the period of manufacture 600 freight wagons were in constant use.[95]

[42] The West Texas-Houston Rancho pipeline system, on stream in April 1953.

— Pipeline

The pipeline from Oloibiri to Port Harcourt in Nigeria was only 70 miles long, but heavy rains and swampy terrain meant that it took 15 months to complete.

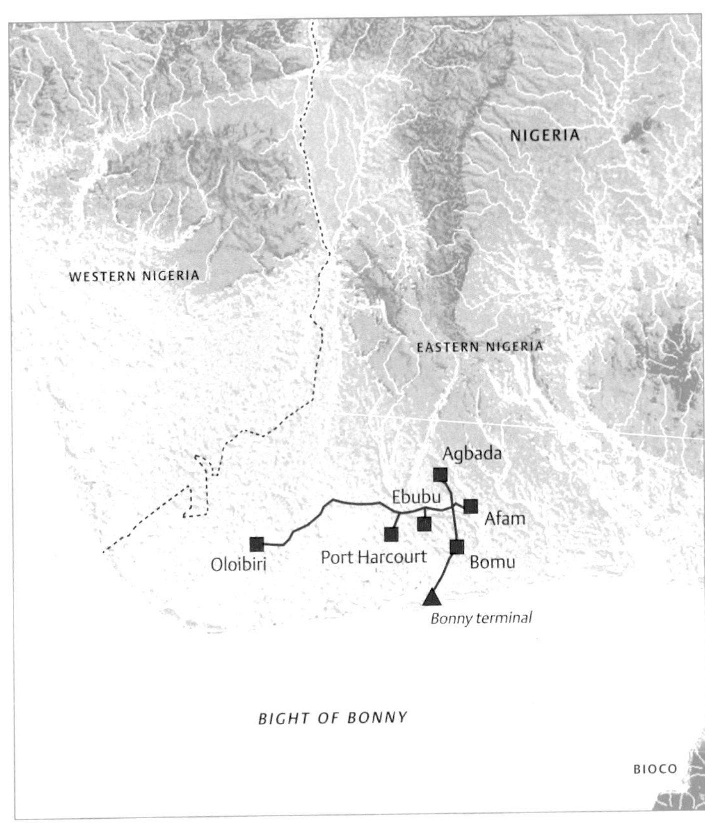

[43]

Elsewhere in the world different pipeline problems were encountered – typically climate and politics. In the US both were fairly benevolent, and the main challenges were more simply those of scale and distance. But in Nigeria, for example, a pipeline from Oloibiri to Port Harcourt had to be built across swamps, and although it was only 70 miles long, with the country's difficult terrain and exceptionally short dry season it took fifteen months (April 1957 to June 1958) to construct. On the other hand, when completed it immediately enabled Oloibiri to improve its production ten-fold from 70 to 700 barrels a day, and led to the completion in the summer of 1961 of the oil terminal and associated installations at Bonny in the Niger delta.[96]

In small crowded developed countries such as Britain and the Netherlands the difficulties tended more towards parochial politics; pipelines could not be simply laid on the surface but had to be buried (although not yet very deeply), and before work began in the spring of 1959 on six new UK pipelines from Stanlow to Carrington – four for Shell Chemical Company, transferring feedstock, chemical intermediates and finished products, and two for the marketing company Shell-Mex and BP – permission had to be obtained, under the authority of an Act of Parliament, from all local authorities and farmers, with the assurance that after the pipeline's installation they would be able to go about their business as normal.[97] Similarly, it was essential for the 24-inch Rotterdam-Rhine pipeline to be buried throughout, except where it crossed the Hollands Diep; there it was laid on the bottom and coated with a layer of reinforced concrete to make it heavier.[98] International politics figured in continental Europe as well as in the Middle East Eastern Mediterranean area, since in both it was impossible to avoid crossing national boundaries, and in 1967, for example, operation of the mammoth 40-inch Transalpine pipeline (Shell 15 per cent), running 285 miles from Trieste in northern Italy to Ingolstadt in southern Germany, was delayed by the Bavarian authorities' 'extreme – indeed, unreasonable – attitude' on the question of pollution risk, until the joint owning company accepted permanent liability of up to $6 million.[99] And with these very large collabo-rative construction projects, inter-company politics were equally inevitable, especially with Esso, a frequent partner. Sometimes Shell's exasperation is almost audible, as in 1956, when it held a long debate with Esso, BP, Gelsenberg, Ruhrchemie, and Ruhröl on 'the most economical and preferable route for a crude-oil pipe-line from the North Sea to present and future refineries in the German Ruhr and contiguous territory'.[100] In this there was a serious difference of opinion among the major participants on the question of the best North Sea harbour. To Shell it seemed plain that Rotterdam was a far better starting point than Wilhelmshaven, Esso's preference, which did not have a developed harbour and was twenty hours' extra steaming time from the sources of oil, and of

[44]

[45]

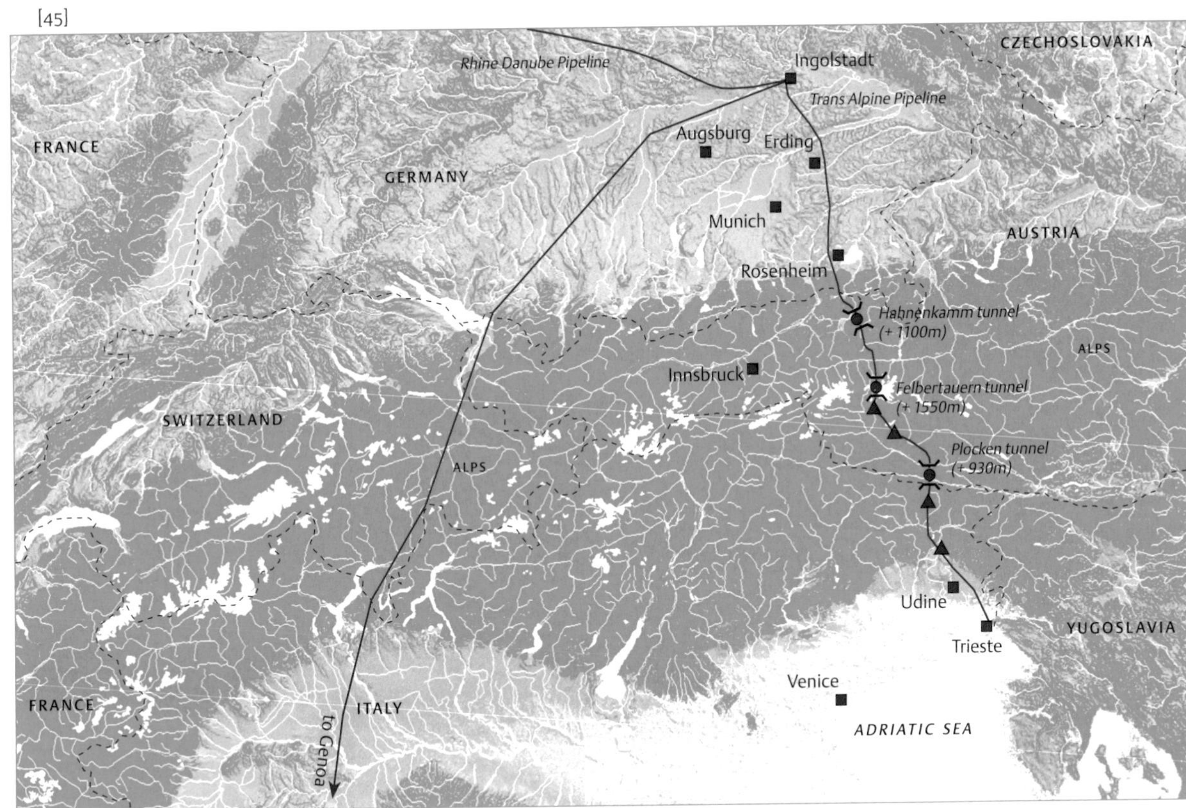

— Pipeline

Overground and underground, the route and construction of the Trans-Alpine Pipeline, 1966. This line was built to rival one from Genoa to Ingolstadt, an initiative from Mattei.

Chapter 4

[46]

course comfortably close to the Group's main continental base at Pernis. As a well-established port the necessary initial capital expenditure for Rotterdam would be 10 million Deutschmarks less, and at least another 10 million Deutschmarks would be saved each year in shipping costs, but, Shell noted, 'nothing presented so far has shaken Esso's apparent determination to proceed with the Wilhelmshaven project'. This was partly because, whereas Shell had 'full confidence in the competence of the Rotterdam Authorities', Esso did not believe they would be able to perform as promised. After a two-day meeting in Hamburg Shell found even the draft minutes, prepared by Esso, to be 'unacceptable' and rewrote them, returning them 'to Esso and BP for review, comments and acceptance'.[101] This somewhat imperious attitude did not go down well with Esso, and when the Rotterdam-Rijn Pijpleiding Maatschappij was formed to see the project through, it was funded by Shell's Bataafse (40 per cent), Mobil, Gelsenberg, and Caltex, with Esso nowhere to be seen. BP's absence was just as noticeable.

Figure 4.3
The Group's cut-of-the-barrel, 1945-1975.

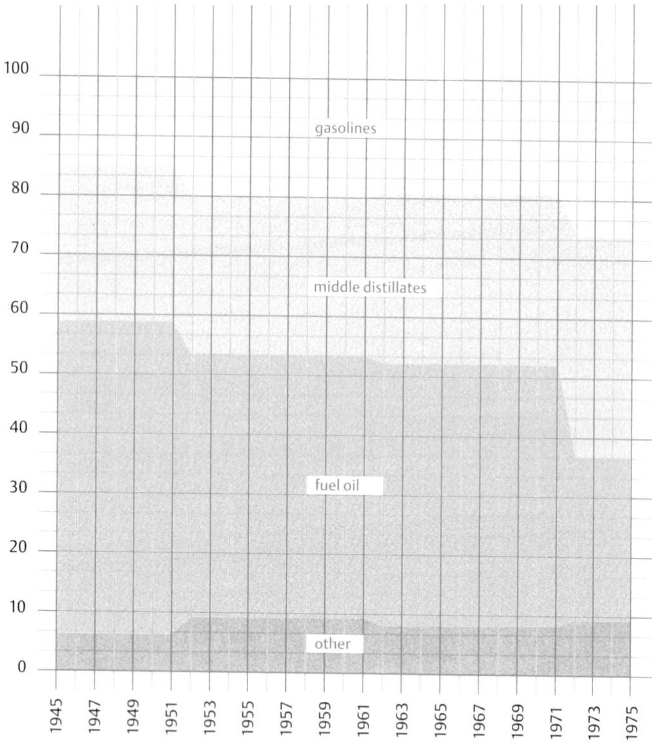

Integrating fragmented markets

During the late 1940s one of the most serious difficulties facing the Group, and the oil industry as a whole, was the enormous increase in demand world wide at a time when access to supplies was still restricted by post-war shortages of material. In Canada supply was so tight immediately after the war that Legh-Jones actually counselled Shell Canada against its proposed advertising initiatives: 'I recognize that aggressive marketing campaigns and programmes of expansion for refining and marketing facilities by competitors can be disturbing,' he wrote, but added that it was impossible to provide the extra supplies Canada wanted.[102] In much of Europe rationing continued, sometimes more stringently than in wartime, and in some countries which had given it up it had briefly to be reintroduced; but while it increased in Britain, there was no rationing in Belgium, and consumption of petroleum products in the Benelux nations exceeded that of 1938. In Britain the Petroleum Board remained in existence until its official dissolution on 1 July 1948, followed by a twelve-month transitional period; likewise in Italy the Government's oil distributing organization was dissolved, and local Group companies resumed competitive trading. As fleets were replaced, pipelines extended and refinery capacity enlarged, the flow of oil products to markets increased, and with the industry's discovery and exploitation of more sources – especially the massive fields of the Middle East – the supply-demand pendulum swung in the 1950s, moving past equilibrium to the point where the industry's ability to supply outstripped demand, even though demand was still growing.

Of all business functions, Group marketing appears to have benefitted most from the managerial reorganizations of the 1950s. The function emerged from the war as fragmented as before, scattered over thirteen regional departments, product departments such as Benzine, Bitumen, Fuel Oil, Candles, Liquid Gas, Lube Oils, or Chemical, each with its own separate accounts department, and finally a few departments dedicated to a particular market segment such as Aviation. Fuel oil even had th[ree] departments, Fuel Oil General, Engine Fuel, and Fuel Oil Accoun[ts], but then it was a product of major importance.[103] Unsurprisingl[y] gathering information was not a simple matter. In September 19[..] Jan Oostermeyer, head of Shell Chemical from 1942 until his retirement in 1953, wrote ruefully to Kessler: 'The difficulties confronting the various overseas marketing organizations in satisfactorily meeting any calls which you make upon them for s[...] data are fully appreciated by me, and in requesting information [...] such nature we have known in many cases it would be difficult, i[f] not impossible, for you to provide us with more than a rough gu[ess] as to the potential market. Nevertheless, it has been felt that not [to] approach you for such information as you might have available o[r] be able to obtain would be to overlook an opportunity of securir[ng] useful data which could assist us to obtain a more complete pict[ure] and possibly open up avenues for later outlets.'[104]

From 1946 the Group's marketing operations were gradua[lly] reorganized with the general aims of achieving shorter lines of communication, abolishing artificial or illogical management

The blessing of a service station somewhere in Japan, probably in 1963. The poster in the window shows that customers' cars would receive a different kind of blessing: Shell with ICA.

[47]

boundaries, and ensuring that reports were full without being unnecessarily frequent and detailed. Building a coherent function from fragmented units with staff scattered over various locations posed a great challenge to managers. To give just two examples, in 1952 a team from Shell Oil was sent to study the organization of Shell's Indonesian marketing operation, and as a totally separate matter, in September 1951 Hopwood pointed out that Group marketing interests 'in the western hemisphere south of the United States' were divided between the St Helen's Court office and Eagle Oil's office at Finsbury Circus.[105] The intention was to place it all under one management at St Helen's Court, which sounds a fairly simple operation; but more than two years later there were still two sources of advice on South American marketing, namely western Area Department and Product Administration, and it was not until 1 December 1953 that western Area took over.[106]

An important step towards closer integration was taken in 1950 with the appointment of Platt and Guépin as joint coordinators of Group marketing outside the US and Canada. Having two managing directors in charge was not an ideal situation, of course, the more so since both continued to have other responsibilities as well, Platt for Marine, Guépin for Chemicals and Research. Moreover, practical considerations rather than any business rationale appears to have guided the assignment of the individual departments, now counting fifteen, to one director or the other.[107] The Organisation Study Unit which scrutinized them during the early 1950s may have effected some internal rearrangements but, as with the other business functions, the Coordinating Committee which began functioning in January 1956 marked the fundamental transition from a virtual marketing department composed of scattered staff clusters loosely held together at the top into a formal administration holding devolved responsibilities, such as drafting the marketing capital budget and piloting it through the Coordinating Committee and the CMD.[108] To give the department sufficient weight in its dealings with the operating companies its director, A. Hofland, received the title Group Coordinator Marketing, and on his appointment he became a director of Shell Petroleum and of Anglo-Saxon.[109] Even then the centralization was not complete. When Marketing began organizing a worldwide review of sales departments in 1956, the operating companies in Argentina and Indonesia still received their instructions from The Hague.[110]

Further down the line the administrative arrangements were sometimes no less confusing. Writing to London about the organization in Indonesia, the Bataafsche general representative

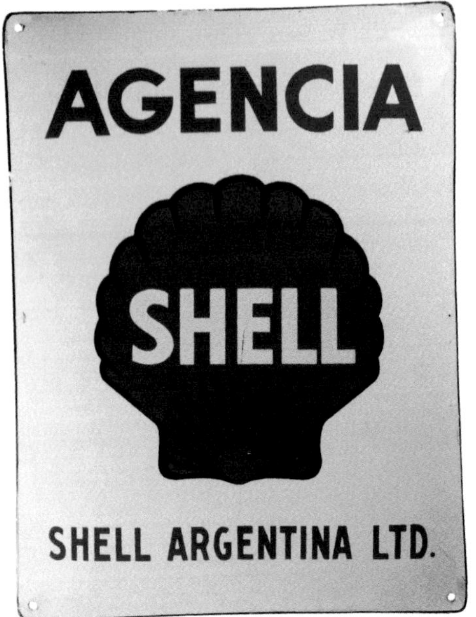

[48]

there produced an explanation which was a classic of its kind. After marketing administration had been integrated with 'some other functions', he acknowledged that 'it is not easy to express the relationship between the Manager Marketing Administration and the Marketing Branches very clearly on the organization chart.' He therefore attempted some clarification: 'As you will be aware the Marketing Branch in East Indonesia has now been integrated under the General Manager for East Indonesia, and the Palembang District has also been incorporated in the sphere of the General Manager Sumatra. The three remaining branches of Medan, Djakarta and Semarang still report directly to the Marketing Manager Administration, whilst the relationship between the latter and the marketing concerns [of] East Indonesia and South Sumatra is on the "staff" concept. Functional advice to the Medan, Djakarta and Semarang branches for Finance, Finance and Personnel, Legal etc are now given directly, also on the staff concept, from the manager of the Administrations concerned. For the time being no other major changes in the Marketing Administration are being considered, except for the possibility that the activities of the Semarang Branch may be split between East Indonesia and Djakarta with Semarang becoming a district office of Semarang Branch.'

He added, without a hint of irony, 'We trust that the organization chart, together with the above explanations, will be of assistance to you.'[111]

As already noted above, heavy oils such as fuel oil and bitumen dominated the Group's overall sales pattern, making up more than half the product volume until the early 1970s, when Coordinator Manufacturing Oil J.H. Choufoer inspired a new policy of 'whitening the barrel', i.e. increasing the share of higher-value gasolines and middle distillates in the products made (Figure 4.3).[112] Shell's geographical spread of sales remained very stable overall. The Eastern Hemisphere took around 20 per cent, Central and South America 10 per cent; North America and Europe changed places, however, the US and Canada dropping from nearly two-fifths in the 1950s to one-third by 1970, and Europe rising from 30 per cent to 40, reflecting the Group's determination to reinforce its marketing position there in the face of mounting competition. The spread of service stations differed slightly from the overall sales pattern, reflecting differences in the type of products sold and in turnover per station. In 1968, Shell had a world total of almost 81,000 service stations, of which 46 per cent were in Europe, 26 per cent in North America, 22 per cent in the Eastern Hemisphere, and 6 per cent in Central and South America.[113]

Lubricants remained a key product for the Group, requiring much technical knowledge and support for customers, notably in the industrial lube oil segment. Consequently, lubricants research usually absorbed the highest amount and the highest percentage of manufacturing oil research costs, 44 per cent on a budget of £3.1 million in 1968. Very high proceeds per barrel compensated for the extra expenditure. The Group had a lube oil market share of 17.5 per cent outside North America, by far the biggest of the majors, during the mid-1960s. The product had its own dedicated supply

[49]

[50]

network, tailored to get lube oil as close to the customer in bulk, to save on the expensive moving of packaged products. The three main export refineries at Curaçao, Cardón, and Pernis, together produced 1.8 million tons of lubricants in 1968, with three additional refineries at Stanlow, Petit-Couronne, and Singapore in the process of being built. Nine special tankers transported the products around the world to local refineries making finished products or, in small markets, to blending facilities. The know-how needed to make lubricants helped to protect the market share in this product segment from the onslaught of the independents and government-supported companies. In addition to its own retail trade of some 17.5 million barrels annually, Shell sold 1.3 million barrels of lubricants wholesale to other oil companies, in the form of either finished products for resale, base oil plus necessary additives for blending, or as Shell products under licence. Another 200,000 barrels were sold to the Eastern Bloc countries.[114]

One of the strong products in this segment was Shell X-100 lubricating oil, launched in the US in 1949. Knowing the strength of its brand name, Group marketing policy emphasized lubrication as distinct from lubricants, an essential feature being the technical service offered to the consumer, and the word 'Shellubrication' was coined to fix the concept in consumers' minds. Shell X-100 was well received wherever it was introduced, and by 1952 was available in almost every country where Shell products were sold. Another

Never shy of using top names in its advertising, Shell gained the endorsement of racing ace Stirling Moss for Shell X-100 (above left) and employed John Castle (above right) and Koen van Os (right) respectively to design its British and Netherlands advertisements.

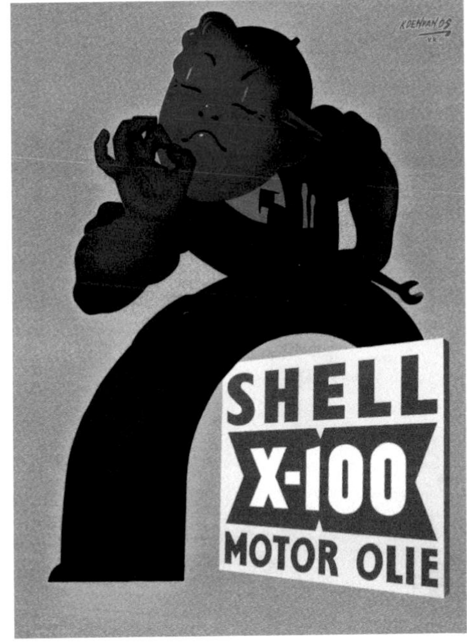

[51]

Taking lubrications technology to the edge: below, a Shell X-100 advertisement in Holland, 1951; right, Schiphol airport in the 1950s; inset right, Aeroshell.

notable new lubricant, an oil that was unaffected by nuclear radiation, was discovered in 1953 and led to a full range marketec under the name Shell APL (Atomic Power Lubricants). From this t Group gained highly specialized contracts, supplying almost all t lubricants for the world's first commercial nuclear reactor, Calde Hall in England, and soon after another in North Wales. Further research, undertaken on behalf of the US Air Force, produced She ETR Grease H, ETR meaning Extreme Temperature Range: the grease could not only withstand radiation but also functioned frc minus 100 to plus 700 degrees Fahrenheit.[115]

Aviation fuel was another strong product area for the Grou In 1954, Shell supplied an estimated 28.9 per cent of the US mark – nearly as much as its two largest competitors, Esso (17.2 per cer and Phillips (13.5 per cent) combined.[116] The following year, Shell Oil supplied the first aircraft turbine fuel in the US, called Aeroshe to a Vickers Viscount turboprop airliner.[117] And in the UK, using a radio-controlled system, from 1953 until at least 1959 Shell-Mex & BP (SMBP) supplied the fuel for eight out of every ten airliners at London Airport, providing at peak periods 150,000 gallons (680,000 litres) every hour. But compared to existing transatlant airliners, 'aircraft of the future such as the Boeing 707 and Dougla DC8' required three times the amount of fuel, and after remodelling at a cost of £600,000, an even faster fuelling system was inaugurated on 26 November 1958, foretelling the eventual need for 'some kind of hydrant system to enable airliners to pull i and be filled up rather like motor cars'.[118] To promote uniformity design, London central office coordinated the development of jet refuelling trucks, hydrants, and turbine fuel bulk installations.[119] Seven years later the Group sold 160 million barrels of aviation fue

[53]

RLANDS GOVE
aeroshell®
OIL
SHELL
ONE U.S. QUART
Shell Aviation

[54]

outside North America, of which 86 per cent was kerosene and naphtha-type jet fuel. Continuing investment in facilities at over 200 airfields around the world provided a firm base for its position as dominant supplier with 30 per cent of the market, but here as elsewhere competition had started to make itself felt. Aviation gasoline had been a specialized product which only the majors could make, but any general purpose refinery could produce jet fuel, so oil companies around the world scrambled to get into the act. As a consequence Group Marketing forecast a market share erosion of 0.5 per cent a year, since 'the international airlines are today more than ever price conscious and it is now comparatively uncommon for Shell to acquire business except at competitive prices'. Typically, the report recommended a remedy not in cost reductions and price competition, but in providing 'good service of all kinds and good personal relations with the comparatively small group of individuals who recommend or decide on contract awards'. In 1970 Shell claimed a 25% market share in aviation fuel.[120]

In the motor gasoline market the Group enjoyed considerable success during the 1950s with its premium fuel 'Shell with TCP' (tricresyl phosphate), known outside North America as Shell with ICA (standing for Ignition Control Additive). Designed to combat spark plug fouling, this product demonstrates to good effect how research findings enabled the Group to turn a supply handicap into a marketing coup. The research into additives which led to TCP began at Wood River in 1948, following complaints by the US Air Force and intercontinental airlines about spark plug fouling on long-distance flights. A patent application for the substance was filed two years later and in 1951 the US armed forces started using it with great effect.[121] At that moment in time, Shell had difficulty in keeping up with the octane race which, after a lull imposed by the Second World War, had resumed with increasing ferocity. Manufacturers competed to develop ever higher octane ratings for their premium grade petrol, but upgrading gasoline required either elaborate and expensive mixing, which yielded acceptable but uncompetitive results, or substantial investment in new catalytic crackers and catalytic reformers. The Group did not possess sufficient modern cracking and reforming capacity to sustain its market position for high-octane gasoline and accordingly started a comprehensive building programme in 1953.[122] However, by launching Shell ICA, the Group succeeded in changing the nature of the gasoline market by switching consumer attention from octane ratings to enhanced performance additives.

Launched with a concerted marketing campaign in North America during 1953 and in most other markets in January 1954, gasoline with TCP/ICA became an immediate success. Shell's US sales of all grades of gasoline, led by the premium grades, increased in 1953 by 11 per cent, while the industry average was only 4 per cent.[123] Sales in South Africa rose by five to 20 per cent.[124] Shell

Shell with ICA (known in the US as Shell with TCP) was launched in the 1950s with a worldwide publicity campaign, as demonstrated by the vehicles below in Perth, Australia, in 1954, and by the Netherlands poster (left) from 1960.

[55]

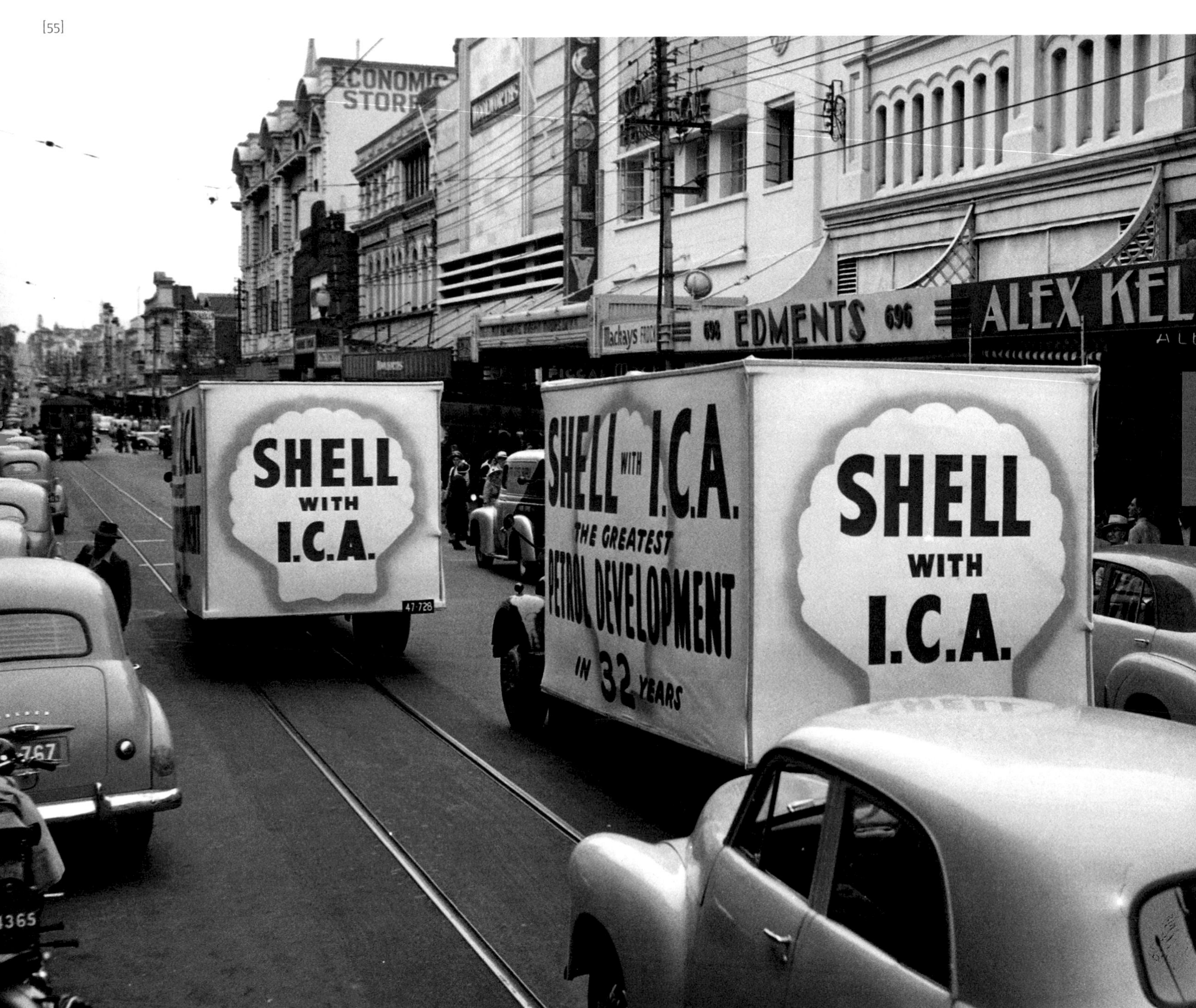

ICA helped Deutsche Shell and Shell Italiana to overcome their competitive disadvantage of being unable to produce high-grade gasoline in sufficient quantities, focusing managers' attention on the need to bring forward the building of up-to-date manufacturing facilities.[125]

The Group fully expected 'stiffer competition (...) from the introduction by major competitors of new premium gasolines'.[126] The expectation was fulfilled: competitors launched substantial advertising campaigns for their own additives and when these did not meet with sufficient success, increased the octane rating of their top grades. Jersey Standard introduced a 95-octane premium gasoline under the slogan 'Total Power'. Moreover, in 1958 the company cheekily introduced to the Venezuelan market a gasoline additive which proved to be identical with Shell's TCP/ICA. Shell's guess was that Venezuela had been chosen 'because this was one of the worst areas in which to defend our ICA patent rights (...) we would need to decide what action to take'.[127] Gulf also produced a 95-octane premium gasoline, and Cities Service one at 94.5; some smaller companies started adding TCP to their gasolines. Deep Rock named its product ECP, but Shell protested and Deep Rock agreed to drop the initials in future advertising. Clark Oil similarly called their additive TTC and used advertisements in a format that closely resembled Shell's. Within the Group, Shell with TCP was described with palpable satisfaction as 'a knock-down blow on the competition', but it was a temporary respite.[128] By July 1956, catalytic manufacturing capacity had expanded sufficiently overall for the Group to return to the octane field with Super Shell, the first 100-octane gasoline,[129] and when Jersey Standard announced a rival premium gasoline under its Esso brand, the Group's sampling of Esso Extra showed it was 'not (...) in the same class with Super Shell'. Knowing beforehand of the launch date, the Group was

Although Shell's marketing often produced beautiful imaginative images, there was also a strong tendency to rely on scientific fact. After the introduction of Shell with TCP/ICA, in 1953 Burns sent Hopwood an indignant cable (above right) about copycat competitors, followed in 1958 by another to Loudon (below right) playing down Esso's 'large headline' method of advertising. However, in 1963 *Shell News* (below) placed the pecten – perhaps symbolically – at the bottom of a multitude of competing logos, and in 1964 Esso introduced (far right) one of its most successful and enduring campaigns, enjoyed by customers all over the world.

[56]

COMPETITION: the pace quickens

Competition has been described as a cold shower – uncomfortable yet invigorating. In an era in which vast and swift changes are almost the normal course of events, it is no surprise that the force of the shower has increased sharply.

The growing intensity of competition is a fact of life for business and industry in general throughout the United States and, indeed, in many parts of the world. This is especially true for the U. S. petroleum and chemical industries. In the petroleum industry, it is caused largely by a slower rate of growth in over-all demand, against a background of excess supply and capacity; in the chemical industry, a major problem is overcapacity for some chemicals. With the growth of industrialization in the United States, competition has been the key factor spurring the American economy to the accomplishment of providing the highest standard of living in the world. To help insure that the vigor of competition would be

This modernistic design highlights a few among the hundreds of major and independent companies which compete in the gasoline market. The "direct" competition represented here is just one of the many forms of competition faced in the oil and chemical industries.

FROM MR BURNS NEWYORK DEC 31/53 1241P
TO SIR FRANCIS HOPWOOD LONDON REG.NO.162985
CONFIDENTIAL

Archief AW

517 -4 JAN 1954

YOUR 564 (FUEL & LIGHT OILS/DEPT. re TCP in Motor Gasoline
(.... patent protection etc.)
TIRE MART INC. HAS INCORPORATED SEPARATE COMPANY CALLED TRI-C-P INC. AND IS ENGAGED IN SELLING GASOLINE ADDITIVE WHICH WE ASSUME CONTAINS TRI-CRESYL PHOSPHATE TO COMMERCIAL TRUCK AND MOTOR FLEET OPERATORS.
HAVE ADVISED THEM OF OUR OBJECTION TO THEIR USE OF TRI-C-P IN THIS COUNTRY BECAUSE OF ITS CONFUSING SIMILARITY WITH OUR TCP.
HAVE ALSO ADVISED THEM OF SHELL DEVELOPMENT'S PENDING PATENT APPLICATION THIS COUNTRY AND THAT YOUR ASSOCIATES HAVE MADE COMPARABLE APPLICATION FOR PATENT PROTECTION THROUGHOUT EUROPE AND THAT THEY PROPOSE TO TAKE SUCH STEPS AS NECESSARY TO PROTECT THEIR POSITION . . .

[57]

Archief AW

RECEIVED

21 MEI 1958

USA/Marketing Policy

HH. Schepers
Brouwer
Boot
Gattiker
Huiskamp
Feith

FROM MR. BURNS NEW YORK MAY 16/58 410P
TO MR. LOUDON LONDON
REG NO 95532

P AND C
595
ESSO STANDARD ANNOUNCES -NEW FORMULA ESSO EXTRA- WITH FULL-(PAGE AND DOUBLE-PAGE NEWSPAPER ADVERTISEMENTS TODAY, MAKING THE UNSUPPORTED CLAIM -OUTPERFORMS GASOLINES COSTING 1 (1) CENT AND 2 (2) CENTS MORE-.
WE HAVE BEEN SAMPLING THEIR PRODUCT FOR THE PAST SIX WEEKS AND FIND THAT THEY ARE NOW ADDING PHOSPHORUS TO BOTH PREMIUM GRADES, IN AMOUNTS UP TO TWO-TENTHS THEORY.
WE DO NOT FEEL THEIR MIDDLE GRADE PRODUCT IS QUITE IN THE SAME CLASS WITH SUPER SHELL, ALTHOUGH OCTANE NUMBERS HAVE INCREASED SLIGHTLY OVER THE PAST FEW MONTHS.
WE HAVE FOUND NO OTHER QUALITY CHANGES COINCIDENT WITH THE ADDITION OF PHOSPHOROUS, SAMPLING CONTINUES.
THEIR ADVERTISING PRESENTS LARGE HEADLINES PRIMARILY, RATHER THAN PRODUCT FACTS.
THEY DO NOT MENTION PHOSPHORUS SPECIFICALLY BUT THEY DO CLAIM -PROTECTION AGAINST SURFACE IGNITION-.
KNOWING THAT THEY WERE PLANNING TO ANNOUNCE NEW PRODUCT TODAY WE PREPARED AND INSERTED IN SELECTED NEWSPAPERS IN ESSO TERRITORY SPECIAL FULL-PAGE ADVERTISING HEADED -TCP --UNIQUE AND ORIGINAL PHOSPHORUS-BASED GASOLINE ADDITIVE--- AND RETELLING THE TCP STORY OF THE PAST FIVE YEARS.
(CONTINUED)

[58]

dismissive of Esso's advertising, which relied on 'large headlines primarily, rather than product facts' and inserted 'in selected newspapers in Esso territory' pre-emptive full-page advertisements retelling the facts of Shell with TCP: 'Our advertising includes a solid product story, and (...) has effectively countered much of the competitive advantage they hoped to achieve.'[130]

Even so, Esso Extra was not about to lie down and die; in advertising matters, 'large headlines' often work better than 'product facts'. In 1958 Group Marketing also acknowledged that the time had passed for 'any outstanding new quality claims calling for special campaigns for automotive products co-ordinated on a global scale'. Consequently, advertising moved to brand imaging in international opinion magazines such as *Time* or *Life* on the one hand, and localized promotional schemes on the other.[131] Jersey Standard developed an ingenious way to do both. As early as 1936 Esso in the UK had used the image of a svelte tiger to advertise its gasoline, with the slogan 'smooth, silent power'.[132] In the 1950s, the tiger returned as a roaring, snarling beast, but was used only in the UK and Oklahoma. However, in the 1960s, with the slogan 'Put a tiger in your tank', it progressed into the famous and attractive cartoon of 'a good-humored, friendly, helpful animal', used to

[59]

Chapter 4

Practical and reliable: Shell's touring service was a real benefit to motorists in many parts of the world, helping to promote customer loyalty. The Shell Touring Service was started in 1955.

A sign (right) indicated service stations throughout Europe where motorists could obtain help and information for trips taking them across borders.

great effect in many countries.[133] Group managers continued to favour promotional campaigns, for their ability to push up gasoline sales by as much as 20 to 70 per cent and for their effectiveness across frontiers and cultures. In 1967, the very popular 'Match and Win' campaign gave customers torn-up Shell 'banknotes' to match and win the amount printed on them. The campaign doubled sales at some stations and inspired the 1967 marketing report to claim industry leadership for Shell in this particular part of the marketing mix.[134]

The switch to brand imaging and promotional campaigns reflected a more general transition in central marketing policy, also evident in the break with long-running and successful publicity tools such as the Shell Guides issued by SMBP in Britain and edited for over thirty years by the poet John Betjeman.[135] This shift followed a thorough overhaul of central marketing. For all its importance, the centralization under a single Marketing Coordinator in 1956 had fallen short of a fundamental reorganization. Hofland's department, looking so purposeful on the organization charts, really remained a motley collection of diverse offices scattered around the City of London. Following the 1959 McKinsey reorganization, marketing was split into Marketing-Oil and Marketing-Chemical, each with their own coordinator.[136]

The marketing reports of the late 1950s and early 1960s suggest that, during those years, the central function nursed a strong orientation towards general conditions and products, taking customers more or less for granted because this fell to the operating companies. The Marketing-Oil reports discussed the technical details of sales such as price movements, transport costs, sales trends of individual products, and the competitive situation, but showed no understanding of individual market segments or customer wants. Following the move to Shell Centre in 1963, Marketing-Oil finally became the focused and coherent organization that the Group needed. The loose assembly of regional and product offices was rearranged into seven functional departments, i.e. Planning and Economics, Product and Application Development, Market Development, Regional Marketing, Special Products, Central and International Sales, and finally Training and

[61]

Personnel, resulting in a staff reduction of twenty-eight people.[137] Outsourcing also helped to achieve savings; by 1965 it had become Group practice to effect substantially all advertising through the medium of advertising agents instead of company staff.[138]

This reorganization marked a watershed in marketing. Together with the concentration of all departments at Shell Centre, the new arrangements enabled managers to integrate marketing with the other business functions and to conceive their activities as part of a supply chain with opportunities for cost reductions between the links. Until then, the department only had figures of overall marketing costs, but could not give a breakdown into separate functions such as supply, distribution, selling, and office costs. Clearly the productivity teams which, during the mid-1950s, had toured around the world to harmonize accounting methods in exploration and production and in manufacturing had not extended into marketing.[139] By adopting the market segment approach, Marketing-Oil could now at last begin to spot demand trends and market opportunities at aggregated customer level rather than at product or regional level as before. It was a surprisingly new departure for the Group, though marketing managers in South Africa had already developed this approach of their own accord.[140] From 1964, Group marketing reports discussed segments and customers.

The belated conversion to customer orientation at central offices is all the more surprising since, from the 1950s, operating companies had crept ever closer to their customers in a material way by the rapid expansion of the service station network. One of the most striking developments in gasoline retailing during those years was the widespread switch to 'solus trading', that is, gasoline service stations tied to a single company. The international companies were all familiar with this system in other parts of the world, particularly in North America. Elsewhere, the pre-war system still prevailed. In the UK, for instance, 'a typical garage presented an array of competing pumps jostling each other in a confined space and each having only a limited storage tank underneath. To supply the tanks came the lorries of all the competing companies at frequent intervals, delivering a comparatively small number of gallons on each occasion, often getting in the way of cars trying to use the pumps, and having to travel quite a distance between each small delivery. (...) Not only was this system of distribution uneconomic but the standards of appearance and service at many British stations fell short of the best practice in other countries.'[141]

[62]

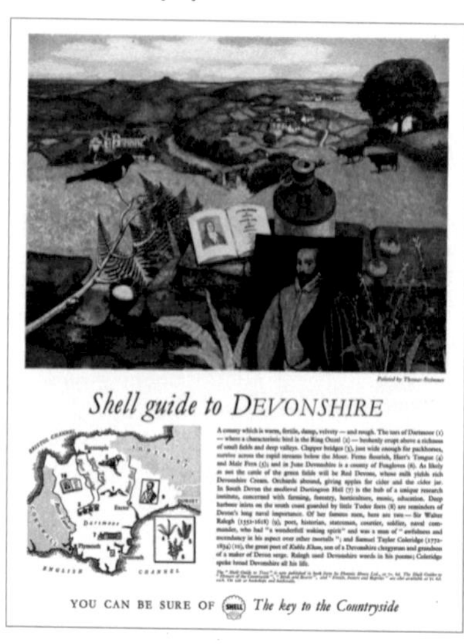

[63]

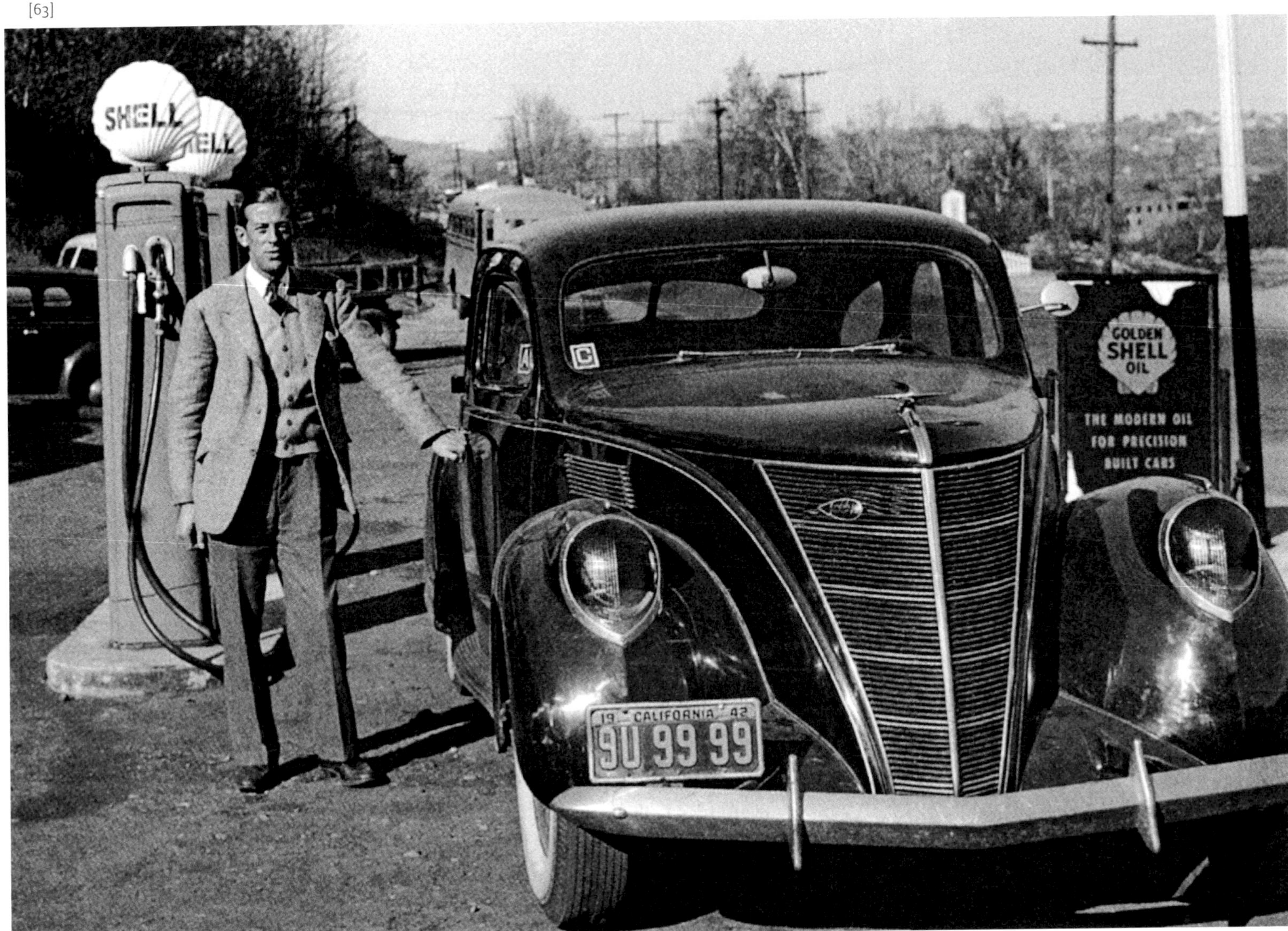

In 1950s Britain, far from being quaint or archaic (left), Shell's charming county guides and monthly nature guides accurately depicted a countryside that was still full of biodiversity, much of which would be lost over the next half-century as new farming practices were introduced. New practices in motoring habits were also imminent, particularly 'solus' service stations, already familiar in the Second World War to American motorists such as the journalist Alistair Cook (above).

[64]

Long accustomed to stocking small quantities of products from many different manufacturers (left and below), independent operators in the US often saw the spread of the oil majors as a serious threat. In Britain (right), a sign for an SMBP depot exemplified the introduction of solus trading.

Roy Copping "RE-TIRE WITH COPPING"

ELEVENTH & OAK . . . . PHONE 4-1151

EUGENE, OREGON

VANWEGE H. M. DE KONINGIN TER BEHANDELING IN HANDEN GESTELD VAN DE MINISTER VAN

March 21, 1955

Her Royal Majesty
Queen Juliana
The Netherlands

Your Royal Majesty,

I have a problem that I would like to submit to you. Perhaps it seems strange that a citizen of the United States and not one of your own people is making this appeal to you.

I have been advised that you are one of the principal stockholders of the Royal Dutch Petroleum Company, which controls 65.44% of the Shell Oil Company of the United States. Since you are known as a very benevolent ruler, I am sure that you are not familiar with the deplorable conditions existing with the small independent Gasoline Service Station Operator in this country.

It seems that Shell Oil Company of this country has decided to eliminate the Independent Operator of this country. Certainly, we the small operators, would not object if the oil companies would lower the wholesale prices on gasoline so that we could still make the profit necessary to operate a business. But the method they are now using is to lower the retail price of gasoline below wholesale; and in many cases below our costs, which is a very simple means of eliminating us from business.

One of our larger Independents advised the public that if this condition continued he would close up and go fishing. He is one of the better-off operators in our business. With me it's different. I can't afford to go fishing because I have a large family to support and would have to try to obtain some other means of livelihood.

We, the people of the United States, have given, and are still giving, extensive financial aid to your country; which we do not regret, because we feel that the Netherlands has a very vital position in world affairs today.

With the above explanation of existing conditions for the small business man in the service station business, I am confident that you will do everything in your power to help eliminate this oppression of small business.

Yours truly,

Roy Copping

Roy Copping

[65]

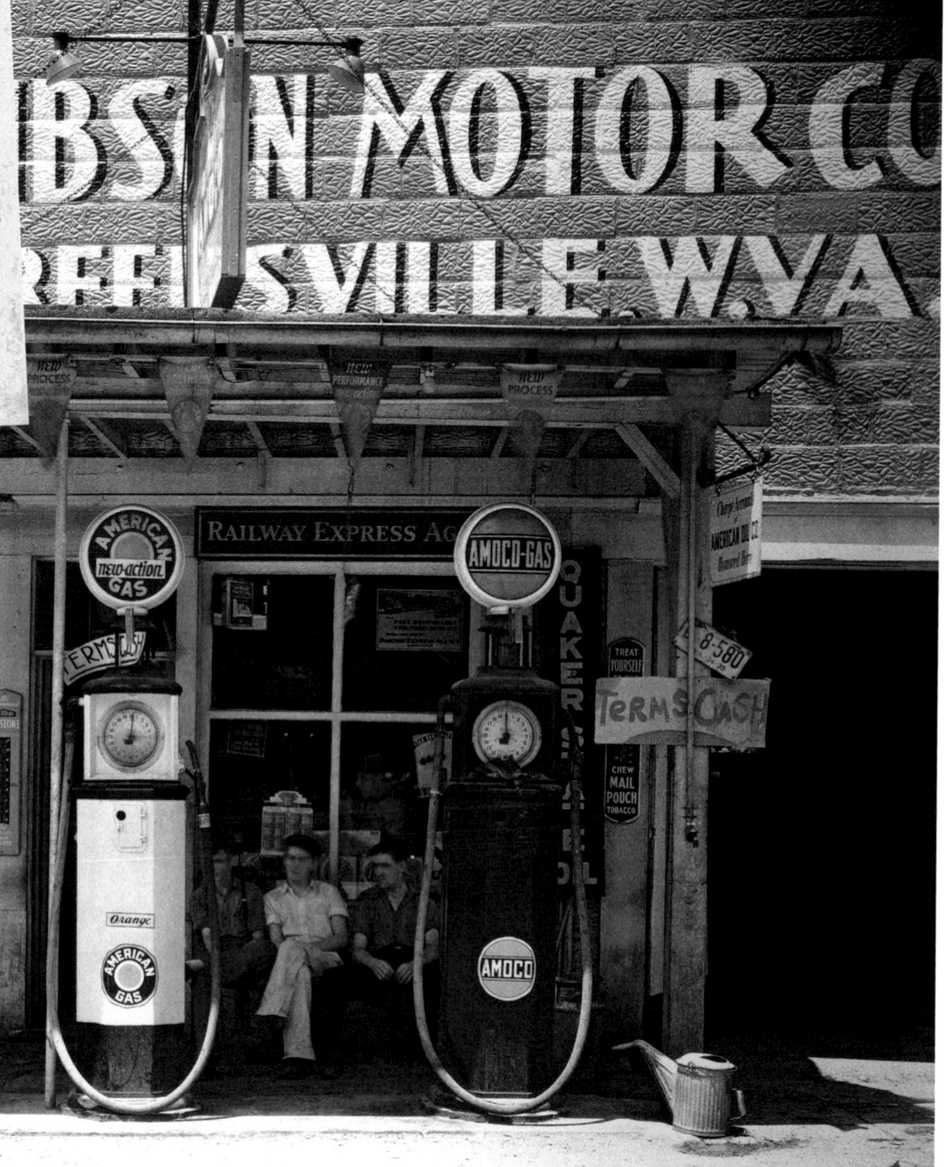

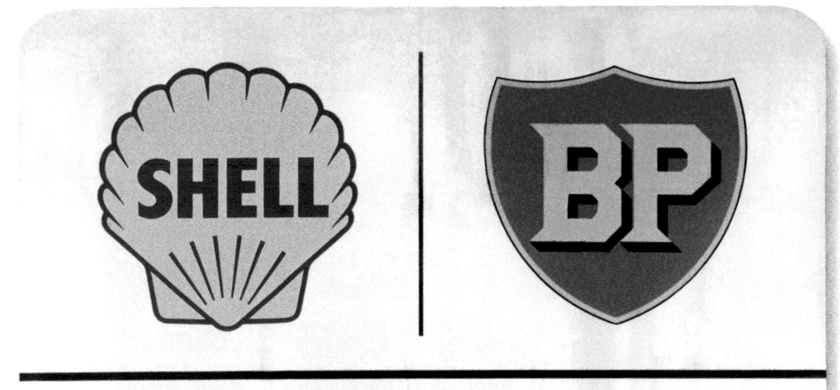

[66]

In 1949, anticipating the ending of government price and brand controls, Esso and SMBP separately began considering the introduction of solus as a means of securing assured outlets for their new refineries, and of providing efficient distribution arrangements at a time when demand was expected to rise very rapidly. Esso was the first to act, introducing its 'Dealer Co-operation Plan' in July 1950, just three months after the abolition of gasoline rationing. SMBP had some doubt about the wisdom of the move, but by the end of 1951 all leading suppliers had started similar arrangements, although to begin with, SMBP's scheme entailed the retailer's agreement to stock not just one brand but the brands of the company's parents and associates – that is, Shell, BP, National Benzole, Power Petroleum, and Dominion. In 1953, branded gasoline was reintroduced in the UK and price controls ended, and by then, enticed by financial incentives and much easier accounting, at least 80 per cent of existing retail outlets had exclusive agreements with one or other of the main suppliers.[142]

This market dominance led to a reaction. The SMBP group was by far the largest gasoline supplier in the UK, supplying in 1964 just over 50 per cent of the market from a network of 15,147 solus outlets and 2,330 company-owned stations. Esso, as Jersey Standard's European operations had become, and its associate Cleveland had about 30 per cent, with 9,148 solus outlets and 1,281 company-owned stations. Moreover, SMBP insisted that its solus retailers should also stock its lubricants exclusively; and with such a dominant position it is not surprising that the legality of this action was challenged by specialist manufacturers of lubricants. The outcome was an enquiry by the Monopolies Commission into the whole solus system, which came to the conclusion that SMBP in particular had been acting in a manner restrictive to competition as far as lubricants sales were concerned, and that it and the other large gasoline suppliers should not be allowed to open any more of their own stations.[143]

The Board of Trade accepted the Commission's first recommendation, allowing all solus retailers to sell other manufacturers' lubricants and tyres, batteries and accessories (TBA), but modified the second, allowing suppliers to open new stations as long as they closed a like number of existing ones. This was scarcely a handicap: theoretically it meant that while maintaining the same number of owned outlets, the large suppliers could increase their market share by closing weak stations and opening new stronger ones. But that share had in fact slipped somewhat already with the entry of new competitors, notably Jet. Starting from nothing in 1958, its simple and successful policy was to expect a high turnover by offering lower prices to the consumer and accepting lower margins for the retailer, and by 1964 it had captured a respectable 3.5 per cent of the market, while SMBP's group share had slipped to 45 per cent and Esso's to 27.4 per cent.[144]

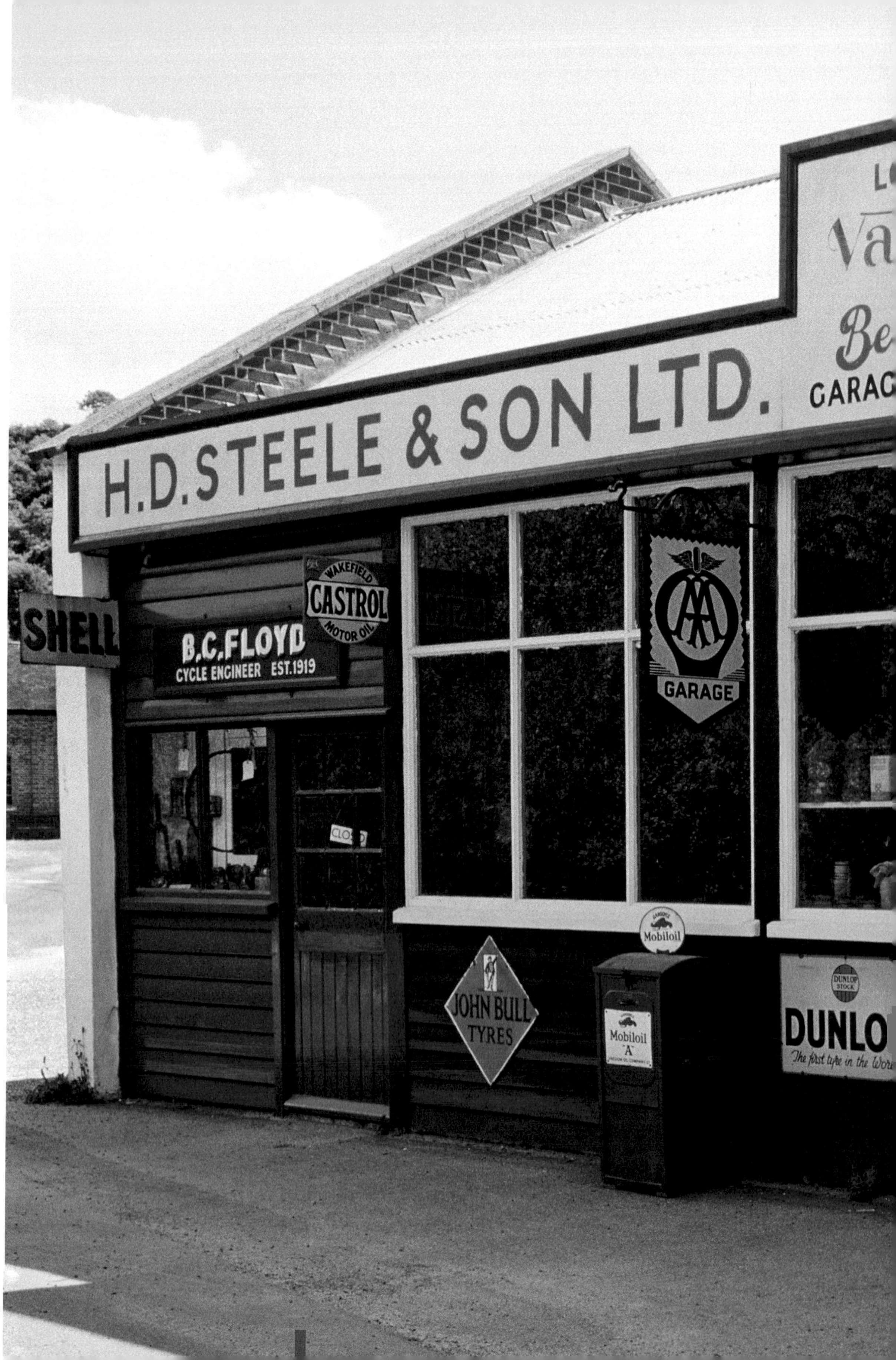

Before solus stations became the norm, independent petrol retailers in the UK commonly sold products from more than one company. Here, Pratt's and Shell stand side by side.

AGENTS FOR
CARS
COMMERCIAL
VEHICLES
ORTHING & GORING
MOTOR ENGI
FILL UP HERE
WITH "SHELL"
PETROL
NO SMOKING
SWITCH OFF ENGINE
NO NAKED LIGHTS
SHELL MOTOR OIL
SHELL MOTOR CYCLE OIL
Pratts
FILTERE
SHELL
MOTOR
Pratts
OIL

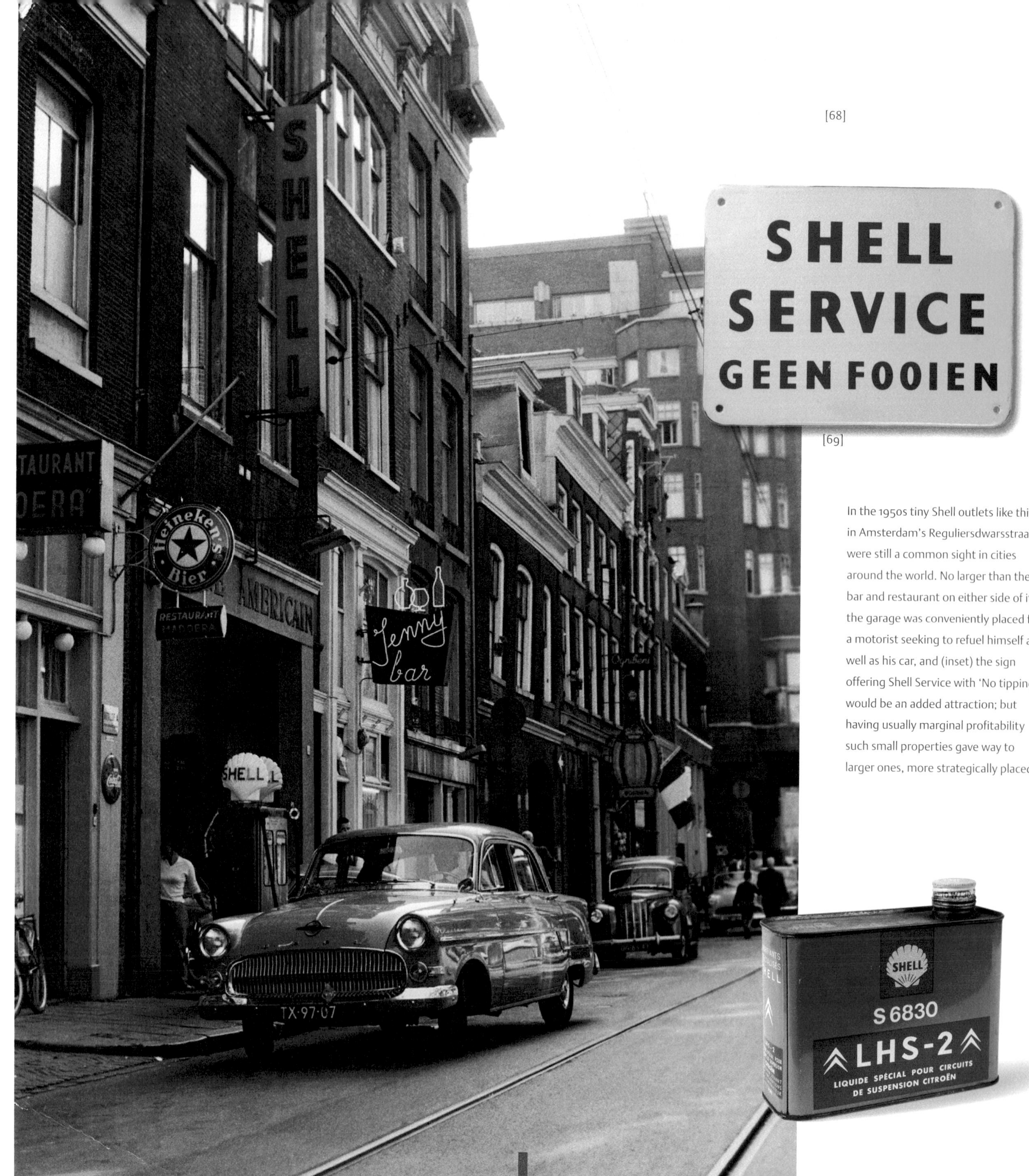

In the 1950s tiny Shell outlets like this, in Amsterdam's Reguliersdwarsstraat were still a common sight in cities around the world. No larger than the bar and restaurant on either side of it, the garage was conveniently placed for a motorist seeking to refuel himself as well as his car, and (inset) the sign offering Shell Service with 'No tipping would be an added attraction; but having usually marginal profitability such small properties gave way to larger ones, more strategically placed.

Very solus! A seemingly lonely pump at Coober Pedy, South Australia, the 'opal capital of the world'. Although the pump is isolated, there are plenty of people there: most of the opal-mining population live in caverns underground to avoid the heat.

[70]

SMBP was also being affected by an internal change. In 1956 it had abandoned its initial multiple-brand policy for solus stations and moved to single-brand. Instead of pumps representing Shell, BP, National Benzole, Power Petroleum, and Dominion all on one site, the motorist now found just one brand; and most of the retailers wanted Shell rather than BP, viewed by motorists as old-fashioned and dull. Those who accepted BP found they had much lower average sales. SMBP attempted to support BP by concentrating its owned stations on BP, but this meant that Shell was effectively subsidizing the weaker brand. In 1960 Shell's takeover of Eagle, a minority partner, raised its shareholding in SMBP to 60 per cent, and in 1965 the marketing company decided to separate the Shell-Mex and BP brands more, establishing separate brand managers and sales campaigns for them within SMBP, in the belief that if both were more strongly promoted, the company's overall position would improve.[145] However, at the end of the year it was forecast that by 1970 SMBP's high market share would fall to 40 per cent, as it was becoming too costly to maintain.[146] In 1966, irked by BP's unconstructive attitude, Shell proposed that SMBP should be dissolved, and after prolonged negotiations the company was gradually dismembered in the early 1970s.[147]

In the late 1950s a new phrase crept into Group terminology: 'visible manifestations', defined as being 'every aspect of Shell's outward appearance except impermanent forms of presentation such as press advertising'.[148] An organization was set up in 1957 to coordinate world-wide studies, both immediate and long term, of Shell's visible manifestations, with the aim of being 'not only abreast, but wherever possible ahead, of public taste while preserving in the public mind the integrity of the Shell image of power and stability.'[149] Consequently, in 1958 a new colour scheme for Group service stations and other properties was adopted for introduction around the world. In the UK, this facelift probably contributed to SMBP's achievement that year of its highest sales to date, an achievement that was certainly aided by the fact that they had a new product for the market, plus a new method of paying for it in the form of a credit scheme for domestic oil fuel which helped to make oil-fired central heating easily available to every home owner.[150]

A concept which rooted only slowly was the self-service station. Pioneered by Shell Oil during the 1950s, self-service stations appeared in Europe during the early 1960s, but they made slow headway on both sides of the Atlantic. Shell Oil relaunched the concept in 1971, claiming that it 'has proved popular with motorists in other countries and is gaining acceptance here'.[151] Similarly, setting up diagnostic repair centres alongside service stations was tried at least twice by Shell Oil, but did not prove an outstanding succes. The Group also experimented with them in Europe.

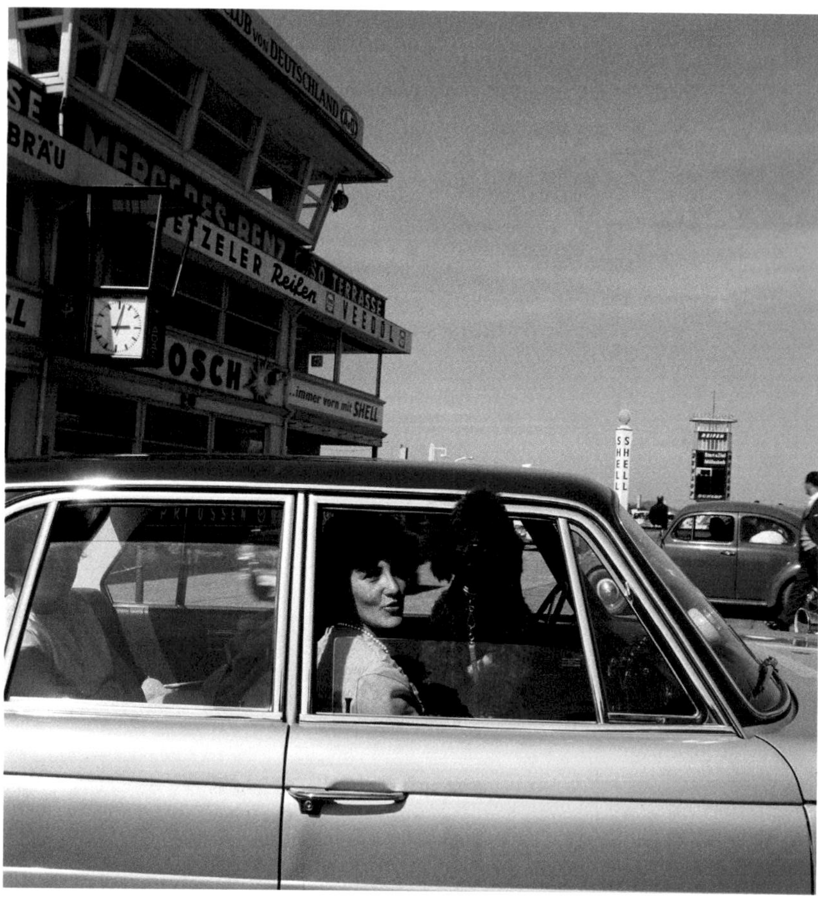

Dog days: Jackie Stewart, the British racing ace, described Germany's Nürburgring (left) as 'the greatest and most challenging race circuit in the world'. With Shell in the background, some are content just to enjoy the sun.

Introducing self-service in Sweden: with an intercom system between each pump and the control panel, the customer presses a signal, the attendant starts the pump, and after filling up, the customer pays in the office. Such do-it-yourself methods are usual today, but at the time, for customers used to attended service, they could seem difficult and were not always welcome.

By the 1960s the pattern of competition in marketing had changed considerably from the pre-Second World War model. During the 1930s 'jobbers' (small-time dealers with little financial backing) entered markets with surplus products, and although their activities were not large in terms of volume, they had a serious effect on prices. Thirty years later, and despite the availability of cheap crude oil, the cost of building and running a refinery meant that new competitors generally had substantial backing, intended to stay in the business, and so were less inclined to act in a manner disruptive to the market. Overall, there seemed to be less threat to 'orderly marketing' than before, but the advent of new price-cutting independent companies was a constant worry.[152] Self-service had been introduced as way to reduce costs and compete a little better with the pricecutters, and in some countries such as Denmark the difference in prices was not very great; but in others, such as Germany, cut-price operators had anything up to 20 per cent of the market, and there the difference was so great that if Deutsche Shell were to match the lower prices, its proceeds would be reduced by £8 million a year.[153]

Strong price competition was almost inevitable, given that outside North America about 35 per cent of refinery capacity was owned by non-major oil companies, but up to 1963 the Group followed 'a policy of restraint' in its marketing, with the rather naïve hope that its competitors would do likewise and so help to prevent the erosion of prices.[154] Not surprisingly the result was simply that it lost markets to competitors, with no appreciable effect on prices, and in 1964 the CMD decided on a change of policy in order to restore Shell's position.[155] However, it was not seeking to regain market share simply for the sake of it, because by then it had recognized another policy flaw. When prices were reasonably stable, as they had been for many years before the appearance of the price-cutting independents, the Group's success or failure

[72]

could be judged fairly accurately by reference to its market share. Thereafter the cost of supplying affiliate companies with discounted crude became so steep that in some places – Italy was one – little or no profit was being realized, and the return on investment was quite insignificant. Consequently, the theory of market selectivity was devised, which essentially meant disposing of unprofitable business and searching actively for more profitable ones, to be exploited aggressively without the customary fear of spoiling the price structure.[156]

The Group's experience in the US appeared to confirm the validity of the theory. There, compared to its competitors, its gasoline sales had been in sixth place in 1958-60, when a very aggressive marketing programme had been adopted. By 1963 it expected to be solidly established in third place, and working on second. This was indeed achieved, though it only represented a relatively small percentage increase in market share.[157] Shell Oil's marketing strategy depended heavily on advertising. In 1968, the company spent $26.3 million on advertising; the second largest spender, Gulf, had a budget of $19.3 million, and the third, Jersey Standard, only $17.1 million. In line with the rest of the industry, Shell Oil preferred television ads, which absorbed two-thirds of the budget.[158] However, marketing outside North America was unfairly expected to compensate for Group cost disadvantages over which it really had no control: the higher cost of crude landed due to distance to source, the unfavourable cut of the barrel, and the fragmentation of markets along national boundaries. With manufacturing costs on a par with Group best practice, marketing received the blame for not cutting staff rigorously enough if costs per barrel sold remained uncompetitive.[159]

Moreover, the North American companies could achieve more economies of scale. Shell Oil could introduce branded TBA at service stations across the United States, because car specifications

[73]

THE MEMORY ... IS ENSHRINED FOREVER
MAN CANNOT LIVE BY FREEDOM ALONE.

[74]

were the same nationwide, but in other countries local operating companies lacked the scale necessary to compete successfully with established manufacturers, whilst the wide variety of national preferences and specifications rendered undertaking this particular diversification at central level impractical.[160] Shell Oil could offer a roadside car breakdown service, insurance, and adopt, in 1958, its own credit card scheme with the administration centralized in just two computerized back offices for the entire US.[161] The absence of nationwide banking even gave Shell Oil a competitive edge in credit cards. Consumers elsewhere were slower to adopt credit cards, however, and until the 1990s high costs seriously hampered financial transactions across national frontiers. As with the branded TBA, a credit card scheme for Europe was considered, and rejected, simply because it made no sense under the circumstances. In 1972, however, Shell introduced the Euroshell Card for truck drivers, who could use the card to pay for fuel at 1,800 Shell service stations in 16 countries. A centralized administration billed the haulage companies using the scheme in their own currency, thus helping to simplify the financial complications of international traffic. A special sign showing a yellow truck cab on a red background identified the stations where the card could be used.[162] America's sophisticated financial markets also enabled Shell Oil to enter into an ingenious scheme to finance its service stations off-balance. The company would lend the sums needed for buying and maintaining the

[77]

[78]

[79]

uw wagen is in
veilige handen
bij:
SHELL
shell smering
... VEILIGE SMERING !

[75]

Shell can help you get better mileage from your dog.
A dog was never meant to carry passengers.
Yet he picks them up all the time.
Unwanted passengers.
Parasites (worms) who are just looking for a free ride.
And who cause needless wear and tear on your best friend.
Shell has taken these worms to task, by creating a product called TASK.*
It eliminates these unwanted dinner guests, and in just one dose.
We have also created a special collar for your dog that knocks off fleas.
What finishes the fleas is a chemical. Shell's VAPONA inside the collar.
It gives off a vapor that a flea finds impossible to live with.
Our work does not stop with dogs. We make ATGARD* to help pigs get fatter faster.
Urea to help forests grow.
Special agricultural products to produce healthier plants.
Shell wants to help you get better mileage from your dog, your farm animals, your trees and plants.
All the good things you enjoy.
And Shell people enjoy.

[76]

How to succeed as a son-in-law.
When George first decided to become a Shell dealer, his mother-in-law wasn't exactly overjoyed.
But eventually the dear lady came around.
After her daughter started to enjoy little things like a beautiful home. A big car. Even a Florida vacation with George.
And George's mother-in-law liked the fact that she could tell her friends all about her successful son-in-law. With 4 people working for him.
About his being on the town council.
About his latest safety speech at the Rotary.
Now she's nuts about George.
If a man feels a little unsure about his mother-in-law's affection, he could do worse than become a Shell dealer.
It's a business that offers a man a good income. Plus the status that comes with being a good guy helping his community.
But when a Shell dealer does something to improve his town, it's not just to impress his mother-in-law.
And it's not just because it's where he does business.
It's because it's where he lives.

Aimed at other factors that could contribute to commercial success, corporate advertising was a complement to product advertising. Influenced in part by the civil rights campaigns of the 1960s, Shell's 1968 corporate advertising in the United States (above) showed considerable lateral thinking. In contrast, these stills from contemporary cinema advertisements in the Netherlands (below) were bright and jolly, but were basically modernized versions of traditional designs.

[80]

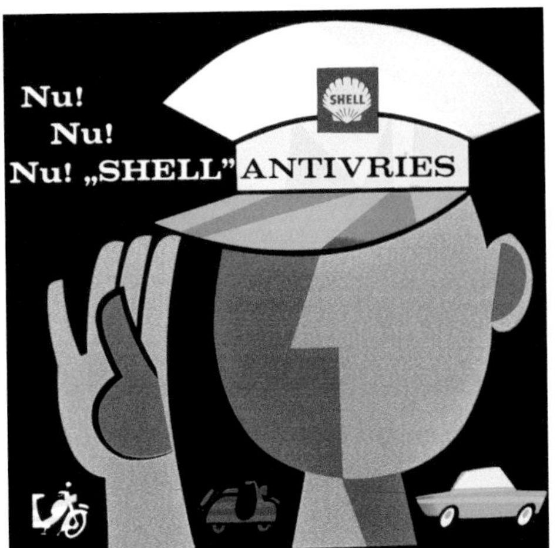

station to the station lessors in the form of a mortgage on the premises. Shell Oil then sold this mortgage to an institutional investor with which it had a long-standing arrangement covering this type of transaction. During 1956-64 Shell Oil financed $93 million through the scheme.[163]

The American companies also had more scope for cost-cutting than companies elsewhere. For example, the inclusion, in 1968, of Canadian fuel marketers in the European marketing figures caused a drop of 3 US cents in costs per barrel to $1.29.[164] Shell Oil also allowed its marketing divisions and districts considerable authority for independent action, but it could, and did, rearrange them as and when needed, something regional organizations elsewhere were unable to do.[165] The spread of refineries to suit national markets rather than optimum supply routes saddled marketing with high and inflexible distribution costs, in Europe with 44 per cent of the total, by far the biggest slice of marketing costs.[166] Cross-boundary rationalization studies achieved some savings, but national frontiers and perhaps also local chauvinism in operating companies prevented substantial transnational integration. The first tentative step in that direction was the appointment of a Benelux marketing chief executive in 1965, followed four years later by the opening of a computer centre serving the three countries.[167]

Moreover, the Group's strategy of operational decentralization gave the operating companies great independence, most visibly manifest in the Shell Mex & BP bastion across the Thames from

The French revolution: Shell's earlier advertising in France had frequently depicted a wooden doll-like model used by artists, but in the mid-1950s its publicity became much more imaginative and appealing, with an emphasis on fun. The aircraft and racing car (below right) were part of the campaign 'Petrol is everywhere in people's lives' which in 1954 gained Shell Berre top prize from the advertising industry, an achievement repeated in 1962 with the lapel badges and cartoons of the 'I love Shell' campaign (below), launched in 1958. Deutsche Shell ran a campaign with the same theme.

[81]

Shell Centre. Marketing-Oil could suggest, not dictate. Visits by service teams of marketing experts to improve performance had to be negotiated and mutually agreed, not imposed.[168] Consequently as late as 1966, and despite an eighteen month cost-calculating harmonization programme, methods for calculating marketing profitability still differed between countries.[169] Group managers monitored investment budgets and overall profitability, but left the operational budgeting and funds management to the operating companies themselves, so local managers possessed considerable financial leeway. Together these factors meant that an aggressive marketing approach to raise market share could work in America, but elsewhere it made sense to concentrate on profitability and accept a gradual market share decline.

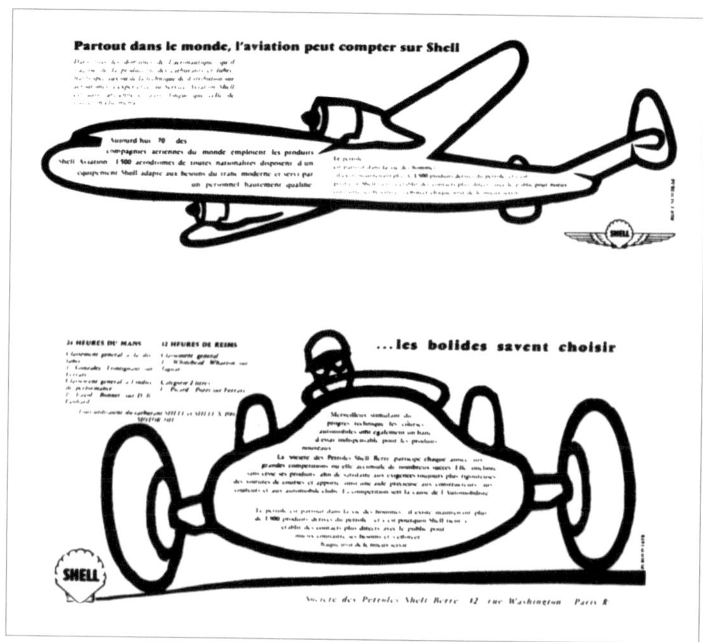

[82]

The Group did apply the selective marketing approach in a thorough review of its service stations. Shell's service station network formed its main customer interface and one of the most profitable operations. The automotive retail trade sold 20 per cent of the volume but generated about 30 per cent of Group inland trade proceeds, whilst absorbing 45 per cent of marketing investment.[170] Still too many stations were dealer-owned and uneconomic, so Central Marketing developed benchmarks for performance and prospects, with the aim of closing down unprofitable stations and rebuilding or extending the profitable ones, since 'our greatest profit opportunities lie in better use of our extensive existing facilities', rather than seeking to increase volume through expanding the number of stations.[171] Expanding service stations with the Shell Shops or Shell Boutiques found on a modest scale in the UK, France, and Germany, remained for the moment under review. Reinforcing consumer loyalty through increasing brand identity was a key ingredient in the exercise. Consequently, Marketing had commissioned the Compagnie de l'Esthétique Industrielle, the Paris agency of the American designer Raymond Loewy, to develop a complete overhaul of all aspects of the service stations, from the Shell emblem, roadside signs, the forecourts, lighting, staff uniforms, and delivery vehicles, to packages and the graphics of publications. Prototype stations were being built by selected operating companies in 1969, to try out the design. Perhaps the Loewy project went too far, and some managers considered it too expensive, but it did mark an important evolution of corporate design. In a bold decision, the name Shell was dropped from the pecten, because it was now considered sufficiently strong as a brand image not to need the name. Only Shell Oil continued to use the name on the pecten.[172]

By 1970 the Group had regained its competitiveness, but Marketing-Oil, which had contributed so much to this recovery by streamlining its operations, had become somewhat dispirited. The year 1969 was disappointing; the operating companies had finally received permission to concentrate on volume growth rather than profitability, but Shell's market share had gone down all the same. Inflation and wage pressures pushed up unit costs, reversing the downward trend that had been achieved with such effort, whilst the effect of the most obvious savings had already worked through. To locate opportunities for further economies, the department had embarked on wide-ranging studies looking at its operations from a systems approach, widening the focus from details like optimizing lorry utilization to the logistics of supply. One such project aimed at integrating the distribution of oil products with manufacturing, finance, and information and computing, to yield the optimum refinery design and siting for both manufacturing and marketing. This approach already appeared to be both very promising and posing new dangers. A reorganization in May 1971 had taken the switch from product orientation to market segment approach one step further. Only lubricants remained a department on its own; all other product groups were now subordinated to one of the new departments, Automotive Retail, Domestic and Rural, or Industry, which together used to make up the Market Development Department. Planning and Product Integration was renamed Marketing Economics and Services. Whilst thus aiming to get closer to its markets, Marketing-Oil felt increasingly isolated at Shell Centre, far from the real world: 'It is extremely difficult to foster a keen businesslike atmosphere in a functional marketing organisation in a central office. We are no exception and being two steps away from the actual market activity aggravates the problem. Marketing staff of top quality – and these are the ones we need – often have difficulty in finding job satisfaction at the Centre. They feel cut off and frustrated, lose touch after a few years and longer-term are in danger of ending up knowing all the answers to yesterday's problems. In such a climate a tendency to excessive criticism of the field effort is likely to creep in. This is a major problem.'[173]

At last you know, my Valentine,
The news I've longed to bring.
Now let the petrol flow like wine,
Let joyful engines sing.
See! On the road each Pool has fled,
The sun has dried the showers;
We've many a carefree mile ahead,
The best of spirits ours.
So off we'll go, and—this we know—
We'll motor far and well.
For every day is . . . . . Day now;
You can
be sure of . . . . .

[83]

My Valentine, I love you with a 'V',
With vehemence, with vigour and with verve.
So various your virtues that to me
You're void of vice, a verdict few deserve.
Most vividly your beauties vex my brain;
I vow that Venus vies with you in vain.
When, with an 'M', we've motored many a mile
'Mid myrtle groves, past meadows massed with may,
Marking the map of love, I've mused awhile
Upon the merry motion of our way.
And, meditating so, my mind has gone
To marvel at this might that draws us on;
Saluting, as with 'S' we seek the Spring,
This supple strength which sends us where we will,
Which speeds us to the stream where swallows wing,
Then surges swiftly up the sunlit hill.
We're seldom sure today: but all is well
When still we can be S for SURE of . . . . .

Every year in the UK, Shell sent out Valentine cards to its female customers. In true Valentine tradition, they were unsigned – and so could sometimes spark the anger of a husband jealous that his wife had apparently received an anonymous love-letter. But if he read all of the verse, the situation would become clear: the penultimate line always ended with a word such a 'bell' or 'well', and the last line, which had to rhyme, would end 'You can be sure of…' This light-hearted annual advertising campaign must have been popular and effective in its time, because it ran for 37 years, from 1938 to 1975. It ceased only when changing social customs meant that women no longer liked being identified as 'lady motorists'.

At the same time the department faced an entirely new challenge of very wide ramifications, i.e. the environmental impact of the oil industry. Exhaust emissions had become a major subject in the US; legislation was being debated and Esso already marketed a gasoline type with an anti-smog additive 'which, while there are reservations about its real value, captured the public imagination to an unusual degree, following strong promotion.'[174] However, Shell Oil's much-trumpeted roll-out, in 1970, of the lead-free Shell of the Future gasoline, served from special white pumps, proved a costly flop because, despite mounting public concerns about the environment, the market would not accept the product. It was replaced by a low-lead Shell Super Regular.[175] Even so concerns about conservation and pollution looked set to change the department's focus on the economics of cost: 'Shell marketers, if asked to define the most intractable problem faced today, would increasingly answer "community acceptance". In the past months this has taken on added meaning. For Marketing-Oil alone, the problems are diverse and we have already been faced with external action in the fields of road vehicle emissions control, selective burning of low-sulphur fuel, protection of inland waters, limitation of adverse effects on the soil and its produce by seepage of oil, and possible carcinogenic hazards from use of earlier cutting oil formulations. (...) The attraction of conservation as a theme of universal appeal, suited equally to student protest, governmental and political pressure, social belligerence, editorial eloquence and consumer motivation, is not only affecting us in the ways mentioned. It is likely to spread to lubricants selling, to matters of service station design, the disposal of old oil cans, unsightly advertising, the appearance and emissions of distribution plants and vehicles and many other facets or marketing. We believe that a co-ordinated and imaginative (not merely defensive) approach is essential, related to all aspects of Group activity, with a strong impingement on related PR activities. Marketing-Oil is represented on the Shell Committee for Environmental Conservation and advice is being given to operating companies on many aspects of this increasing problem.'[176]

In this respect Marketing-Oil showed itself as still having its ear to the ground, identifying both the importance of the trend and the need for a pro-active response at Group level. Consequently, the department of Scientific Development and Experimental Effort became Experimental Effort and Conservation in the 1971 reorganization, a sure sign of things to come.

[84]

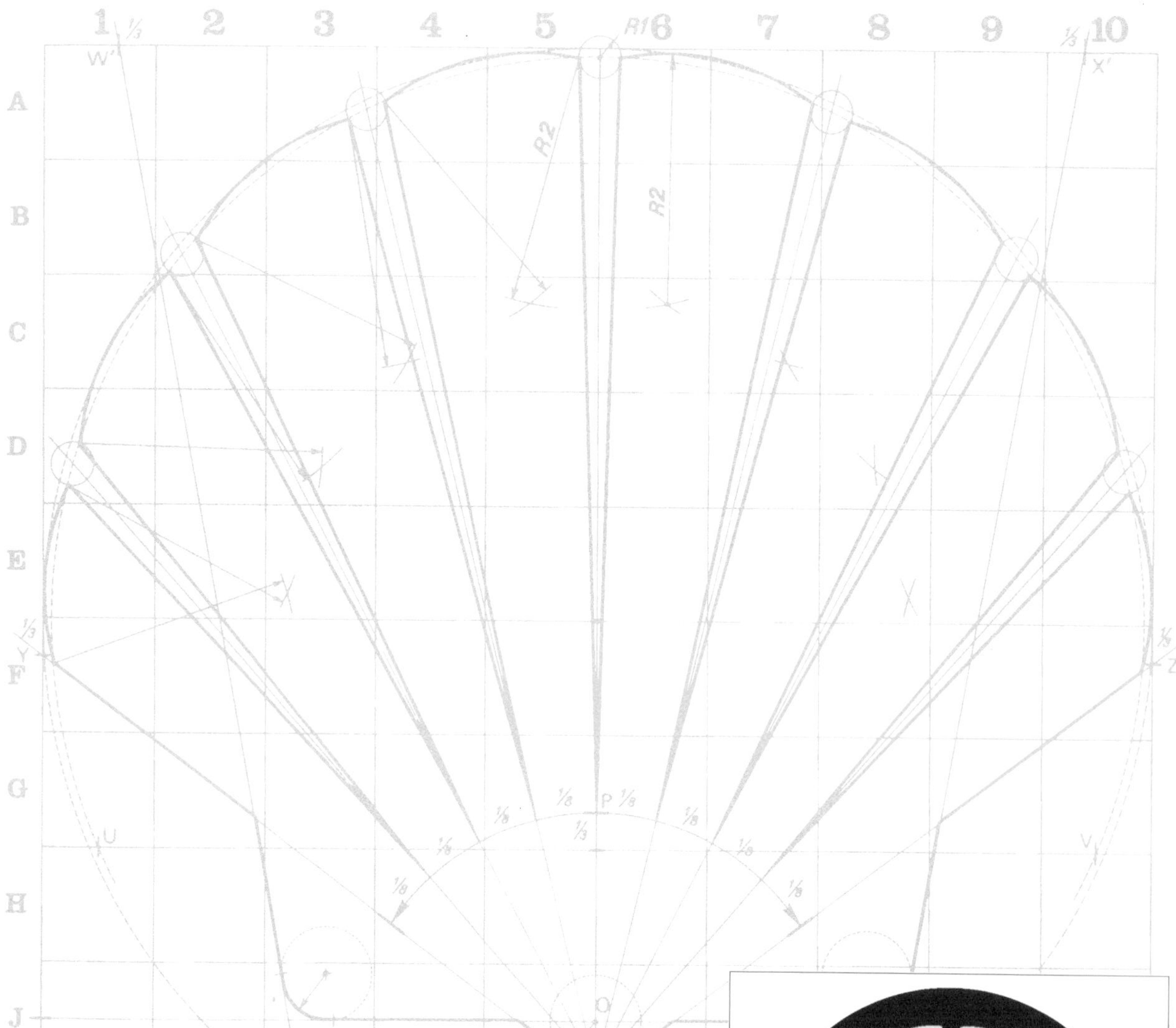

Born in France, Raymond Loewy (1893-1986) became a naturalized American citizen in 1938. In his long and exceptionally creative life, many of his designs became integral parts of modern society's visual language. The classic shape of the Coca-Cola bottle was his, and in the second half of the 20th century, probably no other individual person had more impact on the marketing of the international oil industry. Three of the 'seven sisters' commissioned new enduring logos from him: BP (1958), Shell (1971) and Exxon (1972). The picture on the right shows Loewy with his wife and daughter alongside the Studebaker sports car which he designed. On the left, the succession of logos which the Group used from (top to bottom) 1900, 1904, 1915, 1930, and 1963. Alongside the earlier logos, the diagram shows the precise measurements of Loewy's pecten design. Main picture: before the logo Loewy redesigned the service stations such as this one in Germany, which shows the new design with the new logo.

Diesel
Shell
selbst
tanken
Shell
Super
64
Auch
Super
billiger

[86]

[85]

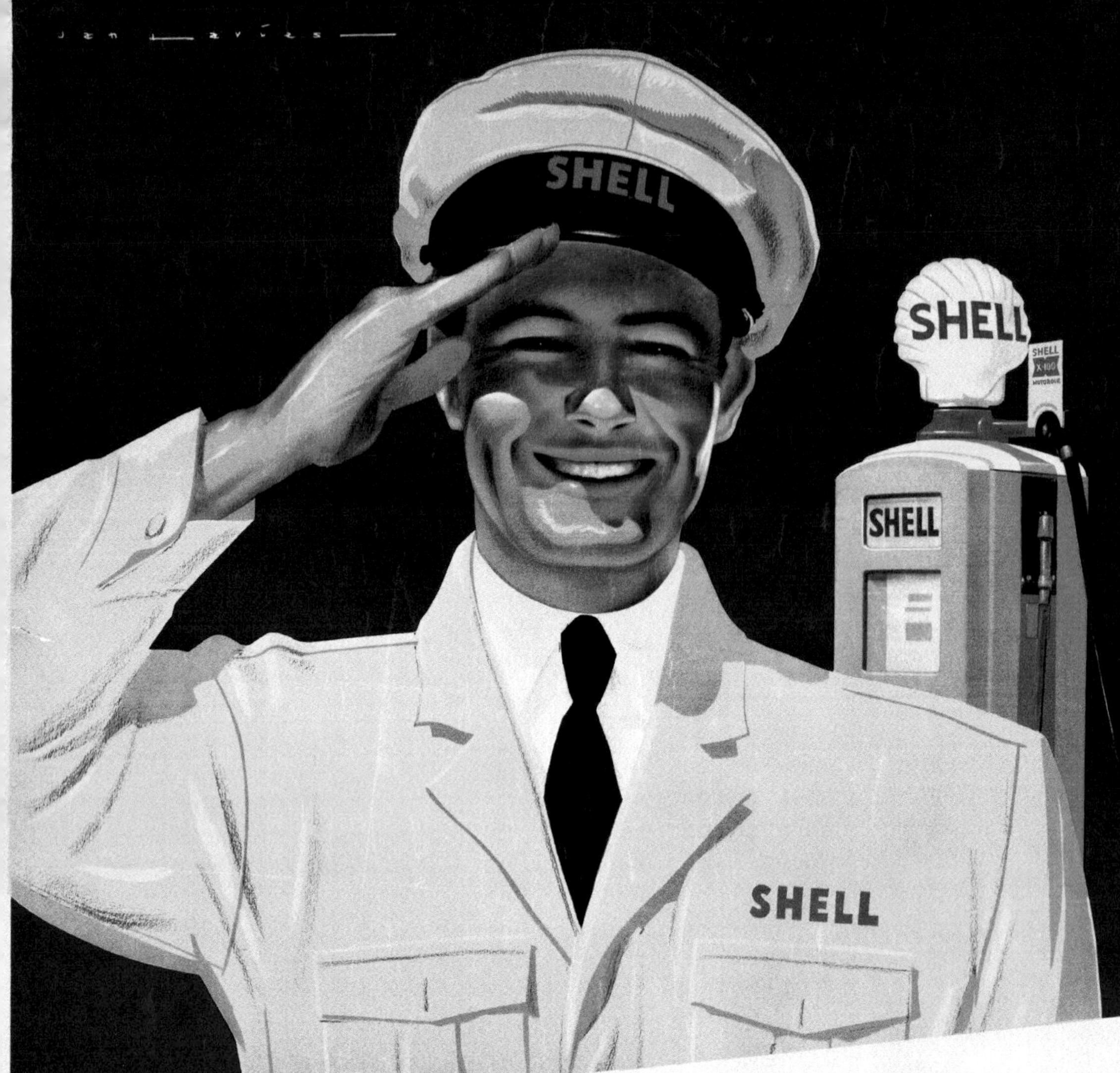
SHELL
SHELL
SHELL
SHELL
SHELL
service

Poster designed by Jan Lavies, in use from 1946 to 1949.

Conclusion The long-term contracts with which the majors divided the Middle Eastern oil between them succeeded in establishing a considerable degree of stability in upstream, at the cost of creating sharp downstream competition because the surplus of cheap crude made oil companies treat downstream as a zero-margin business.

The Group responded with some delay to the challenges in downstream, for four reasons. First, the matrix organization greatly improved the integration of individual functions, but at the same time created demarcation lines between them, while the push for greater autonomy of operating companies and the weight given to regional versus functional interests may actually have led to a certain degree of functional disintegration, notably in marketing. These disadvantages were gradually eliminated during the second half of the 1960s, resulting in a clear improvement in the Group's return on assets until rampant inflation cut across the trend.

Second, it took time before managers could bring themselves to trade the ingrained attitude favouring profitable stability over aggressive competition, price over volume. This was not really surprising. With surging demand creating continuous capacity bottlenecks, keeping pace was difficult enough. There may also have been a generational effect, younger managers who had risen to the top by 1960 having a keener awareness that the cosy pre war arrangements had changed for ever, and pressing for an appropriate response. Third, general policy considerations and requirements handicapped radical cost-cutting in each of the downstream functions. In manufacturing the flexibility of crude supplies remained a sine qua non, and the building of defensive refineries or other small installations near consumers also added cost. Similarly, in shipping it was the high proportion of owned over chartered, and in marketing the autonomy of local operating companies which reduced the overall cost-effectiveness.

Finally, accepting downstream as a zero-margin business would have meant throwing away two hard-earned assets: the premium brand image, and the product differentiation achieved through painstaking research and tailored manufacturing processes. This was not something that the Group wanted to do, because the company considered its downstream business a stronghold in the expected upstream struggle with the OPEC countries. The recovery of competitiveness during the second half of the 1960s showed that sacrificing these assets would have been unnecessary, too; and, as with the far sighted change in E&P policy in 1965-66, the time was near when Shell would profit from having safeguarded its intangible assets.

[1]

Chapter 5

Chasing the rainbow

While the oil business expanded rapidly, the Group embarked on a vigorous diversification into petrochemicals, first along the lines established before the war, but from the early 1950s also in new directions, notably epoxy resins and plastics. Prominent motives driving this diversification were the desire for a second core business, the urge to join a fast-growing and dynamic sector feeding off oil, and Shell's enduring love for new technology. However, for all their allure, chemicals proved to be rather less profitable than oil, which stimulated a renewed and ultimately also fruitless search for a business which could generate the level of profits and growth which Shell required of its business.

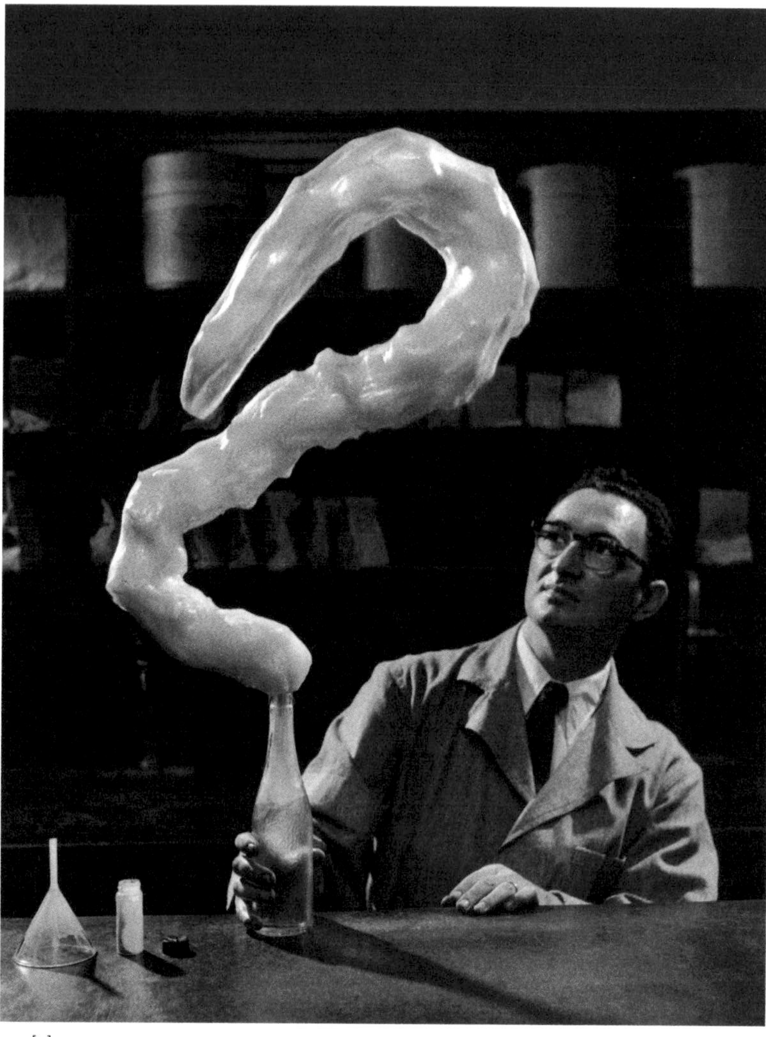

[2]

A spectacular experiment in making rubber from bottled butadiene gas.

Growing a second leg As we have seen in Volume 1 Chapter 6, the pattern of the Group's pre-war chemical operations already testified to an underlying intention towards diversification, with production of fertilizers, solvents, pesticides, and butadiene for synthetic rubber under way, plants for Teepol detergents under construction, and the opportunities for making synthetic resins under active consideration. The expansion programme launched immediately after the war showed that the original intentions had been raised to a higher level, with the building of several large petrochemical complexes centred around fluid bed cat crackers. As with other petrochemical companies, the Group's expansion in manufacturing chemicals was intimately linked with the expansion of the refineries, the production of the former benefiting from design and engineering advances in the latter.[1]

Several motives may be identified for this diversification. First, there was a distinct fashion element. The petrochemical industry seemed to have everything going for it: very fast growth, exciting new synthetic materials, gorgeous technology, and all centred on oil. Manufacturers rushed to make the new chemical products using it; if Shell stood by, it stood to lose a potentially very valuable field to the competition. The urge to join was probably reinforced by the lure of the new hardware, always a great attraction for the Group's large contingent of science-oriented managers. Second, Shell somewhat conceitedly believed itself to have strong advantages over the competition: superior research capabilities, a global marketing network, and a great ability to generate internal finance. When in 1955 Shell Oil moved back into synthetic rubber, for instance, this was clearly driven by an illusion that company scientists could easily outflank the established tyre companies.[2] Planners in Shell believed that 'we can undertake any type of venture anywhere in the World', an optimism based on the company's presence in more countries than any other petrochemical business.[3] As for finance, in discussing the outlook for chemicals in 1971, the CMD opined that 'While therefore prudence

[3]

[4]

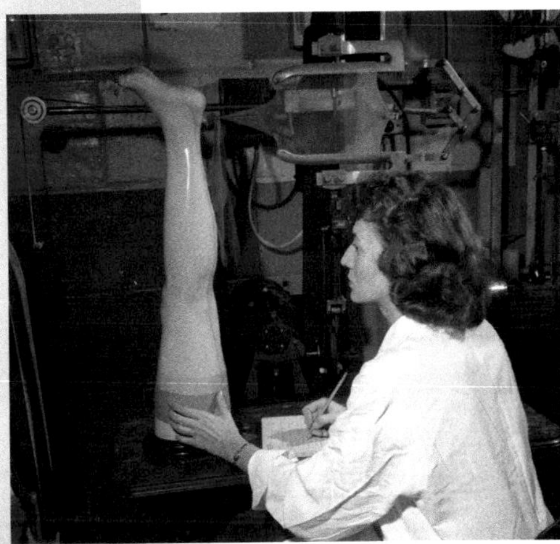

[5]

Chapter 5

The invention of nylon stockings offered the comfort of silk at a fraction of the price. Top, new stockings are drying on heated formers in a UK factory, 1949; above; wearability is tested in a US factory, 1955; left, style and glamour combined.

in capital investment is especially necessary at this time, the Group's principal competitors, relying as they must on the raising of external funds, are presently at a disadvantage and this should not be overlooked.'[4]

Third, senior managers worried about strategic concerns. Notably Kessler was anxious for Shell to have a 'second leg' with chemicals and thus be less dependent on its oil business.[5] This opinion looked persuasive and, as time went by, it appeared to become more so. Like all new industries, the chemical industry grew very fast, outpacing oil. By leaving the petrochemical field to the competition, the Group might find itself sooner or later attacked in its core business, as IG Farben had threatened to do in 1927 (see Volume 1, Chapter 6). Moreover, by the late 1950s Shell's return on assets declined. This could be interpreted as a maturing of its traditional business, rendering the need to diversify all the more evident. Consequently the CMD started to devote increasing attention to this matter.

Fourth, the rapid rise of the petrochemical industry was primarily feedstock-driven, that is to say, powered by the availability of cheap oil in Europe.[6] During the 1950s the chemical industry massively switched to oil. By 1956 petrochemicals were well on the way to ousting, and in some instances had already completely displaced, older sources of carbon-based chemical production.[7] In 1949, only 9 per cent by weight of organic chemicals produced in the UK came from petroleum, but by 1962 the figure had risen to 63 per cent of a greatly increased total production.[8] The German chemical industry, that bastion of carbochemicals, switched somewhat later, the share of coal as feedstock for organic chemicals dropping from 76 per cent in 1957 to 37 per cent in 1963.[9] It appeared to makes sense for the Group as feedstock supplier to try and capture a share of the industry which emerged on the back of it. Finally, several European countries gave tax advantages to the petrochemical industry. In part because of the high depreciation common to capital-intensive industries, Shell's chemical sector as a whole paid only 16 per cent tax compared to a chemical industry average of 40 per cent.[10] In 1954, Australia's tariff protection for the chemical industry helped to persuade the Group to form a joint venture with Union Carbide for building a petrochemical unit next to Shell's Geelong refinery.[11] Until the mid-1960s, feedstocks also created a fundamental difference between the petrochemical industry in the United States and in Europe. The US market was strongly tilted towards gasoline,

[6]

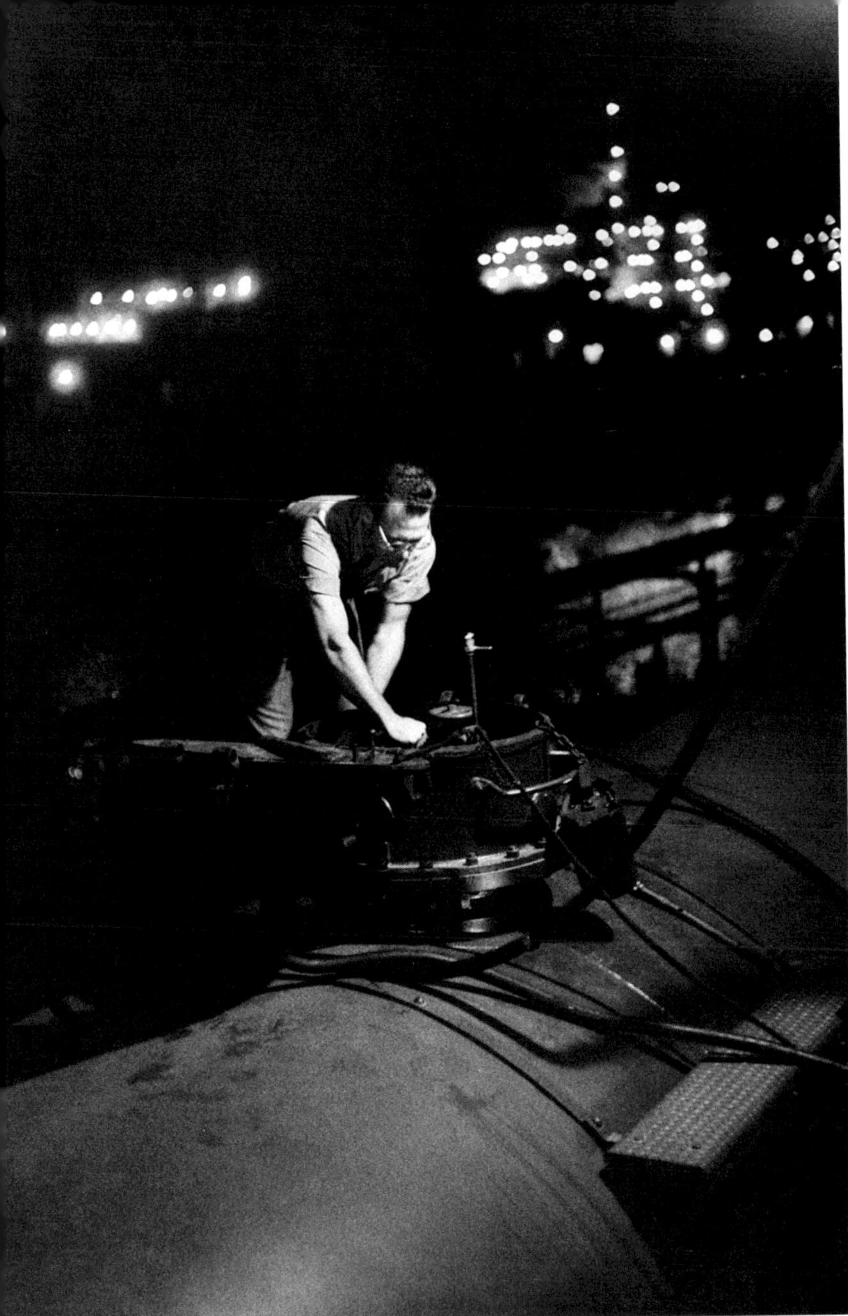

[7]

Hot water bottles made of synthetic rubber (left) might have been completely lacking in glamour, but they brought at least some night-time comfort to many thousands of people. Above, in 1957 a Union Carbide employee loads the raw material, butadiene.

the production of which generated abundant quantities of waste refinery gases. Associated natural gas was also widely available, and building-block petrochemicals could fairly readily be manufactured from both. Europe by comparison had initially only small known sources of hydrocarbons; its major carbon resource was coal, which had formed the basic feedstock of the organic chemicals industry since the nineteenth century. For the same reasons cokes-oven gases had been an important source in inter-war Europe of ethylene and benzene, two of the fundamental constituents of the petrochemical industry, yet such gases were not readily available in the United States.

A further distinction lay in the traditional chemical industries on either side of the Atlantic. In the United States chemical companies were able to continue their progress unimpeded by physical war damage; and although in post-war Britain ICI remained the dominant member of the traditional industry, in Germany the once-mighty IG Farben was dismembered by the victorious Allies into its component companies, with its patents and technical know-how being made freely available to the chemicals industry as a whole. Rather like the break-up of the old Standard Oil Trust in 1911, the dissolution of IG Farben did little long-term harm to its component companies, and the three biggest (Hoechst, BASF and Bayer) soon recovered their positions.[12] In the short term, however, and linked with the physical destruction of so much of the continental European chemicals industry, their functional absence left an inviting gap for others; yet although the patents and knowledge that they had been forced to give up were mainly for coal-based chemistry, it was not a gap that other coal-based chemical companies – themselves recovering from the war – felt able to fill.

For petroleum companies, it was possible to take another view, not least because of the nature of the combined British and

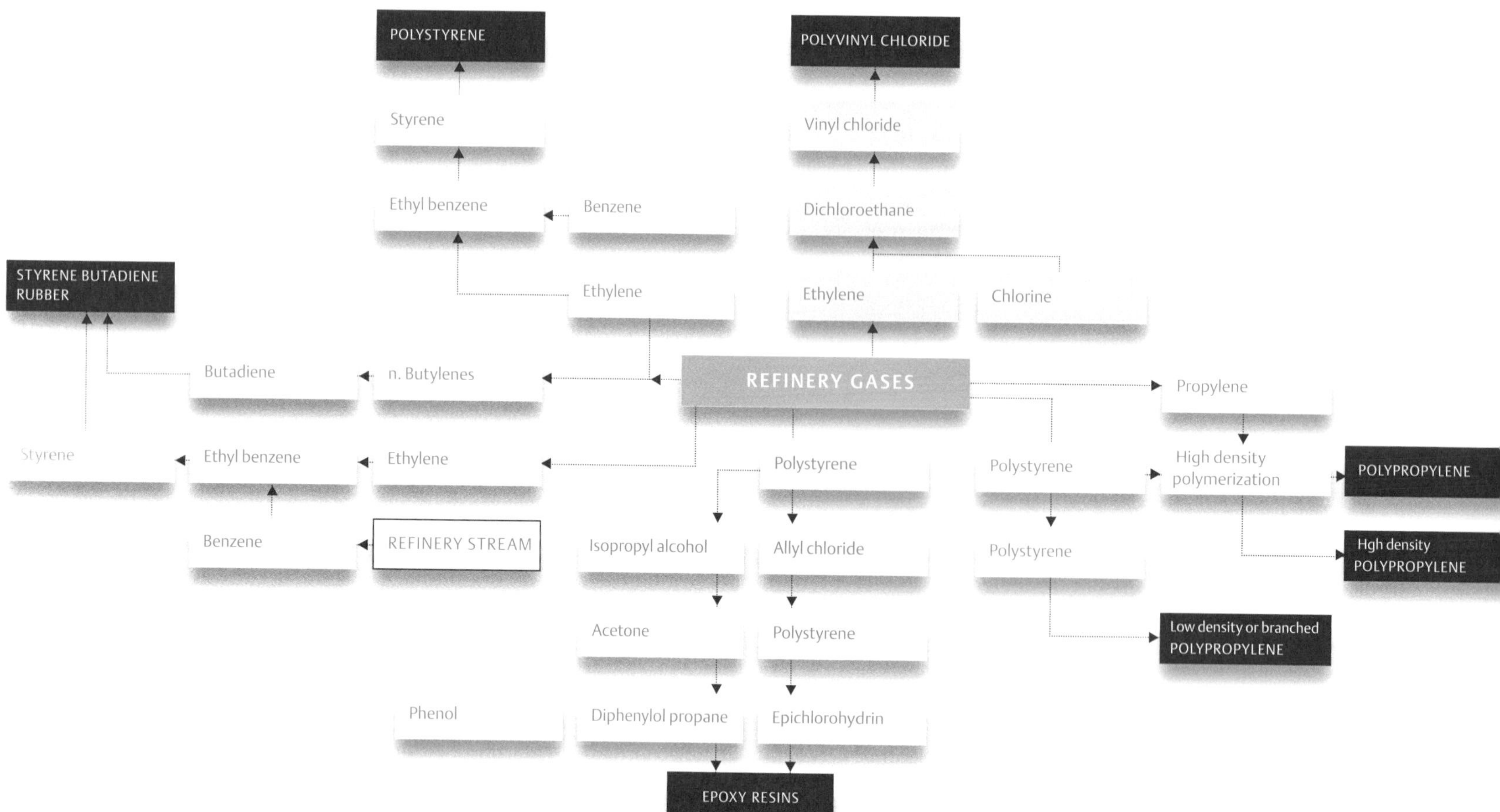

continental European market, taken as one region together.[13] There, there were far fewer automobiles than in the US, and gasoline was in far less demand. Instead Europe wanted more of the refinery products that could help it to rebuild itself (asphalt for roads, runways and roofing, and lubricants, diesel fuel, and fuel oil for its engines) and in producing these, the industry would create large quantities of naphtha from the crude it processed. Naphtha, a basic petrochemical feedstock, is not a specific compound but a generic term for a range of distillates between kerosene and gasoline. Normally it would be used to make gasoline, but, if gasoline were not much wanted, one alternative was to find a market for the naphtha as a petroleum product, even perhaps exporting it to the petrochemical industry of the US, where it could be converted into the lower olefins benzene, butadiene, propylene, and ethylene – some of the most useful building blocks. However, that would imply conceding a large part of the global chemical market to the States, perhaps permanently.

Having naphtha available in Europe thus opened the opportunity to develop and claim a market that coal-based chemicals companies shied away from and which otherwise might be lost to American companies. It was precisely to counter the likely entry of Phillips Petroleum and Jersey Standard into the European synthetic rubber market that the Group embarked, in 1958, on the construction of a rubber plant in France.[14] The economic circumstances immediately after the war provided further incentives: the forecast falling price of naphtha and the possibilities first of reducing the expenditure of scarce dollars on imports, and later of building surplus production for export. Anglo-Iranian came to a similar decision but chose a different route, going into partnership with the Distillers Company, although with some delay after an attempted partnership with ICI had foundered on restrictions imposed by an existing ICI-Du Pont agreement. By contrast, Jersey Standard initially did not start up petrochemical

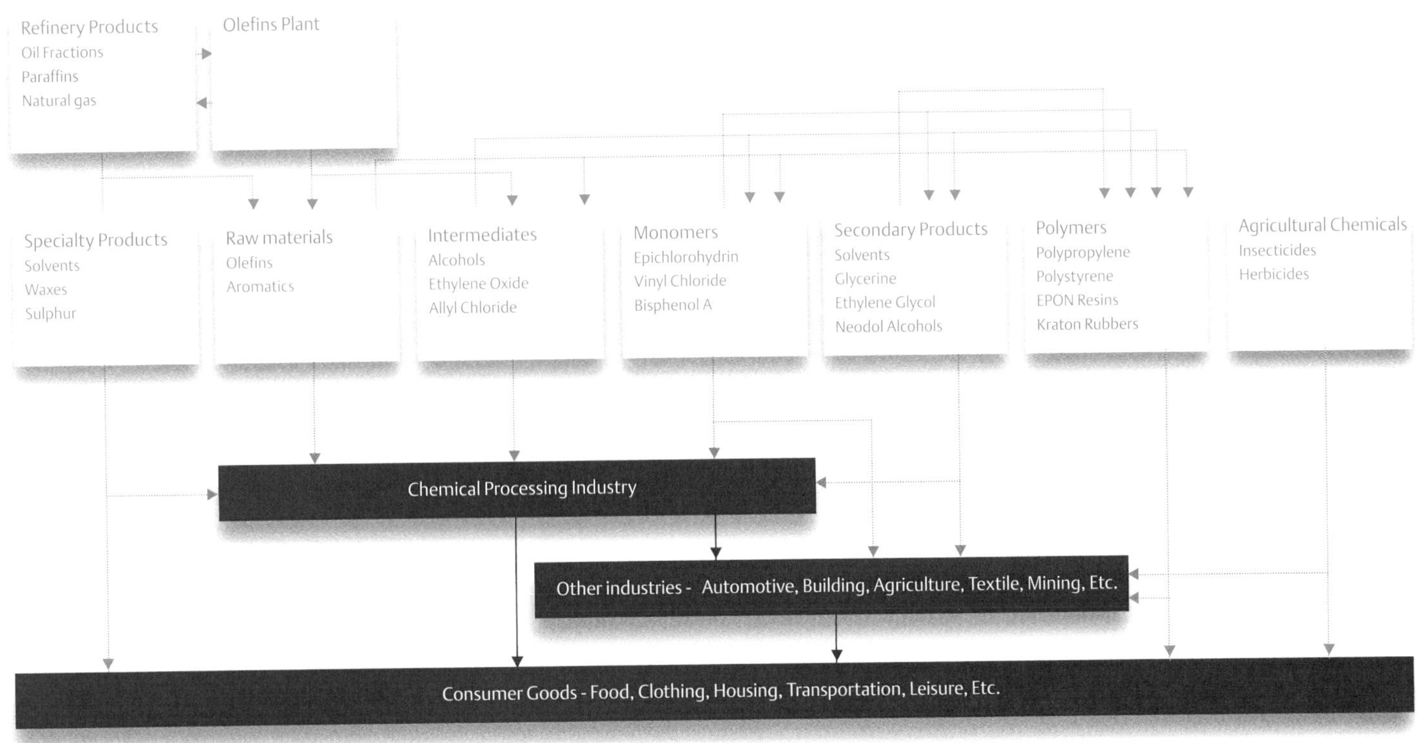

The starting point of the diversification into petrochemicals was the opportunity to derive a large number of useful materials from refinery gases, as the chart on the left shows. The various processes were also closely interlinked (right), streams of intermediates and waste from one process going on as feed to another, rendering it increasingly difficult to stop making particular products if they were unprofitable.

operations in Western Europe.[15] Like Shell, Jersey also had its veteran champions of what it called 'oil-chemicals', but in the immediate aftermath of war their influence was less and Jersey moved much more slowly than it could have done in expanding petrochemicals, even at home.[16]

In expanding its chemical operations, the Group mostly opted for joint ventures with other chemical companies, in order to limit the risk and capital involved, and to get access to specific know-how. The American installations and the two main petrochemical units in Europe, Stanlow/Partington (UK) and Pernis near Rotterdam, were fully owned, but the other ones were nearly all co-owned. Some of the latter evolved over time into 100 per cent Shell companies, but quite a few remained minority stakes. In 1972 the Group's chemical operations employed a total capital of £562 million, of which £200 million was in companies where Shell had a share of 50 per cent or less.[17]

The French subsidiary, for instance, was one which over time evolved into a fully-owned company. In 1947 the Group entered into a 60-40 partnership with the French chemical company Saint-Gobain. Named Shell Saint-Gobain, the company took charge of Shell's Pauillac and Petit-Couronne refineries, and Saint-Gobain's Berre-l'Étang refinery near Marseilles, where the French firm already had a limited production of petrochemicals. The Berre installation was designated as an all-round plant encompassing the full range of petrochemical products which the Group developed: Teepol detergent (1951), isopropyl alcohol (1952), acetone (1953),

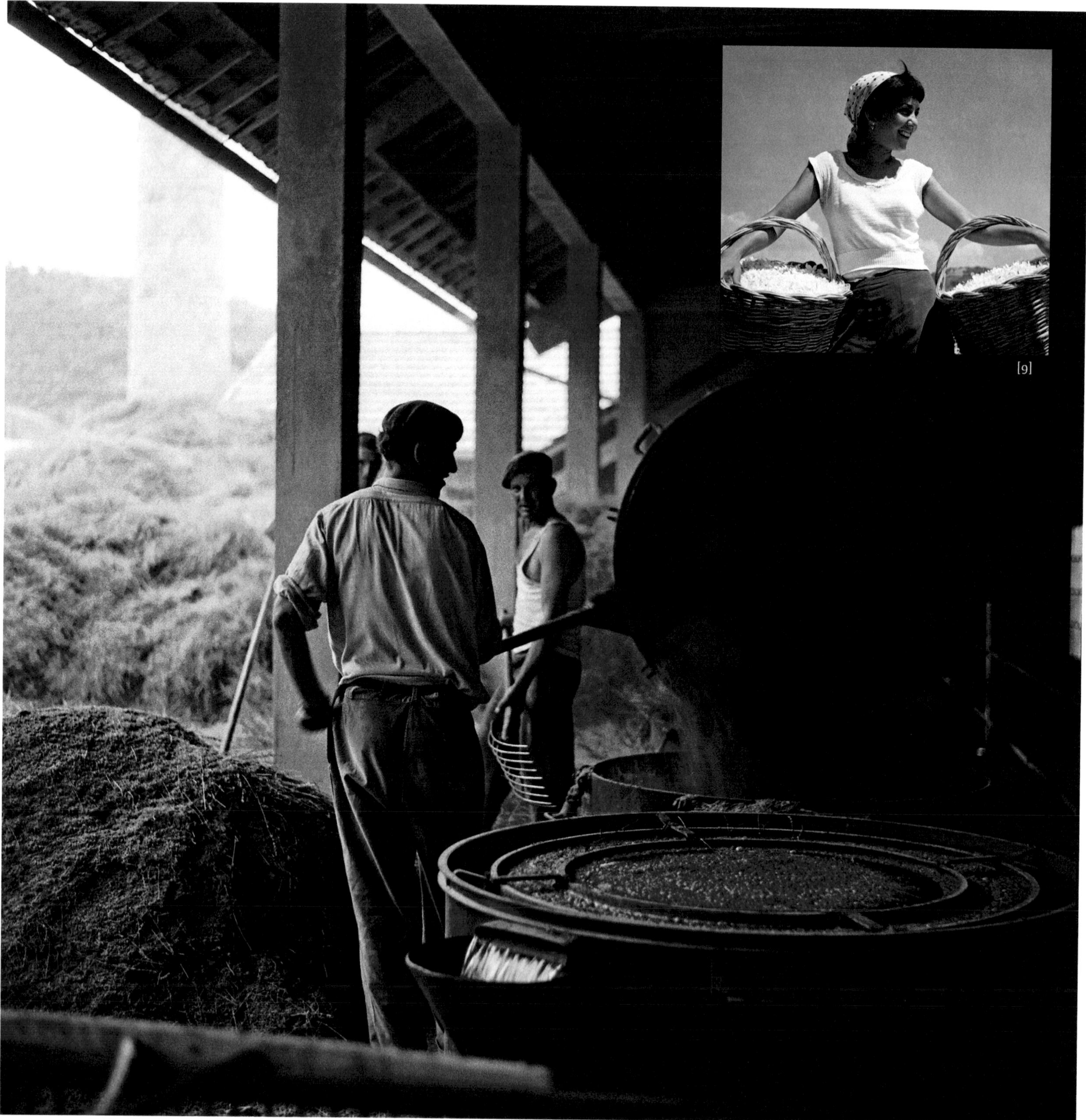

[9]

[8]

Few people who did not know about it would guess that the oil industry and the perfume industry are linked. Left, a still is loaded with lavender flowers to extract their essential oil; inset, under a blue sky in 1955, jasmine for the perfume industry is harvested in the hills at Grasse, France. Volatile solvents derived from crude oil proved extremely useful in the perfume industry, enabling the continuous extraction of perfume from flowers. At first ethyl ether was used, but this was replaced by a spirit known as Essence B, some of which was produced in the Shell St Gobain chemical plant at Berre l'Étang (right).

The plant also produced acetone of extreme purity, used in the manufacture of synthetic perfumes.

[10]

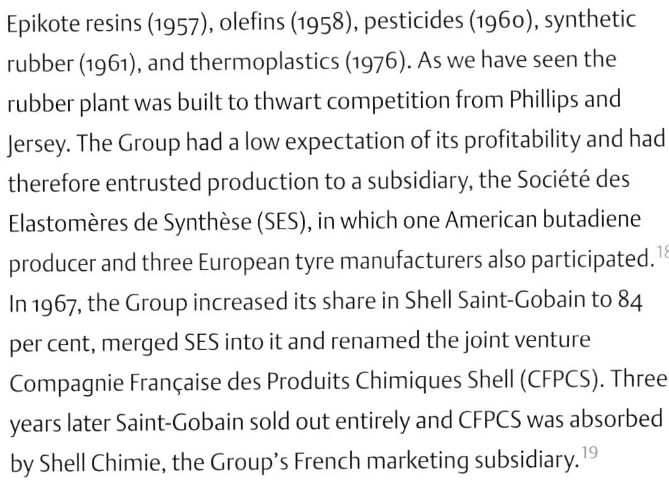

Epikote resins (1957), olefins (1958), pesticides (1960), synthetic rubber (1961), and thermoplastics (1976). As we have seen the rubber plant was built to thwart competition from Phillips and Jersey. The Group had a low expectation of its profitability and had therefore entrusted production to a subsidiary, the Société des Elastomères de Synthèse (SES), in which one American butadiene producer and three European tyre manufacturers also participated.[18] In 1967, the Group increased its share in Shell Saint-Gobain to 84 per cent, merged SES into it and renamed the joint venture Compagnie Française des Produits Chimiques Shell (CFPCS). Three years later Saint-Gobain sold out entirely and CFPCS was absorbed by Shell Chimie, the Group's French marketing subsidiary.[19]

By contrast, Rheinische Olefinwerke Wesseling (ROW) near Cologne, Germany, remained a 50-50 venture with Badische Anilin-und Soda Fabriken (BASF) and focused on ethylene and its by-products. Wesseling was one of the hydrogenation works built under the Hitler government to make synthetic gasoline. In 1948, Shell started to supply crude to Union Kraftstoff, which topped the crude in a destillation unit at Wesseling, hydrogenated the residue into gasoline and diesel fuel, and bottled the butane and propane hydrogenation offgases. Four years later, the processing capacity was expanded with a cracking unit and a reformer. With the ethane and ethylene offgases from the cracking unit, Wesseling became the hub of petrochemical operations. BASF and Shell had explored the opportunities for a joint venture to produce polyethylene and ethyl benzene from refinery gases since 1949, but legal complications delayed its formation until 1953. The plant came on stream two years later and supplied base materials for BASF's successful and rapidly widening range of plastics and polystyrenes. Over the years, ROW also expanded into the production of synthetic rubber and resins.[20] In 1967, BASF and Shell expanded their successful collaboration with the construction of a high-pressure polyethylene plant at Berre-l'Étang run by a 50-50 company named Cochimé.[21] Other joint ventures included small shares in petrochemical manufacturers in Japan, India, and Spain.[22]

The venture with Montecatini in Italy turned sour at a very early stage. The collaboration between the two companies began when Shell, after 'difficult negotiations and a considerable price', had gained a licence from the Italian company to manufacture polypropylene in the UK.[23] Shell and Montecatini then founded the the Rotterdamse Polyolefinen Maatschappij, which in 1963 started making polypropylene using this process as well.[24] That same year they formed an agrochemicals marketing company called Monteshell Agricola. In 1964 Shell acquired a half stake in Montecatini's Ferrara and Brindisi petrochemical plants for £58 million, which were then reorganized into Monteshell Petro-

[11] For all their sophistication, the arrangements of chemical installations such as the Berre-l'Étang works here, can acquire a certain timeless beauty when reduced to shapes and shadows and combined with the distinctly old-fashioned technology of a delivery tricycle.

chimica. Both the CMD and the Conference had serious doubts about taking this step. Monteshell Petrochimica's poor results led to quarrels between the two partners; and then, in December 1965, Montecatini suddenly announced its intention to merge with an important rival, Societa Edison.[25] Doing so would force Shell either to accept an undesirable minority position, or to raise its investment and accept a disproportionate stake in the Italian market – or to withdraw from the partnership altogether. Disappointed with the intentions of its partner, Shell wanted to withdraw from Monteshell Petrochimica immediately and by the autumn of 1966 had agreed terms about the separation with Montecatini.[26]

The Group acquired a different set of subsidiaries and affiliates, with one or two exceptions 50 per cent or majority owned, through its efforts to build market support for its chemical products. In 1960 Shell entered into a 50-50 joint venture with the American company National Distillers and Chemical Corporation. Called Shorko, this company was to manufacture and market polyolefin and polypropylene products, mainly plastic films and packaging. Based in the UK, Shorko quickly expanded its operations into Switzerland, the Benelux, Germany, and Japan, and the company also bought interests in can manufacturers. In 1965 the Shorko group ran into heavy losses and it was dissolved two years later, probably because the two partners could not agree on a plan to reconstruct the business.[27] Meanwhile Shell had acquired, in 1962, 51 per cent of the Dutch polyvinyl chloride (PVC) pipe manufacturer Wavin, largely to keep it out of the hands of rival PVC producer Solvay. Wavin subsequently embarked on a successful European expansion.[28] Two years later Shell Oil bought a Hollywood plastic utensils manufacturer to safeguard its raw material sales, and in France the Group also acquired end-users of plastic ranging from tub-makers to boat-builders and plastic bottle factories.[29] In another key product segment, agrochemicals, Shell acted similarly but for different reasons. When during the mid-1960s the 'drin family of pesticides attracted mounting public criticism for suspected toxicity, Shell started buying up agrochemical distributing companies, notably in the United States.[30]

By 1961 the Group definitely possessed an impressive second leg. Compared to other leading oil companies, Shell had by far the largest aggregate investment in operational petrochemical plants: a total of £208 million, with £91 million in the US and Canada, and £117 million in the rest of the world. Jersey, the nearest competitor, had a total of £128.1 million, with nearly one-third (£41.2 million) in the rest of the world; Phillips Petroleum, next, had invested a total of £80.2 million, almost all inside the US and Canada; and BP, next, had a total of £52.7 million, solely in the rest of the world. The Italian ENI had nothing in the US and Canada and was the only other organization with a substantial 'rest of the world' investment, but that was barely half BP's.[31]

From interface to battleground From the start, chemicals developed spectacularly. Sales rose very rapidly indeed, from the late 1950s even faster than oil sales (Figure 5.1), and by the end of the 1960s they generated nearly 9 per cent of revenues (Figure 5.2). As we will see, growth became the prime justification for chemicals, a goal in itself, which put other business considerations in the shade. This growth was achieved by sustained heavy spending. Manufacturing usually absorbed about a third of investments, second only to exploration and production. Oil tended to receive by far the most, three or four times the amount spent on chemicals during the 1950s and early 1960s, but rising expenditure on chemicals narrowed the gap considerably. By 1970 chemicals received about half the amount spent on oil (Figure 5.3). Notwithstanding the functional separation between oil and chemicals following the McKinsey review, the Group always treated the manufacturing budget as a whole, so spurts of investment in one side of the business curtailed the flow of funds to the other (Figure 5.4). The coordinator's budget meetings must have been very animated, because in proportion to its sales chemicals required far more investment than oil, usually 10 to 15 per cent of sales, against only 3 to 5 per cent for oil. The graph also delineates the three distinct phases of the Group's chemical operations. Until 1961, the sector was considered to be in a start-up phase. From that year, Shell regarded chemicals as having matured and the aim became to raise the return on investment to 8 per cent by way of selective investment. This was very ambitious given the fact that, during 1959-61, chemicals had shown zero return on assets.[32] The new policy was quite successful, however; in 1963 the net return had climbed to 6.6 per cent. When, five years later, it touched the desired target of 8 per cent, the CMD decided to go for the highest number of profitable opportunities, in the expectation that chemicals would yield a sufficient return from then on and thus provide a sound basis for the planned operational diversification.[33]

Though ostensibly complementary, oil and chemicals soon proved to be very different industries, each requiring its own management approaches, investment strategies, marketing policies, and profit yardsticks, whilst competing for shares of the

[12]

[13]

Emphasizing its scientific nature was another theme in Shell's advertising, providing information for an intelligent and curious community. These images from the Dutch publication *De Journalist* also show the growing public familiarity with scientific ideas: the near one (1958) describes research, while the far one (1965) specifies the 'tailoring of molecules' – a nice term for petrochemical manufacture – with a design echoing Brussels' famous 'Atomium', centrepiece of Expo 1958. But with increasing public awareness of environmental matters, the intelligent and curious community was developing into an intelligent and questioning community.

same Group resources. Shell Oil was first to recognize the need for a separation and, in 1945, re-established its petrochemical division as a separate company, Shell Chemical Corporation. This gave Shell Chemical the impetus to switch its focus from upgrading hydrocarbons to exploring the opportunities for making money in the chemical business.[34] The Royal Dutch Annual Report for 1949 shows that Group managers had also percieved the need for a formal separation of manufacturing activities into oil and chemicals within the ongoing reordering of the business along functional lines. The report's customary country-by-country summary of operations, used for over forty years, was discarded in favour of a survey by business functions, highlighted by developments in particular countries around the world, with the manufacturing section having separate subsections for oil and for chemicals, the latter treated as an activity on its own.

To some extent the office organization mirrored this split, the chemical industry administration being part of Guépin's odd assortment of marketing, regional, and product departments, with manufacturing remaining in The Hague. On the formation of the 1955 Coordinating Committee, W. F. Mitchell became Coordinator. Chemicals and J. van Krimpen Coordinator Manufacturing, the rider 'oil' to Van Krimpen's title being omitted because he was in charge of all installations, oil and chemical. Even then it was not until 1958 that the transfer of chemicals to separate operating companies had become the norm throughout the Group.

At the central level the last step was only taken following the McKinsey reorganization in 1959, when the gradual separation between oil and chemicals received its logical conclusion with the formation of identical and fully-fledged central departments for both, each having its own array of sub-departments covering finance, planning, manufacturing, research, marketing, personnel, balanced by regional coordinators.[35] One aim of this separation was to make the chemicals managers directly accountable for their results.[36] The reorganization did not, however, put oil and chemicals on an equal footing. Both departments were headed by a Director of Coordination, but the DCO (oil) was a managing director, usually the most powerful person in the CMD, and the DCC (chemical) was not, though from November 1961 he attended the CMD on a monthly basis. It was only in 1970 that the Director of Chemical also obtained a seat in the CMD. Nor did chemicals become truly independent, tied as the function was to the rest of the Group with

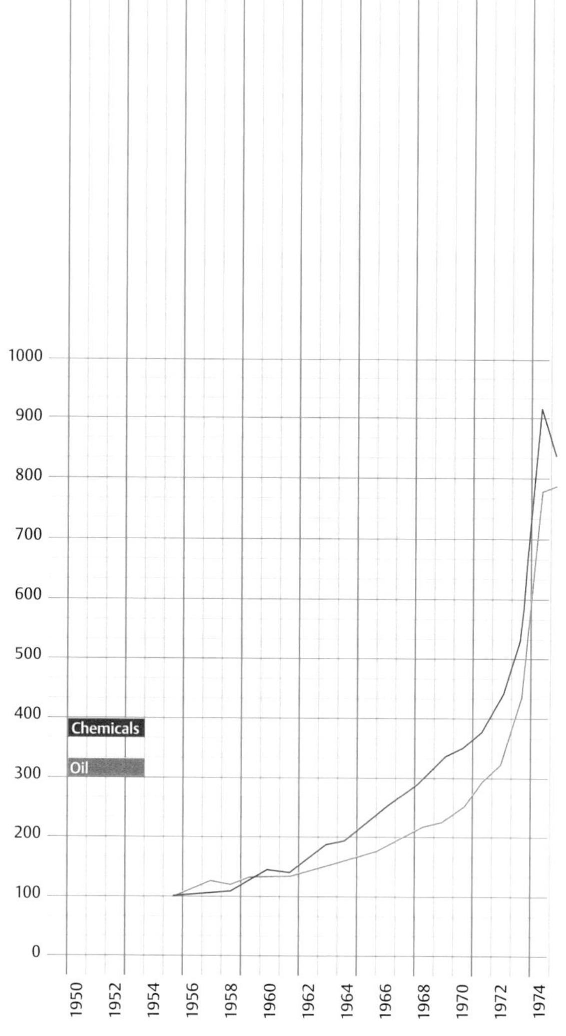

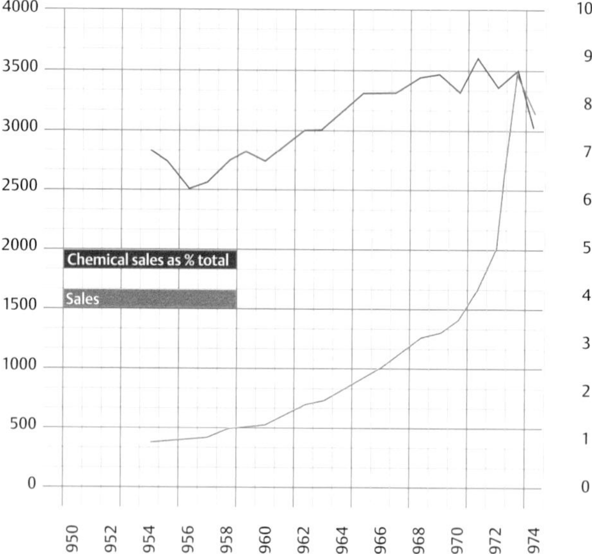

Figure 5.1
The growth of Group revenues, 1955=100.

Figure 5.2
Group chemicals sales, total (left scale, US$ mln).

[14]

[15]

supply contracts, service agreements, and central office overheads. The chemical sector considered the separation from oil as a deserved emancipation. The managers responsible for chemicals thought that taking on the practices of the chemical industry would lead to sounder expansion of the Group's chemical business than sticking to the view that chemicals were simply valuable by-products of a barrel of oil. According to DCC W. F. Mitchell, there should still be 'a mutual interest and understanding between these two great enterprises', but, as an example of the distinctions between them, he cited volume and value. The chemist and chemical engineer were required to use very much smaller volumes in more exacting operating conditions, and with higher value. If a petroleum stream should go off its specification in the course of refining, it could easily be reblended or redistilled to specification, but few chemical products could be salvaged so easily or cheaply. If a ton of heavy fuel were lost through a fractured pipe its value was only about £8, but 'if in a chemical plant a fracture occurs in a cyclopentadiene feed line (...) each ton of that cyclopentadiene is worth a few hundred pounds'.[37] It made sense, given the highly specialized nature of the petrochemical industry, to end the subordination of chemicals to oil and have the function's managers focus on the demands of their industry. Moreover, the two functions had the same raw material base but they shared very little otherwise, so the synergy between them remained tenuous, indeed highly doubtful. Shell Chemical Corporation bought only one quarter to a third of its supplies from Shell Oil and the rest from outside companies.[38] As late as 1970 the Chemical Coordinator E.

In the late 1950s Shell's top men in manufacturing included (from left to right) F. A. C. Guépin, J. L. van Krimpen and W. F. Mitchell.

[16]

Lelyveld reported, somewhat hopefully, to the Conference that the Group was 'uniquely well placed to exploit the growing interdependence of chemical operations with oil operations', evidently not something which he thought had been a central feature before.[39]

Consequently, the functional separation created as many problems as it was supposed to solve. At plant level, it meant having people in charge of the oil and the chemical installations whose interests did not necessarily coincide. At managerial level, it meant constant disputes about the costing of fuel, materials, and secondary processes such as the recycling of spent sulphuric acid. Crude oil itself became a bone of contention, the oil side striving to get the best price for its barrel, leading to the chemical side complaining about the high price of feedstock. From a shared business platform, the interface between oil and chemicals turned into a battleground.[40] Once again Shell Oil identified these problems first. In 1959 Shell Chemical Corporation was merged back into Shell Oil and continued as a division of that company, which in effect meant that chemicals only retained a separate marketing organization. However, central offices remained committed to the ideal of a complete separation for several decades to come.

As a diversification, the chemical sector was thus constrained by a dual compromise: the managers running it were neither free to act in the best interests of the function, nor were operations sufficiently integrated with the rest of the Group so as to maximize synergies and profits. The product range was also quite heterogeneous.

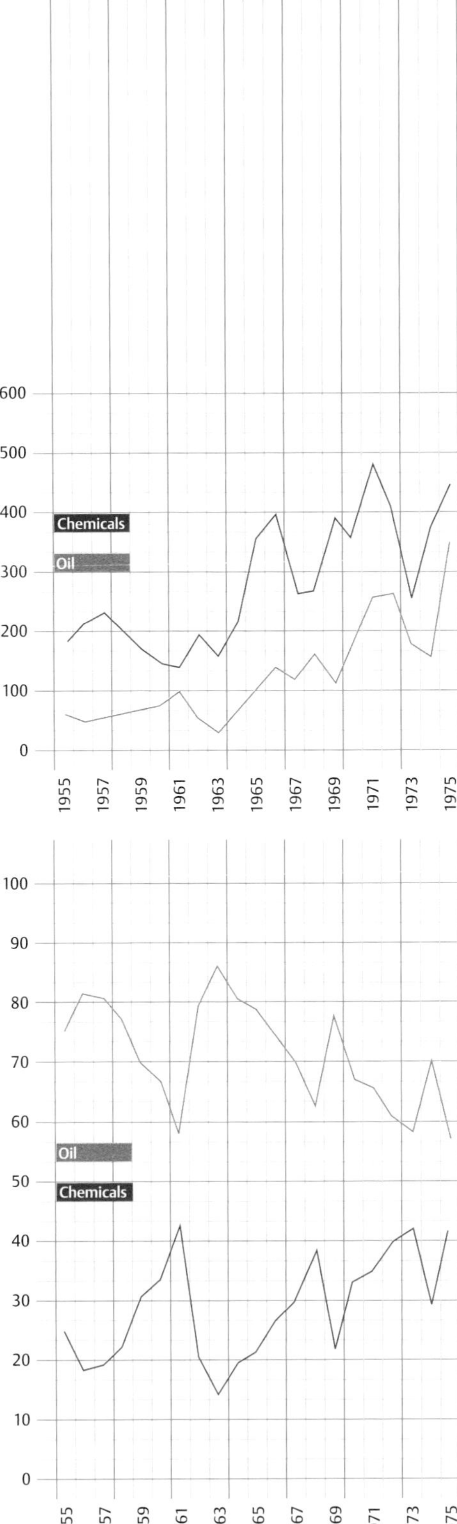

Figure 5.3
Group investment in manufacturing, 1955-1975 (US$ mln).

Figure 5.4
Oil and chemicals investment as a percentage of the manufacturing total, 1955-1975.

Chapter 5

Products and processes During the immediate post-war era Shell, fully engaged in its expansionary drive to keep pace with booming post-war demand, did not push to widen the variety of chemical products offered. Solvents, fertilizers, and, in Europe, Teepol, were the mainstays. To cater for an acute shortage of soap, Group managers increased the production of Teepol in Europe with the construction of new manufacturing units at Stanlow and at Pernis. Stanlow's intake capacity and its production of polypropylene and the pesticide DD were also substantially increased. The additional quantities of butylenes would be used to make more methyl ethyl ketone, with surplus olefins being sold as bottled gas.[41] Fertilizer production was expanded as well, with new capacity at the Mekog works and a new plant at Shell Haven (1959).[42] Over the opposition of the Group's Amsterdam laboratory manager, the Bataafsche board decided to transfer the small pilot plants for asphalt, grease, and pesticides from Amsterdam to Pernis, to serve as the basis of the new and very large integrated petrochemical complex there, but no initiatives were taken towards launching new products.[43] A keen interest in the new kinds of pesticides such as DDT which had emerged during the war did not get a follow-up; the 1946 decision to start making polyvinyl chloride (PVC) at the IJmuiden Mekog works showed the shape of things to come, but the initial volume was not very significant.[44]

Shell Oil similarly expanded along established lines, with alcohols and ketones as its main money-earners.[45] During 1946-50, Shell Chemical spent nearly $55 million on new plant at Houston and elsewhere. In 1948 the Houston synthetic glycerine plant came on stream, followed in October by its synthetic ethanol plant and in November by another new plant, this one being for synthetic ethyl chloride.[46] In the case of ethanol, or ethyl alcohol, Shell's forward step was to devise a method of combining ethylene gas directly with water, rather than through any of the former longer processes.

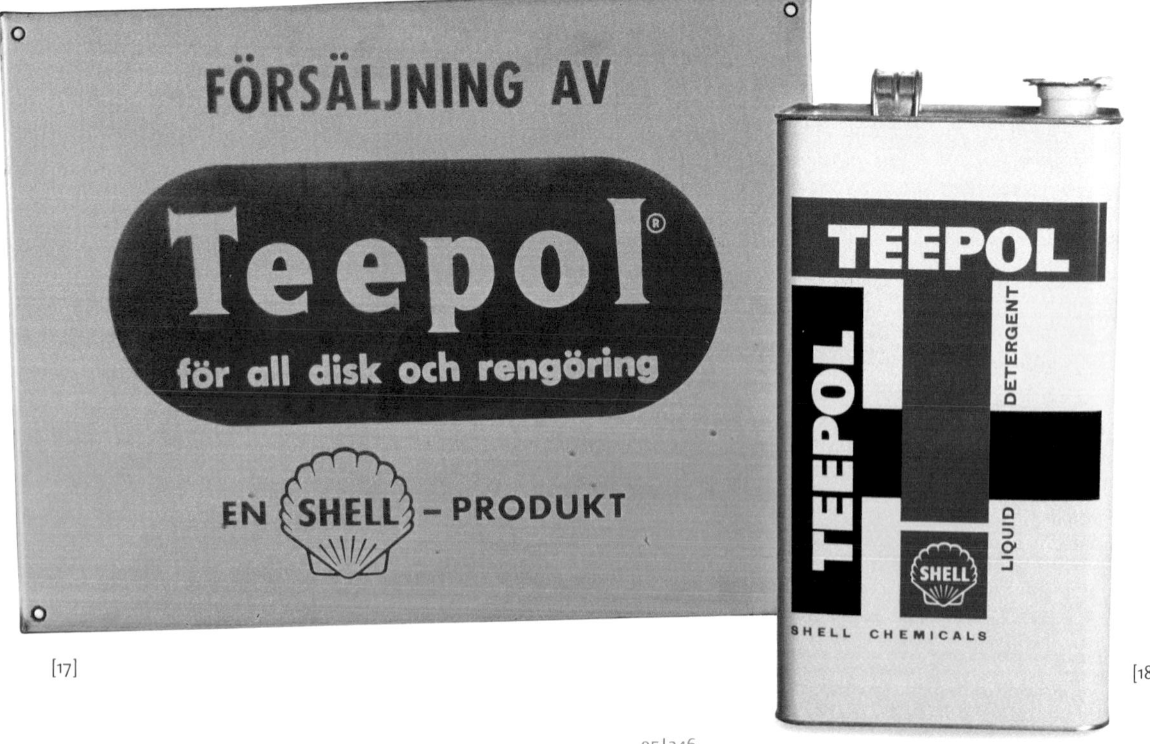

[17]

[18]

Domestic laundry, a dull and routine chore for housewives in every part of the world, was made far easier with the advent of washing machines, and Shell's detergents were part of the formulation of popular soaps such as Tide, seen here being advertised (right) on Hawaiian television.

[20]

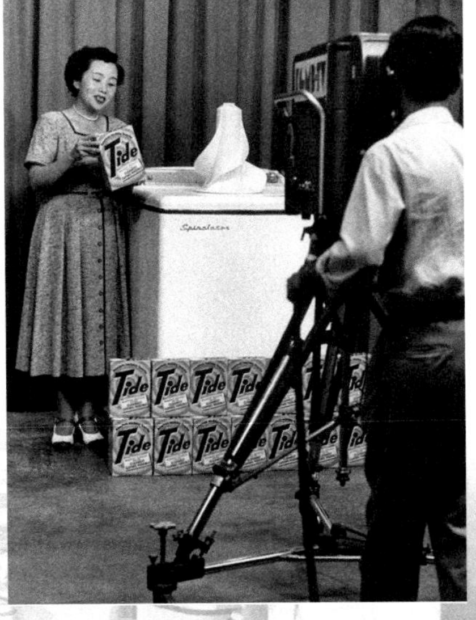

[21]

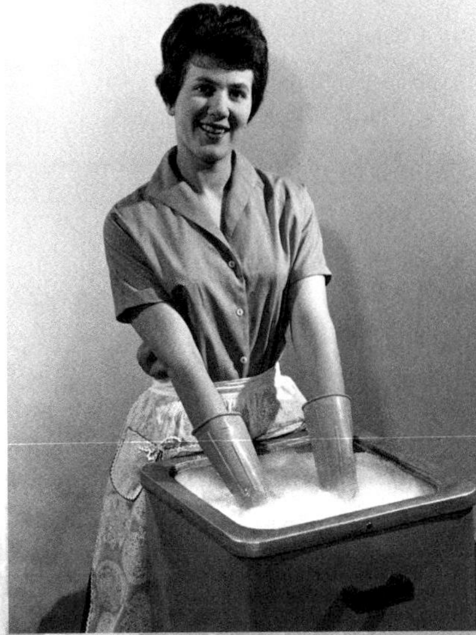

[19]

[22]

Although the different units were indistinguishable to most people, an expert looking at Shell Chemical Company's plant at Carrington in the UK in 1961 (general view, far right) would quickly identify the ethylene oxide plant (left) and the ethylene glycols plant (right).

First, though, the gas had to be made, and at $16 million the Houston ethylene plant was Shell Chemical's largest single post-war item of expenditure, its cost justified by the large number of products that could be derived from ethylene – amongst others, synthetic rubbers, vinyl resins, polythene, PVC, and polystyrene, all of which Shell subsequently manufactured. Ethanol is actually the alcohol in alcoholic drinks, but US federal law insisted that alcohol for drinking must be made from agricultural products.[47] Shell's ethanol was consequently shipped under extremely close controls, and used for the manufacture of printing inks, dyestuffs, plastics and cosmetics. Shell also sold denatured ethanol as a solvent, under the trade name 'Neosol', while its ethyl chloride (made from the same ethylene base) was used as a starting point in the manufacture of the anti-knock gasoline additive, tetraethyl lead. A further step along established lines was the construction of a new fertilizer plant at Ventura in California (1953). On the other hand, Shell Oil's chemical programme contracted when, at the end of the war, the company stopped butadiene production and returned the Torrance manufacturing complex to the US government.

Thus, though the Group's sales of chemicals rose, the variety of products sold was still small. The available figures for 1952 show that in Europe agrochemicals and industrial chemicals, including Teepol, dominated with each around 40 per cent; base chemicals such as solvents yielded nearly 20 per cent, and plastics, resins, and rubbers were negligible (Figure 5.5).[48] Just then the mould had started to break as a consequence of a determined drive to expand the Group's petrochemicals business into five new, or nearly new, directions: epoxy resins, chlorinated hydrocarbon pesticides, synthetic rubber, plastics, and industrial chemicals. The expansion completely changed the Group's chemical product sales (Figure 5.6).[49] Between 1955 and 1972, agrochemicals and solvents dropped from nearly 60 per cent to about 40 per cent of the total, largely because of Shell's gradual withdrawal from fertilizers. Conversely, plastics, resins, and rubber plus industrial chemicals rose from 40 to 60 per cent.

Acquisitions played a substantial part in the new petrochemical strategy. In 1955 the US government decided to sell its rubber-producing assets, and Shell bought the Torrance factory, resuming rubber production at the point where it had left off after the war. Jersey's affiliates Esso Standard Oil Company and the Humble Company did the same.[50] As a consequence, Shell Oil concentrated on developing synthetic rubbers and left plastics to

[23]

[24]

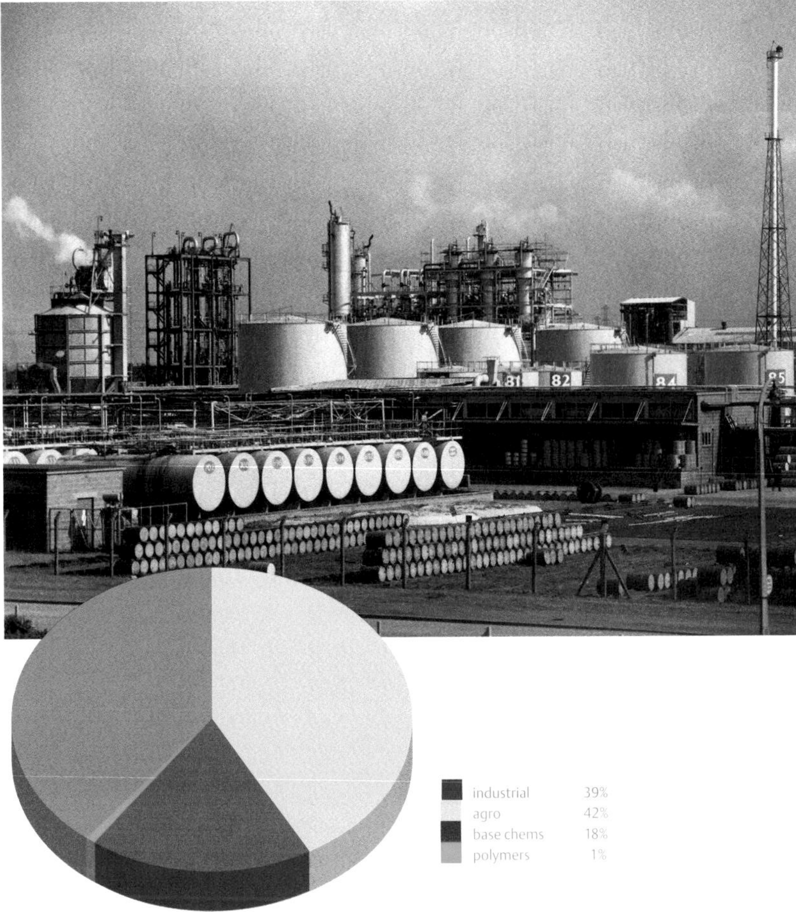

Figure 5.5
Group chemical sales revenue in Europe by product segment, 1952.

the European companies.[51] To get a foothold in plastics, Shell took over a British manufacturer, Petrochemicals Ltd. and its subsidiary Styrene Products, in 1955. Based in Partington near Manchester, the company used the Catarole cracking process, invented by Dr. Chaim Weizmann, to produce ring-like aromatics, such as nitration-grade benzene and toluene, rather than the chain-like aliphatic types of product produced by high-temperature cracking. In addition the company held the exclusive licence for the UK of the valuable Ziegler patents for plastics and synthetic rubber. Petrochemicals Ltd. also possessed new and modern research laboratories and its Partington works could easily be linked by pipeline with the Group's Stanlow complex. Shell transferred Shell Development's process of direct catalytic oxidation of ethylene to Partington and combined the installation with Stanlow into an integrated unit for making synthetic rubber, styrene monomer, polyethylene, and ethyl benzene.[52] In 1962, the Group expanded its UK interest in industrial chemicals by buying a minority stake in the Lankro company.[53]

The resins and pesticides were acquired by purchasing American know-how and technology. The resins sprang directly from the manufacture of synthetic glycerine. One of their most important components, epichlorohydrin, was an intermediate product of that process, and was bought in large quantities by Devoe & Raynolds, a paint manufacturer. Shell Chemical acquired an exclusive licence for resins patents owned by Devoe & Raynolds, and in 1950 began production of resins at Houston, selling them under the trade-name Epon. Houston's production capacity was repeatedly increased over the next few years and Epon manufacture soon spread to the Group's new petrochemical establishments worldwide, including Pernis (1953), Stanlow (1955), Wesseling (1957), Berre-l'Étang (1959), and Clyde in Australia (1962). Outside America the resins were called Epikote, and

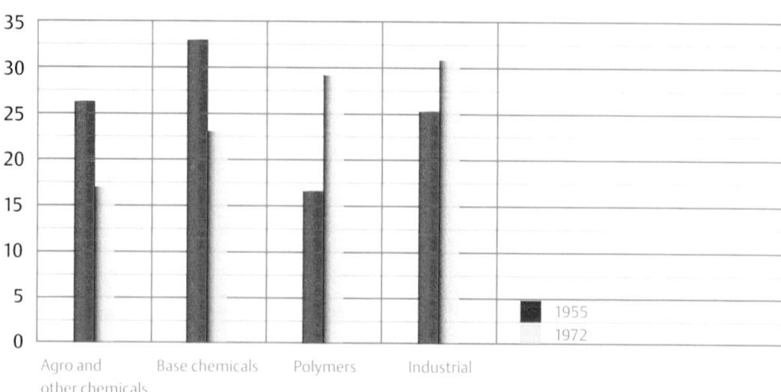

Figure 5.6
Group chemical sales revenue by product segment, 1955-1972 (million pounds sterling).

From the early 1960s Clyde in Australia was one of the sites (right) for the manufacture of Epikote (named Epon in the US), Shell's branded epoxy resin used in paint, varnishes and engineering applications. Facing page below, the formula for Shell's Epikote resin.

everywhere they presented the question of marketing a product that to begin with was entirely new: customers had to be persuaded that this was something they should not ignore.

Whether called Epon or Epikote, the resins were highly successful, proving very versatile as the basis for surface coatings, from paint to airport runways: mixed with asphalt, they could withstand hydraulic fluids, the corrosive effects of cleansing solvents, and even the heat from jet engines. They were also extremely good as a bonding agent, capable of sticking unlike substances together, such as concrete and steel; and entirely by accident, a senior mechanical engineer at Houston discovered that with the right proportion of polyamide curing agent, Epon could easily be applied underwater to the hulls of ships and to fixed metal structures, providing at low cost a very high degree of resistance to corrosion.[54]

The second significant bought-in process was also extremely successful but became the source of very serious controversy: chlorinated hydrocarbon insecticides, of which the main ones were aldrin, dieldrin and endrin, known together as the 'drin family of pesticides. The Group's involvement with pesticides went back a long way and had originated in the search for the right kind of light oil with which to mix and spray them. Selling pesticide sprays remained a modest business until, in 1950, Shell Chemical Corporation started marketing aldrin and dieldrin as sole agents for Julius Hyman & Company, based on the Rocky Mountain Arsenal military base in Denver, Colorado. A lost patent litigation suit over the 'drin formula with a company called Velsicol forced Hyman to sell his business, which Shell Oil acquired in 1952 after having secured exclusive rights to the patents from Velsicol.[55]

In their ability to eradicate not only mosquitoes but also crop-destroying insects such as locusts, ants, grasshoppers, boll weevils, nematodes, and wireworms, the chemicals at first gave the impression of being little short of miraculous: for example, in 1951, when the Iranian government appealed for international help in combating its worst locust plague for eighty years, 13 tons of aldrin were flown out, and it took only two ounces (less than 57 grams) per acre to kill 98 per cent of the pests.[56] The purchase of Hyman's chimed well with the 'war against hunger' declared by the United Nations' Food and Agriculture Organization at the beginning of 1952. In that year, dieldrin saved the $75 million Californian rice crop, after 50,000 acres had been damaged by the worst infestation of rice leaf-miners in three decades.[57] In a hungry

[26]

With pesticides, herbicides and other products, the petrochemicals explosion seemed to offer highly promising solutions to the world-scale problems of disease, hunger and starvation. *The Golden Lands*, *Food or Famine*, *The Land Must Provide*, and *The Threat in the Water* were just some of the associated educational films produced by Shell in the 1960s.

world, it was firmly believed that insecticides of such potency could only be good; for the first time in the history of humanity, the 'drins and other pesticides appeared to offer the real possibility of an end to, or at least a massive reduction in, starvation. With a proposal that seemed so morally praiseworthy as well as commercially sound, production soon followed in Group facilities outside the US. Plants for aldrin and dieldrin came on stream at Pernis in 1954, and within three years production of insecticides there had tripled.[58] By 1972 the 'drins were also produced in the UK, Belgium, Germany, France, and Italy. Contemporary marketing was done in good faith and with missionary enthusiasm, and repeated field trials in customer countries were particularly effective as demonstrations. Yet views changed in the light of experience. By the middle 1960s certain of these agricultural chemicals were the subject of hot international debate, and in many parts of the world their use was restricted or even banned (see Chapter 6).

The 'drins represented the Group's ideal chemical product. They were well protected by patents, hydrocarbon-based, and research-intensive; they were modern and glamorous in offering an instant, technical solution to ancient scourges of mankind; they fitted in well with other manufacturing processes; and they combined low volume with stupendously high margins. During the 1960s, Shell pesticide sales outside North America rose from 66,000 to more than 170,000 tons a year, resulting in a claimed market share of 10 per cent.[59] Profits averaged £43 pounds a ton; only epoxy resins, which had the same allure of modernity and victorious technology, were more profitable, but in much lower volumes.[60] Compared to such spectacular products the rest looked positively mundane. The sales volume of Shell's next best-earning product group, industrial chemicals, amounted to nearly 1.3 million tons with average profits of £8 per ton. Solvents yielded less than half that, fertilizers barely broke even, and until the late 1960s plastics and rubbers were losing money heavily.[61] Moreover, with around 1 million tons a year each, solvents and industrial chemicals were by far the biggest segments by volume. In 1968, the product

[27]

Night without nets
Night has a thousand small and secret sounds. Wind whisper and creak of board, sudden skitter of lizard feet, click and tap, slither and rustle, the ceaseless *crik-crik* of cicadas under the great glittering moon. And among those sounds, the thin, wavering whine of a killer: unseen in the darkness, riding on diaphanous wings, bringing each year disease and misery and death to millions in many parts of the world. It is estimated that in South East Asia alone, before malaria control was introduced, at least 50,000,000 cases occurred annually and that of these, half a million died as a direct result of the disease.
Today the menace is being driven from the scene by eradication campaigns like that in the Philippines.
Slowly but surely. Progressively. By degrees—and by insecticides like dieldrin. Used as a residual spray to kill malaria-carrying mosquitoes (chiefly *Anopheles minimus flavirostris*) and also as a larvicide, this powerful insecticide developed by Shell is playing a major part in a nation-wide house spraying campaign to eradicate malaria completely in the Philippines. Already results are greatly encouraging; in the *barrios* typical of the rural areas, sickness has fallen sharply, in some cases by as much as 75%, and infants are growing up free of the malaria menace. One day soon, it is believed, the night will be made safe for man, without nets. *And not only in the Philippines, but throughout the world.*
dieldrin
DIELDRIN, ENDRIN, ALDRIN, D-D AND NEMAGON ARE SHELL PESTICIDES FOR WORLD-WIDE USE

[29]

[28]

group plastics, resins, and elastomers finally made a small profit, entirely due to resins and elastomers cancelling out the continued loss in plastics.

Thus after some fifteen years of determined diversification, Shell sold an impressive range of new chemical products, many of them very sophisticated and modern. The company had the biggest chemical operations of the oil majors and in terms of sales Shell ranked tenth amongst the world's chemical companies.[62] The return on assets for the Group's chemical sector as a whole had climbed from zero to around 8 per cent. However, by other standards the diversification was not so successful. The bulk of the Group's chemical sales were still in the segments closest to the oil side, solvents and industrial chemicals, both of which yielded steady if unspectacular profits. Among the much-heralded new products were two blockbusters, pesticides and epoxy resins, of which the former suffered from declining public acceptance, while the latter only sold in small volumes. Two further showpieces, plastics and rubbers, had improved but were still losing money, not such a good record given the sums invested. Some operating

[30]

In days of hope and optimism, petrochemicals appeared to provide the answer to many perennial challenges. The insecticidal 'drins could remove the threat of malaria (inset, below left), and in the middle 1960s, whether applied in vapour form in an American tobacco warehouse (far left) or a cattle shed (centre left), or hanging in strip form in a grocery store (near left), Shell's trade-marked Vapona was effective in destroying insect pests. Plastics from petrochemicals also offered protection to people in an entirely different way (below).

companies performed very well, notably Rheinische Olefinwerke. However, the sector's return on assets remained structurally lower than oil and chemicals never generated sufficient cash for its own needs. By the end of 1971, the Group had a cumulative gross investment in chemicals of £585 million. Of this amount, the sector itself had generated only £350 million or 60 per cent; the rest, a staggering £235 million, had come from central funds and thus represented a cross-subsidy from oil to chemicals. And the storm clouds were gathering again: overcapacity, rising inflation, structural labour unrest in the UK, and a global recession hit the chemical sector badly during 1970. Previously 'heroic financial decisions in the face of risks, and patience in waiting for the return' had seemed admirable and stimulating, but now it seemed time for one of the Group's recurrent policy debates.[63]

[31]

Oh! What a beautiful day!

No, they didn't see the little black cloud coming up behind the trees. It was a beautiful day when they set out and, so far as they are concerned, it still is. . . . Young love makes a better umbrella than a newspaper! Of course, if anyone wanted a waterproof paper umbrella it could easily be made and probably has been, for paper is the most versatile of materials. Symphonies, best sellers and the world's correspondence are written on it. It wraps practically everything under the sun, from food and other goods that need perfect hygienic protection to undersea cables that need perfect electrical insulation. In a thousand ways, all day and every day, paper in all its variety is making an invaluable contribution to modern living.

Useful in paper making and essential in paper converting are chemicals from petroleum. The paper industries rely extensively on Shell for these chemicals, and their uses are as wide-spread and various as those of paper itself. Shell is one of the nation's biggest producers of high quality solvents, petroleum waxes, detergents, glycols and polyglycols, and a host of other chemicals. If you would like more detailed information it will gladly be given. If you have any problem involving the industrial uses of chemicals, Shell would be pleased to help you find the right answer.

Write to the Advertising Manager, Shell Chemical Company Limited, 170 Piccadilly, London, W.1.

YOU CAN BE SURE OF SHELL CHEMICALS

Chapter 5

Debating the chemical sector's performance In line with the Group's other business sectors, the CMD monitored chemicals closely, and conducted repeated and fundamental debates about investment decisions and performance. The pace of expansion in oil and chemicals during the 1950s and 1960s stretched Shell's financial power to the limit, necessitating a careful allocation of resources. As early as 1957 the CMD voiced doubt about the viability of the Group's whole chemical business. While losses at Petrochemicals Ltd. continued to mount, managers had to decide on the proposed erection of a synthetic rubber plant at Pernis. The payout time would be only three and a half years at the most; it was, moreover, 'one of the most attractive projects we can expect from our chemical business and presents an opportunity which should not be missed if we are to realise the benefits of the large sums we are spending (...) The need to proceed is urgent, for although we alone have available the raw material, site, research and know-how which together make the project so attractive, it is evident that competitors have only held off building such plant themselves because they believe we have actually decided to go ahead, and because there is only room for one such plant to meet the requirements of the Benelux countries and Scandinavia'.[64]

Against these arguments in favour, the problem was that the project would cost £11.5 million. If this sum were spent then it was uncertain whether funds could be also found for other projects 'more essential fundamentally to the successful continuation of our business'. Coming as it did from the CMD collectively, the description of chemicals as not 'essential fundamentally' was strong stuff, and Mitchell and his his manufacturing colleague J. van Krimpen were asked to set out 'exactly the risks we face in delaying (...) and also the price of any reasonable steps towards construction we might take in the meantime to prevent competitors losing the impression that we were going ahead'. It was agreed that in such an attractive venture, equity partnership would not be a satisfactory solution, but there was a very strong sting in the tail: 'If eventually it should be decided that we cannot proceed with this project for financial reasons, then we shall have

[32]

[33]

to face the fact that we cannot afford our chemical business and that a complete review will be necessary'.[65]

Despite the apocalyptic language, the decision to proceed was made and the plant came on stream in 1960.

That same year, at a key CMD meeting introduced by Group managing director Lijkle 'Skippy' Schepers on Shell's future chemical business policies and programmes, the Group's finance director Chester Peet challenged the sufficiency of chemicals' projected cashflow – 12.5 per cent on gross investment, against 16 per cent 'from the Oil side'.[66] DCC Mitchell defended chemicals on all fronts. While accepting that 'unless the Chemical side (...) could afford to pay for the money it required, they were not justified in asking for it'. Mitchell nonetheless considered that, given time, chemicals would pay. In marketing, he acknowledged there was severe competition, but the Group's sales force was well balanced,

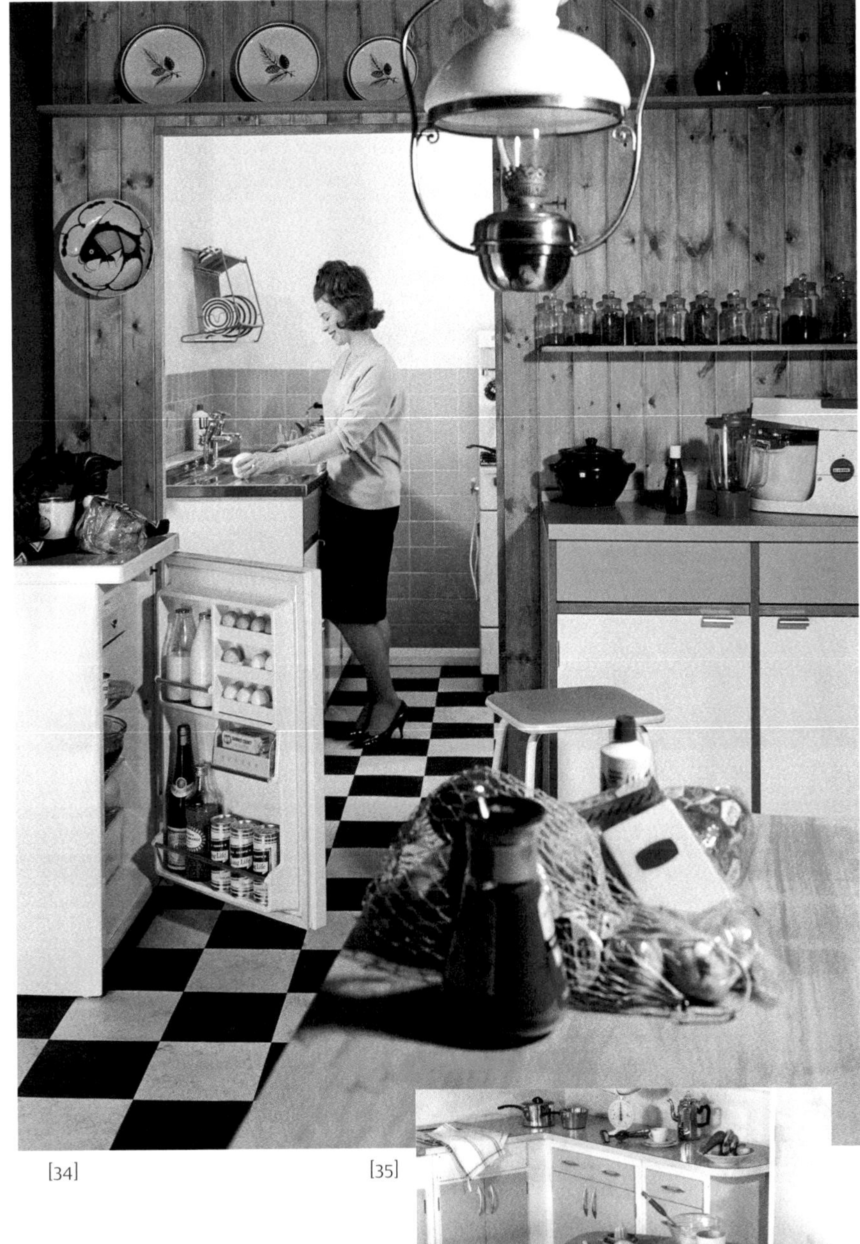

[34]

[35]

O brave new world that has such products in it! On 9-12 June 1959 at Shell's central offices in The Hague, an exhibition entitled 'Plastics and Shell' generated considerable excitement (far left), even though Henk Bloemgarten appeared somewhat sceptical on first seeing a new design of electric kettle (left). By the middle 1950s, wipe-clean Formica was widely used for table-tops and other surfaces (right), and by 1962 a typical kitchen (above right) could include products made from PVC, polypropylene, polystyrene, terylene, nylon, cellophane, polyethylene, Epikote resin, polyvinyl and polyurethane.

[36]

In the late 1950s and early 1960s, Hula-Hoops became a worldwide craze. Named after the swaying Hawaiian hula-hula dance and whirled around the body by movements of the waist and hips, they were used both for exercise and fun (as seen in Rhodesia in 1960, above), and were made from Shell Chemical Company's high-density polyethylene, extruded as 5/8-inch tubing (right).

[37]

and with operations in eighty-six countries it had unique marketing opportunities and could use plants to their full capacity. In manufacturing, older product lines such as detergents were still good bread and butter, but the main opportunities for expansion were in synthetic rubber and plastics, and considerable investment was needed to keep pace with demand. He also put down what seemed a daring option on the future rate of growth, asking the meeting to consider the possibility of no further authorized capital expenditure on chemicals. If that occurred, then in order to complete existing projects, chemicals' drawing from central funds would be £20 million in 1960 and a mere £1.2 million in 1961, after which the operations would move into surpluses of £16 million, £17 million and £21 million in 1962-65. Some of those present who were more sceptical about chemicals must have been tempted to call Mitchell's apparent bluff, but he went on to add that in contrast, if drawings from central funds continued, they would permit disproportionate expansion budgets. An annual average £15 million drawn would permit £20 million expansion in each of the next three years; £20 million drawn would permit £27 million expansion; £25 million drawn would permit £35 million expansion, with respective possible growth rates of 9, 12, and 15 per cent.

The message seemed clear: the greater the investment, the greater the anticipated return, only by growing fast enough would chemicals one day make enough profit. At Shell Oil board meetings debates about the chemical sector followed very similar lines, with identical outcomes. Some directors often criticized the disappointing performance of chemicals compared to oil. One such was Doolittle, who habitually asked whether the division had made any money. On one occasion his questioning stung Spaght, a staunch supporter of chemical and a great admirer of Kessler, into replying that low returns were characteristic for a growth industry, and if chemical's rate of return was lower than oil, he hoped that 'it always stays that way because that means we're growing'.[67]

[38]

Shell was willing and able to make highly specialized chemicals on demand. The semi-liquid material being poured at left was an adhesive made to the customer's specified qualities of strength, heat resistance and curing, as well as with particular flow characteristics, and was devised to overcome a production problem on an assembly line in Utah. In this case the ultimate customer was the US Government, and the product was needed for the third stage of the manufacture of Minuteman inter-continental ballistic missiles, 'stalled' (as *Shell News* reported) 'for the lack of a few gallons of "glue".'

[40]

[39]

[41]

[42]

[43]

In the 1960s more and more daily articles were made of plastic.

When everything from laundry baskets to egg cups and spoons, plates, boxes, cups and saucers, bowls and even juicers could be made of cheap, cheerful, durable and easy-to-clean plastics, there seemed no end to the possibilities in the home.

Growth did matter, of course. During the 1960s, as the Group's plastics production in Europe rose from 50,000 tons to nearly 300,000 tons, the loss of £78 per ton dropped to £11. Moreover, Shell was fully prepared to accept low profits and for a remarkably long time too, if the investment promised rewards in the long term. As Royal Dutch president Brouwer put it, when informing Conference about the reasons for withdrawing from Monteshell: 'In answer to a question it was confirmed that the relatively low expected return on capital during the first ten years had not in itself been a reason for serious concern. We had had other major ventures which had started similarly but which in the second decade of their existence developed most satisfactorily. It was confirmed that we had decided to start our joint venture with Montecatini because this was at the time the only way to obtain a significant stake in the Italian petrochemical market. This market was, and still is, considered a sizeable one. We accepted the comparatively low return on capital during the first ten years in exchange for an established position which could ultimately develop into a very acceptable profitability.'[68]

Size gave the ability to maintain optimism by thinking, as was customary, in the very long term indeed: not many commercial organizations could refer so blithely to investments that performed well only in their second decade.

In addition, the Group probably recognized that the problems encountered in chemicals were to a considerable degree part of the game. During the 1950s and 1960s, the petrochemical industry as a whole suffered from low entry barriers, overoptimism and unrealistic growth projections creating surplus capacity, forcing companies to scramble for market share with cut-price tactics before they had had a chance to recoup their investment. Meanwhile the rapid dissemination of technology shortened product cycles, causing companies to diversify beyond their capacity to manage.[69] Other oil companies equally groped for directions. Jack Rathbone came to office as Jersey's chief executive officer in 1960 and was an enthusiastic exponent of petrochemicals. At that time Jersey earned 4.8 per cent of its gross revenue from chemicals, and Shell 6.5 per cent. Rathbone was convinced of the need for much greater investment, and after three years (1961-63 inclusive) with an average investment of $53 million a year in the sector, there was a massive increase in 1964 to $257.1 million, most of it channelled into fertilizers. At the same time Jersey began buying other companies in Denmark, Germany,

[44]

[45]

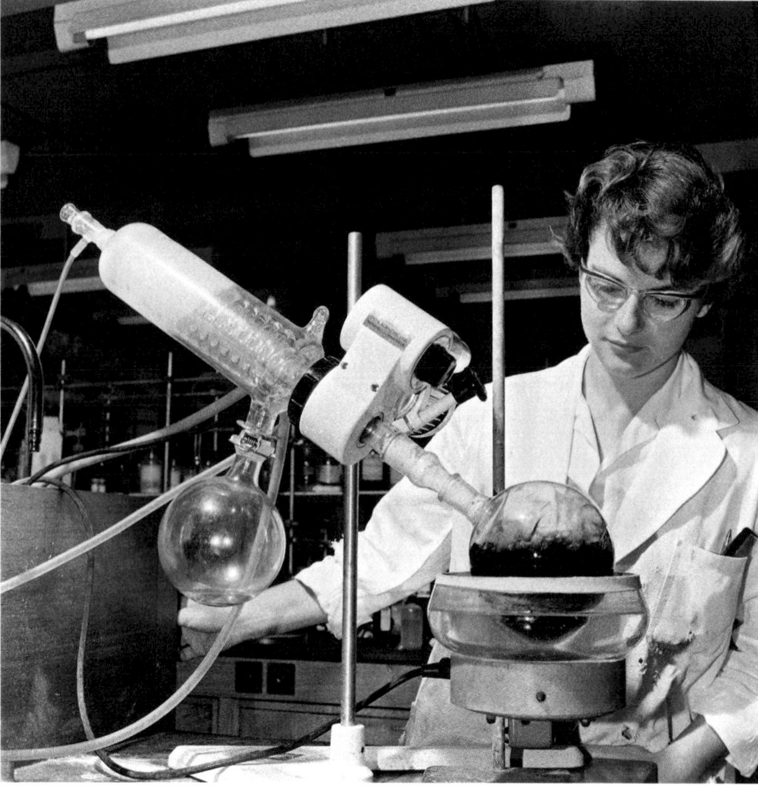

[46]

At Woodstock in Kent in south-east England, Shell's Agricultural Research Centre was a farm with a difference: it undertook most of the research into agricultural chemicals needed by Shell companies outside the US. Covering 440 acres (178 hectares), it included 55,000 fruit trees and ten large glasshouses with breeding facilities for 100,000 insects, including houseflies, mosquitoes, fruit flies and cockroaches. It also reared 20,000 snails a year for the study of liver-fluke and bilharzias. Clockwise from left: inside one of the glasshouses; testing the effects in soil of fungicidal and nematodicidal chemicals; and a rotary evaporator used in organic synthesis.

and Italy to assist its forward integration. Later described by Jersey's own historian as 'the fertilizer binge', all these purchases were eventually abandoned as ill-conceived.[70] BP's joint petrochemical venture British Hydrocarbon Chemicals also slipped badly during the latter 1960s.[71]

Consequently, Spaght's riposte to Doolittle characterizes a general attitude in the petrochemical industry, i.e. the celebration of growth for its own sake leading to a cheerful subordination of profits to growth and the suspension of a reasonable return on assets to an uncertain moment in the future. Tied to a clear strategy, this attitude might have mattered less in Shell's case, but the evolution of the product range appears to have been driven more by a response to appetizing opportunities than by a sense of direction. Mitchell expressed the prevailing attitude well when he said that he wanted chemicals 'to work back from a finished product for which we could see a profitable market and in the manufacture of which, by virtue of our special know-how, we would be in a better position than our competitors'.[72] Each product segment had its own product planning committee, without coordination between them.[73] The result was a rather heterogeneous product range.

Nor does there appear to have been a global policy for individual product segments. In Europe, for instance, the Group tried to improve the overall profitability of chemicals by reducing its exposure to the low-margin fertilizer business, successively amalgamating the operating companies concerned into larger units with majority ownership passing to outside participants. In 1961, Mekog merged with another Dutch producer to form Verenigde Kunstmestfabrieken Mekog-Albatros (VKF). Four years later the Shell Haven fertilizer plant, dogged by ill fortune and underperformance from its inception, was transferred to a 50-50 venture with Armour & Company of Chicago and renamed Shellstar. Then during 1971-73 VKF and Shellstar were merged with another Dutch fertilizer company into Unie van Kunstmest-Fabrieken (UKF), the Group ending up with a 25 per cent share. While the commitments in Europe were thus being reduced, Shell Oil increased its exposure by opening, in 1966, an entirely new fertilizer factory at St Helen's, Oregon. The moment was very inopportune. Only six years later, in 1972, the unit had to close again when the Shell Oil board decided to withdraw completely from fertilizers.[74] Such conflicting policies did little to raise the performance of chemicals.

The CMD definitely recognized the continuing problems in chemicals. One policy debate began in 1961 when Monty Spaght raised serious questions about the performance of chemicals: 'It is probably fair to say that the Group's chemical people have never been prevented from undertaking research or capital expenditures in directions that they really supported, but it is also fair to say that there have been over the years certain doubts and worries relating to the soundness of our directions and the philosophy of what we were trying to do in these ventures. As a Group, we seem to have come to the position of regarding this as a good business opportunity, of considering it one that makes sense to a petroleum-based venture, and as one where by now we have the skills and the organization to carry forward rapidly. At the same time, we don't seem to know just how good it is as a money-making venture. This doubt leads quite properly to thoughts that we might better proceed in some other fashion.'[75]

These doubts were taken up by Schepers, who then spent his summer holiday doing 'some philosophizing about our future in Chemicals'.[76] In a long paper submitted to the CMD, Schepers criticized Shell's strategy in chemicals and recommended three

Photographed in the 1960s, this mountain of barrels at Shell Nederland Chemie's polyisoprene complex at Pernis gives a small indication of the great quantities of chemicals produced there.

possible paths of evolution. The first was that the Group could maintain production of large-volume chemicals based on its own cheap raw materials and using a great deal of specialist know-how, but move away from product lines where it had neither exclusivity through patent protection nor specific know-how unavailable to competitors – especially plastics, rubbers, and other 'performance' chemicals requiring costly technical services in sales.

Schepers' second suggestion was that much more emphasis could be given to pioneering, focusing planning on the discovery of new products which would have small initial volumes but high prices and patent protection. Expensive plants would not have to be built at first; instead the base materials could be bought in until such time as a captive market had been created. If it then became cheaper to build plant and produce the base materials, then that could be considered, while weighing the consequences for previous suppliers; and in support of focused pioneering he gave a vivid example. Researchers at Woodstock, the Group's British agricultural laboratory and farm, were trying to find a new herbicide, but 'We have three scientists working on this subject. How can we hope to be able to compete with Bayer who have 50 working on herbicides alone?'[77] Better, surely, to have a planned list of specific diseases and pests in order of priority. As his third idea Schepers proposed an unorthodox method of developing markets by starting to manufacture end-products, such as plastic utensils, in small-scale fashion in under-developed countries. In this he foresaw a double benefit, both commercial and moral, because 'Once going, the business could be handed over to local manufacturers (...) and apart from having created an outlet for our products (in unfinished form), the Group in the form of our local organisations, also on the Oil side, may benefit from the good-will resulting from our having fostered local industry.'[78]

Schepers's policy paper presumably contributed to the decision, in 1961, to raise the return on chemicals to 8 per cent by selective investment, though the minutes do not spell this out.[79] Contrary to his suggestion, however, Shell did continue with plastics and rubber. Indeed, the Group's weakness in chemicals lay precisely in the fact that the company practised both the first and second options listed by Schepers: it had products across the entire petrochemical range. And, for all the money lavished on research and product development, for all the occasional successes in pioneering new products, the Group could indeed not hope to beat dedicated companies working in particular sectors, because it could not muster similar resources in every sector, and probably also because patent protection in chemicals turned out to be as difficult as in oil. Particular processes and products could of course be patented, but there were nearly always alternative routes, or alternative products.[80] Synthetic rubber had a tough time competing with natural rubber. And sometimes even perfect patents and processes failed to make commercial impact, for one reason or another. From the 1930s, Shell Oil carefully assembled a perfect patent position in acrolein based on a cheap way of making this useful and versatile compound from propylene. Firmly believing in a future for acrolein, the company even built a unit producing it at Houston, but could not find suitable commercial applications before the patents expired.[81] Bad luck took the shine off the commercial introduction of synthetic glycerine. Prices depended on natural glycerine prices which fluctuated very

[47]

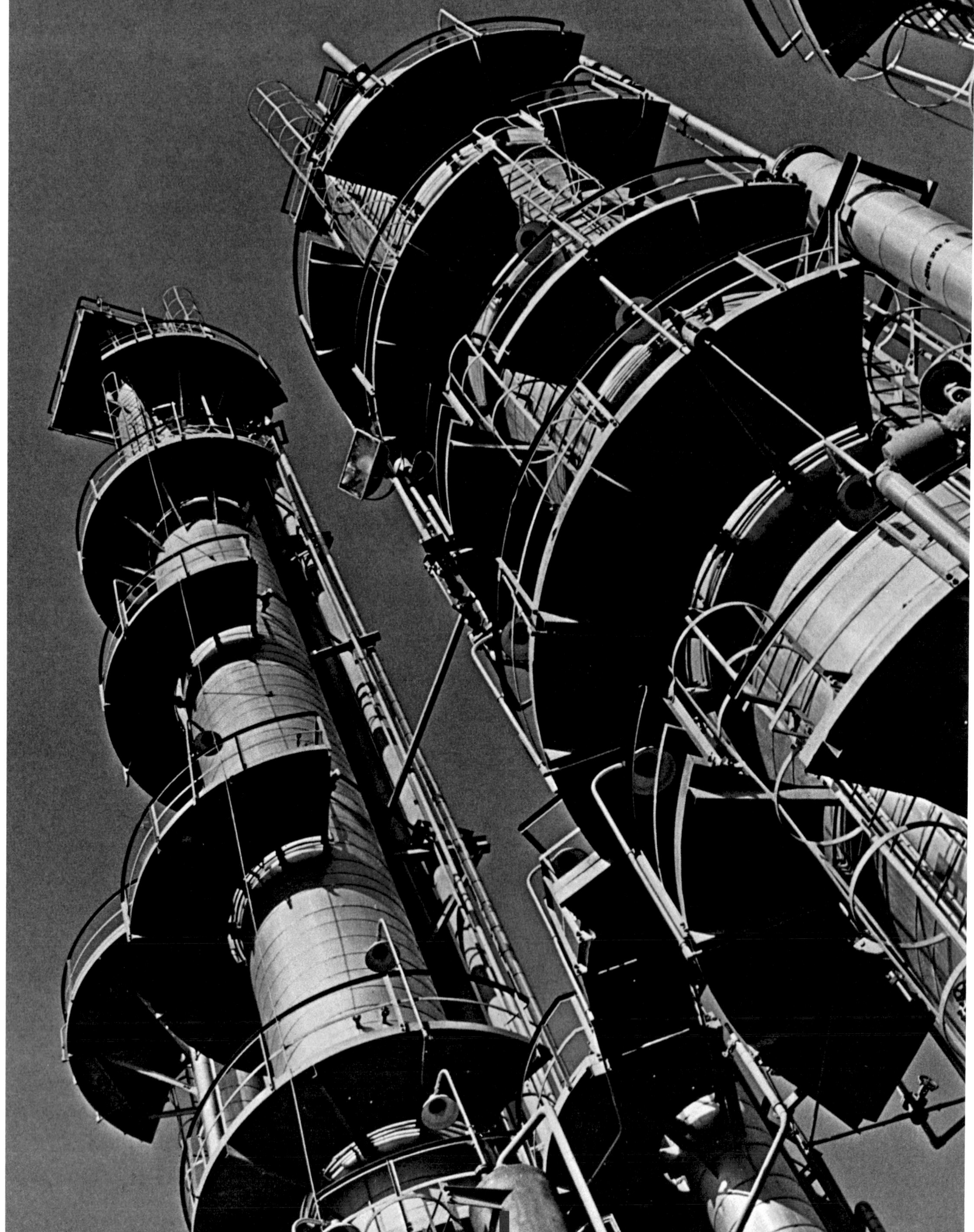

[48]

strongly. Shell Oil countered this volatility by entering into long-term supply contracts at fixed prices based on estimated production costs and a calculated market average. First an unexpected rise in start-up costs overturned this basis, but when tight glycerine supplies pushed prices up, Shell Oil could not profit from them because of the long-term contracts.[82]

The absence of a clear focus on particular products or segments made the vicissitudes of chemicals appear like so many unfortunate mishaps. In November 1961, two months after the CMD debate on Schepers's paper, Mitchell was at the crease again. Without any warning at all, the third quarter's chemical profits were much below expectation, with the exception of Deutsche Shell Chemie. Shell was not alone in this, since chemical companies in general had lower profits. Worst of all was Shell Chemical Company, which had lost nearly half a million pounds in that quarter alone. Mitchell nonetheless gave a sturdy defence to a year of poor returns. There were many contributory reasons, but as a basic cause he singled out economic recession in the US, surplus manufacturing capacity there, and the consequent dumping of American chemical products on the European market.[83] In doing so he was right, up to a point, and for several years Shell made representations on the matter to the Dutch and British governments, asking ineffectively for the imposition of countervailing import duties.

Skyscrapers of science: a part of Shell's 800-acre (324-hectare) chemicals plant at Geismar, Louisiana, on the east bank of the Mississippi. The plant is one of the world's largest producers of surfactants and other chemicals used in detergents.

But dumping was only one aspect of the structural problems affecting the industry. Generally speaking the Group's chemical operations performed on a par with the rest of the industry, and probably a little better than the petrochemical operations of other oil companies. All companies struggled with the erratic nature of the new industry, its cyclical swings, the recurrent overcapacity, and the inexplicable product price falls. While groping for some form of pattern, managers gave a library of reasons for the disappointing performance of Shell's chemical operations. In 1958, losses by Petrochemicals Ltd. were put down to delays in completion of plant, the unexpected need for more staff, and severe competition; in 1959, losses by the new chemicals service companies dragged down profits; in 1960, recession in the US harmed operations elsewhere; in 1961, the cause was pre-operational expenses at the new Pernis styrene butadiene rubber plant and severe operational difficulties at a new Shell Haven plant.[84] All of these were true, as was the observation in 1962 that negligible net income for chemicals in the previous three years had been because of successive reductions in selling prices and the fact that many new plants were not yet covering their operating costs; and a similar lament came in 1963, when there would have been a considerably improved result over 1962 – *if* the Indonesian results were excluded.[85] From the mid-1960s the 'large plant syndrome' reared its head. To achieve economies of scale, petrochemical complexes were increasingly built as large, integrated installations with raw materials and intermediates exchanged between different production processes. The uncoordinated building of such very large plants worsened the industry's already marked cyclical instability. Large plants also turned out to have practical disadvantages. Minor mishaps in one unit could have a domino effect in other, connected units and even cause total shutdowns.

Ground plan for the large expansion of Shell Nederland Chemie's complex at Moerdijk, the Netherlands, in 1968.

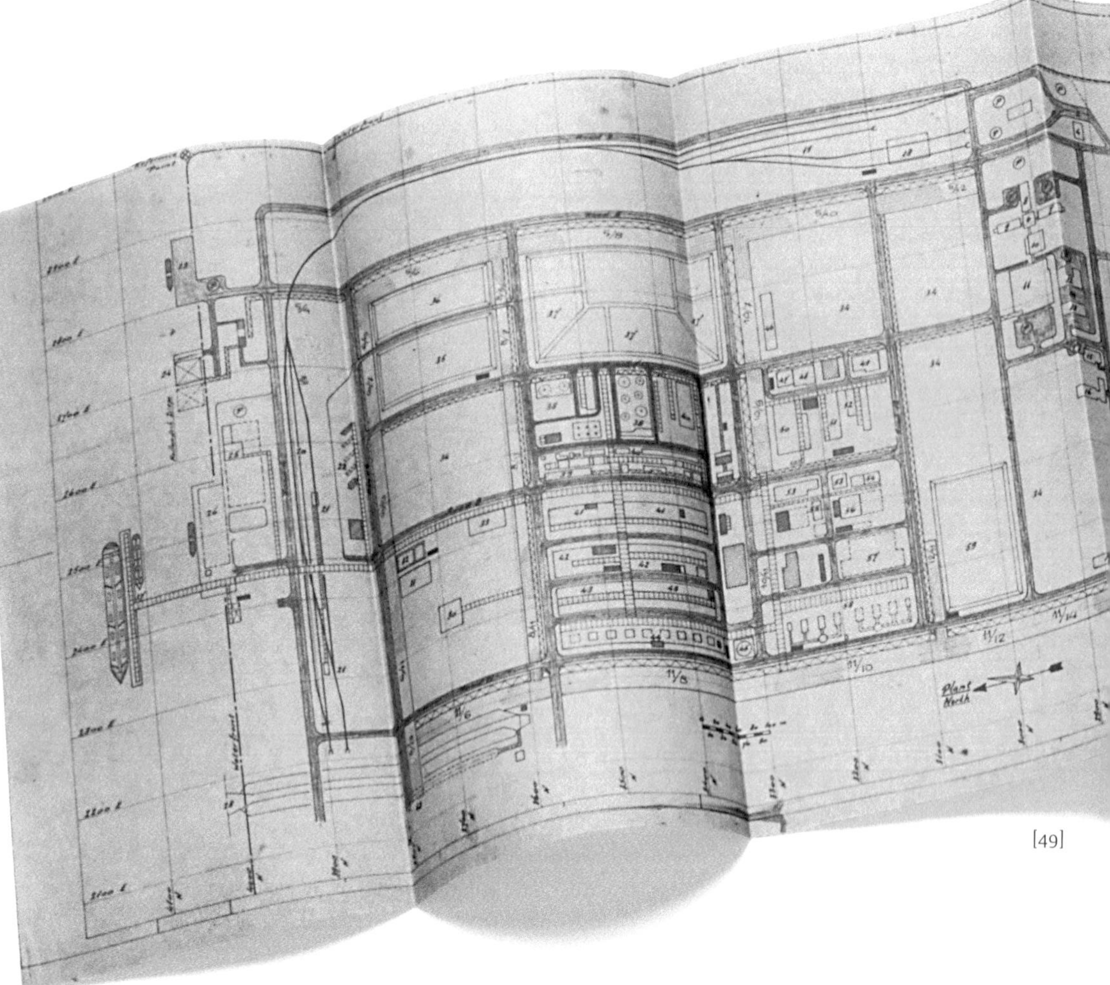

[49]

Moreover, integrated production raised the exit barrier for individual products, as the Group found to its cost.[86]

At last, by the end of the 1960s Shell's 'patience in waiting for the return' appeared to pay off. The losses in plastics and elastomers lessened and the chemical sector's return on assets had risen steadily since 1964: 8.4 per cent was reached in 1968, and 8.5 per cent forecast for 1969, a year that opened with a mood of high optimism and impressive plans, even for Shell. Now that the target of an acceptable return had been reached, the Group went for all-out growth and accepted the maximum number of opportunities. Again, the decision does not show up in the minutes of the CMD or of the Conference, but the outward signs were unmistakable.[87] Construction had begun on a huge new petrochemical complex built around a naphtha cracker at Moerdijk in the Netherlands, linked by an extensive pipeline system with both Rotterdam and Antwerp harbours and with the chemical industries working in the Scheldt and Meuse delta.[88] Chemicals' objectives through to 1974 envisaged an annual average increase in investment of about 12 per cent, maintaining a return on capital of about 8 per cent. According to its own figures, the Group's chemicals proceeds exceeded those of any other oil company engaged in the business, and its net income from chemicals was expected to more than double in 1970-74, while the net income from oil was not expected to grow by more than 40 per cent. Proposed capital expenditure for the period totalled £327 million, and it was anticipated that the chemicals sector was on the brink of thirty years of really serious

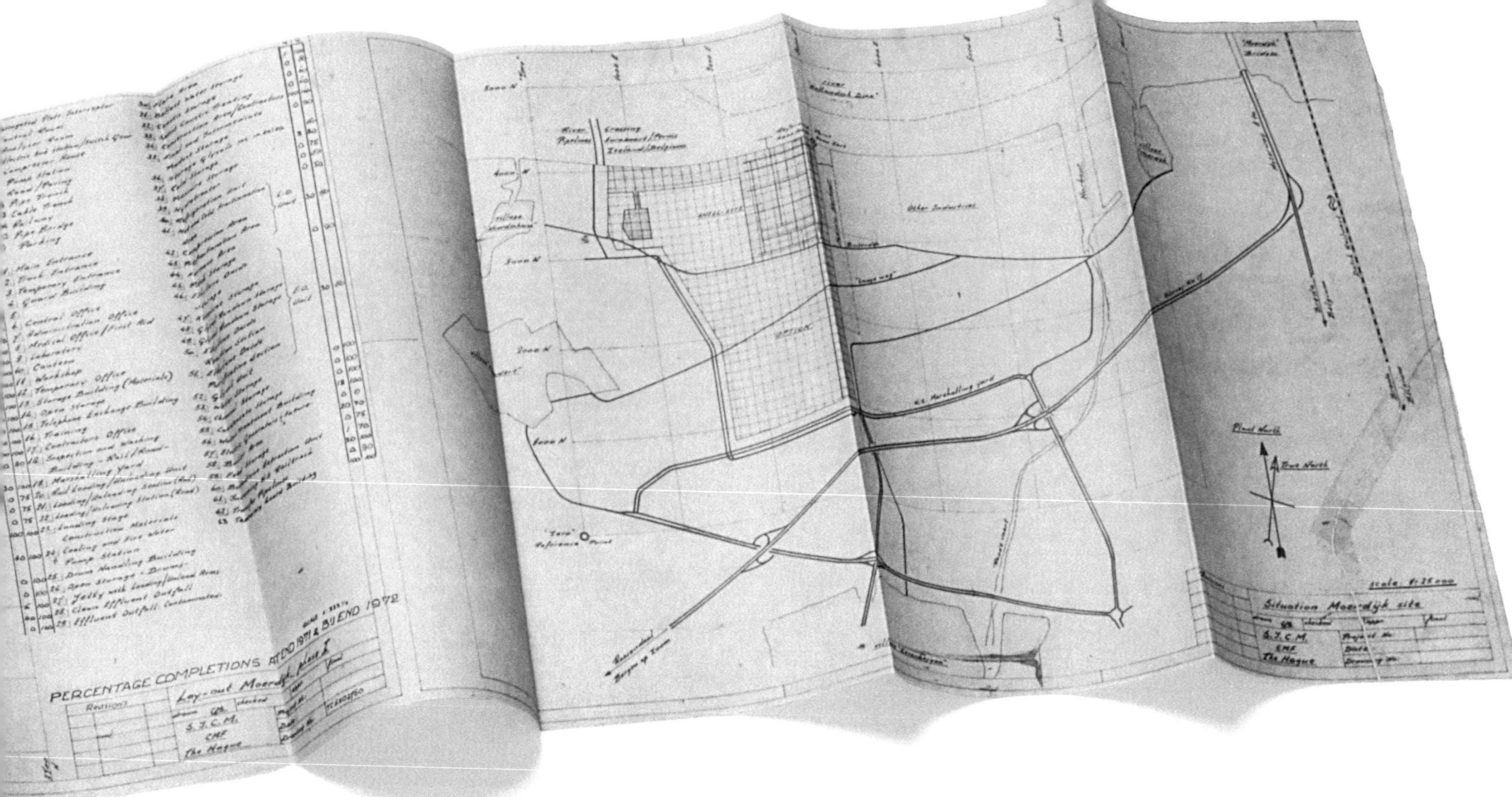

growth, especially as a proportion of total Group business, of which it already represented about 14 per cent. That seemed certain to increase, possibly to 20 per cent of the total in ten years, and even to one-third by the end of the century. The question had actually arisen as to whether the chemical business should attempt to grow even faster, because finance was no longer a great constraint. But skilled manpower and technically qualified managers were hard to find, so it seemed the plans were ambitious enough as they stood.[89]

In presenting these plans the new Director of Chemicals, E.G.G. Werner, made a realistic caveat to the Conference, pointing out some of the industry's vulnerable aspects; yet the mass of optimistic forecast was so much greater and more exciting that the cautionary aspects were far outweighed and were ignored by the Conference. After that came the fall. Chemicals' return in 1969 comfortably exceeded the anticipated 8.5 per cent and touched 10.2 per cent.[90] The CMD received these figures cautiously, commenting that they were 'rather exceptional due to a relative freedom from serious difficulties, and that it would therefore be unwise to assume that the same rate of progress could be maintained'.[91] And indeed, once again the pot of gold disappeared over the horizon; the following year income rose, but capital expenditure and costs far outstripped it. Inflation, labour unrest, and the chronic lack of skilled manpower all combined to create a slipping return of barely 7 per cent in 1970 and less than 3 per cent in 1971 – a level unknown since 1962, and almost the worst the sector had ever experienced. Group chemical net income fell by two-thirds, from £24 million in 1970 to £7.8 million. At Moerdijk, work was slowed down to a minimum, in hopes that its completion might be put off until 1974; but even after the greatest possible deferment, the costly project was still destined to come on stream in 1973, before anyone any longer wanted it to do so.[92] The situation was particularly bad in the UK, with one-third of total investment the Group's biggest chemical operations in Europe.[93] Efforts to leave a planned joint venture for a chemical plant in Thailand seemed unlikely to succeed.[94] Plastics had performed particularly badly, but Shell found itself locked into volume production, because the close integration with other manufacturing processes meant that plastics could not be

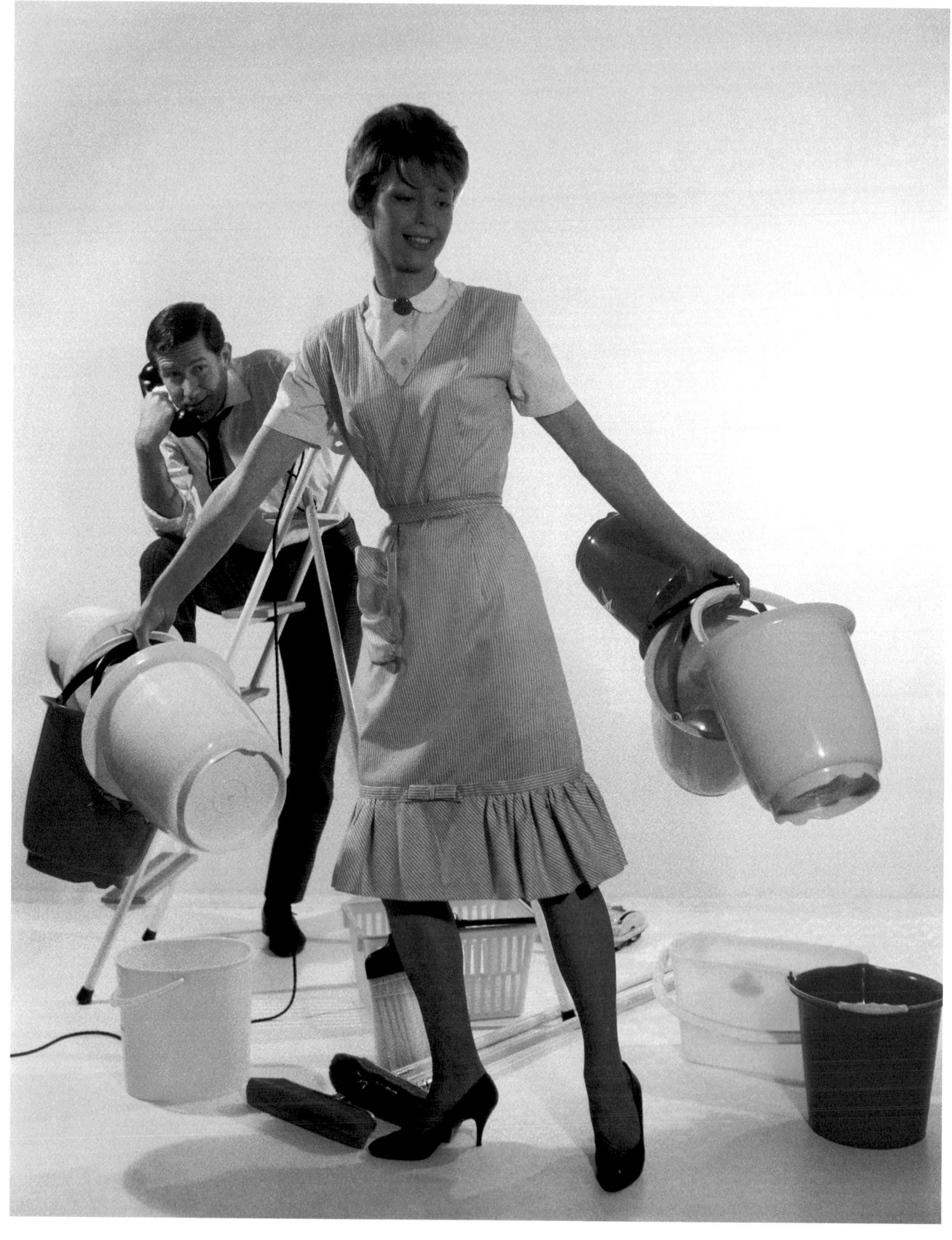

The joy of suds, or glamour by the bucketful. In the minds of Shell's marketers in the 1960s, providing for the domestic needs of housewives had a central position.

'Above all, the idea must be abandoned that rapid growth in sales volumes is in itself an objective to be pursued at almost any cost.'

cut back without seriously harming other production processes.[95]

The Group's sorry results in chemicals during 1971 prompted some of the closest scrutiny that the sector had ever received from the CMD. The period of rapid technological development and sustained growth seemed over, but nevertheless Werner's steadfast analysis was that overall Shell had got it right, with a broad-based chemical business that minimized the volatility of any of its parts. The mistake, he said, lay in the chosen rates of growth. A series of low-return years could be substantially altered by changing the rate of growth: a return of 10 per cent could be achieved with a capital growth rate of 8 per cent, 'but with the function growing at some 15 per cent a year, the return on capital can scarcely be expected to exceed 6.5 per cent .'[96] This was in contradiction of Schepers's view in 1961 – a full decade earlier – that 6.2 per cent was not something for Chemicals to be proud of, and with figures from the UK Werner showed how massively capital-intensive the business was. At one extreme, the car manufacturer British Leyland employed 196,000 people, with a gross investment of just £2,000 per man.[97] British manufacturing industry as a whole employed 8,922,000 with a gross investment of £3,000 per man. At the other extreme, the chemical and allied industries as a whole employed 473,000 with a gross investment of £11,000 per man – and while Shell Chemicals UK employed just 5,000 people, its gross investment was £23,000 per man, largely spent in the pursuit of growth. Addressing the CMD directly, Werner declared, 'The profitability of the Group's chemical business is largely a matter of choice as to the rate of growth desired, but growth rate is not something that can be altered at short notice as lead-times are long and planning complex,' and a little later, in a formal briefing paper to all managing directors, he was adamant: 'Above all, the idea must be abandoned that rapid growth in sales volumes is in itself an objective to be pursued at almost any cost.'[98]

Perhaps Mitchell's apparent bluff in 1960 was less of a ruse than it seemed. Although naturally he had sought the highest possible expenditure on chemicals, he did not have the last say either on that or on the Group's chosen rate of growth, but presented the evidence for the CMD and the Conference to decide. Werner did the same, but by then the Group was ten years further down the road, and although discussion in the Conference was as vigorous as in the CMD, it was virtually fated to fail in reaching any definite decision.

The first main argument given in favour of maintaining the Group's commitment to chemicals was that although the agreed expenditure programme was ambitious, it was the minimum required to uphold the percentage of chemical sales in relation to total sales, and that compared with oil, a lower return had always been expected as a corollary of chemicals' high growth rate.[99] Moreover, the returns were 'not all that unfavourable' compared with similar manufacturing industries. Against this, the more pessimistic line of reasoning was that very high capital costs and an apparent inability to raise prices were very unpromising in a period of rapid inflation; the industry as a whole seemed 'fated to repeat cycles of underestimated demand, huge investments in new plants with excess capacity, and then price weakness'. For a business it was a tragic triangle: first there was too little capacity to meet demand, then much money was spent to provide additional capacity, which turned out to be superfluous when demand slackened and prices fell. The truth of this was acknowledged by the supporters of petrochemicals, and some hopeful comments were made; but even they could end with no more positive a suggestion than that 'there was little practicable alternative to the programme now in hand.'[100]

The dichotomy had always been between growth and returns, with good arguments available either way, but, it is clear, with incomplete understanding between the two sides, the oil men and the chemists. That much is emphasized by the fact that there were two sides, in what ideally should have been one team, and J. F. K. Hinde, Secretary to the Conference, concluded his minutes with the melancholy reflection that 'Although there was a full debate, it did not prove practicable in the time available to resolve these opposing lines of argument, which appeared to represent basically differing views on the strategy for investment in the chemical industry.'[101]

A chemicals strategy meeting convened in October 1973 showed little progress in thinking. The meeting began by assuming as a working hypothesis that the sector would not spend more cash than it generated for the rest of the 1970s, which, given past performance, was not a very reasonable assumption. But even this limitation did not bridle ambitions. Chemicals wanted to widen the geographic spread of operations and also maintain a wide range, combining products having a high added value and a high technological content, with the commodity type of heavy organic chemicals. In addition, the strategy meeting recommended that 'An active, but not over-ambitious acquisition policy, preferably aimed towards the high margin field should be followed.'[102] This sounded sensible, but it showed the hypnotic power that petrochemicals continued to exert, and further expensive lessons – even more than hitherto – remained to be learned.

Beyond the core business: venturing into other industries

As if the drain from chemicals was not enough, from about 1960, the Group started considering further diversifications. Once again a strong impetus was given by Kessler, a powerful figure in the Group until his retirement as chairman of the Royal Dutch supervisory board in June 1961. Once he considered the Group's second leg to be strong enough, probably around 1960, Kessler began pressing for a third one. According to him the oil industry, which he had always regarded as a rather simple business, had lost its excitement, squeezed between producers and consumers, so managers now needed something else to exercise their technical brain.[103] This unusual motive for diversifying a business chimed well with a 'statement of business aims' for internal use which the Group drafted around the same time. As listed in the document, Shell's first aim was to keep the business strong, prosperous, and progressive. This could only be achieved with an enthusiastic staff, so 'Every manager must set out to create a climate where ability, initiative, and ideas flourish and are accepted, and where staff feel that they are sharing in the excitement of competitive business and in our achievements, whether large or small.'[104] This was considered more important than creating shareholder value, which came second on the list.

Was Shell getting bored with oil? Perhaps, but some of the motives which drove the diversification into chemicals probably played a part as well: the concern about the oil industry maturing, no doubt reinforced by the awareness that chemicals, for all their growth and promise, looked unlikely to yield high profits for the foreseeable future. There was also the fanciful confidence quoted above that, with its superior resources, Shell could undertake any business anywhere in the world. Finally, there resided a distinct fashion element in diversification.[105] During the 1950s the United

[51]

[52]

Two of Shell's grand old men: Lijkle ('Skippy') Schepers (far left) and J. B. A. Kessler in 1960, the year of Kessler's retirement from the chairmanship of the Royal Dutch supervisory board.

States witnessed a merger wave characterized by the formation of large conglomerates, companies with a great diversity of businesses not necessarily related to each other and supposedly held together and propelled by a single management vision. Diversification appeared a success, indeed a must for modern companies. By 1963, Spaght reported that in Shell Oil, 'There is an opinion rampant that a company is backward if it doesn't continually enter new fields of activity, particularly through buying other companies. Even in going around the country talking to our Shell people, one encounters some of this feeling here and there.'[106] Diversification thus looked the natural way forward for the world's third-largest company – more natural, in fact, than a continuing expansion in the oil business. During the 1960s the Group spent an enormous amount of time, attention, and money on discussing and effecting diversification, far more than on opportunities for growing through take-overs within the oil industry. Shell Oil did some important acquisitions, most of all the El Paso deal in 1964. Three years later, Barran talked with British officials about the government's keen desire for a merger between Shell and BP, a bright idea from an eminent economic adviser which, on second thoughts, officials realized was 'a non-starter at present', largely because they could not find a way around Royal Dutch.[107] Otherwise, discussions about acquisitions within the oil industry are conspicuous by their absence in Group records, probably because the endogenous growth was considered enough.

In January 1962 a series of meetings with senior managers about diversification revealed widespread interest amongst the staff, coupled with 'some looseness of thinking which is not altogether surprising'.[108] Despite the view among some senior personnel that the Group's petrochemical business was quite sufficient diversification, the mood at the top was to encourage innovation in thinking and practice and to channel and control the enthusiasm for diversification, so as to avoid too great a diversity of activities and the corresponding risk of spreading technical and financial resources too thinly, at the expense of the Group's basic business. Practical considerations added impetus to the need to define a diversification policy. Net investment in Venezuela decreased because Shell considered the terms for new concessions insufficiently attractive and thus shifted its E&P operations elsewhere. At the same time the Venezuelan government wanted 'industries likely to show a return which, coupled with long-term outlets for products, might hold prospects of being commercially attractive'.

the oil companies to assist the country's industrialization policy. To sustain the Group's goodwill, the Compañia Shell de Venezuela (CSV) proposed to form an investment company with the purpose of financing 'industries likely to show a return which, coupled with long-term outlets for products, might hold prospects of being commercially attractive'.[109] At this stage diversification was clearly still seen as being linked to Shell's wider business purposes, in that the companies concerned would have to be customers as well, but this requirement soon disappeared.

By the middle of November 1962 a policy document was ready for discussion in the Conference.[110] Kruisinga was 'designated to act as the prime mover and focal point', allowed to discuss the matter freely at suitable levels within the central offices organization, but with two provisos designed to rein in the wilder reaches of staff imagination: 'we have in no way reached the conclusion that Group finance is available for all manner of new investments. Indeed, in view of the wide ramifications which could be associated with "diversification", it is considered desirable that one of Kruisinga's objects could be to suggest and identify, for illustrative purposes and with reasons some types of activity in which Group investment is *not* to be recommended (an example being ship-repairing, because it is not sufficiently closely related to the Group's main business).'[111]

In January 1963 the Conference considered a paper on diversification, explaining what it meant in general and how it might be applied to Shell in particular. Diversification was categorized into 'offensive' projects – those with high return or growth potential – and 'defensive', contributing only indirectly to Group business by helping to establish the best political or economic climate in a given country. The latter were seen as coming up for consideration in the normal course of business and being largely on the General Managers' initiative. For offensive projects, a fairly cautious initial policy was devised: 'with fresh in mind the many problems, organisational as well as managerial, which the Group had to cope with when entering the chemical field, it is very much in doubt whether breaking out into entirely new industrial fields should be considered, at least at this stage (...) In general, a few large projects are to be preferred to a large number of minor projects (...) but should include new ventures stemming from, for instance, our own research.'[112]

In addition, existing profitable businesses could be taken over if this meant gaining good management and technical know-how that combined economically with the corresponding elements of the Group, and although minority interests were seen as generally undesirable, they could be included if such a holding gave a tied outlet for Group products. The Conference took note of the CMD's emerging policy on diversification, though not without Frank Hopwood, who had meanwhile succeeded to the family title and become Lord Southborough, sounding a note of caution as to the amount of money to be diverted from the Group's traditional business towards purposes unconnected with that.[113] A few months later a single-sentence definition of diversification was adopted for use throughout the Group: it was 'Profitable investment in which the Group has some exploitable competitive advantage over others, preferably in fields not immediately affected by the commercial climate of the oil or chemical industries', i.e. something outside Shell's traditional remit.[114]

Armed with this definition, the CMD set up a team known as the Diversification Studies Division and later renamed Special Trading Activities Division to investigate opportunities.[115] The study team came up with various proposals, including one for manufacturing and marketing oxygen and other industrial gases as a defensive measure against oxygen companies affecting the Group's LPG markets. A joint venture for this purpose was started in France.[116] Retail property development also appeared to offer attractive potential. Shell Nederland decided to cooperate with the Zwolsman Group, the largest property and real estate development enterprise in Holland.[117] Because the Group already had experience in developing shopping and service centres in the US and Canada, exercising influence on the petroleum aspects of projects (such as the use of bitumen, chemical products, and heating), the CMD endorsed a proposal to cooperate with the Swiss company Intershop, a provider of finance for large-scale property development schemes, as sole adviser on petroleum matters in similar projects in Western Europe.[118]

However, the Conference still had serious doubts about the desirability of diversification. When discussing the CMD's new policy document in July 1963, Hopwood was again quite critical. What was the purpose of diversification, he asked, to make the Group bigger or to make it more profitable? Shell's previous diversifications had not been particularly successful. Shares taken in various industries after the First World War had added management burdens for no significant gain; similarly, the financing of office buildings, service stations, and ships for clients or customers had only been a financial drain. If the Group had large cash holdings which tended to increase rather than diminish, these 'could technically be regarded as due for distribution to the shareholders and not for the purpose of capital investments in entirely new fields'. For his part, 'he felt that anything which might either immediately or eventually involve the Group in the management of ventures involving a different technique should be avoided'.[119] After some discussion, the Conference concluded that any intention to diversify 'would be to seek opportunities incidental to and arising from the Group's existing business in order to obtain a higher return on investment'.[120] The Conference thus clearly interpreted diversification more narrowly than the CMD did, as directly tied to oil or chemicals in the way that, for instance, the PVC pipe manufacturer Wavin was, and not aimed at exploiting some unspecified competitive advantage. The minutes make no further mention of Hopwood's suggestion to use spare cash for raising the shareholders' return, but his concern returned seven years later when the Conference asked whether the Group had a moral right to divert the funds at its disposal to purposes other than its traditional activities. The CMD thought that it did: 'in a modern society, the large public corporation had a "life" of its own, directed by its management; and that in exercising this direction management had responsibilities not only to shareholders but also to employees and to the community. (..) All three of these responsibilities would be properly discharged by applying the corporation's asset of sound and long-experienced managerial skills to some additional good purpose, always assuming that this other purpose was prudently selected having regard to the talents and experiences possessed by the company.'[121]

The selection principles laid down by the Conference left a lot of leeway. After all, Shell's operations touched so many other businesses that it did not take a great deal of ingenuity to forecast some economic gain from a deeper involvement with companies that at first sight looked pretty remote from oil. In November 1963, for instance, Monty Spaght had drawn up a catalogue of suitable sectors listing pharmaceuticals, real estate, precision engineering, packaging, specialty metals, general mining, gold mining, building materials, and water supplies. According to him they all merited study as being sufficiently close to the Group's core business.[122] Keen to diversify, the CMD stretched the guidelines laid down by

[53]

[54]

[55]

The Second World War ended with America's nuclear annihilation of the Japanese cities of Hiroshima and Nagasaki in 1945. In 1949 the Soviet Union became the world's second nuclear power, followed by Britain in 1952, and the Cold War between communism and capitalism took shape. Nevertheless, in the summer of 1957 Schiphol Airport was host to a large, ambitious and inspiring exhibition called 'Het Atoom' ('The Atom'), extolling the potential of nuclear power for civil purposes. Beautifully designed by many famous artists, the exhibition was part of the social context in which Shell began thinking about nuclear power. Everyone did, and it looked very promising, creating a widespread and high expectation that the energy needs of future prosperity could be met. At the Schiphol exhibition, visited by many thousands, one of the main attractions was an actual nuclear reactor, pictured here (main picture and above). Shining on the nurses' faces (above), the reactor's glow showed that it was functioning.

the Conference from 'incidental to and arising from the Group's existing business' to 'possibilities which have or could have a close affinity with the Group's existing business'.[123] Even that criterion proved too stringent. Nuclear energy was ruled out at this stage, though it was later to return with a vengeance. The Group had a small team monitoring developments in that industry; its regular reports to the Conference invariably argued that nuclear energy was unlikely to compete with hydrocarbons for the foreseeable future.[124] In 1966, after two years of studying the UK building industry, Shell took a 49 per cent stake in a specialist contractor so as to widen the application of Group products like epoxy resins, plastic linings, and rigid polyurethane panels, but within months the CMD regretted its decision because, despite close prior investigations, the company turned out to be insufficiently profitable.[125] An Australian project for the transportation of solids by pipeline was given a small stimulus, but no more, managing director Berkin noting about this proposal that 'affinity (...) is possibly more tenuous than even the building industry'.[126]

These discouraging results did not lead Shell to drop the idea of diversification as unsound in principle; instead, the CMD redoubled its efforts, probably because in 1965 Jersey Standard had set up a subsidiary, Jersey Enterprises, devoted to the purpose. The company was said to develop interests in cryogenics, real estate, detergents, building materials, and proteins.[127] Shell's intentions had also changed slightly. Doubts were arising whether there would be enough oil in the medium to longer term, so it seemed wise to diversify the Group's operations in order to prepare for that eventuality.[128] Moreover, Shell considered that in the longer term the oil and chemicals business would be unable to maintain the high investment growth rate which shareholders had come to expect.[129]

As a consequence, in 1968 the CMD appointed a task group under the direction of Karel Swart to study the available options. Presenting his initial findings to the Conference the following year, Swart began by outlining the selection principles. As usual, new opportunities ought to be large and have an above average growth potential, i.e. they should be products or services in the stages of development or early market application. They should also be in a sector where the Group could contribute know-how or make some other special contribution, 'in other words, it should not be wholly remote from our current activities'.[130] Shell's diversification policy remained caught between two conflicting requirements. On one hand, prospective acquisitions needed to be substantial or at least fast-growing, but without costing too much and diverting resources away from oil and chemicals. On the other hand, Shell looked for new and innovative projects with a role for Group expertise, i.e. small and risky ventures with uncertain growth prospects. In the end, Shell sought opportunities both big and small, and did so more or less independently from each other, thus compromising the coherence of the diversification policy. Moreover, by turning 'affinity with' into 'not wholly remote from', managerial thinking about diversification distanced itself further from Shell's core business. The first study into potential sectors showed this, by including further analysis of nuclear energy, along with metals, 'leisure' (in its widest sense, and including the wines and spirits trade), and food industries as options.

Nuclear energy was still not considered worthwhile, because of doubtful prospects for private enterprise there. According to the CMD, 'Nuclear power, with all of its ramifications, has of course been and continues to be extensively considered. At this moment we have not been able to find in it as much attraction as in many other fields or indeed as much as apparently some other oil companies apparently seem to find in it'.[131]

Uranium mining did not look attractive either because it was already a crowded field, and technical developments appeared to threaten that sector.[132] Although the wines and spirits trade was a high growth industry, it had the disadvantage of a conflict of interest with the automotive business. Tourism provided a better combination with the transport business and deserved more study. The food business, however, was not considered attractive, because of an expected slump in world agriculture trade and serious over-supply conditions in the developed countries.[133]

Metals, however, and particularly light metals such as aluminium, magnesium, and titanium, were considered very promising, though the CMD realized Shell had no expertise in metal extraction and therefore should proceed by acquiring an existing business.[134] In 1970, Shell took over the Dutch mining company NV Billiton Maatschappij, which it preferred to the much larger Rio Tinto because it was thought desirable to start the learning curve with a small company and then expand gradually. The idea was that Billiton would keep its own identity and become the bridgehead of the Group's non-ferrous metal operations.[135] Billiton appeared a very good fit. Like Royal Dutch, Billiton had its origin in colonial Indonesia, where it had started mining tin in 1860. During the 1930s the company took up bauxite mining and alumina production. After the nationalization of its Indonesian business, Billiton had to find new directions; the company entered into a joint venture for the building of an aluminium smelter with Hoogovens and Alusuisse, and also diversified into a wide range of other activities, such as shipping (forwarding) and sea transport, packaging, steel construction, exploration of oil and gas in the North Sea, and refrigeration systems for food and for flowers.[136]

Open-cast extraction, such as here in one of Billiton's mines in Surinam, was safer and cheaper than deep mining by shaft and tunnel, but it left blatant and lasting scars.

[57]

[56]

Through its ownership of Billiton, Shell's involvement in metals mining in Surinam and Indonesia included mining for tin and bauxite, the chief source of aluminium. Right, in 1966 Surinam's export minister Dr. J. F. E. Einar inspects the first 1,200-ton delivery of aluminium from Surinam to Rotterdam. Billiton's tin-miners in Indonesia used both new and old techniques. Modern equipment contrasted strongly with an update of the methods of the previous century, where high-pressure hoses were used to blast and separate the tin-bearing soil.

[58]

[59]

Within a year of its acquisition by Shell, Billiton's projected profits turned into losses, forcing a thorough reorganization. That disappointment did not cure Shell's itch for a third leg. In 1971 the CMD formed a Panel for Non-Traditional Business (NTB) to monitor the diversification efforts. Individual committee members now rode out their hobby horses, as often as not connected to the part of the business in which they had made their careers. A. Bénard pointed to the underutilized strengths of the Group's retail network. Brought under independent management and with some organizational changes at central office level, this could be given the task of realizing growth outside the oil business. As an example, Bénard mentioned 'Shell Travel Services', which might include a travel agency, car hire, restaurants. Another possibility was 'Shell Domestic Services', which in addition to the traditional heating services could provide plumbing, maintenance, and other domestic services.[137]

In May 1971 the internal debate took a surprising turn. Barran had stated only the year before that there was no need to look at diversification from the perspective of either finding new opportunities for surplus cash, or from a likely competition from other fuels, but meanwhile opinions had changed.[138] Presenting a new diversification policy statement, Swart now argued that oil and gas had reached their maximum penetration of the energy market. After 1980 their share would remain constant, or decline, pushing the likely growth rate of oil companies down to 4-6 per cent a year. Shell thus needed a vigorous diversification effort to maintain its present growth rate. Swart mentioned three attractive sectors: nuclear energy, tourism, and pharmaceuticals. Competitive considerations came into the bargain. According to Swart, other oil companies were already moving into alternative sources of energy, and the Group needed to do likewise if it were not to fall behind further.[139] After a few months of intensive study, tourism was rejected because the CMD could not make up its mind whether to diversify inside or outside the energy sector.[140] Task forces were set up to study coal and nuclear energy as possibilities. In November 1971, Barran informed the Conference that a task group had been established to study possible prospects in coal and its international transportation, with the idea of acquiring anything up to five billion tons of recoverable coal reserves during 1972.[141] By February 1972, the attitude towards nuclear showed a perceptible evolution; the CMD considered a major acquisition still highly unlikely, but regarded the likely impact of nuclear on oil so great that it would be a mistake not to have a close look at it.[142]

The Group came closest to a diversification into pharmaceuticals. Werner, who had meanwhile become a member of the CMD, pushed for this because he considered the sector closely related to chemicals. There were two suitable Dutch companies as candidates for acquisition, Gist-Brocades and Philips-Duphar, both growing fast and showing attractive profits. The CMD considered the pharmaceutical industry past its peak, however, and too labour-intensive as opposed to capital-intensive. Moreover, the likely price of the companies concerned would be twenty times earnings, very high and likely to depress both Group revenues and investment in other operations. Exasperated with these criticisms of what he believed to be very sound proposals, Werner exclaimed that it would be impossible to acquire a new business at a cost that would not depress the Group's return in the initial years: 'I do not know of any other important business but Oil combining low labour and high capital with a high rate of return. "There simply is no other business like the Oil business".'[143]

Part of the 1953 photo-essay 'The Reign of Chemistry' by W. Eugene Smith.

Conclusion The petrochemical industry was a characteristic exponent of the world's post-war economic recovery and growth. Spreading from its North American origins and always intimately linked with the development of its refinery complexes, Shell's quick, determined and very expansive moves into the sector in the US and Europe gave it an initial advantage in what was to become an exceedingly competitive industry. Among the oil majors it was soon the largest manufacturer of petrochemicals, but remained still very much smaller than the largest of its dedicated chemicals competitors around the world.

The Group gradually came to recognize the differences between the chemicals industry and the oil industry, and from the late 1950s chemicals moved firmly away from being a mere adjunct of oil to become a second core business, with a wide variety of new products. Heavy spending fostered rapid growth and the sector's profitability gradually improved, but chemicals never attained the levels of return on average capital employed that were taken as normal in the oil industry as a whole. Instead, for a variety of reasons, much lower levels were accepted, partly because they were deemed to be inherent in the strategy followed, partly because the Group's performance corresponded overall with that of the petrochemical industry as a whole which, while growing precociously, suffered from recurrent overcapacity, marked trade cycles, and a general unpredictability. By the late 1960s, just when the rainbow's end appeared within reach once more, chemicals again entered a deep crisis, reopening the debate between supporters and critics about the sector's poor performance. However, the Group's investment in the sector, its interdependence with manufacturing oil, and a lingering belief in chemicals as the way forward for the oil industry were too great to permit a withdrawal.

Once the diversification into chemicals was well under way, Shell started searching for other business sectors to generate growth, partly following the wider fashion for conglomerate building, partly from concerns about the longevity of the oil industry and the need to provide new stimuli for managers and staff. Success here proved even more elusive than in chemicals. No business could match the performance of oil or even provide a profitable extension to the Group's existing portfolio, and companies taken over invariably disappointed for one reason or another. Rather than retracing its steps, the CMD doggedly stuck to its guns, giving the slip to the Group's shareholders, whose interests the Committee studiously put below other priorities.

[1]

Chapter 6

Discovering tomorrow's world

Immediately after the war, the Group confirmed its conviction about the importance of research by centralizing its coordination under the auspices of the CMD. Shell monitored research costs and benefits continuously and closely. Initially there was a broad agreement about the benefits, but towards the late 1960s doubts arose about the overall cost-effectiveness of the function. The diversification into chemicals boosted Shell's research effort and led to the establishment of institutes dedicated to new products such as pesticides and plastics. At the same time the Group's research scope widened to include the environmental concerns which from the mid-1950s began to have an impact on the business. Shell took the lead in building industry platforms that worked towards agreement on voluntary schemes for reducing pollution. Its attitude towards the environment was circumscribed by cost concerns and specific events, and market circumstances largely determined the speed with which the Group took action. However, around 1970 a wider and more pro-active awareness about the importance of environmental measures dawned.

Durability proved to have a downside, and the promise of plastics became the problem of plastics.

[2]

The evolution of Group research As discussed in Volume 1, Chapter 6, the Group already had a considerable degree of cross-border integration in research before the Second World War. During the war research efforts at the Group's four main centres had necessarily diverged. Emeryville and Thornton concentrated on solving problems of immediate concern to the Allied war effort, notably toluene and butadiene. Further work on the synthesizing of glycerine was postponed until after the war. In 1948, Shell Development saw its achievements rewarded by the American Institute of Chemical Engineers with the Kirkpatrick Award for Chemical Engineering Achievement.[1] By contrast, during the war Delft and Amsterdam kept their staff busy with a mix of projects, some of them designed to relieve shortages of food and materials, others continuing along already established lines in oil and in chemicals. One remarkable wartime result from Amsterdam, which went back to the chloride research started in 1928, was a process for making PVC plastics not from the ethane gases created by cracking oil, but from the cokes oven gases feeding into the Mekog factory at IJmuiden. As a consequence, the Group could bypass existing patents and start building a PVC factory at Pernis in 1946.[2]

In rebuilding the Group after the war, research received a high priority, stemming from a general conviction, as voiced by Deterding in his 1929 speech to the API, that innovation and product quality made a crucial competitive difference. As soon as circumstances allowed, in April and May 1946, research representatives from the UK, the US, and the Netherlands, met in London to consider three key issues: how to budget research, how to set priorities, and how to coordinate the efforts so as to avoid duplication and waste.[3] Kessler attended most sessions, and Legh-Jones joined him in the closing session to endorse the conclusions, which Kessler then summarized in a memo to the Bataafsche board. Research was not just vital to the Group's future, he stated, but was profitable as well. An estimated research expense of $35 million during the last decade had produced extra profits of $50 million; since 1931, Shell's chemical business in the US had made profits of $22 million, with $14 million directly attributable to research. Though spending had increased considerably, mainly due to the expansion of chemical research, these costs would be more than compensated by rising profits from chemicals. The profitability of the work on engines, lubricants, and new engine fuels was more difficult to assess, Kessler admitted, but he argued that it was 'an essential part of the programme to maintain and improve our position in the industry and it is expected to bring its reward in increased sales resulting not only from the quality of the products but also from the goodwill created with Government organizations and other customers'.[4] Once again, he showed his clear understanding of research benefits in terms of intangible competitive advantages and long-term growth, and less in terms of immediate profits. When, at the London researchers' meeting, Legh-Jones showed a keen concern for tangible results, Kessler had countered by emphasizing 'the danger of allowing short-term considerations to become a threat to research. He felt that great care had to be exercised in any paring of research expenditure needed for future prosperity; he believed that in times of depression this type of expenditure may become vulnerable, and

Great expectations! As new boundaries were broken in the 1950s and '60s, research in the Group's Amsterdam laboratory could be genuinely exciting work.

[3]

[4]

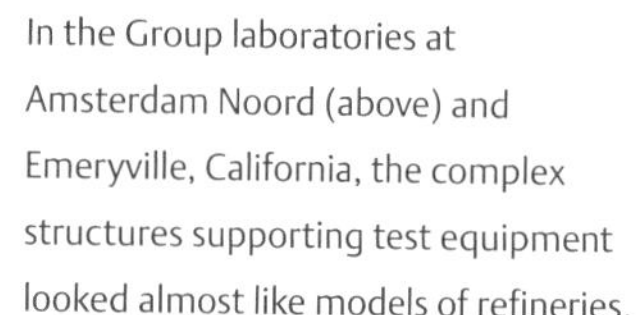

In the Group laboratories at Amsterdam Noord (above) and Emeryville, California, the complex structures supporting test equipment looked almost like models of refineries.

so was anxious that in the future the seeds from which the plants spring should continue to be sown even in years of indifferent harvest'.[5]

The London meeting ended with the establishment of a committee, made up of senior research representatives and two managing directors, to draft a Group research strategy and budget from submissions by the laboratories. The first of these annual exercises resulted in an overview of achievements and plans plus a budget proposal, which were put before the boards of the operating companies in June.[6] Total projected expenditure for 1946 came to $12.4 million, of which about $8 million went to the large research centres at Emeryville, Amsterdam, and Delft, and the rest to the Group's smaller laboratories at Thornton (UK), and the American establishments at Wood River, Houston, Wilmington, Martinez, and Salida. Chemicals claimed just over a third of total expenditure, mainly devoted to process research. The 1947 research budget submitted in October 1946 envisaged a sharp rise in spending to a total of $18.2 million for a staff of 3,300. The old Simplex Agreement between Shell Oil and Bataafsche was replaced by a new Research Agreement, which, as before, laid down the guidelines for the sharing of information and of costs. As with so many Group arrangements, the agreement was largely driven by fiscal considerations and aimed to balance research costs and potential returns on an exchange basis.[7] During the late spring and early summer of 1946, product experts from both sides of the Atlantic sat down to the practical business of coordinating the Group research effort.

The research meeting and budget marks the beginning of an important new phase in the Group's evolution towards greater integration and coordination. Almost a decade before the appointment of functional coordinators in 1955, research became the first business function to be considered as a function in its own right, separate from the operating companies and departments responsible for running it, and with decisions on priorities and funding being shifted from operating companies to the CMD. The committee showed its dedication to the subject by attending, at least from 1957 and probably before, the two-day annual conference of research directors in all fields, economic as well as scientific and technical. This conference discussed subjects ranging from the profitability of research and progress made in specific research sectors, to long-term demand, marketing, and production forecasts. The transformation of what had started as a coordination meeting into a top-level strategy debate underlines the increasing importance of research results in Group policy.[8] By the mid-1950s the annual Group research meetings had become too large to manage and the format was changed. The main session remained, but now subcommittees along functional lines and product specializations provided a forum for detailed information exchange and programme coordination, feeding into regular printed summaries of programmes and achievements for dissemination amongst Group experts. Specialist bulletins also helped researchers to keep in touch. Dedicated meetings fostered the integration between research and the functional departments, for instance helping to coordinate the research into gasoline blends, projected sales, and the planned output of particular refineries. These meetings resulted in annual Research Planning Conferences, bringing together researchers and representatives from the operational departments, beginning in 1960.

In 1954 Shell appointed a Group research coordinator, J. H. Vermeulen, and his efforts seem to have been successful: a study team screening Group research efforts in 1962 was surprised to find duplication of work very rare.[9] This team recommended effecting a closer integration between research and the other business functions, so that same year a Group Research Council was set up composed of the top business executives plus the research directors under Dr. Harold Gershinowitz as chairman. The perceived importance of the Council meant that from 1 January 1963 he was appointed to the Bataafse board as well. Moreover, from 16 January that year until his departure for personal reasons at the end of 1965, Gershinowitz was a regular attendant at the Conference; and in the summer of 1964 he reported 'that there was complete harmonisation of the Research effort as between Shell Oil and the rest of the Group'.[10] Whether this goal had in fact been conclusively achieved is open to doubt: four years later a further submission to the Conference stated that '1967 had for the first time seen a truly coordinated international Shell research programme', with toxicology and pollution receiving intensive and continuous attention.[11]

Group managers constantly debated the level of research expenditure, rates of return, and the optimum ratio between fundamental and applied research, without ever arriving at clear and consistent yardsticks. Kessler had repeatedly stated that research needed to earn its keep, but that it was impossible to find clear measures.[12] The ratio of research cost to sales, to advertising, to employee benefits, to the rate of capital investment, to net income, were all weighed and found wanting.[13] Spending levels and rates of return mattered only for broad comparisons with the competition, but the specific character of research had a direct impact on funding. Projects in applied research were charged to

[5]

Whether by practical experiment or theoretical analysis, laboratory work was always (in the words of the celebrated Dutch photographer Carel Blazer) *Verkenning in het onbekende* – exploration into the unknown.

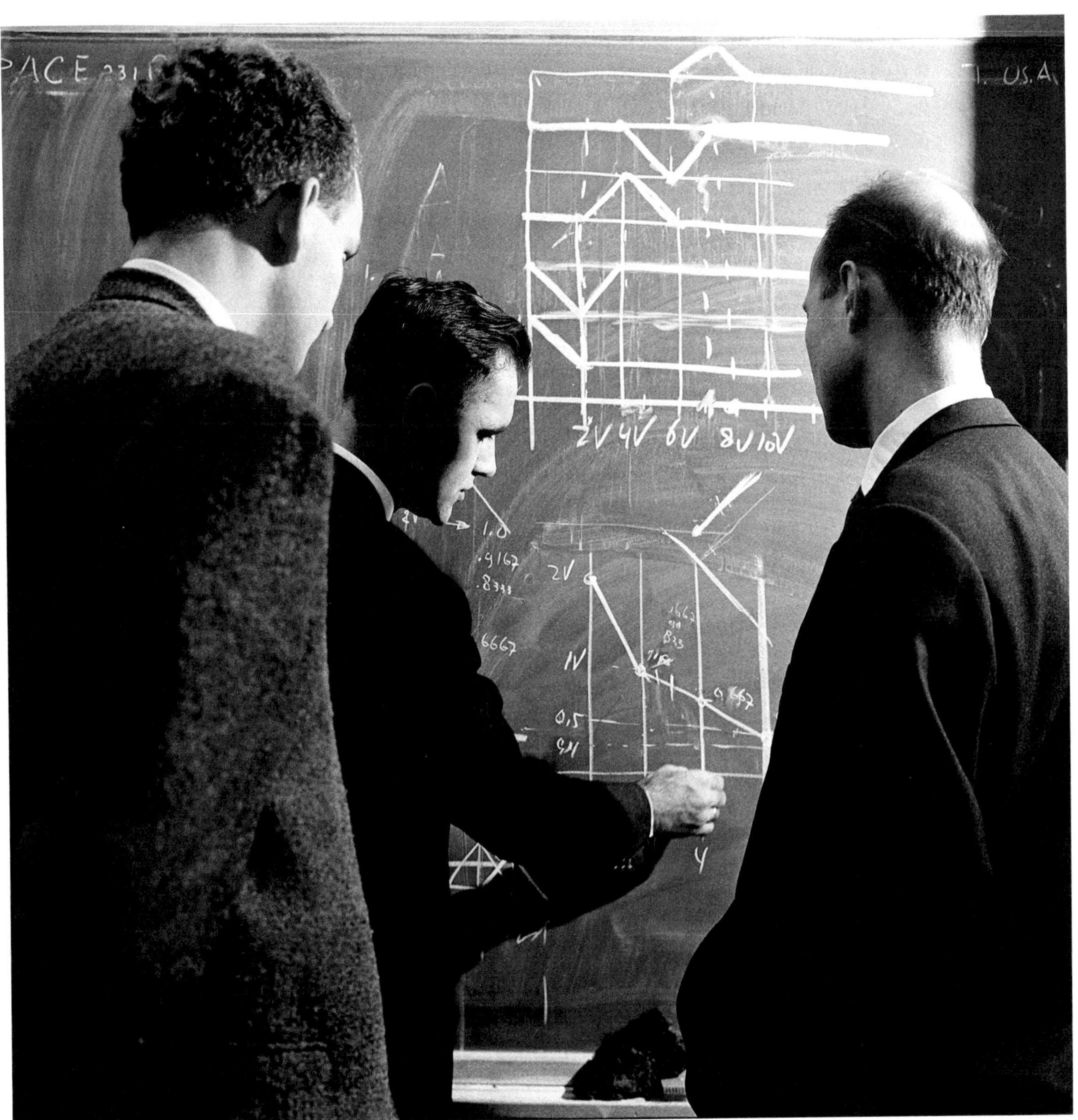

[6]

Building of the Rijswijk laboratory began on 5 January 1959. Here guests are seen walking across the Plaspoelpolder after the ceremonial driving-in of the first pile.

[7]

the business functions, whereas Group funds paid for the fundamental research. At this moment in time, patents hardly figured in the discussions about research result yardsticks, for they were seen as ways to secure market position and not as a mark of distinction. Revenues from patents were used to underline the importance of research, but not treated as a cash resource. Incorporated in 1954 and wholly owned by Anglo-Saxon, Shell Research Ltd. was largely responsible for filing applications for patents and the transfer of foreign patents. Refineries produced more patents than the laboratories, underlining the importance of experimental work done outside the formal research establishments.[14]

Spending cuts and research staff redundancies introduced by Legh-Jones and Bloemgarten after Kessler's retirement in 1948 sparked immediate complaints about the scarcity of new ideas, leading to an increase of the fundamental research budget from 13 to about 20 per cent of the total.[15] Subsequently the importance of fundamental research was taken as a fact and rarely debated, despite the difficulty of gauging results. Profitability indications varied widely. In 1957 Gershinowitz, at that time vice president of Shell Development in the US, estimated that the Group earned $4-14 for every dollar spent on research, but five years later the Group Research Coordinator H. W. Slotboom, formerly head of the Amsterdam laboratory, was more cautious. Reporting to the Conference, Slotboom presented detailed profitability studies of research into asphaltic bitumen and lube oils indicating research costs of $6 million generating additional profits of $30 million a year. In doing so, he did emphasize, however, that research had made substantial contributions to profits only 'in close and constant co-operation with Manufacturing and Marketing operations'.[16] Manufacturing Oil, however, confidently estimated, in 1968, that its annual research budget totalling £6 million yielded benefits of £12 million in cost savings and royalty income.[17] Conversely, research departments were requested to assist marketing by providing information for advertising campaigns. In 1958, a Group estimate claimed that about £11 million was spent on advertising, most of it for gasoline, which does not seem a very high amount on world sales of £2.4 billion. The research laboratories had separate service budgets to process and supply data for further use by Group departments, though it was agreed that 'while they could not be expected to substantiate every claim made by the advertising experts, wherever possible they should use every legitimate technique to provide data'.[18]

Considerations about the Group's competitive position were clearly predominant in the allocation of funds. Thus the 1950 budget overview defended a reduction in research funding for aviation gasoline and lubricating oils with the argument that the Group now had a good competitive position, but no guidelines or targets for evaluating spending and results survive.[19] For the overall budget management directors presumably kept a certain proportion of revenue to costs in mind. Throughout the late 1950s and 1960s, research spending hovered around 0.8 per cent of gross revenues, about 1 to 1.3 per cent of total costs. Spending ratios varied widely, however. Chemicals topped the league with 4.7 per cent of turnover, whereas oil research amounted to only 0.6 per cent, and exploration and production less than 0.3 per cent on the value of crude produced. Keeping pace with turnover, the budget leapt from $34 million to almost $120 million by 1970. By its own account, the Group consistently spent a lower overall amount, but a larger percentage of sales on research, than Jersey Standard.[20] Shell spent more on fundamental research, chemicals, and oil products; Jersey concentrated on exploration and

[8]

[9]

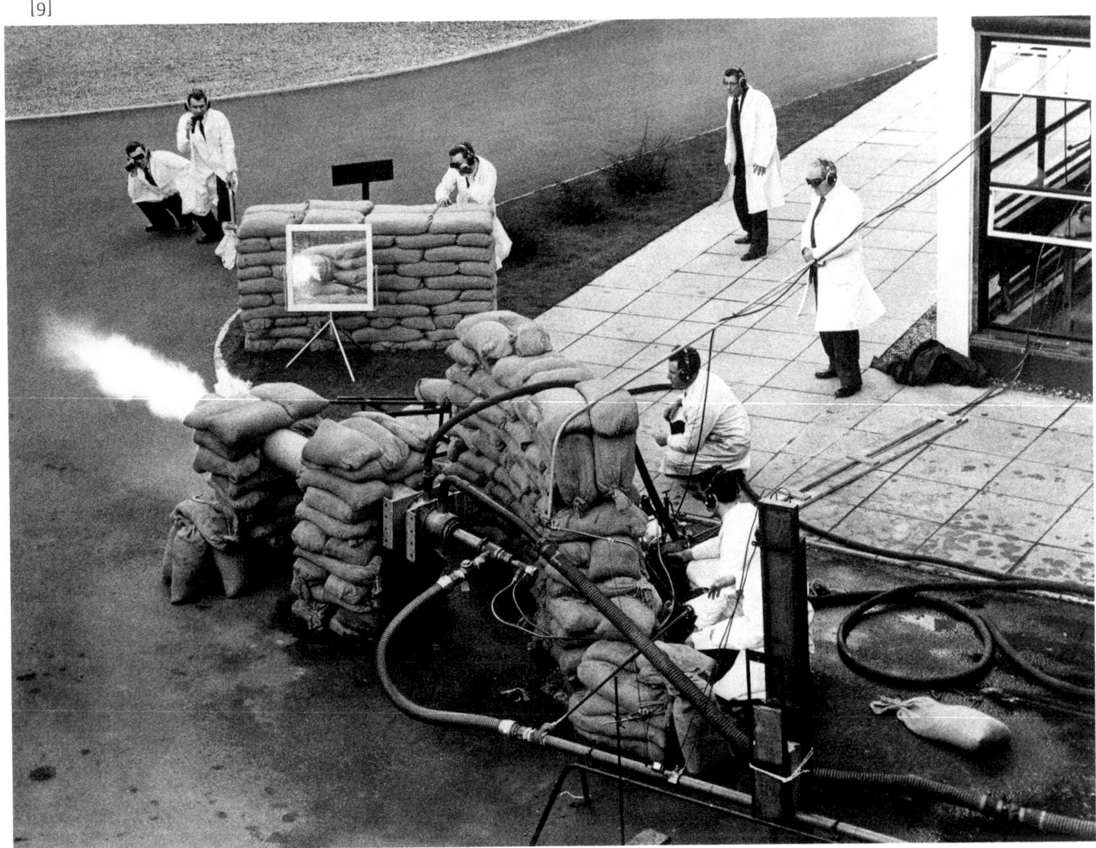

Experimentation was not always a matter of fragile glass test-tubes and distillation columns. At Rijswijk in 1969, this massive tank (left) was used to replicate aspects of an oil field, and at the Egham laboratory (UK) in 1956 (right) an early model of an oxygen/gas oil toroidal burner is tested.

production and oil processes. However, Shell's chemicals research absorbed an increasing share of the total budget. Spending rose from about 20 per cent during the late 1940s to over 30 per cent in the 1950s, to level out at around 30 per cent during the mid-1960s.[21] Consequently the Group, compared to the competition, really underspent on oil research in proportion to its production volume.[22] As with the unfavourable barrel cut emphasizing the middle distillates to the detriment of the more profitable gasoline fractions, the research spending pattern suggests that the decision to adopt chemicals as a second core business had hidden consequences for the rest of the company.

Expansion also meant a continuing specialization following the evolution of business functions and product segments. In 1945, Shell Oil set up a research group for exploration and production to complement its Houston geophysical laboratory. Merging with Shell Development's Emeryville production research team, this new centre had a staff of 200 people by 1950. Following the work done there during the war, Thornton became the prime centre for engine and fuel research. The engine testing done at Delft was transferred to Amsterdam and Thornton, but the E&P research remained and in 1957 Delft was also chosen as the location for a research centre concentrating on plastics and rubber. Two years later the E&P research moved to a new laboratory in Rijswijk.[23] Immediately after the war, Shell Development set up an agricultural research centre at Modesto in California, which worked alongside Amsterdam laboratory in the development of products such as weedkillers, insecticides, fertilizers, and fungicides from petroleum derivatives. In 1955, the facilities at Amsterdam were transferred to a new site at Woodstock, near Sittingbourne in Kent (UK).[24] Two years later Shell Oil expanded Modesto by building a new laboratory of 50,000 square feet to accommodate 110 laboratory staff including the researchers formerly based at the Rocky Mountain Arsenal, where the company manufactured the 'drin pesticides. An organic chemistry group was primarily concerned with the synthesis of new compounds, and a physical and analytical chemistry section developed new formulations and conducted residue studies. The number of staff at Modesto matched that at Woodstock, which in 1958 was also expanded with a toxicological unit, bringing the total scientific staff to 110. The toxicological work in Europe had previously been entrusted to an outside consultant, which was clearly no longer considered satisfactory. A candid memorandum to the Royal Dutch board summed up the need for the new unit as follows: 'The Group is manufacturing and marketing products which may have potential health hazards, particularly agricultural chemicals. Therefore it has been decided to erect a toxicological unit adjacent to Woodstock which would give data with adequate speed and secrecy and would enable the experts concerned to talk on equal terms with Government officials.'[25]

The number of toxicologists at Woodstock expanded gradually and the laboratory became the Group's centre of agricultural research outside the US.[26]

The Group's 1955 acquisition of Petrochemicals Ltd., its Ziegler patents, and the Partington laboratory went hand in hand with a substantial increase in the chemicals research budget and the foundation of dedicated research institutes for these materials.[27] From the late 1940s, regular specialist conferences along product lines or specific functions proliferated, save for where research was carried out under the aegis of governments and therefore regarded as secret, in which case information exchange remained tied to the format of informal discussions during laboratory visits.[28]

By 1960, the Group had fifteen formal research units with a total staff of 5,500, but there were another seventeen laboratories performing some kind of research which was not officially counted as such.[29] The formal research budget amounted to $62 million or 1.1 per cent of sales, but actual spending was probably in the region of $80 million.[30] At that year's research survey for the CMD, Gershinowitz concluded that the Group oil research was insufficient, because 'compared with our competitors, we are not the leaders and even after stepping up our effort in Exploration and Oil Process Research we will still not be up to the leaders. In Chemicals we are amongst the leaders but the business will need more research to be able to expand at the forecasted rate.'[31] The research budget fails to show an immediate and substantial extra effort until the late 1960s, however. The seven laboratories located in the UK and the Netherlands, which employed 3,100 people in all, found it difficult to attract sufficient talent. Consequently a special programme was set up to subcontract research to universities in Germany, France, and Italy.[32] Hiring consultants did not bring the desired results. Expert knowledge was hard to find, expensive, and best built up in-house. One report commented that 'good

[10]

[11]

West Coast experiments: In 1953, the Group expanded its ammonia production with a new plant at Ventura, drawing its supplies of natural gas from Shell's local oil and gas field. At far left, anhydrous ammonia is injected directly into the soil to improve its fertility, and at near left, in 1958 a soil fumigant is applied at the Group's 142-acre (57-hectare) farm in Modesto.

consultants are few and far between, and with some we have retained there has been a disposal problem. Another danger is to rely too long on a consultant when we ought to develop our own specialist knowledge.'[33]

The importance attached to research efficiency meant that the Group adopted computing as early as January 1955, when Slotboom ceremoniously switched on a huge Ferranti machine, which reportedly performed one man-year of calculations in a day, at Amsterdam laboratory.[34] Eventually and inevitably computers supervened in accounting as well as in research, but meanwhile, their advent not only greatly increased the speed and efficiency of research but also helped to create the new discipline of operational research, i.e. the use of scientific systems analysis to determine the optimum solution for any operation, be it the location of installations, the programming of refineries, or process control.[35] The marketing report for 1968 commented favourably on the sophisticated forecasting techniques which operational research had by then developed, which enabled closer control of inland distribution systems and product stocks. Shell-Mex & BP claimed savings of £50,000 a year as a result of these applications.[36]

Nevertheless as research costs rose, net earnings per barrel of refinery throughput fell, presenting the Group managers with a fundamental dilemma. To remain competitive, the Group depended on research either to create new products, to add features to products which had become standard, or to re-engineer products in order to remove undesirable qualities. Increasing official requirements for oil products also led to higher research costs. The first X-100 motor oil introduced in 1949 required only one engine test before its official market introduction; ten years later a new ashless X-100 needed nineteen different tests for the same purpose.[37] At the same time developing and introducing new products was becoming very expensive, with lead times from research project to markets ranging from about two to four or even eight years in the case of pesticides.[38] In 1960, losses on new

DANGER
PERSONNEL
TO
HEAP

Pictured in 1955, this storage chamber, 90 feet (27.5 metres) high, could accommodate up to twenty thousand tons of fertilizer. The giant machinery could shift several hundred tons a day from the store to the bagging plant.

When first introduced, detergents seemed a real boon, but a shocking and unexpected result of their early formulation was they did not easily degrade after use. Here, seen somewhere in Holland, huge quantities of foam float on the waters of a canal.

From 1957 Shell researchers worked to develop biologically soft and degradable detergents (right), but well into the 1960s the same disgusting sight appeared in places as far apart as creeks in Kentucky and the River Seine in France.

[13]

products claimed 20 per cent of gross revenue on chemicals, and research another 10 per cent.[39] Declining margins on commodities cut into the profits needed to maintain research levels there. During the 1950s consumers in Western Europe and the US widely adopted washing machines using synthetic detergents with hydrocarbon-derived active ingredients such as Shell's Teepol. Unlike ordinary soap, these ingredients did not break down in drainage systems and sewage treatment works, but caused malfunctions and created froth floating around on canals and rivers. From 1957 Shell researchers were closely involved in the search for biologically soft detergents, and succeeded in altering the molecular structure of alkyl aryl sulphonate (the base synthetic detergent) in such a way that much more rapid and complete biological breakdown could be achieved. They named the product Dobane JN, and under prolonged post-laboratory tests in actual rivers they were able to confirm that 85-90 per cent of its compounds broke down, compared with only 15 per cent of the industry's earlier products.[40] Moreover, they persevered with the research to improve the product still further towards 'the ultimate goal of rapid and complete biodegradation under all realistic circumstances', using the British government's Directorate of Scientific and Industrial Research to ascertain the biodegradability of ionic and non-ionic samples (sometimes as many as fourteeen different ones in a week); and with the British Association for the Advancement of Science, Shell Chemical Company established an award to help the industry's progress through the systematic exchange of information.[41] As a result of the rapid reaction to the detergent problem, the Group built up a strong position in biodegradable ingredients which went unchallenged by competitors until 1965. Two years later soft detergents became the norm for the European detergents industry.[42] However, the market did not accept a higher price for the new product.[43]

[14]

A similar phenomenon manifested itself in the production of agricultural chemicals. After the Hyman takeover in 1954, the Group spent a total of £3.3 million on research in agricultural chemicals, 3.4 per cent of sales in these products, during the next six years. By 1959 it had a strong position with two products on the market, but it faced growing competition from other insecticides plus the prospect of the 'drin patents beginning to lapse from 1962. Consequently, margins came under pressure but research efforts could not be curtailed since the Group needed new products, and those products were estimated to bring in lower revenues than the exclusive Hyman products were doing at the time. Indeed, when aldrin, dieldrin, and endrin, which generated 90 per cent of income from agricultural chemicals, were discovered to form potentially harmful toxic residues on crops, increased research spending became necessary, hence the formation of the toxicological unit at Woodstock.

Finally gasoline. During the later 1950s, the oil majors switched their marketing strategies from octane numbers alone to additives such as Shell's TCP (tricresyl phosphate), the development of which had taken four years and road-testing covering 70 million vehicle miles. Anti-knock additives remained an active research field, but as part of the continuous search for the optimum fuel blend, combining high performance with the best

Figure 6.1
The number of patents obtained by the Group (left scale) and research costs in US$ (right scale), 1950-1975.

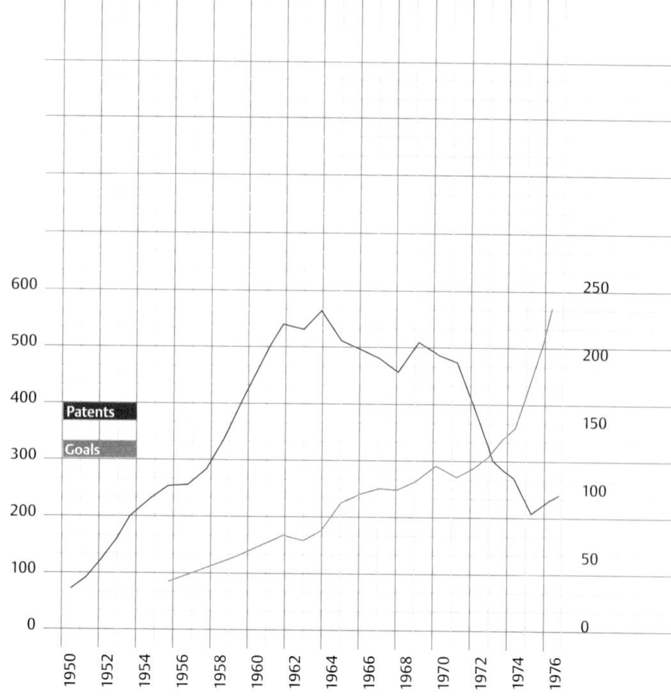

use of Group refinery output. In 1960, atmospheric pollution from gasoline in general and additives such as tetraethyl lead in particular did not yet figure as a pressing concern, but the Group had already begun to investigate refinery processes to produce high-octane low-sensitivity gasoline without additives. Five years later researchers had noted the writing on the wall and increased their efforts.[44] By 1968 two processes had been developed towards this purpose and research continued at a high level. In 1970 one of these processes went commercial at the La Spezia refinery in Italy, where the necessary investment was expected to earn a return in just four months. The other process had reached the stage of pilot plant testing. With legislation banning the use of leaded gasoline in motor cars on the horizon in the US, the Group had more than one option available for marketing the new products required, but when they came to market competitive circumstances made premium pricing difficult.[45]

Thus during the 1960s the Group's research activities were increasingly confronted by a new and fundamental challenge. Top management fully accepted research in all its varieties as a vital competitive necessity, an acceptance exemplified by the appointment of Monty Spaght as managing director of Royal Dutch in 1965. After studying at Stanford and Leipzig, Spaght had joined the Martinez laboratory in 1933 as a research chemist and managed research and development at the San Francisco manufacturing department during the war. In 1949 he was appointed president of Shell Development, then moved to New York to become executive vice president of Shell Oil in 1953, succeeding Burns as president in 1960. As one of the Group's most prominent scientist-managers and a keen advocate of closer integration between research and operational functions, Spaght had given keynote speeches at all research briefings since 1946. During his directorship Group research became more fully internationalized as well, leading to the appointment of an American, Dr. S. A. Ballard, to head the Amsterdam laboratory in 1969.

However, at the same time the gap widened between research costs, lead times and introduction losses on the one hand, and shrinking margins on the other, threatening the rising tide of funds on which past developments had depended. In addition, wage inflation sharply pushed up research spending, while the number of inventions for which the Group filed a patent application showed a marked downward trend (Figure 6.1).[46] The drop in patent filings was largely the result of polymer technology reaching maturity, but it could of course also be interpreted as a sign of declining research effectiveness. These two developments would lead to a profound overhaul of the research reorganization in the 1970s (see Volume 3, Chapter 2).

From cost conscious to environmentally conscious The Group's environmental policy is one of those areas which for the better part of our period cannot be satisfactorily reconstructed because of the rigorous delegation of authority which the company practised (see Chapter 2). Until 1967-68 the coordinators handled most issues, and only isolated cases went up to the CMD or the Conference. As a result, records for the earlier years are patchy, allowing only a glimpse of what went on. We will first discuss the evolution of the organizational structure of environmental matters within Shell, before analysing the company's changing attitude towards them by looking at the major issues.

During the late 1950s, the Group appears to have realized for the first time a need to devote some structured attention to environmental matters. In 1956 the British government passed the Clean Air Act, restricting industrial emissions into the atmosphere, and about the same time Shell set up its own 'Atmospheric Pollution Committee'. However, by 1960 that committee was 'moribund and without an appointed heir' and it was 'by no means clear who are the arbiters of Group policy in this matter'.[47] Apparently no new committee was appointed. Dealing with environmental matters was the remit of, on the one hand, the Technological Department in The Hague, which advised Group installations on how to deal with air and water pollution; and of the London Trade Relations Department on the other.[48]

In 1960, prompted by the emerging concerns about the toxicity of pesticides, Shell did set up a toxicological working committee in the UK. Later that year Shell Development Company in New York and Shell International Research began to exchange toxicological data. By 1964 the toxicological working committee had become a Steering Committee, and that year a Toxicological Division was created under Dr. H.G.S. van Raalte, with one office in The Hague and one in London, each advising its local central office functions and liaising with the other as necessary. The Division was responsible for all contacts with outside bodies, such as the UN-affiliated Food and Agriculture Organization and the World Health Organization, in matters to do with the toxicological aspects of air, water, and soil pollution problems, and provided advice to other Group companies.[49]

Then, in August 1963, Shell and BP jointly established CONCAWE (Conservation of Clean Air and Water, Western Europe), a foundation under Dutch law designed as an industry body to disseminate information on the environmental effects of oil refineries, and to act as a joint representative with governments. At the same time the organization was meant to keep the playing field level by ensuring that environmental regulations would apply to all companies concerned. Within a few years it had thirteen oil companies as members, representing 75 per cent of Western Europe's refining capacity.[50] Following the formation of CONCAWE, Shell set up an environmental conservation committee, which issued a sixty-four-point internal document on the environment, now lost, to be used by Group companies in the preparation of individual policy statements.[51] A few years later Shell Development Company in the US created 'a focal point, a "Mr Pollution", who covers all forms of pollution'.[52] In 1969, Shell conducted a Group-wide inquiry about the perceived importance of various environmental issues at operating company level. Atmospheric pollution by sulphur dioxide emissions topped the list as the issue which raised most concern, with leaded gasoline coming second, and pollution of rivers and lakes third. The survey considered the US, Canada, Japan, Sweden, Switzerland, West Germany, France, Italy, and the Netherlands as the countries where environmental concerns were at their most vivid and thus likely to have an impact on Shell.[53] Following this inquiry, the CMD agreed that 'the field of conservation should be placed under the overall supervision of a managing director (as is done with safety matters); this would also have the merit of underlining the importance attached by the Group to these problems, and would have corresponding publicity value'.[54] In November 1969, the CMD approved an internal guide for operating companies on how to tackle pollution problems.[55] Even so the subject remained confined to individual business functions and was not yet considered sufficiently important to warrant a department of its own.[56]

At that time Shell's American companies already devoted more attention to the environment. In October 1971 the Group held its first coordinating meeting on the environment and conservation in Houston with staff from central offices, Shell Canada, and Shell Oil present.[57] The attendance list showed an interesting disparity. Shell Oil sent a delegation of eight people, the vice president Public Affairs and seven engineers and managers solely concerned with the environment. Each operating function had a head office coordinator for environmental affairs plus a line organization for dealing with issues as they arose.[58] Shell Canada was represented by its coordinator environmental conservation. By contrast, central offices did not yet possess environmental specialists at senior management level. From London a trade relations manager and a staff member working in marine safety and conservation attended the Houston meeting; The Hague had delegated a toxicologist, and one manager each from Manufacturing Oil and Manufacturing Chemicals. This suggests that at that moment in time, environmental matters were still primarily a concern for the operating companies, and not for central offices, which had no department for environmental matters and left the matter to the business functions coordinated by the trade relations department. In 1972, however, the CMD appointed a Group Adviser on environmental affairs, who reported directly to Barran.[59]

Discovering the environment The Group's early attitude to environmental problems was one of enlightened self-interest. Officials showed themselves to be relatively open-minded in responding to outside complaints and prepared to invest in remedies provided that these saved money or paid for themselves in other ways. This was, for instance, how Bataafsche reacted to complaints about the Pernis refinery in 1937, when the mayor of nearby Vlaardingen sent an official complaint to the board about the unbearable stench emanating from the works, and he threatened 'serious difficulties' if nothing were done.[60] Bataafsche responded by denying responsibility and by pointing out 'that our plant is surrounded by other chemical plants, and therefore the source of the stench in the municipality (...) should not only be sought at our plant'.[61] However, the company then went a little further. Firstly, as a step in building public confidence, the mayor was invited to contact the Group's representative at any time; secondly, the board authorized a budget of 20,000 guilders or nearly £2,000 for research into a solution, so that if there were more complaints the Group could at least say it was spending money on the problem.[62] The problem was found to be the emission of hydrogen sulphide in offgases, which was remedied by installing, at a cost of half a million guilders, a Claus installation to extract the sulphur. By selling this sulphur Bataafsche could recoup its investment in four to five years.[63] The installation did not work entirely satisfactorily, because the tailgases still contained too much hydrogen sulphide. One solution adopted was selling the hydrogen sulphide to the Albatros fertilizer works.[64] As a further step Bataafsche extensively tested a different installation on a pilot plant and in 1949 installed a redesigned unit at Pernis, which also served as a prototype for similar ones abroad.[65]

[15]

With messy handling, pollution could start in your own back yard. Poor working practices in Curaçao in December 1949 result in clouds of sulphur dust around the workmen.

In a similar way, chemical engineers Pyzel and De Bruyn had earlier solved a comparable (but not identical) question at Shell Point in California, where large quantities of sulphuric acid were used to make ammonium sulphate. At the same time Shell Oil's crude processing generated waste acid that was afterwards thrown away 'by the ton' as part of the tarry residue. Could that acid not be recovered from the residue and used again in the manufacture of ammonium sulphate? Pyzel and De Bruyn were not the first people to have the idea, but unlike others, in 1932 they developed a viable method which was put into operation in 1933 and became profitable in 1935.[66] Where and how Shell Point's waste acid was previously disposed of is not recorded, but once it had been extracted from the residue there was a second benefit, namely that the remaining sludge oil could be used as boiler fuel in the refineries.

This pre-war experience does not seem to have been fully transferred within the Group. In Curaçao, even beyond the war years and despite the presence of an acid recovery unit there, excess waste acid was routinely dumped into the Schottegat (Willemstad's inner harbour) until 1949. When the time came to plan a new lubricating oil plant, it was discovered that the harbour water was too acid to be useful as cooling water, and a sum for the extension of the recovery unit was included in the capital budget for 1950-52. This, it was foreseen, would reduce both the corrosiveness of the harbour water and the amount of dollars that had to be spent on buying sulphur. The move seems such common sense that at first it is surprising to see how little it cost – 1.7 million guilders (£170,000), or just 2 per cent of the 84 million guilder capital budget.[67] The Group also reacted quickly and adequately, and of its own volition, without much public pressure, if a product

Chapter 6

Shell Oil did set a standard for itself, however, by stating in 1958 its wish to act as a 'reliable and forward-looking corporate citizen' in its business operations.

could be shown beyond any doubt to have direct deleterious consequences and if an alternative could be found, as we have already noted in the case of Teepol.

These examples show that Shell could, and did, act when confronted with environmental problems created by production processes, other operations, or by products. However, such actions remained reactive and incidental, i.e. triggered by outside incidents and confined to solving those. They did not lead to a wider exploration of environmental problems, let alone to the recognition of them as inherent in large-scale industrial processes and particular hydrocarbon products. The problem of hydrogen sulphide emissions, for instance, failed to generate a concern for atmospheric pollution. As we have seen, the committee set up for the purpose during the late 1950s functioned only briefly. In 1961 Group policy on that subject was still no more than 'to keep in touch with all developments in connection with atmospheric pollution without becoming actively involved (...) we should merely keep abreast of developments and no experimental work is proposed'.[68] Shell's intellectual curiosity remained confined to subjects of immediate concern, presumably because the pressure for commercial results prevented researchers and managers alike from widening their focus.

The statement of business aims drafted in 1962 made no mention of environmental concerns either. The list was written entirely from an internal perspective, as a motivation for managers and staff, and emphasized the desirability of keeping Shell's global operations profitable, dynamic, and competitive, driven by an innovative spirit and kept lean through cost controls, with an enthusiastic, self-critical, and highly trained staff. The world outside makes an appearance only at the end, with general precepts for staff to strive for the highest standards of business integrity, cultivate social responsibility towards host countries, and abstain from any political involvement.[69] The absence of environmental concerns from the Group's business aims is not surprising, given the fact that such anxieties only began to surface in wider society during the early 1960s, but the inward-looking attitude to which the statement testifies was not propitious for inspiring a constructive response to criticism from outside, as the controversy over pesticides, about which more below, was to demonstrate. Shell Oil did set a standard for itself, however, by stating in 1958 its wish to act as a 'reliable and forward-looking corporate citizen' in its business operations. The company reiterated this ideal in 1965, expressly linking it to 'the need to protect the purity of the air and water and to insure the public's safety in the use of chemicals designed to improve the food supply'.[70]

We do like to be beside the seaside – but not so much when a pleasant walk ends with tar all over our feet...

[16]

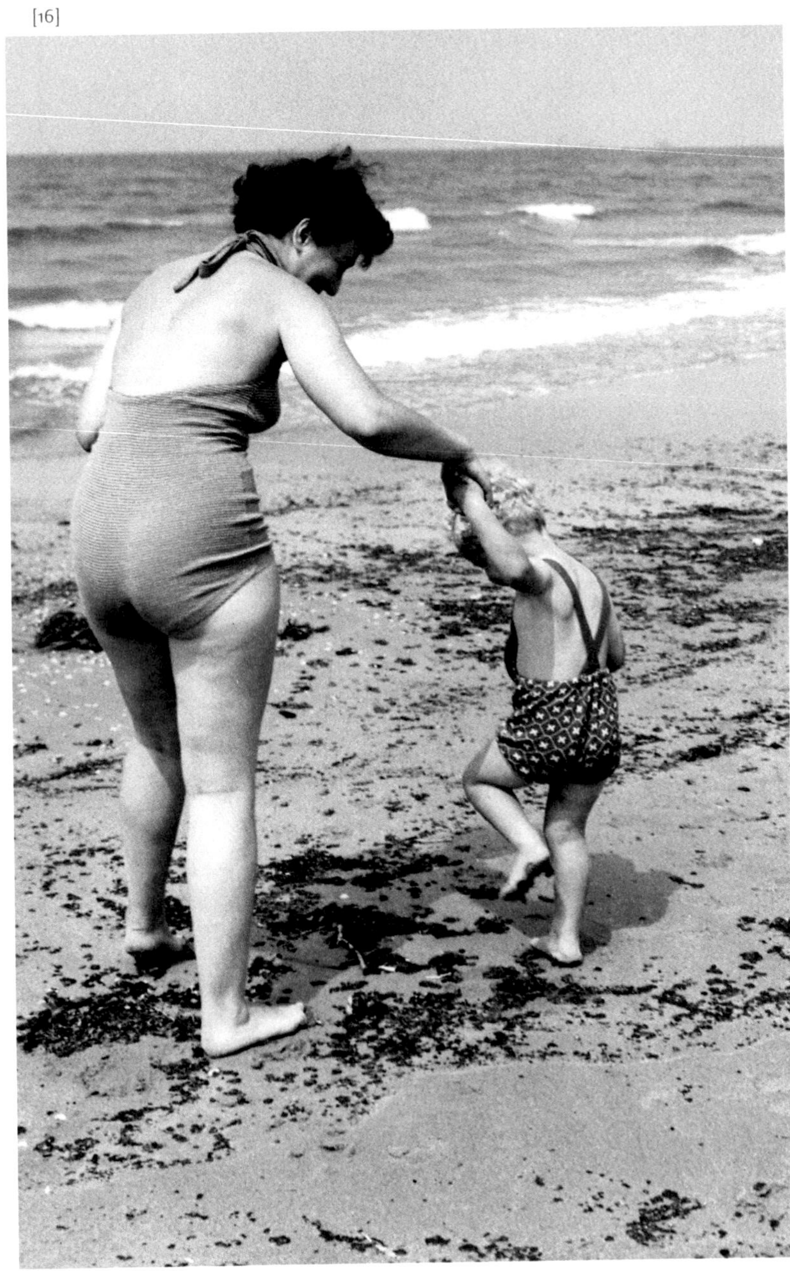

Seeking industry support In combating marine pollution, the Group also showed its constructive side. By the 1950s oil pollution on beaches and shores was worse than ever before, mainly because of the discharge of washings from crude oil tankers, which, with the increased carriage of crude at sea, had grown steadily since the end of the Second World War. The first internationally accepted regulations for the prevention of pollution at sea were introduced in 1954. These regulations originated in an international conference convened by the British Ministry of Transport, and although they did not have the force of law, the Group accepted them. When a second conference, held in 1962, concluded that very little real progress had been made, the Group decided that the basic approach was wrong, and that the oil industry, not the shipping industry, bore primary responsibility and should be the body to solve the problem. Shell therefore set about this, confidently thinking 'that if a practical technical solution could be found, it would then only be necessary to persuade the other members of the industry to follow'.[71]

The Group's self-confidence in taking such a lead was justified when, after practical tests, it found a solution which it called the 'Load-on-Top' system. Using this system, tanks were cleaned in the customary way, but instead of the washings being discharged into the sea, they were consolidated into one tank and a new cargo was loaded on top of them. There was of course a commercial benefit too, insofar as the value of the oil previously discharged overboard was now saved. A temporary hitch arose with the Suez Canal Authority, which took the view that ships carrying even 0.4 per cent of their full capacity were technically loaded and thus liable for full cargo dues, but after negotiation this was altered to a payment of the much lower ballast dues with a 5 per cent premium on consolidated washings. The premium

[17]

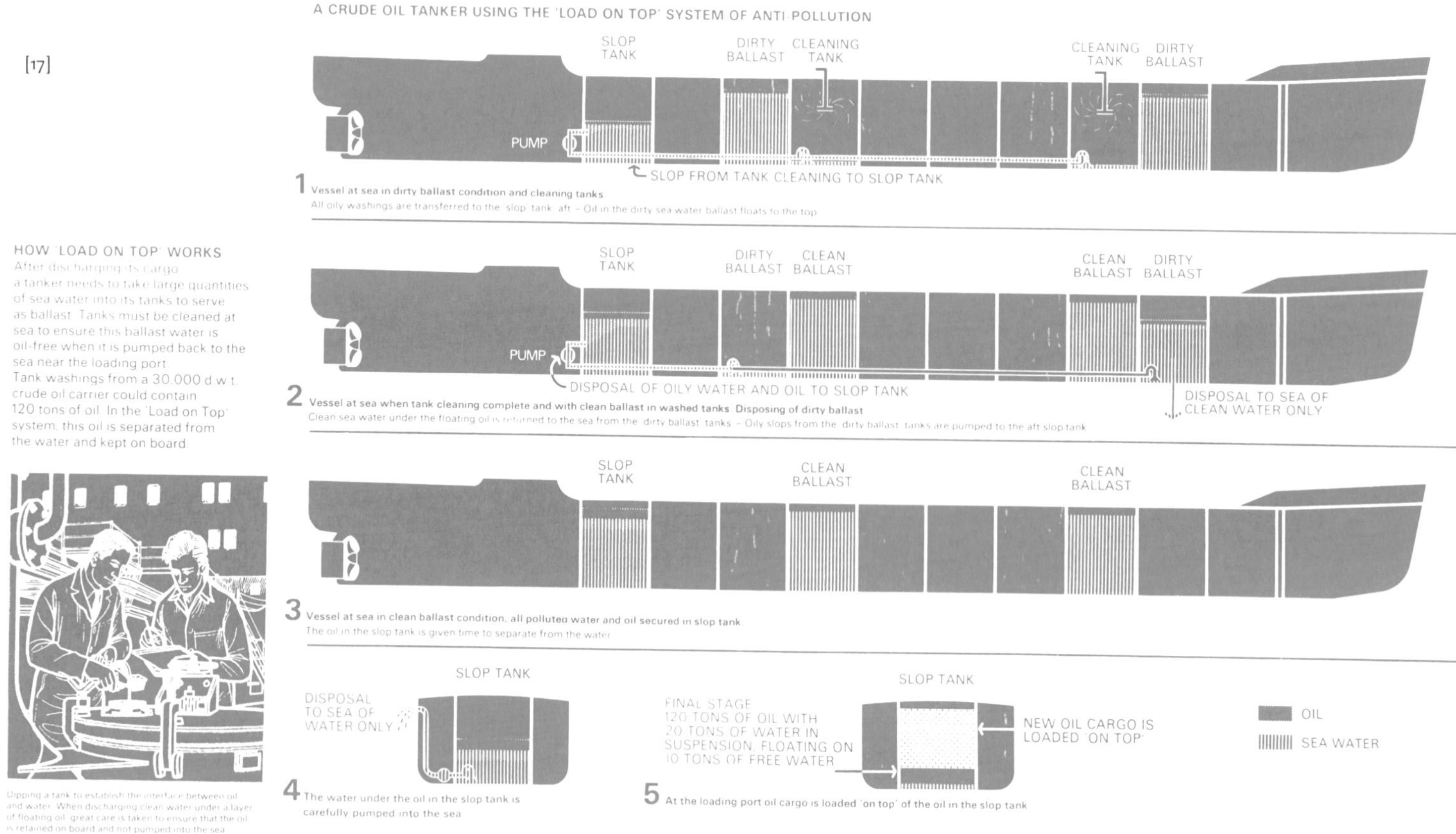

HOW 'LOAD ON TOP' WORKS
After discharging its cargo a tanker needs to take large quantities of sea water into its tanks to serve as ballast. Tanks must be cleaned at sea to ensure this ballast water is oil-free when it is pumped back to the sea near the loading port. Tank washings from a 30,000 d.w.t. crude oil carrier could contain 120 tons of oil. In the 'Load on Top' system, this oil is separated from the water and kept on board.

Dipping a tank to establish the interface between oil and water. When discharging clean water under a layer of floating oil, great care is taken to ensure that the oil is retained on board and not pumped into the sea

represented about 50 per cent of the value of the oil saved, meaning the technique was still economically attractive; so if the industry as a whole would follow Shell's lead, pollution from this source could be largely eliminated. BP was a close and staunch supporter of Shell's efforts, as was Jersey Standard later, but other American companies and CFP still needed to be persuaded. Even at that time, though, the Group was confident enough about the system to draft a code of ethics for general publication, and 'Load-on-Top' eventually became mandatory.[72]

This Group initiative dealt well with the issue of deliberate discharge, but could not cover accidental spillages. Port authorities and others exhibited much interest in this, and Shell and others conducted many experiments for the containment of floating oil. These mainly used floating booms either with flexible weighted skirts or with 'air barriers', consisting of a screen of rising bubbles of air from a perforated pipe installed on the bottom of a harbour. Boats with V-shaped booms attached to their sides were also used. All the devices had their best results in flat calm conditions, but none were completely successful.[73] Moreover, no-one had given very much thought to what might be done in the event of a massive accidental spillage.

That moment came on 18 March 1967, when the 119,000 dwt *Torrey Canyon*, laden with a cargo of Kuwaiti crude belonging to BP, ran onto rocks off England's south coast. Such an accident had been widely feared as oil tankers grew larger, but in the event no one knew what best to do. The Royal Air Force bombed the slick, in the vain hope of setting it on fire; detergents were used to try and dissolve or disperse it at sea and on the coastline, and to clean it from birds' feathers. News coverage was so emotive that Shell did not dare suggest that since crude oil was a natural substance, in the

[18]

[19]

As problems of pollution became acute, Shell developed solutions. Right, technologists at Rijswijk experiment with a 'water wiper' using polypropylene fibre to absorb floating oil, and above, the real thing is seen in action somewhere in Holland.

However, it was clearly better to prevent pollution in the first place, and Shell's 'Load-on-Top' system (above left) became standard practice in the tanker industry.

On 27 March 1967, following a navigational error, the 119,000 dwt *Torrey Canyon* (left) hit Pollard's Rock in the Seven Stones reef west of Cornwall, UK. It was the world's first major supertanker wreck and found the industry and government alike entirely unprepared for coping with such a disaster. As the ship broke up, 120 miles of the Cornish coast and 50 kilometres of the French coast were contaminated, with very great loss of marine life. A flexible boom (right) developed by the Norwegian engineer Trygve Thune was an effective method of containing a patch of floating oil, which could then be sucked away and disposed of, but it could only function in calm waters.

long run the best solution would be to do nothing. But that suggestion surprisingly came in the magazine of the Royal Society for the Protection of Birds, in an article entitled 'Detergents and Wildlife'. A member of Shell Research Ltd. wrote privately,

'this [article] is (perhaps understandably) rather acid and emotional (...) Nevertheless, it is my own feeling that the author's conclusion, i.e. that detergents are a far greater menace to wild life than the oil itself, is probably valid. Despite this, we have to confess that, at the moment, we do not have any other cheap, readily applicable method of removing oil pollution from the sea and from sandy beaches.'[74]

Two years later research efforts intensified after the Santa Barbara oil spill, emanating from a rig off the Californian coast. Shell followed several lines of enquiry such as gathering spilled oil with gigantic sponges, sinking it with sand, or introducing a gelling agent into future cargoes, to reduce outflow from a ruptured tank.[75] Researchers soon found a useful half-way answer which they called Shell Oil Herder. It did not remove oil from beaches or the sea, but it created a barrier that kept the oil away, by forming an extremely thin film (one molecule thick) that was not dissolved by water and had a spreading pressure on water that was greater than most liquid petroleum oils. This enabled it to compete effectively for water surface, and in so doing it could contain and contract an oil slick. The same qualities meant it prevented spilled oil from attaching to beaches (whether of sand, shingle or pebbles), rocks, and structures such as jetties and breakwaters. Moreover, it was practically non-toxic and fully biodegradable. Its drawbacks were that its operators required full protective clothing, the public had to be kept at a distance to avoid any risk of inhalation, and it diluted rapidly, so it was most effective if sprayed on beaches and so forth up to one hour before an oil slick came ashore. But it was a very successful product, and after evaluation by the British government's Departments of Trade and Industry and of the Environment, it was recommended for use by all local coastal authorities.[76]

[22]

With wind, tide and currents spreading the *Torrey Canyon* oil slick far and wide, a floating boom was placed across the mouth of the pretty harbour of St Ives, Cornwall, thirty miles from the wreck, in an attempt to prevent more oil coming in.

The Group took the clean-up of all accidental spillages of its oil seriously; even small ones were dealt with in conjunction with local authorities and their causes investigated and acted on. Such actions might range from physical repairs if damage or wear (such as leaking rivets) was found to be the cause, to disciplinary action against neglectful individuals. The Group would also naturally contest alleged spills when they did not believe the oil to be theirs. Sometimes this was found by independent analysis of the oil to be true; sometimes not.[77] With some other oil majors, the Group was also primarily responsible for the industry-wide introduction of an insurance scheme called TOVALOP (the Tanker Owners' Voluntary Agreement concerning Liability for Oil Pollution). Member companies involved in a pollution incident were obliged either to organize and pay for the clean-up themselves, or to reimburse the cleaning-up costs of the responsible authority. The scheme came into operation in 1969, and by the summer of 1970 had gained the support of nearly 80 per cent of the industry.[78]

The *Torrey Canyon* disaster also helped to convince the Group about the need to open up its affairs to public scrutiny. This and other big oil spills generated a vivid public concern about the dangers of large oil tankers, which deepened further two years later with the occurrence, within a month, of explosions on three VLCCs – two belonging to Shell – during tank-cleaning operations at sea.[79] The direct environmental effect of these accidents was minimal, but the Group decided to hold a public discussion about them in January 1970 so as to counter the growing aversion to oil tankers.[80] The desirability of providing information was one of the big lessons Shell learned from this exercise.

Moreover, like CONCAWE earlier, TOVALOP showed that building industry-wide support remained the single most important factor in combating pollution. Fearful of incurring competitive cost disadvantages, individual companies would resist public pressure or government regulation for as long as possible. With a united front the industry could argue its case to best effect, stave off formal regulation, keep its playing field level, and safeguard costly investments from the impact of sudden legislative changes. For their part, governments liked to collaborate with industry platforms as a way of obtaining otherwise unavailable technical know-how. The Group favoured such schemes for all these reasons, less so for the environmental concerns themselves.

[23]

Three years after the *Torrey Canyon* was wrecked, Shell had a new means of disposing of oil slicks. In the 'Shell Sand Sink' method, sand was treated with a chemical that made oil stick to the grains, then mixed into a slurry with water and sprayed on the oil. A full-scale test 15 miles off the Hook of Holland on 8 April 1970 was judged to be 95 per cent effective, clearing 100 tons of Kuwaiti crude in about 15 minutes. Sticking to the sand, the sprayed oil would sink to the bottom of the sea and remain there, gradually dissipating naturally. Left is a general view of the ship *Geopotes VII* in action with the spray, and below, a close-up of the spraying mechanism.

[24]

On 14 December 1969, *Marpessa*, from Shell's Antilles fleet, exploded during her maiden round voyage. Two Chinese Petty Officers were killed and the 207,000 dwt vessel, over 1,000 feet (328 metres) long, became the largest ship, civil or military, ever to sink. Fifteen days later her sister *Mactra* also blew up and killed two crew-members, but did not sink, and the following day a third VLCC – not a Shell ship, but of very similar design – exploded. All had been undertaking tank-cleaning operations at the time, and the ensuing two-year enquiry established the most probable cause as sparks accidentally generated in the tank's gaseous atmosphere. The solution was to fill the tank with inert gas during cleaning.

[26]

[25]

214

Shell International Petroleum Company Limited

Telex: LONDON 25781
Telegraphic Address:
Overseas SHELL LONDON SE1
Inland SHELL LONDON TELEX
All codes used

SHELL CENTRE · LONDON SE1 Telephone WATERLOO 1234

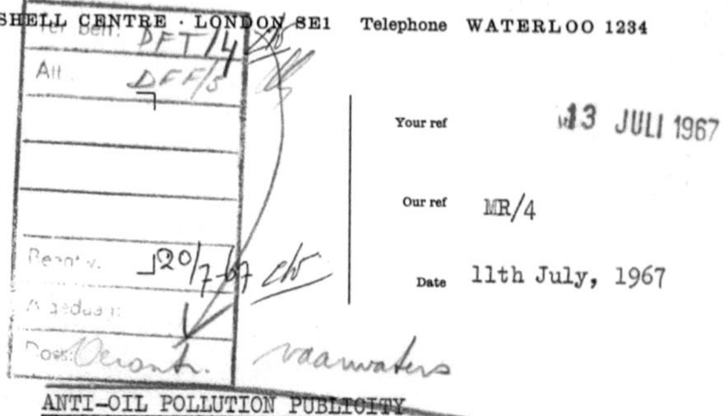

Shell Tankers N.V.
ROTTERDAM

Your ref 13 JULI 1967

Our ref MR/4

Date 11th July, 1967

Dear Sirs,

ANTI-OIL POLLUTION PUBLICITY

LETTER STICKERS

Attached is a sample sheet of letter stickers, which, over the next few months, are to be attached to outgoing mail from Marine and S.I.M.

Whilst the legend of the sticker is aimed at those in a position to "stop pollution", i.e. shipowners, ship's staff, etc., it is the intention to also attach these stickers widely to any mail addressed to firms with a marine interest. The only proviso is that we should avoid dissemination to private persons and to the travel industry.

75,000 stickers are available, and Mr. Kirby has asked that S.T.U.K. should assist in the dissemination via their outgoing mail. We would now like to enquire whether, despite the legend being in English, you feel it would be appropriate and useful for S.T.N.V. to assist also. As you know, some 77% of world-wide crude tanker movements operate load-on-top, and any effort to bring the ramaining 23% into the fold is worth making.

The sticker can be attached to the back of an envelope (in U.K. the Post Office forbid the attachment to the adressed side of the envelope), but here in Marine, the sticker will be attached to the letter itself. This will ensure the sticker arrives on desks rather than Registry waste bins, and will avoid delay, which may otherwise arise, in the despatch of mail from Registries.

If you feel you would like to join in the dissemination of this publicity, we would suggest you having, say, 800 sheets (20,000 stickers) which we estimate would be sufficient for about 6 months. It is not proposed to continue the scheme beyond such a period as otherwise it would begin to lose its point.

Yours truly,
For: SHELL INTERNATIONAL PETROLEUM COMPANY LIMITED

M. P. Holdsworth

[27]

As detailed in the letter (left), these stickers were part of a Shell campaign encouraging all tanker owners to use Load-on-Top, which soon became mandatory. A single use could recover as much as 800 tons of oil previously wasted, and each year the system prevented several hundreds of thousands of tons of oil from being discharged into the sea. But it was not a total solution, and was superseded in the 1970s by Crude Oil Washing, enabling still more recovery of residual cargo.

[28]

In 1959 Pernis started using a mobile laboratory to measure air pollution by hydrocarbons in residential areas.

However, when an environmental crisis emerged around the Pernis petrochemical complex, Shell did not rely on the industry platform alone.

As the Group's Pernis plant expanded, public complaints about the poor air quality became ever louder. In 1956 Shell started taking air samples at the complex and three years later the company bought a mobile laboratory for the direct investigation of complaints from residents living nearby. In collaboration with the Dutch meteorological institute KNMI and the Dutch Royal Air Force, the Group also conducted research into air currents and into the general atmospheric conditions above Pernis. Meanwhile the petrochemical industry in the Rotterdam area grew very rapidly. Attracting this sector was a key plank in the Dutch government's industrialization programme. The companies which located in the area included BP, Gulf, Caltex, and Jersey Standard. Soon the chorus of complaints about the increasing atmospheric pollution rose to a fortissimo.[81] The core issue was the plants' massive emissions of sulphur dioxide and nitrogen oxide, which in specific weather conditions formed a thick smog in the lowest strata of the atmosphere, sometimes as low as 100 metres. Occasional incidents, such as a fire in a fertilizer factory in 1963 and two serious accidents at Shell's Pernis works five years later, reinforced the growing sense of insecurity amongst the population. The chemical industry reacted to the concern by setting up a typical joint body, the 'Stichting Europoort Botlek Belangen', which was chaired by the general manager of Shell Pernis. The oil industry used CONCAWE as a platform for its interests. These forums did not stop company representatives from blaming each other for the pollution rather than accepting collective responsibility.

Initially the industry defended itself in typical fashion, launching a publicity offensive which emphasized the economic benefits of the chemical industry and the resilience of the Dutch climate in withstanding pollution of air and water, and having recourse to palliative measures. In 1964 Shell decided to combine some twenty five smoke stacks into a single stack, 213 metres (700 feet) high, designed to propel offensive gases and particles high up into the atmosphere, beyond the layers in which smog usually formed. This 'practical but costly solution' cost £1.7 million

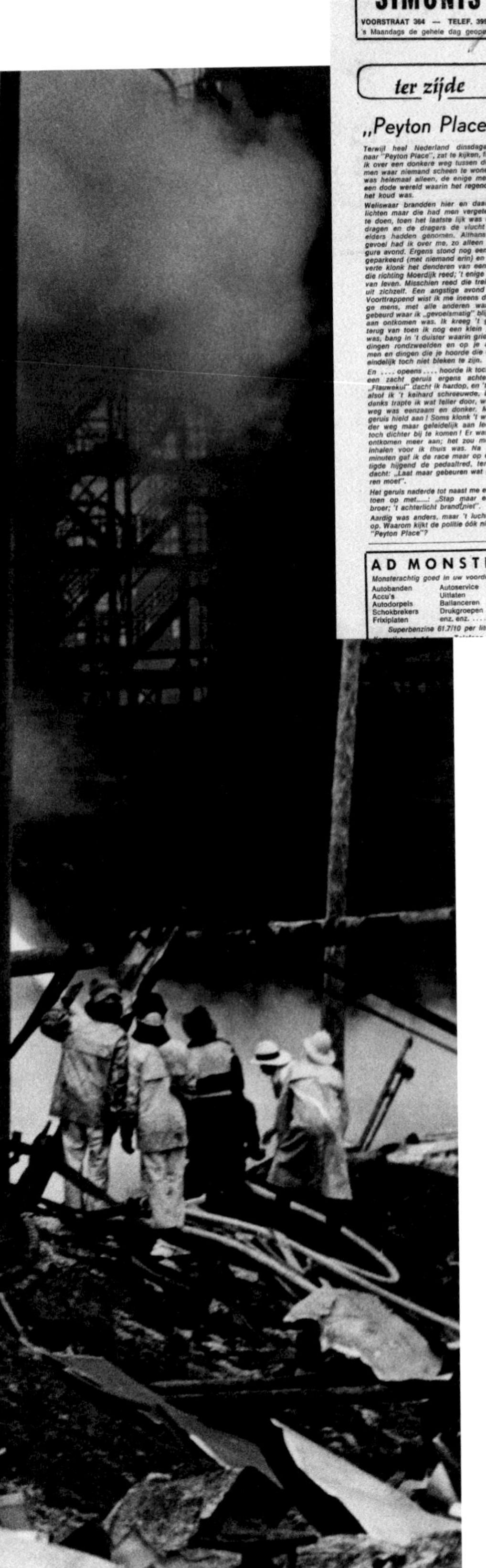

Grandioze Elpee van
HEINTJE
geen 15,— maar
7,90

SIMONIS
VOORSTRAAT 364 — TELEF. 39952
's Maandags de gehele dag geopend

ter zijde

„Peyton Place"

Terwijl heel Nederland dinsdagavond naar "Peyton Place", zat te kijken, fietste ik over een donkere weg tussen de bomen waar niemand scheen te wonen. Ik was helemaal alleen, de enige mens in een dode wereld waarin het regende en het koud was.

Weliswaar brandden hier en daar wat lichten maar die had men vergeten uit te doen, toen het laatste lijk was uitgedragen en de dragers de vlucht naar elders hadden genomen. Althans zo'n gevoel had ik over me, zo alleen in de gure avond. Ergens stond nog een auto geparkeerd (met niemand erin) en in de verte klonk het denderen van een trein die richting Moerdijk reed; 't enige teken van leven. Misschien reed die trein wel uit zichzelf. Een angstige avond dus! Voorttrappend wist ik me ineens de enige mens, met alle anderen was iets gebeurd waar ik „gevoelsmatig" blijkbaar aan ontkomen was. Ik kreeg 't gevoel terug van toen ik nog een klein jochie was, bang in 't duister waarin griezelige dingen rondzweelden en op je afkwamen en dingen die je hoorde die er uiteindelijk toch niet bleken te zijn.

En .... opeens .... hoorde ik toch iets: een zacht geruis ergens achter me. „Flauwekul" dacht ik hardop, en 't klonk alsof ik 't keihard schreeuwde. Desondanks trapte ik wat feller door, want de weg was eenzaam en donker. Maar 't geruis hield aan! Soms klonk 't wat verder weg maar geleidelijk aan leek het toch dichter bij te komen! Er was geen ontkomen meer aan; het zou me toch inhalen voor ik thuis was. Na enkele minuten gaf ik de race maar op en matigde hijgend de pedaaltred, terwijl ik dacht: „Laat maar gebeuren wat gebeuren moet".

Het geruis naderde tot naast me en hield toen op met......: „Stap maar eens af broer; 't achterlicht brandt niet".

Aardig was anders, maar 't luchtte wel op. Waarom kijkt de politie óók niet naar "Peyton Place"?

AD MONSTER
Monsterachtig goed in uw voordeel

Autobanden	Autoservice
Accu's	Uitlaten
Autodorpels	Ballanceren
Schokbrekers	Drukgroepen
Frixiplaten	enz. enz. ....

Superbenzine 61.7/10 per liter

AKTIEGROEP MILIEU-HYGIENE VOLKSUNIVERSITEIT:

Leg „Industrieschap Moerdijk" aan banden!

Dordrecht — De Moerdijkvestiging van Shell is een voldongen feit. Mensen die denken dat de bouw van dit gigantische project nog op de een of andere manier te stuiten is, hebben het bij het verkeerde eind. Wat nog wel tot de mogelijkheden behoort is het drastisch aan banden leggen van de sterke luchtverontreiniging die deze chemische industrie stellig zal gaan veroorzaken.

Een uit 12 mensen bestaande Dordtse aktiegroep, gevormd uit deelnemers aan de cursus „Milieu-hygiene" op de Dordtse Volksuniversiteit, heeft het gouden Shell-kalf bij de horens gevat en is er nu stevig mee aan het worstelen. Wat er tot nu toe uit voortgekomen is, schrijft ons mevr. T. Veenema-Koning, een der 12 leden van bovengenoemde aktiegroep. Mevrouw Veenema is doktersvrouw en tevens moeder van vier kinderen; beslist dus niet iemand die zich uit verveling met dit soort zaken is gaan bemoeien. Haar relaas, dat eigenlijk in onze rubriek „Merwesteyn-lezers schrijven" thuishoort, vonden wij dermate interessant dat wij er onze voorpagina voor ter beschikking stellen. Wij hopen dat haar oproep, om massaal op 25 februari a.s. naar de Shell-hearing in „Americain" te gaan, niet vergeefs zal zijn. Men heeft doorgaans de mond vol over „inspraak" en „medezeggenschap" maar de praktijk heeft uitgewezen dat van deze geboden mogelijkheden tot op heden maar bitter weinig gebruik wordt gemaakt.

PLANOLOGISCHE DIENST

We hebben hier in Nederland een prachtige Planologische Dienst, die plannen maakt voor de eerstkomende vijf jaren (op langere termijn is de ontwikkeling op allerlei gebied niet te overzien). Deze plannen zijn schetsen voor uitbreiding van steden, recreatiemogelijkheden in het groot: openluchtrecreatie buiten de bebouwing, en in het klein: een parkje of speeltuin, clubhuis, enz. op minder dan één km van elk huis.

Er worden groenstroken gepland tussen agglomeraties van steden en/of industrieën, vaak aansluitend aan de grote recreatieprojecten. De omvang van verschillende gebieden wordt berekend, b.v. watersportcen-

bracht; waar voorzieningen ontbreken worden ze geprojecteerd, waar ze al aanwezig zijn, indien nodig, uitgebreid. Vaak worden diverse plannen opgesteld, waaruit de minister een keus kan maken.

Zo is het ook met Zuid-Holland gegaan: voor de openluchtrecreatie zijn er plassen en meren, waar men kan watersporten; bossen hebben we niet veel, maar die komen volgens planning: poldergebieden zullen nog verder geschikt worden gemaakt voor de opvang van duizenden mensen. Het staat allemaal prachtig op papier!

GROENSTROOK

Het is alleen jammer dat de regering niet in het minst gebonden is aan deze plannen. Zonder overleg met de Planologische Dienst en zonder medeweten van de volksvertegenwoordiging werd een industrievestiging toegelaten in het gebied bij Moerdijk, waar toevallig juist een groenstrook was gepland tussen de stedelijke zones, waarin aan de noordkant o.a. Rotterdam en Dordrecht liggen en in het zuiden de Brabantse stedenrij met Bergen op Zoom, Roosendaal, Breda, enz. tot Helmond toe. Gebruikelijk was, dat de omliggende gemeenten tijdig in kennis werden gesteld. Ze konden dan studie maken van de gevolgen van de vestiging met betrekking tot de werkgelegenheid, woningbouw, mogelijkheid tot recreatie zowel binnen als buiten, eventueel vervuiling van bodem, lucht en water, behoud van natuurgebieden, bescherming van spaarbekkens, enz.

Deze situatie is echter veranderd. Er zijn nu CONVENANTEN afgesloten tussen de centrale overheid, de provincie Noord-Brabant en de gemeenten waar de industrie gevestigd zal worden. Deze gemeenten: Zevenbergen, Klundert en Lage-Zwaluwe hebben zich verenigd tot het Industrieschap Moerdijk.

meenschappelijk de voorwaarden op, waaraan de te vestigen industrie zal moeten voldoen. Deze voorwaarden zullen „zoveel mogelijk" op de hinderwet en andere milieuhygiënenische wetten moeten worden gebaseerd en de ondernemingen moe-

de vestiging plaats

FALL-OUT-GE

Dat Dordrecht, da
waaiende zuidw.-
in het fall-out gebi
trieschap Moerdijk,

Stichting voor Sociaal Wijk-opbouwwerk Dordrecht-Zuid
Secretriaat:
Van Gendstraat 29
Tel.: 41557

Het wijkopbouworgaan Dordrecht-Zuid, dat de spreekbuis kan zijn van alle wijkbewoners zet een handtekeningen-aktie op touw. Het is de bedoeling dat de lijsten met handtekeningen aangeboden worden aan de Gemeenteraad van Dordrecht, die op 23 februari vergadert. Neemt U een blocnote-vel en schrijf bovenaan de volgende tekst:

> **Wij, bewoners van Dordrecht, zijn tegen verdere verontreiniging van lucht en water. Ondergetekenden vragen ons gemeentebestuur dringend om i.v.m. de vestiging van bedrijven aan de Moerdijk (o.a. Shell), om voor de strengste eisen en een voor zijn taak berekend controleapparaat**

DE STEM

Principiële discussie Shell-Moerdijk

(Van onze correspondent)

BREDA — „Het is wel een onverkoopbaar standpunt, maar ik persoonlijk (niet het aktiecommité) vind dat Shell-Chemie niet in Moerdijk, maar ook niet in België of ergens anders ter wereld moet komen. Een zekere mate van welstand is welkom, maar de welvaart is niet de enige faktor van het menselijk geluk", aldus gisteravond de sekretaris van het actiecommite Shell-Moerdijk tijdens een discussieavond in Breda georganiseerd door de Linkse Konsentratie.

[30]

On 20 January 1968 Shell suffered one of its worst peacetime disasters when a hot oil and water emulsion in slop tanks at Pernis reacted with volatile hydrocarbon slop, causing an extremely violent explosion. The subsequent fire covered 30 acres (over 12 hectares), with off-site damage occurring as far as 9.5 miles (15 kilometres) away. Two men were killed, nine hospitalized, and 76 slightly injured. Early estimates of the financial loss were put at around $100 million. Together with contemporary fears about acid rain and other environmental issues, the explosion no doubt contributed to public opposition (above) to Shell's large expansion of its chemicals plant at Moerdijk in the late 1960s and early 1970s.

(16 million guilders) to build, yet its value was questionable. Commercially it had the drawback of disproportionate expense – a 200-metre stack cost five times as much as one of 100 metres – and while it succeeded in alleviating the ground-level problems for the immediately surrounding area, this was only achieved by lifting the sulphur dioxide further upward into the atmosphere, from which it would come down again further afield.[82] After the stack had been completed in 1968, evidence soon emerged showing that this was hardly environmentally ideal, with claims from market gardeners at some distance from Pernis and even from people as far away as Sweden that particles washed down on them from the high atmosphere. Group research suggested that 'There might be truth in this, and so control of the gases emitted would be a better long-term solution than high stacks. Unfortunately the removal of sulphur from crude oil or residue was extremely expensive.'[83]

The Pernis stack stands as a symbol of well-intentioned 1960s environmentalism: an attempt to combat pollution by pushing the

VERMEUL
SCHOORSTEEN
BOUWBEDRIJF

Construction of the gigantic chimney stack at Pernis took four years. Its base was so large that it straddled a divided highway with footpaths on either side, and although the stack reduced local air pollution it attracted opposition, expressed in graffiti (above). The Dutch expression 'We gaan de pijp uit', literally meaning 'We're going out of the chimney', is a way of saying 'We're going to die.'

noxious substances out of sight, rather than tackling them at source. The attitude was rooted in a conception of pollution solely in terms of a nuisance to man; a wider and more integrated conception, which looked at pollution in terms of the spread, accumulation, and transformation of chemical substances in nature, would emerge only during the 1970s. Consequently, Shell considered that, with the stack, it had done enough to alleviate the problem. Nor did the industry show itself willing to cooperate with further pollution abatement measures. During 1967-69 the 'centrale meld en regelkamer', a local government monitoring organization tied to a network of smog warning detectors, was set up to monitor public complaints about pollution. Initially the industry rejected any responsibility, but protracted negotiations led to the acceptance of a voluntary scheme to restrict emissions in times of high smog formation.[84] This time the companies failed to forestall formal regulation; in 1972 the Dutch parliament passed a new air pollution law which gave the government powers to introduce a compulsory scheme of emission restrictions.

[31]

Combating sulphur emissions Sulphur emissions were not just a problem for the environment surrounding the Group's own installations, but also for the public acceptance of its largest product by volume, fuel oil. Both Venezuelan and Kuwaiti crude had a high sulphur content, which rendered Shell's competitive position difficult because of the costs to remove it. In 1956 Jersey Standard, which was better placed than Shell with regard to low sulphur crudes, started selling a low sulphur product in the UK. By segregation of crude intakes Shell could make available a certain amount of low sulphur product, but to produce this in large quantities it needed a cheap desulphurization process. The Shell Trickle Hydrodesulphurization process, introduced in 1955, provided some relief, and refineries were successively equipped for it. The units entailed a higher capital cost, but that was repaid in two years through lower operating costs. Better still, the improved units brought an extra benefit: 'from the public relations angle this new combination unit has the advantage that only "sweet" products are obtained and stored. Consequently, no pertinent air pollution problems exist and no disposal problems of waste chemicals have to be solved'.[85]

In 1958 Gulf's laboratories made a technical breakthrough that seemed to offer an economically acceptable process of reducing the sulphur content of residual oils, and to which the Group had royalty-free rights.[86] From the middle 1960s a new serious challenge arose for the Group and the industry, as lawmakers on the east coast of the United States started to insist on ever-decreasing levels of sulphur, to the point of its nearly complete extraction from fuel oil. In 1966, even current levels of extraction (reducing the sulphur content to 2.2 per cent by weight) were extremely expensive, at around 70 cents a barrel. Going far beyond that, a new intermediary target of 1 per cent was to be

[33]

London had long been notorious for its extremely thick 'pea-soup' fogs, and the main picture is a 1953 London scene in daytime. Created by industrial emissions and the smoke from thousands of domestic coal fires, the phenomenon became known as smog (smoke and fog combined) and through to the 1960s smog pollution in developed urbanized nations reached levels that would seem almost unbelievable in the early 21st century. In Los Angeles the city's air pollution officer stated that about 85% of the smog came from automobiles – a claim that was easy to believe when seeing the result (above) of starting just one car there. Below right, the Manhattan waterfront in 1946 was also shrouded in smog. Such conditions caused many deaths from bronchial disease, and although in Britain the 1956 Clean Air Act did much to help, even in 1962 (below left) people would wrap scarves around their faces to avoid breathing the smog.

[34]

[35]

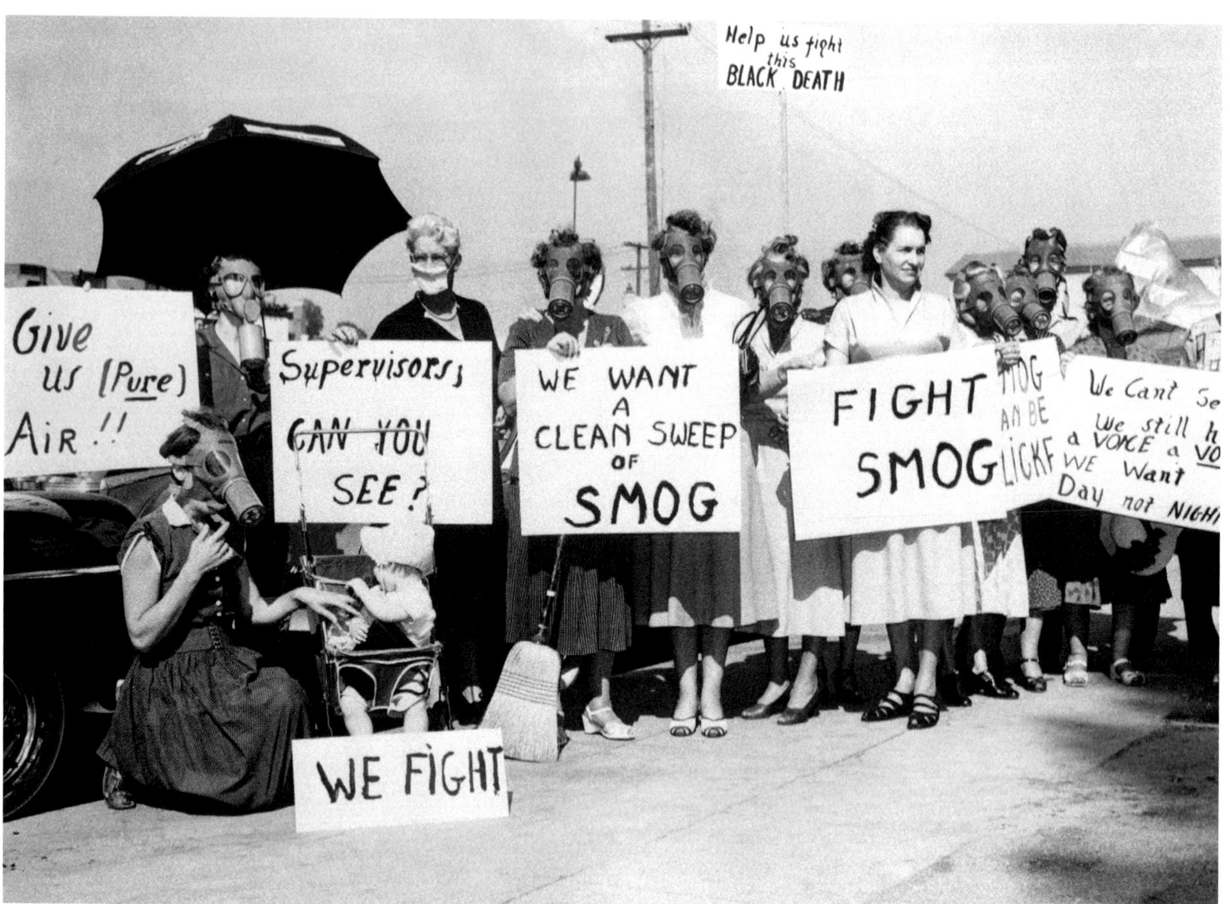

[36]

On 20 October 1954, after a smog lasting for 15 consecutive days, housewives in Pasadena, Texas, had had enough and held a protest against the conditions that blighted their houses, their laundry and their health.

achieved by 1971, with the possibility of still further enforced reduction thereafter. Experiments with deep desulphurization methods resulted in expensive processes without reliable results.[87] Because Group research could not find good solutions, it was suggested in Conference that the basis of investigation could be broadened by involving universities and other research bodies, and by offering a substantial prize for the discovery of an economic process.[88] The suggestion was short-lived: after just two months, following consultation with Monty Spaght and the Group's Research Coordinator Lord Rothschild, who had succeeded Gershinowitz, it was abandoned on the basis of expense and difficulty of administration. Spaght in particular felt the problem would only yield to a sustained and protracted Group effort, since 'the nature of the investigation made it unlikely that an intuitive flash of understanding by an individual could provide the answer'.[89]

However, by 1967 the situation was becoming acute, because after very bad smog conditions in New York City the power company Con Edison was being obliged to switch to 1 per cent sulphur almost immediately, instead of in 1971. The Group assessed its own supply position for low-sulphur crudes as about average, but was cautious about rushing into long-term sales contracts when better prices might be obtained in Europe if equally stringent anti-pollution measures were introduced there, as seemed possible: 'The spearhead of the attack on pollution is in the US (...) it is only a question of time before other countries become equally disturbed.'[90] Reaching the 1 per cent limit could be done by importing crudes from Nigeria and Gabon into Curaçao and blending them with Venezuelan residual fuels in Cardón, but if a recommendation from the US Department of Health, Education and Welfare for an ultimate level of 0.3 per cent was accepted, Venezuelan residuals could not be used at all. No one could yet say

'The spearhead of the attack on pollution is in the US (...) it is only a question of time before other countries become equally disturbed.'

if those levels would actually be imposed, but the Group mood was both glum and realistic, with the ring of an organization that had become accustomed to criticism it regarded as unfair: 'it seemed likely that the pollution question would eventually be pushed to the limits of technical possibility and feasible cost (...) and since desulphurisation of coal was at this juncture out of the question, it could be expected that pressure would be concentrated on the oil industry, and we should be prepared for this'.[91]

Group research into desulphurization continued, and there was some discussion as to whether it might be better to build more desulphurization facilities or buy more 'sweet' low-sulphur crude.[92] But then came a quite rapid and unexpected change, and in Conference, early in 1969,

'The question was asked whether the Group was justified in apparently reversing its judgement as to the seriousness of impending low sulphur fuel requirements. It was agreed that this problem no longer loomed ahead as urgently as last year: this was due partly to our estimate as to the physical availability of the requisite material to us, and partly to the recent developments in the US where the authorities' requirements were proving less stringent than feared – this represented not a temporary respite, but success in achieving a better understanding of the technical and supply problems involved (e.g. the expenses which would have to be incurred in respect of crudes low in sulphur content but requiring special processing or blending).'[93]

This 'success in achieving a better understanding' probably indicates that behind the scenes, Group representatives had once more been lobbying the relevant authorities and successfully explaining the industry perspective. For Shell it meant that this period ended with at least the chance of more realistic and achievable sulphur legislation.

Leaded, unleaded – or something else? In 1964, the Group could still claim publicly that 'Petroleum gases and oil when burnt efficiently are basically clean fuels.'[94] But was this so with lead? During 1961-3 the American Public Health Service published a wide-ranging survey of leaded gasoline and air pollution in three cities, which established that the inhabitants had significantly higher concentrations of lead in their blood. This report opened a sometimes vehement debate about lead pollution in the US.[95] As so often, the scientific evidence long remained open to interpretation. Whether lead-based additives in gasoline were toxic when burnt was a question which kept Shell experts divided for some time.[96] They did agree that lead actually inhibited the efficient burning of fuel, and as such caused an increase in the main pollutants emitted from the exhaust, namely partially burnt and unburnt hydrocarbons, carbon monoxide and nitrogen oxides.[97] They also agreed that market preferences were moving against such additives. As we have seen above, the Group anticipated the impending requirements for premium gasoline without lead additives and, by 1968, possessed two different processes for making it.

At the same time Shell realized that, for optimum performance, lead-free gasoline required new types of engines, so in 1968 the company signed an agreement with British Leyland Motor Corporation to collaborate in an intensive research programme aimed at reducing exhaust emissions from motor vehicles to levels that they envisaged would be mandatory in the 1970s.[98] The Group also looked at a variety of sometimes unusual alternatives to the existing internal combustion engine running on gasoline. LNG, which had good anti-pollution characteristics, could be a potentially commercial vehicle fuel, so Shell devised an experimental programme involving both engine and distribution equipment, for experimental marketing in Osaka in 1973.[99] Other possibilities

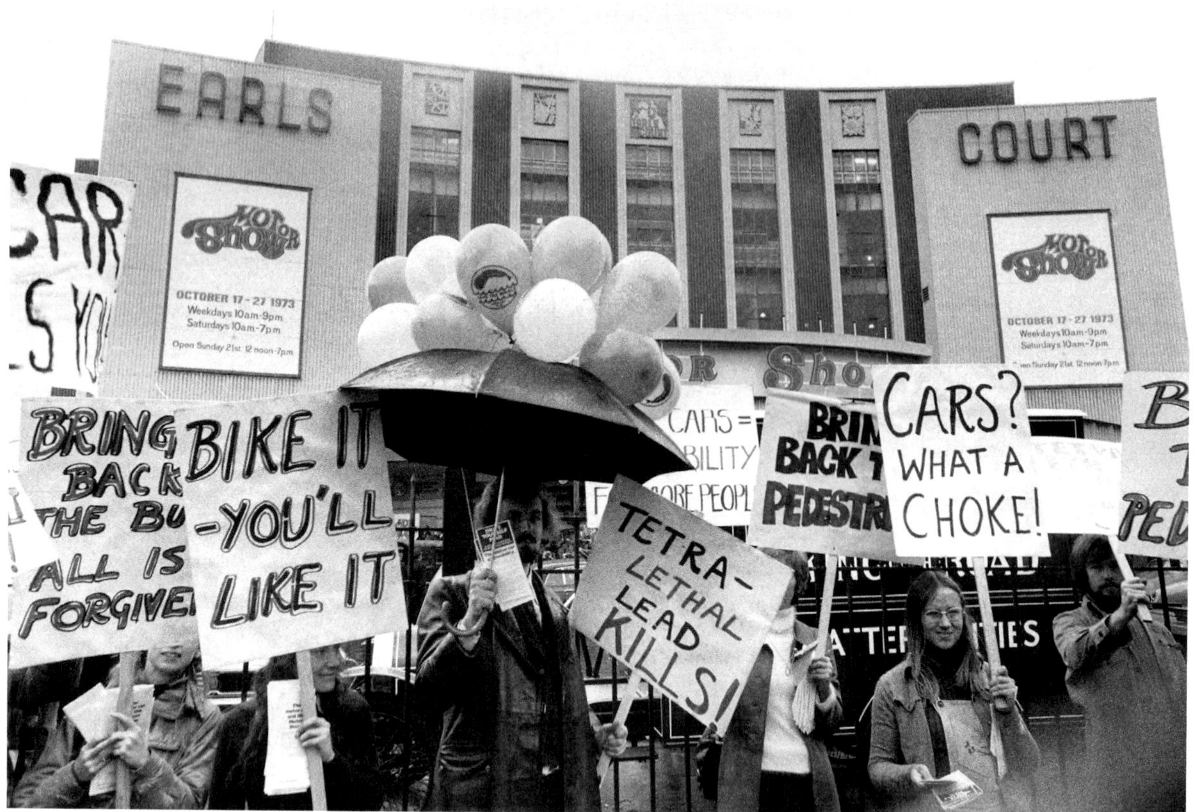

[37]

surveyed included electric cars, fuel-cell cars, and hybrid cars combining a small gasoline engine with an electric battery and motor. Shell did not see any likelihood of electric cars becoming real competitors to conventional cars (a view shared by Jersey Standard), but thought the UK might introduce a discriminatory tax, and so it was experimenting with fuel-cells in a modified DAF. The fuel-cell generator was a highly advanced type but it ran on hydrazine, which even the Group admitted was hopelessly uneconomical. The research was acknowledged to be defensive in character, but knowing more about the fuels for fuel-cells was deemed to be beneficial, and the idea was that if hydrazine proved interesting the Group would try to replace it with methanol. As for hybrid cars, General Motors, General Electric and Westinghouse were all considerably interested, but after an intensive study of the vehicles Shell's Thornton laboratory came down strongly against the Group becoming involved.[100]

Outwardly, the Group reacted diffidently to concerns over the effects of leaded gasoline and worked to downplay the seriousness of the issue. When in 1969 such concerns surfaced in Sweden, Marketing commented that 'efforts are being made to avoid precipitate action and encourage careful study'.[101] That same year, the Group's environmental survey concluded that leaded gasoline was 'an acute but localised problem.'[102] The Group's main concern with gasoline was not whether the product should be leaded or unleaded. Making a completely or nearly unleaded product was technically feasible, and environmentally speaking Shell did not believe there was a great argument one way or the other. However, without much more cooperation from the motor manufacturing industry, the consumers – automobile drivers – would find an unacceptable loss of performance in their vehicles: 'we believe many of the problems associated with exhaust gas pollution could be greatly ameliorated, if not eliminated, if the motor manufacturers could be persuaded to alter engine designs. In

As the environmental movement gathered strength, supporters of Friends of the Earth staged a protest against the pollution caused by leaded gasoline (left) and at the University of Minnesota (right) members of Students for Environmental Defense staged a mock funeral, burying an internal combustion engine.

[38]

principle, there are no insuperable difficulties in producing a "smog-free motor car". There are, however, serious technological problems in making such a car conform with the rigorous requirements of the motoring public. In addition, the cost of modifying engines is significant and, at any rate in the UK, will probably be resisted by the manufacturers.'[103]

Consequently, launching lead-free gasoline would not make sense if the motor manufacturers did not move at the same time: the Group advanced slowly because the market was not ready for it. The public concern with car exhaust emissions translated first in oil companies racing each other to introduce gasolines with super detergents, which removed deposited carbon compounds on engines and enabled a more complete combustion of fuel.[104] This brought back memories of the late 1950s: 'We were conscious of the danger that competitive and generalised claims for detergent additives would destroy credibility, as had happened in the marketing of washing powders.'[105]

In 1970 the moment appeared to have come, with impending legislation in the US stimulating the automobile industry there to start making engines suitable for use with unleaded gasoline.[106] Shell Oil decided to take the plunge and launched the unleaded gasoline 'Shell of the Future' on the US market. Unfortunately the product was ahead of its time. Despite heavy expenditure on additional marketing facilities and on advertising, Shell of the Future flopped, either because customers would not pay a premium, or because they did not believe that the gasoline offered a similar performance to leaded alternatives. Shell Oil rushed to replace it with a low-leaded alternative.

Labelling dangers Public concerns about the health hazards associated with hydrocarbons and allied materials also had their impact on other products. Research into health hazards went on in many centres under several authoritative bodies, and in Shell there was already a growing feeling that this was a problem to be addressed on an inter-company basis by the industry, perhaps through the Institute of Petroleum Research Group, rather than by Shell alone.[107] Many companies were affected by another part of the spate of US environmental legislation in the early 1960s, namely the Hazardous Substances Labelling Act, whose regulations took effect on 1 February 1962. The only instance of mandatory warning labels previously known to Shell International Petroleum Company (SIPC) was in Germany, where lubricants such as AeroShell Fluid 4, whose specification included the highly toxic tritolyl phosphate (TTP), had to be labelled 'Poison' with a skull-and-crossbones emblem.[108] Shell Oil advised that in the US lubricants containers had to be labelled with warning notices about health hazards only if the materials were likely to be used in the home, but SIPC reported receiving a sample of an American motor oil used for small marine outboard engines and marked 'Caution: Contains organic metal compound. Avoid prolonged skin contact. If spilled on skin wash off with soap and water. Harmful if swallowed.' Remarking that 'This is the first time we have seen a note of this kind on an engine oil container', the SIPC view was that 'we are generally against special labelling about health hazards unless there are good reasons for recommending that this be done in the case of a particular product.'[109]

However, the spread of product liability litigation forced companies to protect themselves from possible lawsuits by placing increasingly detailed analytical labels on their products. The Group learned this might be a good idea after a very lively party at Norske Shell's Værnes airfield on New Year's Eve 1962. In this macabre episode, some of the party-goers – perhaps having run out of other alcohols – cracked open the stocks of AeroShell Compound 7 (for de-icing propellers and wings) and drank the fluid, 'with fatal results for one of the participants'.[110] Pending new regulations the Norwegian Ministry of Health suggested that any Shell oils containing additives not covered by existing rules should bear the warning 'Caution. The oil contains toxic additives. Must not be drunk.' This was not quite as superfluous as it might appear,

[39]

Human folly could lead to fatalities. The two tins at left look nearly identical, but the far left is older and carries no health warning. From the early 1960s many Shell products (with two more examples at right) began to carry such warnings.

because even if there were no repetition of the airfield episode, a related product called AeroShell Fluid 12 was used in dentists' drills.[111] Even so, Norske Shell was worried by the wording of the suggested warning, which went on to say 'Avoid contact with the skin.' The company's comment to SIPC in London was that

'The use of such a text is obviously impractical on account of the inevitable sales reactions (..) We are especially concerned about the implications with respect to industrial oils, e.g. Tellus, the reference to contact with the skin more or less precluding the use of such oils in industry.'[112]

A colleague from SIPC was reassuring about the precise example of Tellus oil, which held no toxic additives, and in other instances of nervous inquiries, the point was made that sometimes risks had to be accepted as comparable to the natural world:

'it has to be kept in mind that these carcinogens are ubiquitous and present in smoked food, forest soil and other "natural" surroundings. Moreover, it is known that only part of the multitude of polynuclear aromatic compounds and their derivatives are carcinogenic, whereas many are definitely not.'[113]

The double-edged drins As discussed in Chapter 5, the Group possessed exclusive rights to the 'drin family of chlorinated hydrocarbon pesticides, which were by far the most profitable products from its range of chemicals. Prominent members of the family were aldrin, dieldrin, endrin, phosdrin, and bidrin, used around the world to combat a variety of pests after amazing results achieved in Iran (1951) and in California (1952) had shown their effectiveness. The chemicals were in great demand from gardeners, cereal farmers and fruit farmers, a valuable, varied, and widely extended purchasing constituency. From time immemorial all such producers had fought an unceasing battle against pests, weeds and fungi, but by the mid-1950s petrochemicals appeared ready to put an end to that, as was shown by a complete list of Shell garden products. Based on aldrin, Ant Doom was marketed as 'the best and safest remedy for all types of ants', while Aldrin Soil Pest Killer gave 'all-purpose soil pest control', and was safe and non-tainting. Shell Liquid Derris was rated 'one of the best liquid controls for black fly, green fly, etc,' Derris Dust controlled the majority of horticultural pests, and Shell Weedkill for Lawns was 'most successful (...) water your weeds away'.[114] Other apparent blessings for the gardener included Dieldrin Garden Pest Killer ('non-tainting (...) more persistent than any other preparation'); Pillakiller (a 'positive killer' of caterpillars, apple blossom weevil, etc); 5% DDT Dust, for pea and bean weevil and similar bugs; Sulficide for fungal diseases, with 'perfect mildew control'; and Coppicide as the 'finest preventative for mildew'. Shell Tomato-Set helped in the setting and full development of tomatoes; Shelltox with Dieldrin had 'exceptional residual properties'; and with just one spray of Universal DNC Fruit Tree Wash, the fruit farmer could kill the five major pests that might overwinter on his trees.[115] Yet over time these chemicals turned out to have very serious

drawbacks. Their inherent toxicity made them difficult to handle, necessitating strict guidelines for handling and storage.[116] In the field, the handling instructions were often rendered useless by the indiscriminate spraying of large areas from the air with blithe disregard for neighbouring people, animals, or crops. In the United States, the Department of Agriculture sponsored such programmes in a drive to eradicate particular pests. Mixed with light oils, the pesticide compounds could easily be inhaled and stuck to the skin, through which they could enter the blood.[117] For the same reason produce treated had to be thoroughly cleansed before food preparation. Moreover, the 'drins were not very specific and killed other species besides those targeted. Worse, pests repeatedly exposed to spraying became resistant. Chlorinated hydrocarbons also proved to be persistent, leaving residues in the soil which washed into rivers and seeped into the groundwater. They travelled through the animal food chain via contaminated crops and pests eaten by rodents or birds, showing up in raptors laying eggs with abnormally thin shells or dying from doses built up over time. The 'drins were also said to be carcinogenic to humans.

Some of the negative effects from drins were recognized early on. Bataafsche's Amsterdam laboratory had already

[42]

Insecticides and pesticides derived from petrochemicals were so effective that for many years after their invention they were used almost indiscriminately, with little or no thought for possible side-effects. Despite the presence of a worker on the ground, a farmer's fields are sprayed from the air (1961), and a man sprays DTT against mosquitoes in advance of a conference in Bangkok (1955). Though not generally toxic to humans, the chemical's persistence and toxicity on other species led to a ban on its use for agricultural purposes firstly in the US (1972) and subsequently world-wide. Despite increased resistance to it on the part of mosquitoes, practical experience shows it to be valuable in the fight against malaria, but its use remains highly controversial.

experimented with chlorinated hydrocarbons before the Second World War. In 1947 the biologist C. J. Briejèr, who had done this work, was appointed head of the Dutch government's Department of Plant Pathology, and in that capacity he began to sound the alarm on the effects of pesticides. In 1949 he published a fundamental article criticizing their ongoing indiscriminate use, following that with others highlighting specific effects, including the increasing resistance of insects.[118] Six years later the British journal *Chemical Age* castigated experiments on cattle carried out by Shell with 'drins as possible systemic preventatives of infestation by insects. Oral doses had killed yearling Hereford cattle in less than five hours, but with subcutaneous injection only one animal died among many. The periodical considered that this was going too far: there was no point if residues in the carcass meant the meat was unfit for consumption, and there was a clear sense of disgust: 'we cannot help feeling that on the whole this is a chemical venture that interferes too grossly with natural complexities and had better be abandoned'.[119] It was indeed abandoned, but the criticism was notable in coming from other professionals. Gradually evidence about the wider effects of pesticides reached the public domain. In 1957 two British veterinary doctors established a clear link between bird deaths and the ingestion of dieldrin-dressed grain, which helped to explain the unusual numbers of dead birds in the British countryside.[120] Other effects, particularly the persistence in soil and water and the spread through wildlife, took time before being noted and scientifically proven, partly because of the difficulty in measuring the sometimes infinitely small traces and in establishing the causal nexus between disparate phenomena. A year after setting up its toxicologial unit at Sittingbourne in 1958, the Group started fundamental research there into the mechanisms of resistance to pesticides and into compounds, 'preferably with low mammalian toxicity', which would not show such resistance.[121]

In response to the emerging problems, scientists strove to find safer variants of the 'drins. Proponex was marketed as 'safe and easy to handle', and in January 1958, improved applicators were introduced for Shell garden products including Ant Doom, now available as a dry powder in 'Puffit' packs. A few months later

Netelex was brought in as a non-toxic, non-corrosive hormone weedkiller for nettles, brambles and briars, along with dry Slug Doom and liquid Slug Kill.[122] Phosdrin, claimed to last only three days in the crop, was introduced in 1957-58 as 'a new wonder insecticide [which] should silence the concern about the residues of toxic pesticides in food crops', and was initially hailed as 'yet another example of expenditure on research and development paying handsome dividends'.[123] Its sales at first were good, but by 1961 the figures were already falling 'seriously, principally because of its high cost and high toxicity.'[124] Lastly in this run of new products, early in 1962 Telodrin was brought to the market as the most efficient and versatile insecticide yet discovered, after more than 400 field trials in 65 countries. Acting on the insects' nervous system on contact, it was claimed not to affect the flavour of crops, damage plant tissues, or pose risk to human beings, provided of course that the recommended safety precautions were taken.[125]

In June of that same year, the *New Yorker* magazine began its pre-publication serialization of *Silent Spring* by Rachel Carson. Trained as a marine biologist, Carson had gradually developed an interest in the environmental damage done by pesticides, and she used Briejèr's work to support her scientific analysis of causes and effects.[126] *Silent Spring* marks the birth of the modern environmental movement. Its shocking analysis of the impact of DDT and other chemical pesticides on nature found an immediate resonance. The book had a huge public impact in the US, where President Kennedy ordered his Science Advisory Committee to conduct a special investigation into pesticides. The committee recommended a gradual reduction of chlorinated hydrocarbon pesticides, working towards an eventual ban.[127] Elsewhere, *Silent Spring* was widely noticed and translations quickly appeared; its direct influence on public debate and government policy was

[43]

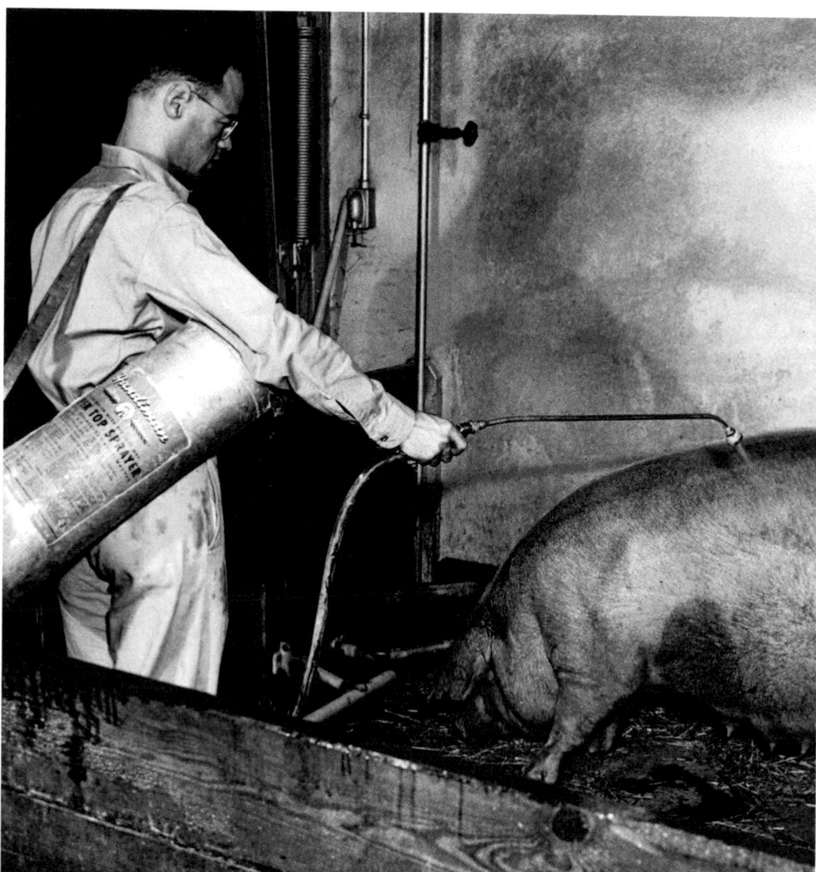

[44]

[45]

The insecticidal qualities of DDT were discovered in 1939. Above left, making life more comfortable for the pig, a man sprays it with a 5 per cent suspension of DDT in water (1945). Although he is wearing working overalls, there was no particular need for a mask or other protection. Above right, transmitted to humans by ticks, mites, fleas and the human louse, typhus flourishes in dirty overcrowded conditions where mortality can be high. When possible it is treated by antibiotic drugs and can be prevented by vaccination, but in Seoul, Korea, a faster and cheaper method of prevention was to spray humans with DDT (1951). Left, Rachel Carson, pioneer of the green movement, raises an admonishing finger (1962).

greatest in Sweden, Britain, and the Netherlands.[128] Carson's message quickly garnered support. In 1963 the British environmental organization Nature Conservancy presented a comprehensive study to the government about pesticide residues found in all terrestrial, freshwater, and marine wildlife species from around the British Isles. The study also established a link between the marked decline of certain birds of prey and the spread of pesticides. From its evidence, Nature Conservancy concluded that persistent organochlorine insecticides were by their nature uncontrollable and the organization therefore recommended a total ban on aldrin, dieldrin, heptachlor, and endrin.[129] That same year the journal *Science* reported DDT and dieldrin contamination in rivers all over the United States.[130] In 1964, the British government's Advisory Committee on Poisonous Substances used in Agriculture and Food Storage also recommended that the use of aldrin and dieldrin should cease except for very limited purposes.[131]

The pesticides industry of course reacted indignantly to the charges levelled by Carson's book and subsequent confirmations. Velsicol, the holder of the original 'drin patents and still a major pesticides producer, attempted to have *Silent Spring* withdrawn by threatening its publisher with legal action to obtain damages for bringing its products into disrepute.[132] The industry counter attacked along three main lines. First, doubts were raised about Carson's scientific credentials, as a single, female, marine biologist allegedly writing outside her sphere of competence, whose emotional appeal to reason ruled her work out of serious academic debate. Company representatives and researchers contested the solidity of the evidence and the causal links presented. Second, scientists were mobilized to reassure the public with an avalanche of statements that pesticides were both vital for world food production and essentially safe if used with the right precautions.

[47]

[46]

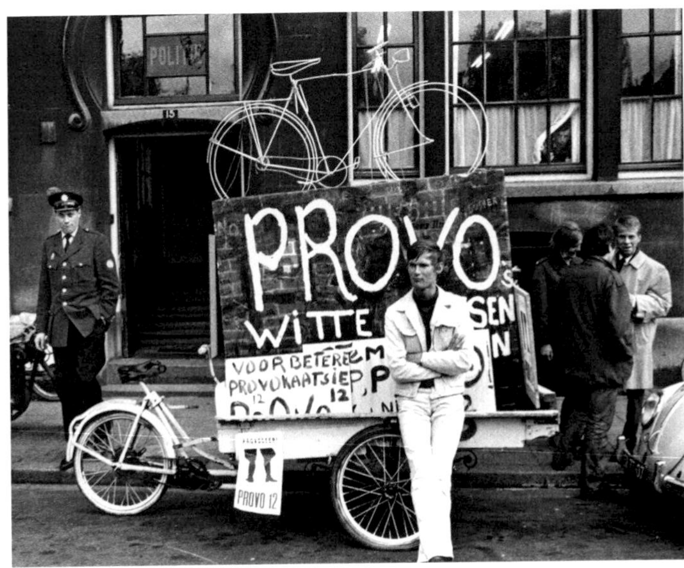

[48]

The environmental movement of the 1960s formed part of wider social protest movements, which included protests against tobacco companies (inset). Some of these groups, such as the Dutch Provo, strove to air their grievances with peaceful and often playful manifestations (main picture) and the promotion of healthy and non-polluting alternatives to cars in the form of white bicycles (left).

Third, the pesticides producers exercised pressure on government departments of agriculture, with which they had collaborated in pest eradication programmes, to obtain support by public reassurances about the safety and necessity of pesticides for modern farming, and to forestall any initiatives for a ban.

As a leading pesticide manufacturer, Shell adopted the same attitude as the rest of the industry, refusing to accept the evidence and retreating into an ivory tower of scientific and technocratic certainties.[133] Rather than recognizing that the problem lay in the essence of the products themselves, the Group considered the furore about pesticides as a public relations problem to be solved by educating the public with better information.[134] Why would the company have opted for a stubborn policy of denial, so markedly different from the more open-minded attitude taken over other issues? First, the controversy appeared to remain isolated, limited to a few countries and to small sections of the population there. The 1969 Group environmental survey noted that pesticides caused substantially less concern than sulphur dioxide emissions, leaded gasoline, surface water pollution, or marine pollution, 'and then only in more developed countries. As with lead in petrol though, there is a kind of submerged neurosis which could easily burst out if telling evidence of harm from residues were produced and publicised.'[135]

Second, 'drin sales suffered very little from the negative publicity. The total volume of aldrin sold in the US peaked in 1966, four years after the publication of *Silent Spring*.[136] The Group's annual sales of pesticides peaked in the same year and subsequently remained more or less stable, but profits per ton rose, so the market continued to accept the products readily. A third, obvious reason lies of course in the profitability of the 'drins. In 1965, a fairly typical year, pesticides generated 14 per cent of the sales and 23 per cent of the profits from the chemical function. However, this reason is a little too simplistic. Gasoline was a very profitable product, and yet the Group responded to the market's demand for eliminating lead while remaining sceptical about the scientific case for doing so. Fourth, as noted before, environmental thinking during the 1960s focused on individual nuisances. Shell simply would not accept the side effects of its pesticides as a public nuisance; from the scientific viewpoint then current, the residues in wildlife posed no danger to human life at all, and they were a small price to pay for the great benefits to food production.

However, the fundamental reason why Shell refused to change its mind lies in the fact that there were really very few incentives to do so. In every case discussed above, environmentally friendly policies were inspired by the discovery of cost savings or other advantages, by the need to keep up with market trends, or because the cost of removing a nuisance yielded a benefit in the

The protest movements of the 1960s culminated in 1968 when students confronted the authorities in many European cities, notably in Paris. On the left, protesters hurl projectiles against riot police, and right, citizens climb over barricades of paving stones erected by the protesters.

[49]

[50]

form of favourable publicity. At this time the pesticides market was not as competitive as the one for gasoline, where a mistake in spotting trends was penalized by an immediate loss of market share. Indeed, the agrochemical industry was united against change, so under the circumstances it made sense to hold out for as long as possible until a joint solution could be found, as Shell had done in the case of the sulphur emissions and the Pernis smog abatement. For pesticides, the overall solution proved as elusive as the panacea to sulphur emissions. Some specific disadvantages could be remedied. In 1960, for instance, after unusual numbers of dead birds had been found in fields round the countryside, the likely cause had been identified in the shape of grain seed dressed with chemicals to prevent attack by insect pests. Shell and other manufacturers had investigated alternatives, and in 1961 Shell announced a positive outcome with the introduction of Kotol, a liquid seed-dressing based on lindane, which was claimed to be 'the least hazardous insecticide to birds and other wild life'. The product was launched after a new voluntary agreement with the British government over regulations on the application of the pesticide.[137]

However, this was very much an exception. Despite a considerable effort to develop new types of pesticides, Group researchers could find no alternatives for the 'drins with similarly strong patents, and with the efficiency of the 'drins but without their disadvantages. In 1967, the Conference noted that it appeared unlikely that the Group would ever get 'an alternative insecticide as ubiquitous – or as profitable – as dieldrin'.[138] Consequently, Shell could not switch to a different formula, as it had with the detergents. By 1970 the company spent £3.1 million annually on research and development of agricultural chemicals, on a budget of £50 million. The Conference queried the item, expressing doubts as to whether the Group should continue to invest such substantial amounts given the paucity of the results.[139]

Nor did governments provide the impetus for change which the market failed to give. When in 1964 an advisory committee recommended that the use of aldrin, dieldrin, and heptachlor should cease, the British government preferred to rely on the cooperation of the industry rather than proclaim an outright ban.[140] Consequently, in the emerging controversy Shell took the

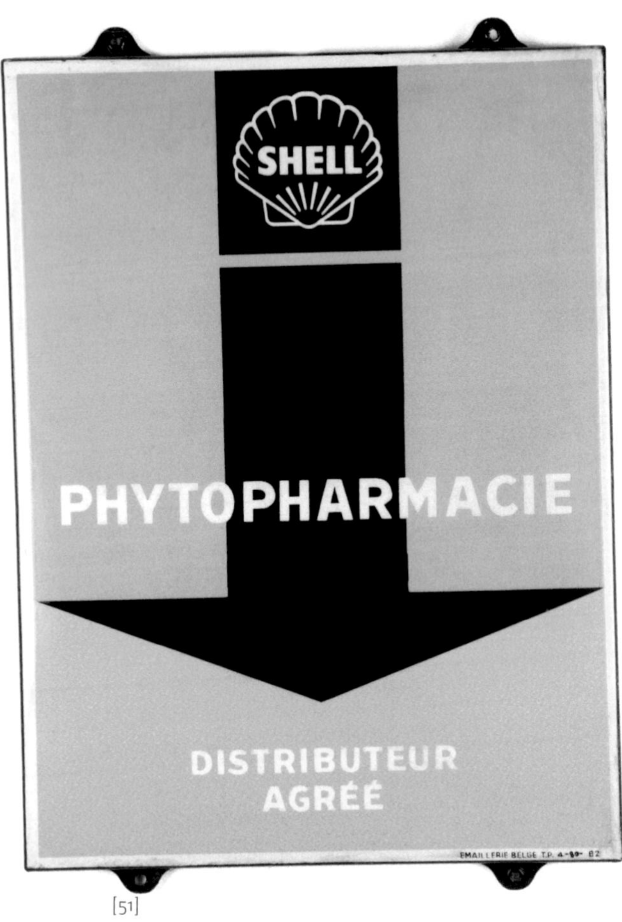

[51]

Phytopharmacy, or the chemical care of plants, is a word that did not exist before the 1960s, but as soon as it came into use Shell was quick to recognize and adopt it.

same line as the industry generally, contesting the evidence, sending out positive publicity, and working on government officials. On the eve of the British publication of *Silent Spring*, the company distributed a leaflet entitled 'For Your Information: Chemicals and Our Food' to staff.[141] This was followed with a special issue from Shell Briefing Services to Group companies, visits of expert UK journalists to the Group's Woodstock laboratory, and training sessions in television techniques for spokesmen of the Association of British Manufacturers of Agricultural Chemicals.[142] A film was sponsored giving both sides of the question, and Shell Transport's Annual Report for 1962 took a firm line: 'Agricultural chemicals play an essential part in ensuring better yields from the land. Fertilizers raise the productivity of the soil, herbicides kill the weeds, and pesticides destroy the insects and other pests that today deny man so much of the yield from all the crops he sows. These vital chemicals are sometimes attacked as "upsetting the balance of nature", as dangers to bird and animal life, and even as a threat to human health. Such claims are greatly exaggerated: man "upset the balance of nature" the first time he grew a crop.'[143]

The riposte came under the general title of 'Progress and Prospects'. Its length (more than one whole page out of three) and the wide range of other Shell responses showed how dangerous the criticisms were seen to be. The approach of *Silent Spring*'s UK publication in 1963 prompted a long memo from the chairman of Shell's Agricultural Products Committee, C. L. Raymond, to the DCC. Pointing out that Carson was working outside her field of marine biology, Raymond noted that US government scientists and independent scientists considered her book useless as a scientific assessment, through its bias and innuendo. But he acknowledged the expected effect of the book and the risk it posed, and he recorded an interesting part of the Group's reaction: 'A large amount of toxicological work has been carried out by, or on behalf of, the Group in respect to Group pesticides. Until recently the Group's policy was to "play down" toxicology, the word itself almost being banned. This policy has now been reversed, but the result of the earlier policy has been that little of the Group's toxicological work has been published and laboratory reports have had a very restricted circulation.'[144]

Raymond also identified a 'powerful need to educate the public and co-operate with Government scientific advisers'. The secrecy formerly associated with the Woodstock laboratory was thus seen to have been a mistake: the industry could defend itself

'... man "upset the balance of nature" the first time he grew a crop.'

better from a critical public by sharing its information, proving it was doing everything possible to protect the public. Should the Group launch some 'aggressive' PR? The CMD thought so, until counselled otherwise: regrettable though it might be, Shell was not actually a name that sprang to peoples' minds when the chemical industry was mentioned, and its purposes might be better served through trade associations rather than directly.[145] Likewise in the United States, where the pesticides controversy was growing and Congressional hearings were about to start, it was decided not to change Group PR policy 'at this rather critical moment. We would not wish to appear unduly on the defensive'.[146] Instead it was felt again that the problem would be better tackled by the industry than by individual companies, but 'Meanwhile studies will of course continue with special reference to the "side effects" of pesticides in humans.'[147]

By the end of the year there seemed some room for hopefulness, when the Duke of Edinburgh, the chairman of the influential World Wildlife Fund for nature who had earlier that year assisted with the public presentation of *Silent Spring* in Britain, visited Shell's laboratories at Sittingbourne in Kent to participate in a conference entitled 'The Countryside in 1970'. The WWF had earlier condemned pesticides wholesale, but in his closing speech the Duke praised the extent to which Shell employed biologists alongside chemists, and he sought a constructive middle path: 'I think we have avoided a head-on collision between the pros and the antis. We have established this very difficult subject I think in the right perspective (...) the scientists and the conservationists now realise that they both have responsibilities, that they both have limitations, and I think that there is an atmosphere in which this immensely difficult problem can be worked out in a sensible way to the benefit of both sides.'[148]

Therefore the publication, only two months later, of the Advisory Committee's recommendation of a ban on 'drins came as an unpleasant surprise.[149] It was not so much the imposition of a UK ban that Shell feared: if accepted and acted upon there, the effect on Group business in the UK would be relatively unimportant, but if that lead were followed elsewhere it could assume serious proportions. The Group made immediate high-level representations to various departments of the British government, as well as interested bodies such as the National Farmers' Union.[150] In preparing thereafter for a meeting with the Minister of Agriculture, a potential dilemma was recognized. If the Minister refused to permit a study of Shell's compromise proposals (which the archives do not specify) and if he instead asked the Group to go along with the report's recommendations voluntarily, then the CMD felt that agreeing to do so would constitute 'an admission that the Group had acted irresponsibly, even improperly, in promoting the sale of the "condemned" substances for its own gain. On the other hand, a negative answer might well be represented as a "defiance" of the Government, the will of the people, etc. and be damaging to the Group in its relations both with the Government and the public.'[151]

The CMD clung to the interpretation that the controversy stemmed from a genuine difference of scientific opinion, and consequently managing directors felt that the Group could not properly accept recommendations which, on the basis of its own research, knowledge, and experience, it considered to be ill-founded. To sidestep this dilemma, Shell accepted an extension of the earlier voluntary scheme agreed over Kotol to provide the government with information on new pesticides or new uses for old ones and on the properties and safety of chemicals marketed. As a result the recommendations of the advisory committee were not implemented at once.[152]

The tide against pesticides reached a peak during the early 1970s. In 1972, the use of DDT was banned in the US, and during 1973-74 the US Environmental Protection Agency (EPA) held public hearings about a ban on aldrin and dieldrin in the US. The hearings concentrated on the charge that these pesticides could cause cancer of the liver. Again Shell tried to fight the case on scientific grounds. Expert witnesses contested the evidence for the alleged carcinogenic properties, and presented counter-evidence including a long-term study carried out on staff at the Pernis 'drin plant, which had observed a lower than expected incidence of cancer.[153] It was to no avail; the Agency imposed a partial ban on aldrin and dieldrin in 1974 and a total ban the following year.[154] Other countries soon followed, and by 1990 they had become the world's most widely circumscribed class of insecticides, with their use either banned or severely restricted in approximately 150 countries. As their step-by-step banning proceeded, the Group closed their production and ceased their sale in countries where it was no longer permitted.

When production ceased in the US, the 'drins left a dreadful legacy at the Rocky Mountain Arsenal, a huge site of 27 square miles (nearly 7,000 hectares) west of Denver, Colorado. The US government had produced chemical weapons there during the Second World War, and in 1947 had leased part of the land and the buildings to Hyman to make pesticides and herbicides, as Shell Oil subsequently did. The 'drins were manufactured separately from the US Army's chemicals production, which included mustard gas, white phosphorus, napalm, and GB1 nerve gas (also known as sarin). But disposal of the large quantities of toxic waste generated by the combined civilian and military production was undertaken jointly, by pumping the wastes into unlined trenches and pits.[155] By the middle 1950s, soil, groundwater, and surface water were contaminated with aldrin, dieldrin, and arsenic; crops around the Arsenal suffered damage; ponds and lakes within the Arsenal were dying. As complaints mounted, efforts were made to contain the situation. An artificial evaporation lake of 93 acres (230 hectares) was dug and lined with asphalt; a well two miles deep was sunk on Army authority for the pressurized disposal of waste below the groundwater level. It was used from 1962 to 1966, when its use was discontinued after a series of earthquakes, possibly caused by the lubricant effect of the accumulating waste liquids. The pollution at Rocky Mountain Arsenal would lead to lengthy litigation between Shell Oil and the US government during the 1980s about the clean-up costs (see Volume 3, Chapter 5). A civil action launched by the Dutch state against Shell about the disposal of drins in the period 1953-60 failed because the company successfully claimed that the method employed had been legal at the time and had been considered satisfactory.[156]

These cases of soil pollution helped to bolster an insight which the Group had started to recognize from the late 1960s: 'Broadly speaking environmental conservation can be brought about by either prevention or cure. It is generally cheaper to stop pollution at the outset than to control it afterwards.'[157] This fundamental insight broke the ground for a change in attitude, from reactive to pro-active. During the 1970s Shell came to view environmental costs as part and parcel of the business and no longer something to be accepted grudgingly in the face of public pressure, so they should be factored in from the beginning. The same transformation happened with regards to the public relations aspect of environmental affairs. In 1968 the Coordinator Trade Relations described his department's task to the Conference as 'to maintain and enhance the freedom of action of Group companies to pursue policies and activities that are commercially attractive by

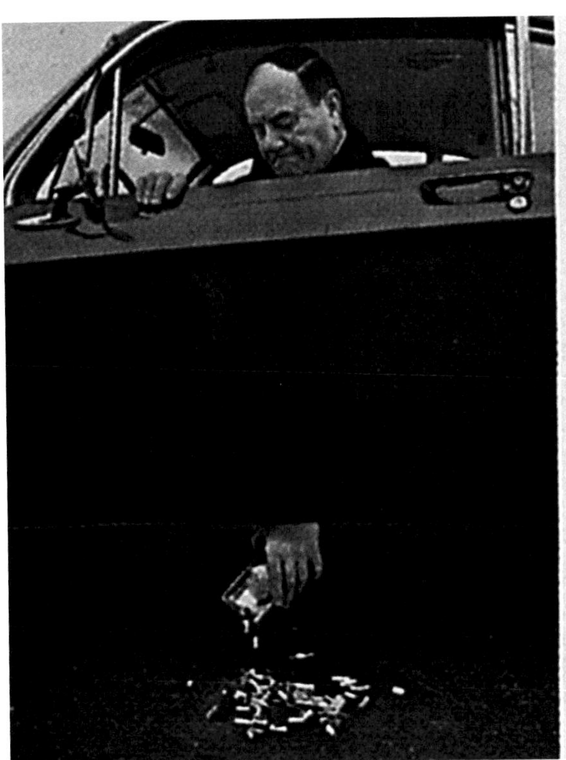

What have you done to your country lately?
Cigarette butts. Gum wrappers. Candy paper. Don't drop them in all the wrong places. Like a sidewalk. Or the highway. Or on somebody's lawn. Or in the gutter.
Every once in a while, make a deposit in a waste can at your Shell station. It's a great way to save. The landscape.
Now you can visit your Shell dealer when your tank is empty, or when your ash tray is full.
That way,you not only keep a tidy car. You get a tidy country to drive it in.
Please keep this in mind: if we keep throwing trash away on the streets and highways, we're throwing something else away.
A nice place to live.

[52]

In 1968 the assassination five years earlier of US President John F. Kennedy was an event still shockingly fresh in the minds of most people around the world, and in furthering its new environmental message Shell adapted one of Kennedy's famous phrases. 'Do not ask', he had said, 'what your country can do for you – ask what you can do for your country.' Shell's adaptation was sure to strike a chord with its public, and remains one of the most powerful single environmental messages it ever issued.

gaining public acceptance through understanding of rapidly growing international operations in a strongly nationalist environment'. Combating public ignorance figured large in his presentation and environmental concerns were conspicuous by their absence from it.[158] Three years later a new coordinator put pollution and pesticides at the top of his department's agenda.[159] Meanwhile Marketing Oil had realized that its entire programme depended not on getting the right publicity, but on Shell achieving 'community acceptance'.[160]

The idea of 'community acceptance' had come a long way. At the end of the 1930s officials had first realized the need for Shell to appeal to a wide public with a positive corporate image, and they had adopted film as the best medium to achieve that (see Volume 1, Chapter 6). Guided by that idea, public relations remained confined to the one-way traffic of disseminating information, but with the rise of pressure groups Shell recognized that 'community acceptance' was needed, which implied an open debate. This did not immediately become a widely accepted view. The 1971 Group meeting in Houston still regarded environmental matters not as issues in their own right, but as primarily dependent on the coincidental balance between public concern, government regulations, and the company's aims, with forceful public relations efforts as the most efficient means to alleviate them.[161] However, in combination with the insight that environmental costs were best taken beforehand, the recognized importance of public acceptance led to the emergence of environmental consciousness. Two events in 1972 accelerated that shift: the United Nations Conference on the Human Environment at Stockholm; and the publication of *Limits to Growth*, Dennis L. Meadows' report for the Club of Rome about the sustainability of the current exploitation of the earth's resources. From localized issues, environmental concerns suddenly achieved a global dimension, with the 1973 oil crisis adding further urgency.

What are these chemicals doing to my world? Pictured in October 1964 at the Mericourt dam in France, the eloquent gestures of a waterman pose the question.

Conclusion

The 1950s and 1960s were in many respects a golden age of Group research, full of exciting new products and processes, full of scope for the building of new laboratories to develop plastics, synthetic rubber, agrochemicals. Research was closely coordinated and integrated with the other business functions so as to ensure a good fit between commercial requirements and scientific enquiry. Judging from the calculations of revenues generated by research results, Shell profited greatly from its dedication to science, but by the late 1960s the research effort came under sustained pressure. Competition in all major markets made it increasingly difficult to charge a premium for new products; research costs rose exponentially, and the yield in terms of patented inventions showed a long-term decline. During the 1970s, the critics of the Group's research spending would gain the upper hand and press for an extended pruning exercise. In future years, periods of research expansion regularly alternated with contraction.

Together with the technical and trade relations functions, research was also the first department to be confronted with the environmental damage caused by manufacturing processes and the products issuing from them. Initially the Group viewed environmental concerns primarily in terms of cost, taking measures if and when they reduced cost or even yielded revenues. From the early 1960s, Shell also initiated the building of industry platforms to negotiate voluntary pollution abatement schemes and achieve an equitable distribution of the costs over all participants. Growing public pressure reduced the need for cost effectiveness, as long as the steps taken alleviated the nuisance concerned. At the same time, however, Group policy was also constrained by the market. Shell was heavily dependent on sulphur-rich crude from Kuwait and Venezuela, but could not charge a premium for developing an effective method to remove sulphur from fuel oil, because other oil companies could supply low sulphur fuel oil from sources without a need for expensive secondary processing. Similarly, trends in the gasoline market could not be forced, as Shell Oil's premature launch of lead-free gasoline showed. On the other hand, with pesticides neither the market nor governments exercised a pressure for change, so the Group continued selling the 'drins until gradually forced to phase them out by legislation.

By 1970 Shell's attitude to environmental affairs had shifted again, from the close focus on public relations efforts and individual nuisances to an increasing awareness that environmental costs should be factored in from the start. Managers slowly came to see that, if the business were to prosper in modern society, it needed to move from the policy of enlightened self-defence by means of corporate information to the achievement of 'community acceptance' of its operations. This implied the need for opening a dialogue; this required a major shift in attitudes which would take some time to complete.

Conclusion

The thirty-four years of Royal Dutch Shell's history covered by this volume fall neatly into two sub-periods. The first, 1938-48, was of course overshadowed by the Second World War. For the Group, however, the importance of that period lies more in the fundamental conflict between the British and Dutch directors over control and its resolution by the creation of an informal committee of managing directors, later formalized as the CMD. Booming demand then dominated Group policy during the years 1948-72, as the hydrocarbon revolution transformed the world's leading economies from coal-based to oil-based.

During the Second World War, Shell showed great resilience. Its competitive position suffered from losing a third of production plus a third of manufacturing capacity, but even so the business kept up remarkably well and the lost ground was quickly recovered after the war. This was in no small part due to Shell's strength in the United States which, in the Second World War as it had in the First, supplied the products that secured victory. The war also reinforced the Group's predilection for upgrading and the component-oriented approach to manufacturing developed in the 1920s. It prized flexibility of supplies and accordingly equipped its manufacturing installations with units capable of processing any type of crude. Consequently, the rapid post-war expansion manifested itself partly in a growth of primary capacity, but to more telling effect in the rise of secondary processing for the upgrading of particular product flows. With secondary processing, hydrogen technologies became the core of Shell's manufacturing concept, bringing to fruition experiments begun in the 1920s.

The compromise between Dutch and British managers achieved towards the end of the war finally succeeded in changing the cumbersome top management structure of companies with overlapping boards – previously split between London and The Hague, responsible to different sets of non-executive directors, and run as separate units by different managing directors – into an integrated board led by a close-knit team of managing directors. Over time the original structure (which was really a birth defect caused by the Rothschilds' understandable refusal to accept the intended liquidation of Asiatic) had increasingly disfigured the Group, but under Deterding change was unthinkable, and once he had gone the antagonism between Dutch and British directors made it impossible. Alas, we do not know how they reached the deal trading a 50:50 ownership ratio in return for a Dutch style of corporate governance, but both sides clearly attached great importance to the change in governance. The Dutch directors gave a staunch defence of the proposed 50:50 to their government, but when their stratagem to force it into accepting the 50:50 failed, the other part of the deal was not cancelled. The episode is remarkable for another aspect, too. It was a rare occasion of one of the Group's home governments successfully thwarting something which it badly wanted, the exchange rate regime giving the Dutch government a veto which it would not normally have possessed. The much belated rationalization of the managerial structure was rounded off in 1961 by the institution of the Conference. This was the realization of something first tried in December 1913 but abandoned when the First World War intervened. The Conference operated in every respect as a joint supervisory board to the CMD's executive, and thus Shell's managerial structure assumed the outward form of other large corporations.

This re-forging at the top of the Group finally enabled Shell to give its central offices a more rational organization. Typically, research was the first business function to be given a firm framework designed to obtain closer coordination between the various research centres and between the laboratories and commercial ends. With the Group's increasing commitment to petrochemicals, research acquired a central importance to its business, but towards the later 1960s competition made it ever more difficult to charge a premium for research-intensive products. As for Shell's other functions, it took about a decade to unravel the Gordian knots tied during forty years of compromises. The matrix structure that emerged, a balance between functional and regional interests, harked back to the reforms introduced by Colijn and Pleyte at Bataafsche during the First World War. It was a very daring managerial innovation, and one of Shell's greatest successes during the period covered in this volume. By providing a framework of checks and balances for delegating authority to the lowest level practicable, the matrix kept the playing field for operating companies sufficiently narrow, giving them some leeway but not too much. This enabled the Group to combine the advantages of a large corporation with the flexibility of local businesses, a factor of supreme importance in a world of fragmented markets and political polarization. Shell's adaptability to local circumstances only failed in Indonesia, where its patience ran out just before the tide turned.

Another important achievement of the 1950s was a fundamental reform of the Group's training programmes, replacing the on-the-job system with formal courses which reduced the former large differences in background between staff. Together with the introduction of more uniform employment conditions, these training programmes helped to draw the corporate culture closer together around a recognizable common identity at a time when the regionalization policy increased the diversity amongst the staff. The high contemporary value given to employees was further demonstrated after the launch of Royal Dutch shares on the New York Stock Exchange, when despite the newly perceived importance of investor relations the staff kept their top position in the Group's table of stakeholder priorities, some rungs above the shareholders.

At first sight the organizational overhaul would appear to have been counterproductive, for the Group's performance as measured by its return on assets dropped from the late 1950s. There are indeed reasons to believe that the demarcation lines between functions were initially a little too sharp, hampering the optimization of supply lines, a problem successfully addressed from the early 1960s. In addition the Group appears to have taken time to adjust to the newly competitive downstream environment that developed during the later 1950s. However, the main cause for its underperformance lay in a number of competitive disadvantages. Compared to the American oil companies with their stronghold in Saudi Arabia and their concentration on gasoline, Shell had a greater distance from source to marketplace and a blacker, less lucrative barrel. The American companies were further helped by tax legislation designed to facilitate overseas investment and, unlike the Group, could treat downstream as a zero-margin business.

This difference in position largely determined the Group's performance and policy during the 1950s and 1960s. Contrary to the other majors, it did not want to put all its eggs in the Middle Eastern basket. Instead, given the steady worsening of terms in the OPEC countries, it strove to maintain a judicious balance of supplies and decided very early on to channel its E&P investment into offshore ventures, accepting the extra cost as a necessary precaution. This very courageous decision was in the Group's best tradition of taking a long-term view, and was fully vindicated by later events. Shell also showed an open-minded and flexible attitude towards the rise of OPEC, preferring an open dialogue rather than confrontation, which enabled it to stand out from the other more combative majors. This was no doubt because it was less dependent on OPEC oil than they were, but its attitude was also formed by its long experience in dealing with economic nationalism, from which it had learned the virtues of tenacious flexibility, standing firm while always keeping negotiations open. The Group's E&P policy stands out, with the matrix structure, as another of its key achievements during the 1950s and 1960s. The growing importance of gas originated in chance circumstances as an offshoot of this policy, namely Bataafsche's position in the Netherlands on top of the Slochteren field, but Shell's rapid adoption thereafter of gas and LNG as a core activity similarly demonstrated its forward-looking attitude.

Despite investing heavily in E&P, the Group also spent very large and increasing amounts on its downstream operations, which expanded at a prodigious pace. Yet it failed to keep pace with booming demand, losing market share throughout the 1950s and 1960s. To a degree, this was unavoidable. Shell could hardly have grown faster than it did, without stretching its financial and managerial resources beyond the limit. Nor could it have met the competition head-on without possessing the same tax and other advantages. As in the past, the Group continued to favour price over volume; cut-price competition also ran counter to the carefully nurtured brand image with its cherished premium. Shell naturally wanted to grow and the business grew very rapidly indeed, but it did not want to grow at all costs, and therefore did not want to enter into the headlong downstream competition resulting from oil companies racing to build outlets at whatever price in order to offload bountiful supplies from the Middle East. Consequently, until the mid-1960s the CMD and the Conference accepted the erosion of Shell's market share with relative equanimity as an ineluctable fate; but when the decision was taken to adopt a more aggressive marketing policy, they found that regaining market share was considerably harder than losing it. The verdict on the Group's downstream policy must therefore be a mixed one. In conserving brand image and in building a sales network focused on profitability rather than volume per se, it was successful as a holding operation, but only at the cost of giving ground to the competition.

The Group's acceptance of market share erosion also stemmed from a nagging feeling that the sharp competition, lower margins, and declining profits were sure signs of the oil industry reaching maturity. Though originally begun for other reasons, the Group's diversification into petrochemicals and later into metals derived part of its motivation from the same feeling. Unfortunately, with petrochemicals the future was always brighter than the past or present. The business grew very fast, Shell built up very substantial petrochemical operations with a wide range of new products, and gradually the profitability of chemicals improved as well. However, chemicals remained a serious financial drain on the oil side and never completely fulfilled the intention of becoming a solid, independent business. Two product groups (resins and pesticides) were very profitable, but two other innovative ones (plastics and synthetic rubber) never came close to repaying even a fraction of the huge investment in research effort and plant. Meanwhile the product groups closest to the oil side generated most of the chemical function's sales volume and yielded steady, if unspectacular, profits. When the Group at last recognized the need to change, it found it could not do so at all easily. By committing itself to large-scale integrated plants, it had locked itself into continuing with loss-making product lines. In petrochemicals, Shell had been a trendsetter for the oil industry, other majors following sooner or later, with equally dubious results. By contrast, in its enduring search for diversification into other economic sectors, Shell followed the business fashion of the day, the formation of conglomerates. The concern for the oil industry's alleged maturing definitely helped to inspire the diversification, but the constantly shifting arguments used to defend it demonstrate that there was really no firm economic reason for doing so, only a tenacious belief in vague virtues.

The dogged persistence with chemicals and other diversifications showed the occasional dangers of taking the long-term view and subordinating short-term gains to undefined future benefits. This was Shell at its most self-absorbed, a technology-driven enterprise keeping its eyes fixed on an ever-receding horizon, with scant regard for common business yardsticks such as return on assets or shareholder returns. Its attitude towards growing public environmental worries was initially similarly inward-looking and carried by a conviction of knowing best, which found its most deplorable expression in the stubborn refusal to accept the toxicity of the 'drins. Both stand in marked contrast to Shell's open attitude to OPEC. But in respect to environmental concerns, despite its size and economic power, Shell had to move with the tide, and could not really move against it. Under pressure from the marketplace, the Group gradually came to understand that it must take the concerns of society seriously if the business were to continue prospering.

Notes

Chapter 1

1 RA Alkmaar De Lange papers (no inventory numbers at time of writing), Godber to Van Eck, 13 May 1940, a copy in SHA 190D/730. Kessler was probably away on business when the incident happened and saw the letter only much later; his main comments are dated November 1940. A last comment pointing to the likelihood of Waley Cohen having inspired the letter appears a later addition, since it refers to Godber in the past tense. Godber had indeed been fully involved in the discussions with the Dutch directors: SHA 65/36-3, memo Godber to Van Eck, 4 May 1940.

2 The appointment announced in SHA 7/507, telegram Oppenheim to De Booy, 6 June 1940, reporting on the outcome of the board meeting held for the purpose. This part of the emergency programme had been extensively discussed before with the British directors: SHA 65/36-3, memo Godber to Van Eck, 4 May 1940.

3 Payton-Smith, *Oil*, 41.

4 Payton-Smith, *Oil*, 61, 63.

5 SLA SC7/92/9/2-2, memo Hill for Godber, 31 December 1936, Godber to De Kok, 4 January 1937; SHA 11/24, memo CI department, July 1939 (Pernis iso-octane plant 1938-39). On the development of iso-octane, see Volume 1, Chapter 6.

6 Payton-Smith, *Oil*, 55, SLA SC46/1, Manuscript J. W. Vincent, *The History of Shell's United Kingdom Oil Refineries 1914-1959*, 23; Nockolds, *Engineers*, 10.

7 Payton-Smith, *Oil*, 56-7; Reader, *ICI* II, 266, 275.

8 On the development of alkylation, see Volume 1, Chapter 6. Contrary to what Payton-Smith, *Oil*, 260, believes, alkylation was not developed by Anglo-Iranian. As so often, several oil companies worked on this process; in the end, Bataafsche's patent application beat Anglo-Iranian by thirteen days. Schweppe, *Research*, 65; Bamberg, *BP* II, 204. In April 1939, the companies concerned formed a patent pool for alkylation.

9 Reader, *ICI* II, 266, 275. According to Reader, the Group never really liked the project and had only joined to please the British Government.

10 SLA 119/11/26-6, board memos Anglo-Saxon, memo 3 August 1938 (plants for slack wax cracking, ester salts, ketones), two memos 5 August 1938 (steam distillery, expansion laboratory).

11 SLA 119/11/27, board memos Anglo-Saxon, 29 April 1939; SLA SC46/1, Manuscript Vincent, 'UK Refineries', 11, 17-8, 35; Nockolds, *Engineers*, 22; Payton-Smith, *Oil*, 53-4, 192. In December 1939, Anglo-Saxon decided to expand the aircraft engine-testing facilities at Stanlow because of the rising volume of work, the increased difficulty in maintaining contact with the Group's Delft engine-testing station, and because the Air Ministry did not want the work to be done abroad: SLA 119/11/28, board memos Anglo-Saxon, memo 19 December 1939. These testing facilities were eventually moved to Thornton.

12 SHA 11/16-2, memo world advertising budget, 6 June 1944. The budget dropped from £1.1 million in 1939 to £760,000 in 1943 and then recovered slightly in 1944, the European share falling from 2.6 to 0.7 per cent of the total.

13 On the Petroleum Board the anonymous *Petroleum at War: The War-time Story of the Petroleum Board* (a series of articles reprinted from *The Petroleum Times*, 1945); Bamberg, *BP* II, 206-7; Payton-Smith, *Oil*, 43-5.

14 Payton-Smith, *Oil*, 112; Nockolds, *Engineers*, 35-6.

15 The Board's original members had pondered whether it should be a limited company, but had decided that 'from the psychological point of view' it would be much better to leave out the word Limited: using it would make the Board sound like an ordinary peacetime company, creating 'an entirely different impression in the minds of the public and possibly in the minds of the personnel concerned'. SLA SC7/P1/1, memo meeting 22 May 1939, legal opinion Henry Willink KC, 7 November 1939. The comparable German pool was formally a limited liability company: Karlsch and Stokes, *Faktor Öl*, 201.

16 Beaton, *Enterprise*, 561-3.

17 Beaton, *Enterprise*, 574-5. It was another one of Doolittle's foresights to have a monthly report compiled about the production capacity for iso-octane. During the war, Doolittle returned to the US air force and rose to become Lieutenant-General commanding the Eighth Air Force stationed in Britain. The following data give a good impression of the volume of US iso-octane production in 1939. Two years earlier, the British Air Ministry had estimated the RAF's need for 100-octane fuel and agreed to buy a total of 74,000 tons of iso-octane for five years from three companies, including the Group. Following the Munich crisis of 1938, the ministry more than doubled its original estimates for fuel demand, so some 150,000-200,000 tons would have been needed: Payton-Smith, *Oil*, 55.

18 Yergin, *Prize*, 371-2.

19 Nouschi, *La France*, 100-1; Payton-Smith, *Oil*, 70-4, 120. Anglo-French co-operation over oil began to assume a more definite

shape only after the Munich crisis of September 1938.

20 SHA 11/21-1 and -2, minutes meeting 17 March 1938 (producing in France not economically viable), Riedemann to Godber, 16 and 20 June 1938, Godber to Riedemann, 17 June 1938, Kessler to De Booy, 6 March 1939; 49/69, OA Department to OD Department, 15 February 1936 (first discussion polymerization plant France), Dooijewaard to Bastet, 23 January 1937 (potential octanes production in France). Nouschi, *La France* 102-3, creates the impression that the French attempts to build a common oil policy with Britain foundered on the appeasement attitude prevalent amongst British government officials and on the uncooperative attitude of the oil companies, which he relates to the links between Nazi Germany and the oil companies, construed from Deterding's well-known sympathies (see Volume 1, Chapter 7). This seems a little unfair on Deterding. By the time of the first preparations for war, he was no longer in overall charge of Group policy; moreover, French government oil policy was not conspicuously more effective after his resignation in 1936 than before it.

21 Anderson, *Standard Vacuum*, 75-6.

22 SHA 11/24, Van Embden to De Booy, 28 December 1939; cf. 11/21, Department AGT/AMN to Van Eck, 9 June 1937; SLA SC7/G1/24, W. R. Carlisle to Godber, 1 March 1940. Anderson, *Standard Vacuum*, 119-20. Japan eventually obtained the hydrogenation technology from Germany, but production of synthetic gasoline remained negligible: Cohen, *Japan's Economy*, 137.

23 Anderson, *Standard Vacuum*, 121-9, 139; Van Mook, *Netherlands Indies*, 26-41, De Jong, *Koninkrijk*, XIA, 692-5. B. Th. W. van Hasselt, the general manager in Venezuela, may have attended the Batavia talks as well, cf. Manning and Kersten, *Documenten* I, 337-8; Yergin, *Prize*, 311.

24 Karlsch and Stokes, *Faktor Öl*, 190; Overy, *Why the Allies won*, 230. Birkenfeld, *Treibstoff* gives an excellent overview of the synthetic gasoline programme.

25 Karlsch and Stokes, *Faktor Öl*, *187*.

26 Payton-Smith, *Oil*, 54 (87 octane with maximum lead addition); Yergin, *Prize*, 383, and Overy, *Why the Allies won*, 234 (*Luftwaffe* on 87 octane, superior performance of 100 octane).

27 Karlsch and Stokes, *Faktor Öl*, 200.

28 Birkenfeld, *Treibstoff*, 123-4.

29 Thus technical difficulties rather than, as often believed, administrative incompetence held up jet figther development in Germany: Green, *Warplanes*, 361-3, 619-27.

30 SLA SC29, Whittle Turbine Project 1938-45; Whittle, *Jet*, 122; Kerr, *Time's Forelock*, 68, 70; *Shell Magazine*, December 1961, 345 (obituary Isaac Lubbock, head of the Fuel Oil Technical Department, who alerted Whittle to the combustion chamber).

31 Karlsch and Stokes, *Faktor Öl*, 199; on this episode, see Volume 1, Chapter 7.

32 Flieger, *Gelben Muschel*, 163-6; Karlsch and Stokes, *Faktor Öl*, 200-1. The oil companies' sales agency was offically a limited company, however, which functioned, like the British Petroleum Board, as a pool.

33 Quoted in Yergin, *Prize*, 387.

34 The average German infantry division required over 4,000 horses to move: DiNardo and Bay, 'Horse-drawn Transport' 132.

35 Van Crevelen, *Supplying War*, 151-4, Yergin, *Prize*, 384.

36 SHA 15/198, undated memo, probably November 1923.

37 Taselaar, *Koloniale lobby*, 487-8 (Sandkuyl in pressure group for pilot training); SHA 15/198, P. F. S. Otten (Philips NV) to De Kok, 17 June 1939, De Kok to Otten, 21 June 1939 (lobby for building navy ships), memo Oppenheim, 7 December 1923 (no collaboration), Sandkuyl to De Kok, 9 November 1931 (no active collaboration), Oppenheim to Sandkuyl, 1 March 1933 (collaboration only if pressed hard), Sandkuyl to Surabaya, 2 June 1937 (no collaboration in time of peace, only oral information), Oppenheim to Sandkuyl, 22 August and 9 September 1938, Sandkuyl to Oppenheim, 2 September 1938 (exorbitant powers *Staatsmobilisatieraad*, the government mobilization council). On the activities of the *Staatsmobilisatieraad*, see De Jong, *Koninkrijk* XIA, 666-75. SHA 190C/179, defence preparations Pernis and 12/13, destructions in France.

38 SHA 15/198, memo Oppenheim, December 1923, Oppenheim to Sandkuyl, 11 November 1931. The concern for the continuity of the business under a potential occupation showed from the repeated instruction not to give written information, in order to prevent occupying forces from finding out the company's complicity from the archives and accordingly treating it with hostility. The dilemma about command and information in Payton-Smith, *Oil*, 132-3.

39 SHA 15/198, Oppenheim to Sandkuyl, 22 January 1940, De Booy to Godber, 26 January 1940, De Kok to Sandkuyl, 26 January 1940.

40 De Jong, *Koninkrijk* XIA, 669-70, 676-7; Boer, *Koninklijke Olie*, 108-13; Keppy, *Sporen*, 28; SHA 11/38, Van Eck to De Booy, 8 December 1940 (no uniform demolitions plan presented by Bataafsche to *Staatsmobilisatieraad*). Cf. Payton-Smith, 132-3 for the dilemmas in Britain concerning the question to whom to entrust a decision concerning the destruction of stocks.

41 SLA 119/11/26-1 and -4, board memos Anglo-Saxon, memo 24 June 1938 (shelter Shell Haven), 30 June 1938 (shelter St Helen's Court), 119/11/27-5, memo 4 July 1939 (shelter at Plymouth office intended for London staff), SHA 49/723, memo 11 November 1937 (air raid precautions Pernis), 8/1861, memos 19 September 1938, 21 March 1939 (air raid precautions The Hague Central Office), memo 7 September 1939 (air raid precautions Amsterdam laboratory and Tank-installatie), 10/503-9, Sandkuyl to BPM, 20 December 1938 (air raid precautions Dutch East Indies); SLA 119/11/27-10, memo 14 March 1939 (strengthening decks), Bundesarchiv Berlin, Reichswirtschaftsministerium (BB RWM) R87/5951, Anhang 1, 13 (tanker transfers). Most of the ships transferred to the British flag appear to have come from La Corona, Wouters, *Tankers*, 97, i.e. 'they were formally Dutch tankers'. This raised the concern of the Dutch government: Gabriëls, *Koninklijke Olie*, 108. For preparing the fleet for war, the Group could call on the experience of Cornelis Zulver, who was still the Marine Superintendent, as he had been during 1914-18. He drafted a tight schedule to have the upper decks of tankers strengthened to carry gun platforms; the first tanker to receive the modifications docked in March 1939 and by August another 83 ships had been so equipped out of a total of 94. The Dutch tankers

received a similar strengthening, paid for by the Dutch government: SHA 8/1870, BPM to NIT, 3 February, 3 May 1939, BPM to Ministry of Defence, 21 June 1939.

42 SLA 119/11/27-1 and -5, board memos Anglo-Saxon, memos 24 January 1939 (three groups, telex lines), 26 April 1939 (Bournemouth flats bought), 16 May 1939 (freehold flats bought for offices or accommodation), 27 June 1939 (freehold house bought for Wax & Candles Dept.), 4 July 1939 (Marine departments to Plymouth); 119/3/11, board minutes Anglo-Saxon, 6 September 1939 (board meetings at Lensbury Club); 141/3/9, board minutes Shell Transport, 7 February 1939. To avoid excessive travelling, the frequency of Anglo-Saxon board meetings was reduced from weekly to fortnightly.

43 SHA 10/617, memo 25 September 1939 (organization Hague Party); 10/441A, Chapter 3 (about 100 staff); Gabriëls, *Koninklijke Olie*, 107; *Olie*, May 1975, 148-50, interview with A. van Noortwijk, who formed part of the first group to go. SHA 12/13, giving vivid reports of the chaotic situation in and escape from France during June 1940.

44 This becomes clear from the long summary of operating companies in BB RWM, R87/5950-5953. The Royal Dutch preference shares, which held the rights of board appointments, were also transferred to London: SHA 65/36-1, memo 16 January 1940.

45 Cf. SHA 65/35-1, including minutes board Bataafsche and Royal Dutch, 1 September 1939, in which the managing directors received full powers to prepare a transfer to another part of the Kingdom of the Netherlands. Following this decision, the companies concerned made all necessary preparations, which were in place by November. In February, Oppenheim travelled to Curaçao with full powers to enact the transfer: SHA 65/36-1, memos 16 and 22 January 1940. The law was signed by the Queen on 26 April and published in the Gazette on the 30th. Cf. De Jong, *Koninkrijk* II, 424-5; Blanken, *Philips* IV, 128-36, 143; Wubs, *Unilever*, 77-81. The Dutch law came late, but only slightly later than in Belgium, where a similar law was passed in a hurry during February 1940, after the capture of papers outlining the German invasion plans from a crashed *Luftwaffe* aircraft: Van der Wee and Verbreyt, *Oorlog*, 48. On May 7, the Shell Transport board had already given Oppenheim proxy to vote for them in Bataafsche shareholders' meetings should the company be transferred to Curaçao: SLA 141/3/9, board minutes Shell Transport, 7 May 1940. The Philips company had long urged the Dutch government for legislation on transferring the legal seat abroad, but preferred setting up trusts, presumably to avoid the risk of the law coming too late. When the Philips management in exile in London belatedly transferred the seats of two holdings to Curaçao on 14 May 1940, A. S. Oppenheim, in charge of the Group's business there, helped with executing this transfer and found his efforts handsomely rewarded with a Philips share: SHA 11/8, Oppenheim to De Booy, 10 October 1940, De Booy to Oppenheim, 18 November 1940. Two of the shipping lines also moved to Curaçao, three others to Batavia.

46 SHA 65/36-1, Oppenheim to Jacobson, 20 May 1940.

47 SHA 11/8, Oppenheim to de Booy, 21 August 1941. There was also a tax advantage to keeping Royal Dutch and Bataafsche operating from Dutch territory, but this only came up during the war and does not appear to have been a preponderant reason for the move to Curaçao: SHA 11/38, Kessler and De Booy to Van Eck, 22 January 1941 (not too many London residents on boards for tax reasons), 11/16-1, Van Leuven to Kessler, 11 September 1941, Koeleman and Van Leuven to Kessler, De Booy and Van Eck, 17 October 1941 (instructions to bank from Curaçao and not London for tax reasons).

48 Only the Royal Dutch administration was conducted in Curaçao, 'for special reasons' as the post war report put it: SHA 10/441A, 27. An undated memo in SHA 65/35-1 even referred to directors actually residing in Curaçao as 'an unlikely event'. Van Eck went to New York to attend to Group business there, but also 'to be available should conditions become such as to make general operations from London more difficult or impossible', SHA 65/36-3, Van Eck to Oppenheim, 9 July 1940. De Booy and Kessler put it even more emphatically, stating that Van Eck's colleagues had begged him to go to New York 'so that the Group will never be without a Captain', underlining Van Eck's stature amongst Group directors at that moment: SHA 11/38, De Booy to Van Eck, 22 January 1941.

49 SHA 16/38, De Booy to Van Eck, 5 February 1941, colonial government to Van Eck, 17 June 1941. When returning from Batavia to New York in December 1940, after the Japanese trade talks, Van Eck suggested to De Booy that the Group might reorganize the Indonesian operations into a separate operating company with some independence, on the lines of Shell Union: SHA 11/38, Van Eck to De Booy, 8 December 1940.

50 We have not found an original written agreement between the Group and the Treasury, but reference to an arrangement is made in SHA 190C/441A, 172, SHA 11/8, De Booy to Oppenheim, 11 June 1940, and Nederlandsche Bank, Amsterdam (DNB) file 2.3/1827/1, memo Ministry of Finance, 22 June 1954. The agreement followed from the introduction of the Defence (Finance) Regulations 1939 in Britain.

51 A lucid description of the need for such an arrangement in Payton-Smith, *Oil*, 60, 147-53. On the position of the Dutch East Indies SHA 11/38, Van Eck to De Booy, 30 December 1940, Bank of England Archives (BEA) OV 86/1, Van Eck to Kessler, 17 July 1941, the latter intercepted by the British secret services and relayed to the Bank of England, which makes one wonder whether Group directors used couriers or the standard mail. The arrangement was officially extended to cover Curaçao by an agreement between the British and Dutch governments dated 25 July 1940: BEA OV 86/1, memo 28 July 1942.

52 SHA 16/2, W. Keppler to A. Fischböck, 18 October 1940, H. Lochner to A. Fischböck, 28 October 1940; Pearton, *Oil*, 237. Thus Gabriëls, *Koninklijke Olie*, 108, errs in suggesting that the Germans repeatedly tried and failed to overturn the transfer to Curaçao. The manager appointed, J. H. W. Rost van Tonningen, was a member of the Dutch Nazi party NSB and an elder brother of the prominent Dutch fascist leader M. M. Rost van Tonningen, later appointed president of the Nederlandsche Bank. Rost's appointment to Astra was made by the first *Verwalter*, Hans Lochner, replaced by Hauptmann Eichardt von Klass in November 1940. Von

Klass had been a director of the Benzolverband, a German oil company.

53 SHA 11/8, Van Eck to Oppenheim, 31 May 1940; *Olie*, May 1975, 148-50 (memoirs Visman); De Jong, *Koninkrijk*, III, 32, 8. The standard version has it that it had been agreed that De Booy would stay and De Kok leave, but on the morning of 14 May, Kessler and Van Eck wired from London to The Hague urging both to come over, so there may not have been a set plan: SHA 11/78, Kessler and Van Eck to De Kok and De Booy, 14 May 1940. The Dutch army had capitulated at ten to five on Tuesday afternoon. German troops took possession of the Van Bylandtlaan office very early on Wednesday morning: SHA 11/42-1, Bloemgarten to De Booy, 30 December 1941. Of the nine truckloads of Group papers to be transferred to London, only six could be loaded on the ship for a lack of time: SHA 65/36-1, Van Eck to Oppenheim, 31 May 1940.

54 SHA 11/8, Oppenheim to De Booy, 4 December 1940, giving a vivid description of the funeral based on a letter from Oppenheim's son, who was there; 11/42-1, Bloemgarten to De Booy, 30 December 1941; *De Bron*, December 1940 for a report of the funeral.

55 SHA 11/42-1, Bloemgarten to De Booy, 30 December 1941.

56 SHA 11/38, De Booy and Kessler to Van Eck, 22 January 1941, Van Eck to De Booy, 30 January 1941.

57 Cf. SHA 12/169, Kessler and Van Eck to Van Goethem, 18 May 1940, stating that 'Experience Netherlands has shown presence NSB members or former members caused unrest in organization.'

58 SHA 11/101, De Booy to Curaçao and Batavia, 20 May 1940; 12/169, De Booy to Van Goethem, 20 May 1940. An exception was made for French nationals born in Alsace and Lorraine during the German occupation and naturalized before 1920: SHA 11/101, Asiatic to Brylinski, 25 May 1940.

59 SHA 12/169 for the lengthy legal complications following the dismissals in Argentina; 190C/30, a report from 1949 on the issue.

60 BB RWM R87/5951, Anhang 3, 29-30. Shell Union refinery runs also rose less than crude production, so the company followed a policy similar to Jersey Standard: Beaton, *Enterprise*, 784-5, 787. SHA 11/16-2, memo Asiatic Corporation balance sheet and sales 1944, giving a total of $38.8 million, against $42.6 million for 1939 in the BB RWM document

61 Nockolds, *Engineers*, 3.

62 SHA 10/441A, 43-4 (production difficulties Venezuela); SHA board minutes Bataafsche, 17 July 1947 (asphalt lake with 2 million tons). Nockolds, *Engineers* 36, mentions 1.5 million tons in the asphalt lakes; we have preferred to use the higher figure as coming from internal sources, deriving from plans in 1947 to get rid of the lakes.

63 Yergin, *Prize*, 379.

64 SHA 10/441A, 51.

65 On the Haifa refinery, see SHA 49/16 and Bamberg, *BP* II, 165, 218.

66 SHA 10/441A, 56-7.

67 SHA 10/441A, 39-40 (production in Egypt); SLA 119/11/26-3 and -5, memo 29 June 1938 (reforming unit), memo 15 March 1938 (butane gas installation). AIOC also had a 32.2 per cent stake in Anglo-Egyptian; the Egyptian government held 5.5 per cent, and 30.1 per cent was in hands of the public.

68 SHA 10/441A, 49-50, SLA 141/3/10, board minutes Shell Transport, 1 January, 13 March 1947.

69 SHA 10/441A, 52-3.

70 SHA 190B/95E, Allied intelligence report 14 June 1945; on the development of the Schoonebeek field, Gales and Smits, 'Olie en gas', 81-3.

71 Yergin, *Prize*, 277 (Iran); Philip, *Oil and Politics*, 48, 54 (Chile, Ecuador); Larson, Knowlton and Popple, *New Horizons*, 479 (Venezuela); NA Kew FO 371/24185/A1304, Godber to F. C. Starling (Mines Department) 12 February 1940 (Colombia).

72 SLA GHC/Mex/B2-2, Wilkinson to Godber, 9 October 1939; GHC/Mex/B3-1, Davidson to Godber, 26 June 1942 (position after US deal), Hopwood to Van Hasselt, 31 August 1942 (reluctant acceptance Jersey); GHC/Mex/B3-2, memos 12 and 29 October 1942 (outline potential deal); GHC/Mex/B12-1, memo 25 April 1944; 141/3/10, board minutes Shell Transport, 10 September 1947 (terms very satisfactory); SHA 10/441A, 40-3 (overview developments). Van Vuurde, *Países Bajos*, 105-28; Yergin, *Prize*, 276-9; Larson, Knowlton, and Popple, *New Horizons*, 426-7, 433-4; Manning and Kersten, *Documenten* III, 319-22, 393-4, 441; idem IV, 536-7, 591-2; idem V, 473-4, 634-6; Van Faassen and Kersten, *Documenten* VI, 237, 338, 438-9, 454-6.

73 Cf. SHA 11/16-1, memo 17 July 1941, Van Hasselt giving as his impression that the Venezuelan government was 'prepared to blackmail the companies into adapting their existing concessions to a new, and less favourable, law'.

74 SLA 119/11/31-6, board memos Anglo-Saxon, memos 30 July, 24 September, 7 November 1942; 119/11/32-6, board memos Anglo-Saxon, memo 2 January 1943; 141/3/10, board minutes Shell Transport, 5 January and 4 May 1943; SLA GHC/Mex/B3-1, Hopwood to Van Hasselt, 31 August 1942 (position Jersey Standard); SHA 10/441A, 49 (survey developments), minutes Bataafsche 21 December 1950, undated memo Curaçao capital expenditure 1951 (Cardón conditions); Rabe, *Road*, 81-7; Yergin, *Prize*, 434-7; Larson, Knowlton, and Popple, *New Horizons*, 479-84; Manning and Kersten, *Documenten* V, 97-9, 130-1, 284, 291-4, 311-2, 342-4, 457-8; Van Faassen and Kersten, *Documenten* VI, 116-7, 141-6, 260-1, 302-3.

75 Figures from SHA 15/15.

76 Royal Dutch shares traded at around 340 per cent of par at the time of issue, Shell Transport at 120 per cent.

77 Veldman, *Voorziening*, 40-1; De Jong, *Koninkrijk* III, 442; Schweppe, *Research*, 91-3.

78 SHA 49/729, during the first year of the war alone, Pernis was bombed on seventeen occasions. The failure to do much damage was known in Britain: SHA 11/16-1, memo 18 September 1940.

79 SHA 190C/179, memo Pernis operations during the war; SHA 11/38, Van Eck to De Booy, 2 January 1942 (report remaining oil stocks at Pernis); BB RWM R87/5951, Anhang 2, 21-2; *Van Rotterdam Charlois*, 42-5, 54-5, 59; De Jong, *Koninkrijk*, III, 327-8. A part of the mixed-up products was processed in Germany for Group account.

80 SHA 10/441A, 93.

81 SHA 12/13, memos 29 and 30 June 1940, 10/574, Brylinski to von Klass, 20 February 1942; SHA 10/441A, 94-5; BB RWM R87/5952, Anhang 10, 3-4; Pradier, *Shell France*, 26-9.

82 Payton-Smith, *Oil*, 136-7.

83 Yergin, *Prize*, 373-7; Nockolds, *Engineers*, 30-2; Payton-Smith, *Oil*, 381, 416; Beaton, *Enterprise*, 604-9. The partners in the Plantation line were Jersey Standard (49%)

and Standard of Kentucky (27%).

84 Shell Transport Annual Report for 1947, 17, for the immediate pre-war figure 'under one House flag', i.e. the flag of Anglo-Saxon, flown in addition to the red ensign showing British registration. Dutch-registered ships flew Royal Dutch's flag and the Dutch tricolour, and Eagle ships flew their own distinctive house flag and the red ensign. Deadweight tonnage indicates the actual carrying capacity of the vessel. Wouters, *Tankers*, 100; Middlemiss, *Tankers*, 48. After a brisk series of post-First World War purchases, the Group's collective tanker fleet was the largest in the world from at least 1927: Larson, Knowlton and Popple, *New Horizons*, 205. Jersey Standard overtook it during the Second World War.

85 Calculated from Lucas, *Eagle Fleet*, 123-36. Compared to modern tankers, contemporary ones were small vessels and the Group's thirty-six Dutch-flag ships particularly so, averaging barely 6,500 dwt, largely because most of these tankers plied the short haul between Venezuela and Curaçao or the inter island trade in the Dutch East Indies and Asia: Wouters, *Tankers*, 100.

86 Allen, *Wartime*, 61.

87 Shell Transport Annual Report 1947, 17. Definitive statistics of loss are notoriously difficult to compile from the archival records, in part because they depend upon definitions which are not always given. Similarly, published records are frequently inconsistent with one another for that reason and because they are superseded by later research.

88 Wouters, *Tankers*, 126-7; Lucas, *Eagle Fleet*, 123-36.

89 Royal Dutch Annual Report, 1947; Bamberg, *BP* II, 216; Larson, Knowlton, and Popple, *New Horizons*, 398.

90 SHA 190C/441A, 203; Wouters, *Tankers*, 128, Middlemiss, *Tankers*, 55.

91 Manning and Kersten, *Documenten*, IV, 485-9.

92 Manning and Kersten, *Documenten*, V, 35-9, Van der Horst, *Wereldoorlog*, 2-4.

93 Manning and Kersten, *Documenten*, IV, 550-2, 577-8, 584-5, 589-90, 598-90, 672-3, 675-8; idem V, 80-1, 137-8, 256-9, 319, 358-61.

94 SHA 11/38, Van Eck to De Booy, 8, 12, and 18 May 1942, De Booy to Van Eck, 23 June 1942.

95 SHA 11/17, De Booy to Engle, 31 March 1943.

96 SLA, Manuscript J. Lamb, *Development of the Tanker, with particular reference to Royal Dutch/Shell's Contribution*, 1957; 'The Shell Group and the War: a summary of Group contributions to the war effort', *Shell Magazine*, vol. 25, July 1945, 154-7; Lucas, *Eagle Fleet*, 80-1, 95; Hope, *Tanker Fleet*, 24-6, 93-5.

97 Lamb, *Development*; 'The Shell Group and the War: a summary of Group contributions to the war effort', *Shell Magazine*, July 1945, 154-7; Wouters, *Tankers* 123-5; Hope, *Tanker Fleet*, 96-101.

98 Wouters, *Tankers* 126-7.

99 SHA 10/441A, 31; Royal Dutch Annual Report 1945, 13.

100 SHA 10/441A, 29-30.

101 Anderson, *Stanvac*, 151-7, adding the crucial detail of tanker shortage and thus providing a more complete picture than De Jong, *Koninkrijk* XIA, 694-5, who considers the outcome of the negotiations to have been a disappointment to the Japanese government on the basis of the reduced volume alone. Unfortunately Anderson creates confusion on the point of actual volumes supplied, giving half of the total on page 156, two-thirds on 167, and 56 per cent on 168. The difference may be due to the volume of crude delivered, which Anderson does not specify. SHA 11/38, for Van Eck's correspondence from the Batavia talks.

102 Anderson, *Standard-Vacuum*, 158-92; Yergin, *Prize*, 313-27.

103 Cf. Cohen, *Japan's Economy*, 141

104 SHA 733, memo Van Diermen 23 April 1942, SHA 11/38, summary demolitions Pladju, 26 February 1943; Fabricius, *East Indies Episode* 13-4, 42, 49-53; Boer, *Koninklijke Olie*, 108-10, 112-3-165; De Jong, *Koninkrijk* XIB, 39-40. The Group was quick to secure its position in rebuilding the installations after the war. On 28 January 1942 Bataafsche wrote to the Dutch government in exile asking for a diplomatic approach to the US government for priority in obtaining favourable conditions in the supply of materials for rebuilding the demolitions effected on government orders after the war. In June, the Dutch ambassador received assurance on that point from the State Department: SHA 11/38, Bataafsche to Minister for the Colonies, 28 January 1942, Van Eck to De Booy, 3 June 1942.

105 SHA board minutes Bataafsche, 13 March 1948, memo 8 March.

106 Cohen, *Japan's Economy*, 134, 140-1; Anderson, *Standard-Vacuum*, 194. According to Cohen, Japan had expected the wells and installations to be totally destroyed, and was 'pleasantly surprised to find little or no damage in other areas', 140.

107 SHA board minutes Bataafsche, 13 March 1946, memo 8 March. The memo does not give data for Ceram and Balik Papan, but estimated these roughly at 1.3 million tons for a total of loss of 9 million during the occupation. This would raise the volume lifted as a percentage of 1941 production to 57 per cent.

108 SHA 11/38, De Booy to Van Eck, 16 December 1942, Bloemgarten to Van Eck, 9 February 1943.

109 SHA board minutes Bataafsche, 13 March 1946, memo 8 March (1,781 European staff); Royal Dutch Annual Report 1938 (1,900 Europeans and 19,000 Asians); SHA 734 gives 24,216 Asians in Group employment.

110 Fabricius, *East Indies Episode*, 140-55.

111 Cohen, *Japan's* Economy, 143-4; Anderson, *Standard-Vacuum*, 194-5; Overy, *Why the Allies won*, 228-30.

112 Pearton, *Oil*, 223-31, 244-58; Karlsch and Stokes, *Faktor Öl*, 205-7.

113 SHA 190D/799, F. Lutter to Zuiveringscommissie, 25 June 1945 (Rost membership NSB), minutes interviews Norton-Griffiths, Kirby, and Wieringa with Rost, 22, 23, 24 January 1946, report Norton-Griffiths and Kirby on Rost, February or March 1946; SHA 190C/309A, report 21 June 1945 (alkylation plant); Karlsch and Stokes, *Faktor Öl*, 208-13 (Konti Öl). Managers in London knew about the alkylation plant: SHA 11/16-1, memo 26 December 1941.

114 Karlsch and Stokes, *Faktor Öl*, 207.

115 Cooke and Conyers Nesbit, *Target*, 81-107. The total of 350,000 people mentioned by them is probably the total used by the Germans to repair all oil installations, including those in Germany itself: Karlsch and Stokes, *Faktor Öl*, 236.

116 SHA 190C/309A, report 21 June 1945 on the war damage at Astra's Ploesti works; Royal Dutch Annual Report 1944, 6.

117 On the damage assessment committee, see Faassen and Stevens, *Documenten*,

707. On the bombing offensive Cooke and Conyers Nesbit, *Target*, 140-53; Karlsch and Stokes, *Faktor Öl*, 232-43; Yergin, *Prize*, 346-8; Flieger, *Gelben Muschel*, 174-6 (damage Group installations Germany); Birkenfeld, *Treibstoff*, 196, 197 (destruction Pölitz).

118 Data on the Group's processing capacity in 1940 in BEA G1/482, memo 28 December 1942, Annex C.

119 Knowlton, Larson, and Popple, *New Horizons*, 499; Beaton, *Enterprise*, 787.

120 Royal Dutch Annual Report 1945, 14; Beaton, *Enterprise*, 759.

121 Beaton, *Enterprise*, 572-4, 586-7; Nockolds, *Engineers*, 39-41; 'The War Years Remembered', *Shell News*, April 1994, 28; SHA 49/19-1, Caland to Murch, 21 December 1935, 49/69, De Kok to Godber, 25 October 1934, 49/74, BIM to Asiatic, 19 May 1938.

122 SHA 10/441A, 65-6; Beaton, *Enterprise*, 567-70; Nockolds, *Engineers*, 40-1; Forbes and O'Beirne, *Technical Development*, 614-5.

123 SHA 10/441A, 31. One of the alkylation units came from the UK, where it had been under construction for the Air Ministry at Thornton. When building fell badly behind schedule due to government vacillation, the Air Ministry wanted to abandon the unit, at which the Group offered to transfer it to Curaçao: Payton-Smith, *Oil*, 274-5, 278.

124 SHA 10/441A, 66-7; Beaton, *Enterprise*, 583-4.

125 Payton-Smith, *Oil*, 279, Nockolds, *Engineers*, 40-1.

126 SHA 10/441A, 64-5; Payton-Smith, *Oil*, 269; Beaton, *Enterprise*, 581-3.

127 SHA 10/441A, 67; Beaton, *Enterprise*, 584-5.

128 SHA 11/22, Braybrook to Kessler, 13 May 1940. For Kessler's wish, see Volume 1, Chapter 6. We do not know what became of the scheme.

129 'The Shell Group and the War', *Shell Magazine*, 25 (July 1945) 154-7.

130 Beaton, *Enterprise*, 588-93; Forbes and O'Beirne, *Technical Development*, 621.

131 SHA 190D/729, Oostermeyer to Kessler, 21 January and 7 April 1942; Beaton, *Enterprise*, 592-8.

132 Forbes and O'Beirne, *Technical Development*, 611-2 (toluene and extractive distillation); Beaton, *Enterprise*, 599-602. On 629-30, Beaton claims that Shell companies preferred to do without government support, but SLA 119/11/31, board memos Anglo-Saxon, memo 17 March 1942, shows this not to have been the case.

133 Beaton, *Enterprise*, 544-7.

134 Beaton, *Enterprise*, 610-5; the portable pipeline could be laid at a speed of thirty to fifty kilometres a day, Royal Dutch Annual Report 1944, 6.

135 SLA 119/11/3-7, board memos Anglo-Saxon, memo 13 March 1944 (Stanlow decision).

136 Beaton, *Enterprise*, 615-6; SHA board minutes Bataafsche, 20 July 1950, memo 21 June 1950; 10/441A, 73-4.

137 SHA 11/16-1, Braybrook to Kessler, 16 August 1943 (growing interest pesticides), 11/16-2, Braybrook to Van Eck, 15 July 1944 (importance DDT); 12/63, various memos and minutes of meeting concerning the Argentina project. Already before the war, the Group had had an involvement with anti-malaria pesticides which originated in the manufacturing of oils which formed films on the waters and morasses in which mosquito larvae grew.

138 SLA GHC/US/D9/2/2, memo 22 February 1943.

139 SLA GHC/US/D9/2/2, memo 22 February 1943.

140 SLA GHC/US/D9/2/2, memo 10 March 1943.

141 SLA Sc46/1, manuscript Vincent, *History Shell Refineries* (1959), 42.

142 SHA 10/549, Van Eck to Van Wijk, 4 May 1938, Van Wijk to Van Eck, 9 May 1938, Van Eck to Kruspig, 10 May 1938, memo Moraht, 25 May 1938, Kruspig to Van Eck, 14 November 1938; BB RWM R87/5953, Anhang 13, 19-22.

143 SHA Country Volumes Germany II, s.v. shareholders; Flieger, *Gelben Muschel*, 166.

144 SHA Country Volumes Italy, II, s.v. Nafta; 10/441A, 96.

145 SHA 10/441A, 92.

146 SHA 11/16-2, note on Bataafsche during the war, undated, giving expenses of 106 million guilders on receipts of 112 million, the revenues deriving from sales (51 million guilders), dividends (18 million), redemption of loans and balances (17 million) sale of securities (9.5 million), ship hire (4.2 million), interest and commission (4 million).

147 SHA 42/1, Bloemgarten to De Booy, 30 December 1942; Gabriëls, *Koninklijke Olie*, 114.

148 SHA 11/16-2, undated memo on Bataafsche during the war; Karlsch and Stokes, *Faktor Öl*, 221.

149 SHA 16, memo 5 March 1943, 190B/95E, Allied intelligence report on the Amsterdam laboratory 14 June 1945, showing 1,200 staff out of the pre-war total of 1,450 still at work; cf. Schweppe, *Research*, 95-7; SHA 190A/179, memo 8 November 1943 (BIM car conversions), memo 3 December 1943 (Pernis stocks May 1940 and December 1943). SHA 190B/95E, Allied intelligence report 14 June 1945 (DDT). SHA agendas BPM 1947, memo PVC plant IJmuiden, 25 October 1946, with appendix, Schweppe, *Research*, 123-4 (PVC).

150 SHA 10/574, memo 15 November 1941; Pradier, *Shell France*, 26-8.

151 SHA Country Volumes Russia, XIII, s.v. black pages WW2, von Klass to Fischer, 15 December 1941.

152 SHA 190C/30, report 16 June 1949. For the ideas about scrutinizing staff after the war, SHA 11/16-1, Kessler to Gray, 17 May 1944, stating that this would be the responsibility of nationals from the country concerned, not of managing directors.

153 Cf. SHA 190C/270 for correspondence on the efforts to get Jewish refugees to safety.

154 Van Tielhof, *Banken*, 59-60; Wubs, *Unilever*, 144-145; Blanken, *Philips* IV, 278-9.

155 SHA 190C/270, circular 10 February 1942, undated memo with staff survey following the circular, memo von Klass, 21 March 1942, von Klass to Rüstungsinspektion, 7 July 1942; 190C/270A, memo RA 20 March 1944, memo 28 August 1991. For legal reasons, Bataafsche refused to pay the sums which Jewish staff had in the Provident Fund to the Lirobank which the Germans used to rob Jewish assets.

156 SHA Country Volumes Russia XIII, s.v. black pages WW2, von Klass to Dihlmann, 17 September 1942, Dihlmann to von Klass, 23 September 1942, memo Schippers and De Klerck to von Klass, 30 October 1942, memo Musehold 31 October 1942, minutes meeting 14 November 1942, von Klass to Kontinentale Öl, 9 December 1942.

157 SHA 190C/179, undated memo Pernis during the war.

158 Bataafsche's first *Verwalter*, Hans Lochner, already found it difficult to justify his

position to his Nazi superiors: SHA 16/2, Lochner to Fischböck, 28 October 1940.

159 SHA 16/1, memo staff situation at Bataafsche, 3 March 1943.

160 *Van Rotterdam Charlois*, 46-53, Schweppe, *Research*, 110-4.

161 RA Alkmaar De Lange papers (no inventory numbers at time of writing), Godber to Van Eck, 13 May 1940 with comments by Kessler, a copy in SHA 190D/730; SHA 65/36-3, memo Godber to Van Eck, 4 May 1940 (appointments Godber and Legh-Jones); 11/38, Kessler and De Booy to Van Eck, 22 January 1941 (Agnew proposing 50:50); SHA Country Volumes Russia I, Haly to Boyle, 1 September 1921 (declining support Russian claims). The British Treasury raised the matter of Royal Dutch's 60 per cent ownership time and again in the talks preceding and following the 1946 Treasury Agreement.

162 SHA 11/38, Kessler and De Booy to Van Eck, 22 January 1941.

163 SLA GHS/3E/1, Group directors; *ODNB*, 'Agnew', and 'Godber'; see Volume 1, Chapters 2 and 6 for more details on Waley Cohen, Agnew, and Godber.

164 See Volume 1, Chapter 6 for more information about Kessler and BEA G1/482, undated memo on his strong chauvinism; on Van Eck, *Shell News*, March-April 1965, 30-1; Wielenga, 'De Booy'.

165 The memos exchanged between the Dutch and British Group directors have not been found; SHA 11/38, De Booy and Kessler to Van Eck, 22 January 1941 refers to them and suggests the decision to leave the position of Director-General open until after the war.

166 SLA 11/16-1, Van Hasselt to Kessler, 24 January 1941, Koeleman to Kessler, 18 February 1941.

167 The following section is, unless otherwise indicated, based on SHA 11/38, De Booy and Kessler to Van Eck, 22 January 1941.

168 BEA G1/482, memo 3 December 1942, Annex D.

169 SHA 11/38, Van Eck to Kessler, 11 August 1943.

170 SHA 11/38, De Booy and Kessler to Van Eck, 22 January 1941, 11/36, Kessler to Van Eck, 12 March, 8 and 21 April 1941, Kessler to Van Eck, 4 June 1941.

171 SHA 11/38, Van Eck to De Booy, 30 January 1941.

172 SHA 11/38, Kessler to Van Eck, 17 June 1941. If necessary, the cash target was to be achieved by suspending the payment of dividends on the preference shares, the company having stopped paying dividends on ordinary shares in 1940: Van Eck to Kessler, 25 June 1941.

173 BEA G1/482, memo Abel Smith 30 November 1942, Montagu Norman to Abel Smith, 1 December 1942, Abel Smith to Montagu Norman, 2 December 1942, memo 28 December 1942, with marginal notes 29 January 1943. Oral Royal Dutch tradition has it that the plan to have Shell Transport take over a majority in the Group went up to, and was turned down by, Churchill, magnanimously not wanting to take advantage of the Dutch government's precarious position, but the 28 December memo would seem to suggest that the proposal did not go further than the Treasury. Bakker and Van Lent, *Lieftinck*, 111, give the Churchill story in the context of the 1947 debate about 50:50, when of course Churchill was no longer prime minister. Earlier in 1942, Godber and Kessler had had a row over a futile question: RA Alkmaar, De Lange papers (no inventory numbers at the time of writing), Kessler to Swaine (New York), 24 July 1942, Swaine to Kessler, 30 July 1942, Kessler to De Booy, 5 August 1942, Godber to Kessler, 6 and 11 August 1942, De Booy to Kessler, 8 August 1942, Kessler to Godber, 8 and 12 August 1942.

174 SHA 11/38, Van Eck to De Booy, 6 August 1940 (Lacomblé appointed Group research director); SHA 11/38, Batavia to Handelszaken, 11 January 1941, Handelszaken to Batavia, 14 January 1941, Sandkuyl to Van Eck, 24 January 1941, Van Eck to De Booy, 26 March 1941, De Booy to Van Eck, 22 April 1941, SHA 11/16-1, Godber to Oudraad, 1 July 1941, SHA 11/17, Godber to Oudraad 1 November 1941 (rationalizing DEI organization); 11/38, Kessler to De Booy, 16 June 1941 (reorganizing research), SHA 11/16-1, Langen van der Valk to Kessler, 9 October 1941 (raising importance product development, research organization Jersey Standard, annual research meetings); SHA 11/38, Van Eck to Kessler, 11 August 1943 (reorganizing reporting lines area coordinators and sales organization).

175 Ministry of Finance, The Hague (MFH) code 1.822.145.3, Kessler to Minister of Finance, 30 April 1945, memo 23 April 1945.

176 MFH 1.822.145.3, memo 6 October 1945 (initiative taken by the Dutch managing directors).

177 SHA 415/1, Bishop to Wieringa, undated, but June or July 1946. It took several years before the Royal Dutch and Shell Transport Annual Reports formally acknowledged the formation of the committee: Royal Dutch, Shell Transport Annual Reports 1951.

178 SHA 11/16-1, memo 29 October 1940 (Bataafsche funds New York), 11/17, memo 24 June 1941 (Bataafsche funds New York); BEA G1/482, Van Eck to Kessler, 17 July 1941 (position Dutch colonies under agreement), memo Foreign Office 1 September 1941, 28 July 1942 (grouse against the Dutch).

179 MFH 1.822.145.3, Bataafsche to Ministry of Finance, undated but probably July 1945 (dollar amounts), Ministry of Foreign Affairs to Ministry of Finance, 8 April 1946 (dollar deficit).

180 MFA 1.822.145.3, W. Eady (Treasury) to Godber, 1 June 1946, shows that the Treasury was informed of the intention to transfer Bataafsche's major assets to Anglo-Saxon in April 1945.

181 NA Kew T236/5875, report 17 July 1952; DNB 2.3/3766/1, memo 25 August 1955 (calculation benefits Britain).

182 NA Kew T236/5875, report 17 July 1952.

183 MFH 1.822.145.3, Ministry of Foreign Affairs to Ministry of Finance, 18 June 1945, Bataafsche to Ministry of Finance, 4 October 1945, memo Ministry of Finance 6 October 1945, Ministry of Foreign Affairs to Ministry of Finance, 25 October 1945, memo Ministry of Finance 6 November 1945, Ministry of Finance to Bataafsche, 14 November 1945. Van Hasselt and Oppenheim had asked the Dutch government not to give a clear formal refusal because this would force them to divulge it to the British government, from which they expected unpleasant repercussions, presumably on the negotiations for the Treasury Agreement: memo Ministry of Finance 6 October 1945.

184 MFH 1.822.145.3, Ministry of Foreign Affairs to Ministry of Finance, 8 April 1946 (objections), DNB 8.2/1650/1, Foreign Affairs to Finance, 29 April 1946 (more objections), Finance to Foreign Affairs, 25 June 1946 (reply to objections), Ministry

for Overseas Territories to Finance, 4 June 1946 (objections), Finance to Overseas Territories, 25 June 1946 (reply); DNB 2.331/3/1, Finance to Bataafsche, 9 May 1946 (assent with conditions).

185 SHA agendas Bataafsche, board minutes 13 December 1945.

186 SLA 120/3/9, board minutes Asiatic 13 February 1946.

187 Meijer, *Indische rekening*, 33, 38-41.

188 SLA HR boxes, Blair report 1959, 4-16 (regionalization); SHA 12/177, Waley to Stephens, 16 March 1944 (public relations committee), memo 26 April 1944 (staff education Argentina and elsewhere).

189 The Bataafsche meeting of 13 December 1945 had adopted a pension scheme for managing directors, who until then had had no such provision. SLA 120/3/9, board minutes Shell Petroleum, 9 July 1947.

190 SHA 11/16-1, memo 16 May 1944.

191 Beyen papers (private collection), memo 15 September 1947; SHA agendas Royal Dutch board meeting 13 March 1948, memo 13 January 1948. We are greatly indebted to Dr. W. H. Weenink for his giving free access to the Beyen papers, then under his care.

192 Beyen papers, memo H. Riemens, 19 April 1948.

193 SHA agendas Royal Dutch board meeting 13 March 1948, memo 13 January 1948, Annex 1.

194 SLA SC7/G1/28, Van Eck to Godber, 12 December 1947 (regret at failure to obtain 50:50); BEA G1/482, Foreign Office to Washington embassy, 31 January 1948 (Dutch directors in favour).

195 Minutes of the September 9 meeting in the Beyen papers. Bakker and Van Lent, *Lieftinck*, 110-1.

196 NA Kew T233/34, memo Beyen 3 October 1947, minute GLFB 15 October 1947, Lieftinck to Dalton, 18 October 1947, memo 23 October 1947; MFA 1.822.145.3, memos Treep 6 November 1947, 21 June 1948. Weenink, *Beyen*, 263.

Chapter 2

1 SLA 141/9/1, board folders Shell Transport, meeting 10 November 1948.

2 SLA 120/3/9, board minutes Asiatic, 23 June 1948, for comments after Kessler's retirement as a managing director of Asiatic.

3 Familiar in continental Europe, the *Raad van Commissarissen* concept was so little known in Britain that in February 1949 *Shell Magazine* felt the need to explain to its readers that 'The Board of Commissaries of the Royal Dutch supervises and advises but does not direct the company's affairs.' For British people the term 'supervisory board' was the nearest equivalent. In September 1948 the Bataafsche board, which the senior Dutch delegate member chaired, voted to have Kessler take the chair when present: SHA board agendas Bataafsche, 16 September 1948.

4 Royal Dutch Annual Report, 1948.

5 SHA 190D/701; Royal Dutch Annual Report, 1951, SLA 141/3/10-1, board minutes Shell Transport, 2 January 1952. After resigning as Director-General Van Hasselt remained a director of the Group holding companies until his death.

6 SHA 190D/760, Godber to W. T. S. Doyle, 27 January 1939.

7 *Shell Magazine*, May 1961, July 1965.

8 *Shell Magazine*, February 1947; SLA SC45/1, interview Loudon, 9 November 1987.

9 SLA SC45/2, interview Loudon, 17 January 1989, 48.

10 SLA SC7/G2/11/1, Godber to Van Eck, 18 December 1947.

11 *Time Magazine*, Vol. 75, No. 19, 9 May 1960, carried a detailed profile of Loudon. See also Howarth, 'Passports are not that important', *Shell World*, April 1996.

12 Sir Peter Holmes (1932-2002), conversation with Stephen Howarth, 28

June 1995.

13 *Time Magazine*, Vol. 75, No. 19, 9 May 1960.

14 SLA SC45/2, interview Loudon, 17 January 1989.

15 SHA190C/393A, Peet talking to New York Society of Security Analysts, 8 January 1957.

16 Royal Dutch Annual Report, 1953.

17 SLA 141/3/15, board minutes Shell Transport, 18 September 1963, details the conditions attached to warrants.

18 Gabriëls, *Koninklijke Olie*, 168.

19 SHA 2/387, memo discussions between Van Hasselt and Fraser, May 1950; according to this memo, Kessler had already argued for some years that Royal Dutch stock needed a New York listing. Painter was senior partner of the New York law firm of Cravath, Swaine and Moore, an adviser to Shell from the start of Group operations in America in 1912. He became a director of Shell Caribbean Petroleum Company in 1948, and in June 1961 was the first American to be a member of Royal Dutch's's supervisory board: *Shell Magazine*, July 1961; Gabriëls, *Koninklijke Olie*, 66.

20 SHA 190C/393A, memo May 1950.

21 The first fully reliable figures were those for 1951, published in the Annual Reports for 1952. SHA, Royal Dutch board agendas, 25 November 1958, memo November 1958 on accounting policy recommendations.

22 Peet had been treasurer of Shell Oil Company Incorporated, 1944-48; subsequently was its 'Financial Organiser' and General Vice President, and one of its directors; and eventually chief financial officer of the Group. Beaton, *Enterprise*, 302, 696, 768, 777, 780; SLA SC45/2, interview Loudon, 17 January 1989; SLA 141/3/11, board minutes Shell Transport , 1 December 1953.

23 *London Shell*, 27 June 1962.

24 Peet's first such presentation was the kind of report any director would wish to hear: 'In conclusion (...) sales, earnings, production, and cash were up, and debts were down', SLA 141/3/11, board minutes Shell Transport, 21 September 1954.

25 Gabriëls, *Koninklijke Olie*, 168.

26 SHA2/3, ANP to Bataafse, 16 June 1954.

27 SHA 190C/393A, Howard H. Banker to Wieringa, 19 April 1955, and reply, 28 April 1955. Quarterly reporting began in June 1955, and even seven years later *London Shell* could still claim smugly, 'In the publication of full quarterly results, the Group is much ahead of general practice on this side of the Atlantic', *London Shell*, 27 June 1962.

28 SHA 190C/393A, Royal Dutch fourth report, June 1954.

29 Gabriëls, *Koninklijke Olie*, 168.

30 SLA SC7/G2/11/2, Godber to F. C. Mitchell, 7 November 1945.

31 Gabriëls, *Koninklijke Olie*, 140.

32 SHA 190C/393A, 'The Royal Dutch/Shell Group Stockholder Relations', 7 May 1954.

33 This was seemingly a national characteristic: the owner of a small independent gasoline service station in Oregon wrote to Queen Juliana of the Netherlands complaining that 'It seems that Shell Oil Company (...) has decided to eliminate the Independent Operator', and asking her to use her influence to stop 'this oppression of small business'. SHA 2/364, Roy Copping to Queen Juliana, 21 March 1955.

34 SHA 190C/393A, G. K. Lind to Wieringa, 14 September 1954.

35 SHA 190C/393A, Haight to Wieringa, undated but after 14 September 1954.

36 SHA 190C/393A, Dupont & Co to O'Malley, November and December 1954; O'Malley to Royal Dutch, 16 December 1954.

37 SHA 190C/393A, H. J. Post to Royal Dutch, 27 December 1954.

38 SHA 190C/393A, Emanuel, Deetjen & Co to Loudon, 24 April 1955.

39 SHA 190C/393A, O'Malley to Royal Dutch, 22 November, 3 December 1954.

40 SHA 190C/393A, memo stockholder relations, 7 May 1954, O'Malley to Royal Dutch, 16 December 1954, H.J. Post to Royal Dutch, 27 December 1954.

41 Royal Dutch Annual Report, 1954; *Elsevier's Weekblad*, 'Gesprek met Jhr. Mr. J. H. Loudon', 24 March 1956, a copy of which in SHA 2/3.

42 *Elsevier's Weekblad*, 'Gesprek met Jhr. Mr. J. H. Loudon', 24 March 1956, a copy of which in SHA 2/3.

43 SHA 2/3, Proposal for alteration 27 April 1954, request for quotation 16 June 1954.

44 SLA 141/3/11, board minutes Shell Transport, 12 March 1957; trading began on 13 March.

45 *Financial and Operational Information* 1970-1979, 1.

46 SLA unnumbered file with organization charts from the 1950s.

47 SLA 120/3/9, board minutes Asiatic/Shell Petroleum, 23 January 1946.

48 SLA 141/9/1, board folders Shell Transport, 12 May 1948. At the same time Shell Française's capital was increased by 50 per cent.

49 SLA 141/9/1-13, board folders Shell Transport, 3 January 1950.

50 Royal Dutch Annual Report, 1948. The new name was adopted at an EGM on 19 January 1949 and took legal effect on 14 February 1949.

51 Beaton, *Enterprise*, 776.

52 SLA 141/3/11, board minutes Shell Transport, 8 November 1955; 120/3/11, board minutes Shell Petroleum, 23 November 1955; SHA board agendas Royal Dutch, meeting 20 October 1955, memo 14 October 1955.

53 SLA 141/3/11, board minutes Shell Transport, 3 November 1953; SLA 141/3/13, board minutes Shell Transport, 12 April 1967.

54 SHA 15/3, De Booy to Curaçao, 4 January 1940 (staff councils Curaçao); *De Bron*, 20 January 1947, *Olie*, February 1956; SLA DCS S12, CMD minutes 6 March 1962, with memo 2 March, minutes 5 April 1960 (quote). See also Volume 1, Chapter 5.

55 SHA board agendas Bataafse, meeting 3 June 1948, memo 28 May 1948.

56 Spaght relates how in 1961 or 1962 he and Loudon assembled all their courage to persuade Forsyth Wickes, who was already well into his eighties, to retire from the Shell Oil board. Wickes had been engaged as counsel in the 1920s, had helped to put Shell Union together in 1922, and had been on the board ever since. When Spaght and Loudon carefully broached the subject with him, Wickes replied that he could not retire because he had never been really appointed to the board, since Deterding had objected to a lawyer on the board and only consented on condition that the appointment be a temporary one: Shell Oil Houston, interview Spaght, 75-6.

57 SLA SC7/G2/11/2, Bearsted to Kessler, 18 October 1945.

58 SHA 190C/271, memo April 1954.

59 *Shell Magazine*, 'I remember...', June 1957.

60 *London Shell*, 21 November 1962. One thing the management tried to encourage but which employees did not like was the wearing of 'Shell' lapel

badges while on holiday, so that if they saw another Shell person they could get to know each other. The scheme never took firm root.

61 SHA 2/1948, Schepers to Loudon, 12 June 1956.

62 For example, Sampson, *Anatomy*, 481-2. On the same pages Sampson referred to Shell's 'internationalisation' and 'regionalisation' as the same thing, but even though he confused the terms, his judgement of their outcome is correct.

63 SHA 2/1948, Schepers to Loudon, 12 June 1956.

64 SLA Boxes HR, Blair report 1959, 4-6, 12-3. There was one exception, Rai Bahadur Govindachari, who joined the Group as a clerk in 1905 and rose to a senior management position in Burmah-Shell.

65 SLA Boxes HR, Blair report 1959, 14-5.

66 SHA 2/1907, report De Graef and Sierhuis, March-April 1956.

67 SHA 2/1948, memo Sinclair 18 July 1957.

68 SLA boxes HR, Blair report 1959, 7.

69 SLA GHC/UK/C4, Personnel.

70 SLA Boxes HR, Long Range Planning of Managers, 1964 review.

71 SLA Boxes HR, Long Range Planning of Managers, 1963 review.

72 SLA Boxes HR, Long Range Planning of Managers, 1963 review.

73 SLA DCS S12, CMD minutes 3 May 1961.

74 SLA DCS S12, CMD minutes 3 May 1961.

75 SHA Royal Dutch board agendas, 25 June 1958, memo 17 June 1956 on proposed sale of shares in Shell Cambodge to local interests; cf. SLA DCS C12, Schepers to Bataafse, 21 January 1959, noting that the Group until then had resisted requests for local participation in India and Pakistan, but now it had 'become increasingly evident that the time is coming when we shall need to identify ourselves more fully with the local scene in order to survive'.

76 SLA GHC/3F, MWG/AAF conversation, 22 June 1959.

77 Shell Transport Annual Report, 1957.

78 Shell Transport Annual Report, 1958.

79 Royal Dutch Annual Report 1969, 15, 1971, 13. Interestingly, The Hague was far more international than London, the former having 340 expats and 2,740 locals in 1969, whereas the corresponding figures for London were 153 expats and 4,307 locals: SHA Group personnel statistics 1969, 21.

80 Sampson, *Anatomy*, 482.

81 Sir Peter Holmes, conversation with Stephen Howarth, 28 June 1995.

82 Engle, 'Organisation', 346. Engle was the head of Shell Petroleum's Organisation Department. SLA DCS S2, memo E. B. Mayne, 13 July 1961.

83 De Jongh's career can be gleaned from SHA 190B/138, memoirs Späth, 10; 190G/245, Annual Reports Rhenania; 190A/146, correspondence Rhenania; 190C/8, transactions with Rhenania-Ossag; 10/523-1, correspondence Germany; 10/525, board Rhenania-Ossag; 10/549, financial reports Germany; *Olie*, December 1948.

84 SHA 2/1911, Loudon to Van Hasselt, 5 August 1949.

85 Engle, 'Organisation', 345.

86 SLA Central Office telephone directory, 1957.

87 Engle, 'Organisation', 349.

88 SLA unnumbered file with London office organization chart, 1953.

89 SHA 190D/707, circular 21 October 1955 announcing the appointment of the coordinators; 2/96, Bataafsche to CPIM, 15 November 1955, explaining the function of the coordinating committee. That the committee was seen as one step in a process of balancing functional with regional coordinators becomes evident from a sentence in the letter to the effect that 'at this stage no changes have been made to the status of the area departments'.

90 *Olie* 1958, 292.

91 The CMD briefly considered introducing the title vice president in 1959, but dropped it because the committee wanted a more profound consideration of the matter which the rush to set up service companies did not allow: SLA DCS S3, memo 26 February 1959. However, Shell Française adopted the title in 1962 as part of a reorganization bringing all Group operations in that country under one umbrella: Pradier, *Shell France*, 3.

92 Engle, 'Organisation', 348-50; SLA DCS S24, report April 1965; *Olie*, June 1949, for a description of the first regional meeting Europe on research and manufacturing, and September 1955, the first exploration engineering research conference.

93 Interview authors with Bill Bentley, 27 July 2005.

94 Chandler, *Strategy and Structure*, idem, *Scale*.

95 SHA 2/330-2, Wilkinson to Loudon, 30 May 1956.

96 SHA 190D/760, memo Looijesteijn; interview Maurits van Os with Hugh Parker, 31 August 2004. Parker was one of the McKinsey consultants who worked on the Venezuela and Central Offices project. Shell Houston, No. 9532919035, Texaco McKinsey & Co. reports from 1954-61, covers the reorganization of that company in detail. The box probably ended up with Shell Oil as a result of a merger of downstream activities with Texaco in the 1990s. On the early activities of American consultancy companies in Europe and their specific competencies see Kipping, 'Consultancies' and idem, 'American Management Consultancies'. According to Kipping, 'American Management Consultants' 212, Loudon was introduced to the McKinsey CEO Marvin Bower through the CEO of Texas Instruments, which appears unlikely given the link with Texaco. The summary of the McKinsey project with the Group given by McKenna, *Newest Profession*, 173, 177-9, is unfortunately rather inaccurate.

97 SHA 2/1418, correspondence Forster, 1949, 1950.

98 SHA 2/1413, Kemball-Cook (Caracas) to Asiatic Corporation (New York), 7 February 1957. The reports were entitled 'Programmed Management', 'Strengthening Personnel Administration to Improve Effectiveness and Profits', 'Increasing the Effectiveness of Marketing in Venezuela', and 'Staff Support for Top Management'.

99 SHA2/1418, memo 3 March 1958.

100 SHA 2/1418, Dunsterville (Caracas) to Stokes (London), 22 April 1958.

101 SLA DCS S10, report De Bruyn, undated; his visits took place in April-June 1958.

102 SLA DCS S3, CMD minutes 10 July 1957, agreeing with the draft terms; cf. SHA 2/1906, memo 18 July 1957, for draft terms.

103 SHA 2/1418, memo McKinsey & Co. to Group MDs, 28 August 1957.

104 SLA DCS S3, CMD minutes 10 July 1957, accepting the terms of reference for the McKinsey study.

105 SHA 2/1418, Dunsterville (Caracas) to Shell Petroleum, 11 October 1957.

106 SHA 2/1418, memo McKinsey & Co. to Group MDs, 28 August 1957.

107 SLA GHC/UK/D1, J. Taitt to Ministry, 16 September 1957.

108 SHA 2/97, Emery (London) to Puricelli (Buenos Aires), 20 August 1957; SHA 2/1418, memo McKinsey & Co. to Group MDs, 28 August 1957; SHA 2/1906, memo R. P. Keegstra to co-ordinators and departmental managers, 30 October 1957.

109 Interview Maurits van Os with Hugh Parker, 31 August 2004.

110 Sampson, *Anatomy*, 485; this was partly confirmed by Hugh Parker talking to Maurits van Os, 31 August 2004, with regards to the dual head office structure and to the exercise being about organization and not about cost-cutting.

111 Kruisinga, *Vraagstukken*; in 1954, Kruisinga had edited a collection of papers by various authors on the same issue under the title *The Balance between Centralization and Decentralization in Managerial Control*.

112 SHA 2/1906, memo 18 July 1957; 2/1418, memo McKinsey 28 August 1957 refers to this document having served as terms of reference for the consultancy, which were taken over in their totality.

113 SLA DCS S3, CMD minutes 7 November 1957, progress report Bower. According to Hugh Parker in his interview with Maurits van Os, 31 August 2004, it was Loudon who told the McKinsey team privately to remove the dual head office structure from their brief.

114 SLA SC45/2, interview Loudon, 17 January 1989, 9.

115 SHA 2/97, Berkin to Schepers, 20 December 1957, with the terms of reference for the team visiting overseas companies.

116 SLA DCS S10, CMD minutes 4 February 1958.

117 SHA 2/1906, report 15 August 1958.

118 SLA DCS S3, draft report McKinsey & Co., 22 December 1958, CMD minutes 5, 6, 7 January 1959, minutes meeting with coordinators, 15 January 1959. The changes to the top of the organization are well brought out by the unnumbered file in London with organization charts from the 1950s and 1960s.

119 The diagram drawn from SLA unnumbered file with organization charts.

120 Stephens' speech of 20 November 1961 printed in *London Shell*, 13 December 1961.

121 SLA DCS S2, memo Brouwer 18 May 1960, Brouwer to Langford, 19 May 1960, minutes meeting CMD with coordinators, 2 June 1960, CMD minutes 5 July 1961, 1 February 1962; DCS S3, CMD minutes 5, 6, 7, 8 January and following, draft memo 8 January 1959, minutes meeting CMD with coordinators, 15 January 1959, CMD minutes 25 February, 5 March, 3 April 1959, Central Offices Organization Guide, October 1959.

122 SLA DCS S2, memo E. B. Mayne, 13 July 1961; Wall, *Growth*, 60-87, for Jersey's reorganization. It appears to have been difficult for Larson to understand how the regional coordinators worked: SLA DCS S4, CMD minutes 27 March 1961.

123 SHA Conference 9 June 1971, Marketing Report 1970; SLA DCS S57, CMD minutes 4 July 1972; Wesseling, *Fuelling*, 30-1, 83-4, 191-2; for a vivid, if biased, description of the atmosphere at Shell Centre in 1972 3-9.

124 Shell Oil Houston, interview Spaght, 71.

125 SHA Conference minutes, Peter Samuel to Conference, 16 June 1965.

126 Information from Pieter Folmer.

127 Interview with Bill Bentley, 27 July 2005.

128 SLA DCS S3, CMD minutes 7 November 1957; DCS S35, unsigned memo optimum location head office, 17 March 1967, memo Brouwer, 10 April 1967.

129 Priest, 'Americanization'.

130 Relying heavily on Spaght's memoirs in Shell Oil Houston's Historical Documents Room and on Bridges, *Americanization*, Priest, 'Americanization' overemphasizes the Americanization of Shell Oil.

131 Beaton, *Enterprise*, 691. Priest, 'Americanization' 193, erroneously ascribes the rise of new managers in the Shell Union/Shell Oil ranks to a putative disruption in recruitment during the Second World War, whereas this would appear to have been a simple generational effect. On appointments SLA DCS C3, CMD minutes 2 May 1958, 10 November 1961. Shell Oil Houston, interview Spaght, 11.

132 Cf. for instance SLA DCS C3, CMD minutes 2 May 1961, Wilkinson reporting to the CMD about recent anti-trust cases in the US giving rise to serious concern 'as to the dangers which we would be facing if we were not to eliminate the risks which we had been content to run in the past'.

133 SC45/2, interview Loudon, 17 January 1989, 43-4. The danger was eliminated after strong diplomatic pressure from the UK and the Netherlands.

134 SHA 49/151, 15/215 (Simplex and Research Agreements); 49/23-1, Van Eck to De Booy, 20 July 1938 for the gist of the agreement and the date of the first agreement; on the function of Asiatic Corporation see Beaton, *Enterprise*, 149 and SLA DCS C27, memo Wagner 8 September 1972.

135 *Shell News*, November 1958, 1-4

136 Cf. Priest, 'Americanization' 194. For the Shell Oil budget cycle SLA DCS C3, CMD minutes 1 February 1961, memo Loudon on talks with Spaght, 23 February-3 March 1961. The limitation to E&P, chemicals, and research suggested by CMD minutes 7 March 1961, report Loudon on talks with Spaght; regular meetings for all functions evident from DCS C19, memo relationships Shell Oil with the service companies, 21 December 1965. Cf. DCS S3, CMD minutes 19 January 1959 about the coordination structure with regards to the US and Canada, also referring to 'the assistance that The Hague Office is already giving to Shell Oil and Shell Oil Canada in the exploration and production field, and intends to give increasingly in the manufacturing field', services to be coordinated by the DCO.

137 Shell Oil, annual review 1999; Shell Oil Houston, interview Spaght 1975, 2; *Shell News*, December 1961, 8-12, overview of the first season of eleven programmes, in which eleven different countries were visited.

138 'Going global', *Shell News*, summer 1998, 8-9; Shell Oil Annual Report 1999 gives the various American pectens. Information about the colour scheme, interview with Jack Doherty, 7 February 2006. On the trade mark issue between the Group and Shell Oil see Shell Oil Houston, historic documents room, legal memorandum, 'Shell Oil Company/Scallop Corporation Trademark Matter; Use of the Pecten Symbol', by James J. Mullen, 3 September 1980.

139 Erb took Dutch citizenship before his appointment as managing director of Royal Dutch in 1921.

140 Shell Oil Houston, interview Spaght, 90-1.

141 Shell Oil Houston, interview Spaght, 78-9. It was quite a large transaction for Shell Oil, which had offered $60 million for El Paso at a time when its own net income was around $200 million; according to Spaght, Loudon consulted his colleagues and then asked only one question, 'Are

you offering enough?', before consenting.

142 SLA DCS C19, CMD minutes 14 December 1965, memo 21 December 1965, CMD minutes 4 April 1966. Interestingly, Shell Oil's position within the Group differed completely from the situation of Unilever's American subsidiaries: Jones, 'Control', 464-76.

143 SLA DCS C19, CMD minutes 13 December 1966.

144 SLA DCS C27, CMD minutes 23 June 1969; Priest, 'Americanization', 199.

145 Priest, 'Americanization', 199, referring to the main evidence being articles and other public statements from oil analysts unconnected to Group companies.

146 Shell Oil, board minutes 28 June 1979; Priest, 'Americanization', 200.

147 SLA DCS C27, CMD minutes 10 July 1970.

148 SLA DCS C27, memo Wagner 8 September 1972.

149 Cf. for instance Shell Oil board minutes, 26 October 1972, 28 June 1973.

150 Priest, 'Americanization', 200; interview Joost Jonker with Alan Lackey, former General Counsel Shell Oil, Houston 9 February 2006. According to Spaght, 'corporate lawyers, corporate attorneys are a sort of a, they're a fraternity unto themselves, and it's the easiest thing for a corporate lawyer to tell his management of all of the risk that they face, and what will happen if so and so, and to warn them off of all of the pitfalls to the point where the poor guys are afraid to go to the bathroom', Shell Oil Houston, interview Spaght, 80.

151 SHA 2/1906, report L. A. Toone about staff selection development, 3 December 1958.

152 SLA DCS S10, memo 11 January 1962.

153 *London Shell*, 13 December 1961.

154 SLA DCS S12, Mullock to Hofland, 4 July 1961.

155 Cf. SLA DCS S12, CMD minutes 4 February 1958, a decision to help a retiree buy a home in France, presumably because due to particular circumstances he had been unable to build up suitable pension provisions.

156 SLA DCS S12, CMD minute, 4 June 1959. During his years in Venezuela Dunsterville had become an expert on orchids and after retirement from Shell he became a Research Fellow in Botany at Harvard.

157 SLA DCS S19, memo Brouwer on Van Bloeme, 13 November 1963.

158 SLA DCS S12, Escoffier to DCO ('defection'), 4 July 1962; Loudon to V. A. J. Jørgensen, 10 July 1962.

159 SLA DCS S12, Escoffier to DCO, 4 July 1962.

160 SLA DCS S11, unsigned memo, 2 October 1957.

161 SHA 190C/271, report McKinsey & Co., September 1956

162 SLA DCS S11, H. C. Langford to all MDs, 15 October 1959.

163 *Ibid.*

164 Shell Oil had introduced stock options in 1956; Royal Dutch and Shell Transport adopted them only in 1966. Shell Oil Annual Report 1956; Royal Dutch Annual Report 1966.

165 SLA DCS S12, memo 30 March 1960, referring to 23 people in the unclassified category.

166 SLA Boxes HR, report long-range planning of managers, 1962-68, and report Van Lennep and Muller, 1970; DCS S26, memo 15 February 1966 (Dr. Hollis W. Peter, Professor Van Lennep), S34, memo 21 May 1968 (CMD test), 29 August 1968 (new staff appraisal system); Muller, *Search* (research results). The separate career planning in London and The Hague had been integrated in 1957: SHA 2/96, circular Schepers 30 October 1957, with circular from Stephens and an organization chart of the new personnel department of the same date.

167 SLA DCS S12, CMD minutes 3 March 1959.

168 SLA DCS S11, CMD minutes 20 April 1960.

169 SLA Boxes HR, PN32, L. C. Whiteley to P. F. Nind, 21 December 1966.

170 SLA DCS S12, CMD minutes 28 January 1958.

171 Cf. correspondence in SHA 12/177, on scholarships for employees from Argentina, Cuba, and Brazil to train in the US. In 1949, the Group set up a secondary school in Praboemoelih, Indonesia, to start training Indonesian staff there: SHA board agendas Royal Dutch, meeting 20 October 1949, memo 7 October 1949.

172 SHA board agendas Bataafsche, meeting 16 October 1947, with memo outlining the scheme; *De Bron*, January and July 1947, *Olie*, July 1948, August 1952

173 SLA 141/11/16, board folders Shell Transport, memo 23 February 1945; SHA board agendas Bataafsche, meeting 13 February 1946, meeting 19 January 1947, memo 7 January 1947.

174 SLA DCS S26, Berkin to all MDs, 8 February 1965, 26 April 1965.

175 SLA DCS S11, memo Group Training Policy, 5 August 1958; Boxes HR, Long-Range Planning of Managers, 1963 review; *Olie*, February 1951, 120-1.

176 *Shell News*, June 1959, 1-5, July 1960, 27, May-June 1964, 10-5.

177 SLA DCS S11, memo Group Training Policy, 4 November 1958.

178 SLA DCS S11, CMD minutes 20 January 1959.

179 SLA DCS S11, note concerning discussions between Kirby and Loudon, 22 January 1959, 29 January 1959.

180 SLA DCS S11, Sir John Walker to CMD, 10 December 1959.

181 SLA DCS S11, CMD minutes 7 January 1960.

182 *London Shell*, February 1957.

183 *London Shell*, 13 September 1961. Although Shell could claim to have pioneered this method in British commercial life, an 'exactly similar' method had been used in 1942 in training young pilots of the Royal Air Force. A. W. Deller, Shell Refining Company, *London Shell*, 11 October 1961.

184 SLA DCS S11, Mullock to CMD, 23 June 1961.

185 SLA DCS S10, CMD minutes and memo 11 January 1962.

186 SLA DCS S10, memos 11 January, 15 March 1962.

187 SLA Boxes HR, Long Range Planning of Managers, 1963 review.

188 SLA DCS S19, Barran to CMD, 31 October 1963. The six points were 1. that the Group was already active in this field and should welcome an outside source rather than continuing to rely too much on its own efforts; but 2. that diplomas or degrees might result in men becoming available for employment only in their late twenties and therefore at higher cost than in their early twenties; there also should be limited numbers at the outset; 3. he would favour an institution attached to an existing university for 6-12 month courses; 4. there should be refresher courses in later years after practical experience; 5. the institution should be able to conduct research into methods and techniques of business and provide advice to outsiders; 6. only one body should be responsible for fund-raising. For this purpose HMG had suggested the Foundation for Management Education. The CMD were willing to subscribe funds provided the eventual scheme was one of which they approved and had financial support from other major business concerns.

189 SLA DCS S26, Berkin to MDs, 8 February 1965, 29 April 1965.

190 SLA DCS S34, report and reaction CMD.

191 SLA DCS S92, Pocock to all MDs, 30 April 1970.

192 SLA DCS S92, Training Review 1972, G. H. Gandy to all MDs, 9 March 1973.

193 Ibid.

194 *London Shell*, 23 November 1960.

195 *London Shell*, 7 December 1960.

196 Ibid.

197 The curious chauvinism displayed by Sampson, *Seven Sisters*, 136, that the Group was run by cautious yes-men until Legh-Jones arrived at the top and led a recovery is belied by the facts; and the notion that people like Godber or Kessler were yes-men is simply hilarious.

198 The good-natured aspect of these jousts is well brought out by an anecdote circulating within the Group which relates how, when discussing the opening of Shell Centre, Barran wondered whether this would not be a suitable occasion to invite the Queen, at which Loudon supposedly asked 'Who do you mean, the 60 per cent Queen or the 40 per cent Queen?'. The building was opened by Queen Elizabeth II. On the other hand, two Dutch CEOs of Anglo-Dutch multinationals are credited with saying that with 60:40 the field is level when playing the British, so much more adept at power games than the Dutch: attributed to Morris Tabaksblat, former Unilever CEO in *Financieele Dagblad* 21 March 2007; for a similar remark by Jeroen van der Veer see Volume 3, page 297.

Chapter 3

1 The term 'economic hostage' and the fundamentals underlying that situation derive from Shell Oil Houston, Historical Documents Room, paper entitled 'Historical Perspectives on the Oil Industry', (Shell Oil, December 1988), 6.

2 Bamberg, *BP* II, 483-511.

3 SLA SC45/1, interview Loudon 9 November 1987, 16; SLASC45/2, interview 17 January 1989, 15, 17.

4 SHA board agendas Bataafsche, memo 14 September 1954 about the formation of the consortium; *Shell Magazine* 1958, 206; Loudon appears to have been particularly keen to ensure that the managing directors of the consortium did not come from BP: interview Stephen Howarth with Sir Denis Wright, 30 June 2000.

5 Bamberg, *BP* III, 75-99, 294.

6 Yergin, *Prize*, 541.

7 SLA GHC/ME/A1, memo 2 November 1961.

8 Skeet, *OPEC*, 15-9; Yergin, *Prize*, 517-8.

9 Quoted in Skeet, *OPEC*, 18.

10 SLA SC45/2, interview Loudon 17 January 1989, 12.

11 Skeet, *OPEC*, 18.

12 Table 3.1: Royal Dutch Annual Report, 1952; *Financial and operational statistics*. Further figures in pounds sterling and guilders can be found in Volume 4.

13 Bamberg, *BP* III, 2.

14 Penrose, *International Petroleum Industry*, 91; Bamberg, *BP* III, 493.

15 Penrose, *International Petroleum Industry*, 121, Bamberg, *BP* III, 493.

16 SLA DCS DCS S16, analysis Jersey Standard, August 1963; Shell Oil Houston 163055274, Industry Studies Report, Analysis Jersey Standard, April 1971; SLA DCS S15, analysis Texaco December 1963, 3.

17 SLA DCS DCS S16, analysis Jersey, August 1963, 11; Jones, 'Control', 437-8.

18 SLA DCS S15, analysis Texaco, December 1963, 7-8; idem S16, analysis Jersey, August 1963, 10.

19 SHA Conference 24 November 1960, Marketing Report January-September 1960; idem 14 March 1962, Financial Report for Q4 1962; SLA DCS S15, analysis Texaco 1963, 3; DCS S16, analysis Jersey, August 1963, 14, 16, giving the Group's production balance between Venezuela and the Middle East for 1962 as 340 million barrels against 399 million, whereas for Jersey this was 502 million against 322 million.

20 SHA Conference 18 April 1962, Marketing Report 1960-62; idem 13 March 1963, Marketing Report 1961-62; idem 15 April 1964, Marketing Report 1963.

21 SHA agenda's Royal Dutch, 22 July 1959, Long Term Survey Summary of Group Supply and Demand Forecast, 1961-65, 30 June 1959.

22 SHA Royal Dutch board minutes, 30 May 1960; idem Conference 12 June 1968, Marketing Report 1967; idem Conference 10 June 1970, Marketing Report 1969.

23 *Petroleum Intelligence Weekly*, 28 August 1967, 7.

24 SHA Conference 20 January 1965, memo Group Objectives; Conference 16 June 1965, Marketing Report 1964; Conference 10 June 1970, Marketing Report 1969.

25 SHA Conference 12 June 1968, Marketing Report 1967.

26 Royal Dutch Annual Report 1964.

27 SLA DCS S16, analysis Jersey Standard 1963.

28 SHA Conference minutes 21 April 1965, 6.

29 Shell Oil Houston, Historical Documents Room, interview Spaght 1975, 36, describing the Group as a classically crude-short business. SHA Exploration Review 1966, for the data.

30 In 1965, supplies from long-term contracts totalled 869,000 b/d, of which 722,000 or almost 90 per cent from Kuwait: *Financial and Operational Information 1963-72*, 15, and idem, *1965-174*, 15.

31 Yergin, *Prize*, 420, quoting the British Foreign Office opinion in 1955 that the Group was 'to all intents and purposes a partner in the concession'.

32 SLA DCS S15, analysis Texaco 1963, 9-10.

33 SHA Conference minutes 11 February 1970, 9.

34 SHA Exploration review 1966, Appendix 8.

35 Bamberg, *BP* III, 206.

36 SHA Conference minutes 9 June 1971, 7, 17 May 1972, 6, 18 May 1973, E&P report 1972, 35.

37 SHA Exploration review 1966, Conference minutes 16 November 1966, 5-6.

38 Bamberg, *BP* III, 204; SHA Conference minutes 18 May 1973, E&P report 1972, 35, gives 42 countries for Shell, 38 for Texaco, 33 for Jersey, 32 for Socal, 29 for Gulf, 27 for Mobil, and 23 for BP.

39 SHA Exploration review 1966, Appendix 10.

40 SHA Exploration review 1966, 13.

41 SLA GHS/3D/1, Marketing Prices and Competition 1886-1959, 11 February 1959.

42 Bamberg, *BP* II, 172.

43 E. E. de Bruijn, Group co-ordinator of exploration and production, in G. Burck, 'Royal Dutch/Shell and Its New Competition', *Fortune*, October 1957 (reprinted 1957 by *Fortune* in conjunction with G. Burck, 'The Bountiful World of Royal Dutch/Shell', *Fortune*, September 1957), 23.

44 Shell Transport Annual Report 1958, 17.

45 On the 50:50 in Nigeria see SLA GHC/NIGE/C1/2/2.

46 Shell Oil Annual Report 1949, 8, 1950, 5-6, 1951, 7, 1962, 7-8, 1966, 8 and *Shell News* November-December 1965; SHA board agendas Royal Dutch, meeting 25 June 1958, Manufacturing Oil report January-May 1958 (Venezuela); SHA Conference minutes 14 March 1962, 2 (Venezuela); Gales and Smits, 'Olie en gas', 85 (Schoonebeek). The scope for secondary recovery appeared to be particularly promising for Venezuela, where the Group estimated to have some 20 billion barrels of heavy oil left in concessions after primary recovery: SHA Exploration review 1960, I-8, idem 1961, II-9-10; Long term exploration and production development review 1964, Graph 7.

47 Burleson, *Deep Challenge*, 37.

48 Beaton, *Enterprise*, 645-6.

49 Burleson, *Deep Challenge*, 40-5; Brouwer and Coenen, *Nederland*, 110-1.

50 *Shell News*, October-November 1962, 5-10, January-February 1963, 8-12; cf. *Shell News* September 1959, 21-3, for the first story about a deep floating platform.

51 Burleson, *Deep Challenge*, 89; *Shell News*, September-October 1967, 4-5.

52 Burleson, *Deep Challenge*, 102-7, Brouwer and Coenen, *Nederland*, 111.

53 Shell Oil Annual Report 1955, 8, 1956, 8, 1972, 9.

54 Cf. *Shell News* 1954, 4-6, 1958, 7-9, 1965, 14-5; according to Spaght, Shell Oil discussed taking up leases in Prudhoe Bay and decided not to, considering the opportunities insufficient; Shell Oil Houston Historical Documents Room, interview Spaght 1975, 18. Shell Oil held Alaskan leases to the maximum permitted under Federal law, the nearest being 15 miles away from the initial discovery on the North Slope: SHA Conference minutes 9 April 1969, 12. As Spaght put it to the Conference, 'Shell Oil are giving urgent consideration to their position and the possibilities open to them', i.e. they were mad as hell at being beaten to such a bounty and heads would roll. A well drilled on the lease closest to Prudhoe Bay and in the same formation yielded water rather than oil: SHA Conference minutes 10 September 1969, 8-9. Cf. Bamberg, *BP* III, 185-95.

55 SHA Long term exploration and production development review 1964, 15-8.

56 Shell Oil Annual Report 1964, 6.

57 SHA Exploration review 1966, 23, Conference minutes 15 May 1968, E&P report May 1967-April 1968, 3.

58 SHA Conference minutes 9 June 1971, E&P report May 1970-April 1971; Royal Dutch Annual Report 1971, 14, 1973, 18.

59 SHA Conference minutes 1 May 1968, E&P report May 1967-April 1968, 8-9, minutes 14 May 1969, 6, plus E&P report May 1968-April 1969, 8, 9 June 1971, E&P report May 1970-April 1971, 10, 21; SHA Longterm exploration and development review 1964, 25-9. Cf. also Exploration review 1966, 23-5.

60 *Financial and Operational Information 1963-1972*, 14; idem *1965-1974*, 14. In 1965, Loudon had predicted that, by 1990, one quarter of the world's oil and gas would come from beneath the seabed: *Shell News* September-October 1967, 5. SHA Conference minutes 18 May 1973, E&P report 1972, 2, also referring to the fact that 70 per cent of the additions to the reserves during the year had come from offshore. SHA Conference minutes 18 May 1973, E&P report 1972, 37, giving Group offshore acreage as 995,000 km^2, with the next largest, Gulf, holding 460,000 km^2.

61 SLA DCS C3, Burns to Loudon, 11 December 1957.

62 This can be deduced from SLA DCS C3, memo 13 December 1962. The CMD began functioning as such in 1946, but the DCS files only start in 1957; we have been unable to find earlier records for the CMD.

63 Shell Oil board minutes, 29 June 1948.

64 Shell Oil Houston, interview Spaght 1975, 34; according to Spaght, the CMD asked Shell Oil around 1950 whether it wanted to go abroad, and the board rejected the suggestion for the reason mentioned in the text.

65 SLA DCS C3, memo 13 December 1962; the amount of US$20 million taken from DCS 15, McFadzean to Loudon, 2 January 1963.

66 SLA DCS C15, McFadzean to CMD, 31 December 1962, McFadzean to Loudon, 2 January 1962; the loss of purpose evident from DCS C19, CMD minutes 30 June 1966.

67 SLA DCS C3, CMD minutes 31 July, 1 August 1957. Though there may have been some truth in the accusations, the margin is unlikely to have been the 10-30 per cent above market prices claimed by the plaintiffs, simply because no Group company would have accepted such a penalty in dealing with another Group company. The plaintiffs also wildly overestimated the volume allegedly bought from Group companies by putting it at 200-300,000 b/d. Shell Oil imported only about 50,000 b/d during the mid-sixties. Asiatic Corporation handled another 100,000 b/d of Venezuelan fuel oil, which the company itself sold along the Eastern Seaboard.

68 Data from Shell Oil Annual Reports.

69 SLA DCS C19, memo 19 April 1966.

70 Data from Royal Dutch *Financial and Operational Information*.

71 SLA DCS C19, memo 19 April 1966. The following month, Shell Oil sounded out the CMD with a proposal to invest US$300 million in the US coal industry, leading to the setting of three criteria which such a proposal would have to meet: SLA DCS C19, CMD minutes 5 May 1966.

72 SLA DCS C19, CMD minutes 30 June 1966.

73 Shell Oil Houston, interview Spaght 1975, 34-5.

74 Shell Oil Houston, interview Spaght 1975, 35.

75 SLA DCS C27, CMD minutes 13 May 1969.

76 SLA DCS C27, CMD minutes 7 December 1970.

77 According to Wesseling, *Fuelling* 104-6, Shell Oil also bid successfully for South Vietnamese offshore concessions in 1973. No oil in commercial quantities was found before the fall of Saigon two years later.

78 Five of the seven ventures mentioned had been terminated by 1975. SHA Conference minutes 24 February 1972, 7; Shell Oil, board minutes 26 October 1972; Shell Oil, Annual Report 1971, 11; 1972, 11; 1975, 9; Priest, 'Americanization', 200. On the Iranian talks Shell Oil board minutes, 27 November and 6 December 1973 (begin) and 24 April 1975 (indefinite suspension). As part of the Halpern vs Barran settlement in 1981, the Group also transferred rights in its Cameroon production company to Shell Oil.

79 SHA 69/74, Horstmann (NKPM) to De Kok, 12 October 1933, De Kok to Horstmann, 27 October 1933, Bolton to De Kok, 21 January 1938.

80 In a 1994 interview, Gerrit Wagner claimed to have done the search and found the document: Kaijser, 'From Slochteren', 338.

81 SHA 69/74, Wilkinson to Van Hasselt, 13 May 1946.

82 SHA 69/74, Wilkinson to Van Hasselt, 5 August 1946.

83 SHA 69/74, Van Hasselt to Legh-Jones, 12 August 1946.

84 Larson, Knowlton, and Popple, *New Horizons*, 733.

85 Larson, Knowlton, and Popple, *New Horizons*, 733, only hint at these arguments, saying in their sole mention of NAM that the participation was 'obtained' from Shell in 1946. Wall, *Growth*, 251 is much more detailed.

86 Gales and Smits, 'Olie en gas', 85. Production peaked in 1965-66 with 22,000 b/d, or 1.1 million tons.

87 On E&P in the Netherlands see Borghuis, *Veertig jaar NAM*, and Visser, Zonneveld, and Van Loon, *Seventy-five Years.*

88 According to Gales and Smits, 'Olie en gas', 86, NAM had explored for gas since 1948 and somewhat against the wishes of the company's two shareholders with their focus on oil. However, at least for the Group natural gas had always been a key concern in the United States and more recently in Canada as well; moreover, it is hard to see how the NAM managers could have executed an exploration programme for gas without the shareholders fully agreeing to it. Moreover, gas exploration was a NAM priority from the start: cf. Stheeman and Thiadens, 'History', 265.

89 Kaijser, 'From Slochteren', 345.

90 Algemene Bank Nederland, 'Oil and Natural Gas in the Netherlands', research report 1968, 1; Brouwer and Coenen, *Nederland* 177-94.

91 Gales and Smits, 'Olie en gas', 86-8; Schippers and Verbong, 'Revolutie'; Kaijser, 'From Slochteren', 346-56; *Financial and operational information 1968-77*, 17. See also Correljé, Van der Linde, and Westerwoudt, *Natural gas.*

92 SHA Conference minutes 18 September 1963, 8-9. Loudon commented that 'the Group would thereafter have or have applied for an interest in the offshore areas of all countries bordering the North Sea'. The Brigitta negotiation was successfully concluded in 1964 (Conference minutes 17 June 1964, 7).

93 SHA Berkin to Conference, Conference minutes 16 October 1963, 15.

94 SHA 190C/202, Typescript 'History of Shell Expro (25 Years)'.

95 By the autumn of 1963 about twenty companies were actively exploring the North Sea. SHA Schepers to Conference, Conference minutes 16 October 1963, 12. The seismic surveys were done by helicopter, because the area was still not free from mines, eighteen years after the war: Royal Dutch Annual Report 1964, 19.

96 SHA Berkin to Conference, Conference minutes 15 January 1964, 12, and 19 February 1964, 5-6; Bamberg, *BP* III, 199.

97 Bamberg, *BP* III, 199-200.

98 SHA Berkin to Conference, Conference minutes 16 September 1964, 8-9.

99 SHA Conference minutes 17 November 1965, 7, 9 June 1971, E&P report May 1970-April 1971, 2, 6-7.

100 SHA Conference minutes 18 May 1973, E&P report 1972, 36, quoting a figure of £1,000 b/d for the North Sea against £100 b/d for the Middle East.

101 Howarth, *Century*, 280.

102 Royal Dutch Annual Report, 1972; SHA, Conference minutes 27 February 1974.

103 SLA SC45/2, interview Loudon, 17 January 1989, 13.

104 SLA GHC/PGULF/A3, Thomas to Hopwood, 8 January 1949.

105 Yergin, *Prize*, 420, terming the contract a 'shadow integration' between Gulf and Shell, and quoting a British Foreign Office memo regarding the Group as 'to all intents and purposes a partner in the concession'.

106 SHA board agendas Bataafsche, meeting 5 June 1947, memo 28 May 1947 for a detailed overview of the initial agreement.

107 The ups and downs of the Gulf/Shell Kuwait agreement are charted in SLA SC91/b/27, history Gulf/Shell Kuwait agreement.

108 SLA DCS C15, memo 8 February 1963, Cummins to Berkin, 8 March 1963, memos McFadzean, 9 March, 29 July, 17 and 20 August, 3 December 1964, C19, memos McFadzean 17 and 26 March, 13 July, 2 September, 19 October, 4 November 1965.

109 SLA SP74/2, Shell Briefing Service 66/104, 1.

110 Shell Transport Annual Report 1946, 10; see also SLA GHS/2B/30 and GHC/EGY/B6/2.

111 SHA board agendas Bataafsche, meeting 20 November 1952, report Consolidated areas August-September 1952 (price agreement cancelled); meeting 18 December 1952, quarterly report manufacturing, drilling, and production (Jersey 1948, staff regionalization); meeting 15 October 1953, quarterly report manufacturing, drilling, and production (curtailing investment).

112 SHA board agendas Bataafsche, meeting 16 July 1953, report Consolidated May 1953. To facilitate such coordination, Anglo-Egyptian had taken a former British ambassador to Egypt onto its board.

113 SHA board agendas Bataafsche, meeting 20 May 1954, report Consolidated March 1954.

114 SHA board agendas Bataafsche, meeting 21 December 1954, report Consolidated, September-October 1954.

115 On the evacuation and sequestration see SLA GHC/EGY/B6/2, LGL 77/1-4, 80/1-2, 88/1-4, and 97/1-4; on the negotiations for a return, GHC/EGY/A1; SLA DCS C12, CMD minutes 30 September 1958 (acceptance monopoly), DCS C3, memo discussion Loudon and Barran with Eugene Black in New York, 28 February 1961, Loudon being quoted as saying that 'things were not working out too badly, although it was not expected that Shell would ever make any money there. Indeed this had not been expected at the time when the return to Egypt was negotiated, and the important thing was that Shell was back and in possession of its properties'. Black then suggested that the Western countries should make an effort to pull Nasser out of the communist orbit, and start by organizing state visits for him, for instance to Germany or the Netherlands.

116 SLA SC/45/2, interview Loudon 17 January 1989, 35-6.

117 SLA TR14, talk Stephens, 8.

118 SLA SC33/2, Wylie to Francis, 12 October 1967; SHA, unknown (probably Brouwer as chairman) to Conference, minutes 20 July 1966, 7.

119 SLA SC33/2, Wylie to Francis, 12 October 1967.

120 SLA Shell Transport Annual Report 1974, 15, 47.

121 SHA Loudon to Conference, Conference minutes, 20 September 1961, 6.

122 Bamberg, *BP* III, 165-6.

123 SHA Barran to Conference, Conference minutes 21 February 1962, 6.

124 SHA Barran to Conference, Conference minutes 14 December 1966, 9; Bamberg, *BP* III, 168-69.

125 SHA Barran to Conference, Conference minutes 11 October 1967, 9-10; Barran to Conference, Conference minutes, 13 December 1967, 15. See also Bamberg, *BP* III, 167-71, noting that in 1969 the Soviets also provided finance and technical assistance in developing North Rumaila, which came on stream in April 1972.

126 Cf. Micheels, *Vatenman* 222-3, for the barrel magnate Bernard van Leer being summoned to London immediately after the war and entrusted with building barrel factories in Indonesia and Singapore.

127 SHA board agendas Bataafsche, meeting 8 May 1946, memo 1 May (Java and Sumatra oil regions still inaccessible), meeting 20 May 1948, memo 10 May (idem for Djambi, Pladju, Cepu, and Pangkalan Brandan), meeting 17 January 1952, memo 7 January (access gained to Djambi, Mangundjaja, and Kawengan), meeting 19 November 1953, quarterly report (Indonesia holds on to northern Sumatra), 20 May 1954, quarterly report (idem), 15 July 1954 (idem); SLA GHC/ID/A2 (negotiations reopening Pladju and Soengei Gerong), EA96/2, Mackay to Bataafsche, 22 December 1952 (completion rehabilitation).

128 SHA 190C/21A, US Antitrust proceedings 1952-59, 183.

129 SHA 2/1739, Barran to Berkin, 29 August 1951.

130 Cf. for examples of the close control exercised SHA 2/1755, Oosten to Scholtens, 13 January 1950, Van Hasselt to Legh-Jones, 20 January 1950. The idea of forming an operating company was discussed and rejected in 1949-51: SHA 2/1755, memo 10 October 1949, 2/1739, memo 21 November 1951. The reason was probably the fear that the Indonesian government might not recognize a transfer of concession rights, or it might tax them: SLA DCS C13, CMD minutes 4 December 1957.

131 SLA DCS C13, memo Schouten 30 October 1957, with a CMD minutes of 23 October noting that Caltex was the biggest producer in Indonesia.

132 SHA 2/321, Bosman to Van Hasselt, 3 November 1948, 2/1755, Scholtens to Oosten, 5 January 1950, Oosten to Scholtens, 13 January 1950, 2/240, Scholtens to Bolderdijk, 29 January 1952, 2/1742, Schepers to Hilling, 20 August 1954, Schepers to Ministry Economics Djakarta, 23 September 1954, 2/1748, memo 9 January 1956, memo 2 December 1955, 2/243, Brouwer to Astley-Bell, 21 April 1958, 2/244, Sediono to Boot and Van der Voorst, 26 August 1958. Bartlett et al, *Pertamina*, 119-23.

133 SLA DCS C13, memo Schouten 30 October 1957.

134 Meijer, *Den Haag*, 581-94.

135 DCS C13, memo December 1957. By the end of 1959 the proportion of Dutch expats versus other nationalities was expected to be 60:40: idem, CMD minutes 9 July 1958; CMD minutes 14 October 1959 (240/550). Cf. *Shell News*, May 1958, for the story of four Shell Oil men assigned to Indonesia.

136 SLA DCS C13, CMD minutes 1 April 1958. Scheffer had followed up this daring feat by obtaining a government document giving protection against such attacks, at a cost of £6,000, half paid up, half to be paid on Group business passing safely through the troubles.

137 SHA 2/1742, memo 23 December 1955; 2/1907, Hilling to Schepers, 29 December 1954, memo 10 November 1955, Hirsch to Van Duursen, 15 December 1955, Van Duursen to Boot, 24 February 1956, Van Voorst to Brouwer, 23 May 1958, memo Brouwer 11 June 1958, memo Astley-Bell, 15 December 1958; 2/1948, memo 16 May 1956, Wagner to Van Duursen, 8 June 1956; 2/281, Schepers to Nederlandsche Bank, 10 July 1958. SLA SC45/2, interview Loudon 17 January 1989, 31-2; DCS C13, CMD minutes 9 and 17 December 1957, 8 and 15 January 1958. Canadian Shell, not to be confused with Shell Canada, was fully held by Bataafsche, so the transfer would have been more cosmetic than real. By 1960, a new arrangement had been hammered out involving Canadian Shell Overseas Ltd., a company specially set up for the purpose, see SLA 141/3/13, board minutes Shell Transport, 9 March 1960; 141/9/5, board memos Shell Transport, memo 19 September 1960; DCS C13, CMD minutes 12 October 1960; GHC/ID/A1/8/3, correspondence and contracts about the new company.

138 SHA 2/243, memo Astley-Bell, 18 December 1958.

139 SLA DCS C13, minutes 1 April 1958 (profits of £7-10 million on an investment of £300 million, investment prospects), minutes 9 July 1958 (market share and annual increase), minutes 29 October 1958 (quote). The Tandjung reserves were reckoned to provide half of the Group's total reserves in Indonesia of 450 million barrels: SLA DCS C13, memo 28 September 1959. In October 1959, the CMD considered estimates of profit in Indonesia for the years 1959-62, giving an average of £6 million a year.

140 SLA DCS C13, CMD minutes 10 August 1960, with undated memo Indonesia general. The Netherlands and Indonesia had broken off diplomatic relations.

141 SLA DCS C13, CMD minutes 22 September 1959, with memo 28 September counting the cost of cancellation.

142 SLA DCS C13, CMD minutes 7 March 1961, S16, minutes 30 April 1963; SHA

Conference minutes, 13 March 1963. After his resignation from the World Bank in 1963, Black joined the Royal Dutch board.

143 SLA DCS C13, CMD minutes 2, 10, 11 August, 15 September 1960.

144 SLA DCS C13, CMD minutes 2 January 1961, memo 16 January 1961.

145 SLA DCS C13, CMD minutes 11 April 1961, memo 5 April 1961.

146 SLA DCS C18, memo 17 June 1963, referring to tension building up in 1960, 1961, and 1962.

147 SLA DCS C13, CMD minutes 2 May 1961.

148 SLA DCS C13, CMD minutes 15 and 28 November 1962; SHA Conference minutes 21 November, 12 December 1962.

149 SLA DCS C13, CMD minutes 28 November 1961, 22 January and 28 February 1963; SHA Conference minutes 12 December 1962.

150 SHA Conference minutes, 19 June 1963. Even after the signing of the agreement on 1 June, the Indonesian government tried to force a split between the American companies and the Group by attempting to have them sign the definitive work contracts separately; this was pre-empted by 'high-level action', SHA Conference minutes 16 October 1963.

151 SLA DCS C18, CMD minutes 22 January, 28 February, 14 and 20 March, 14, 21 and 29 May, memo 17 June 1963; SHA Conference minutes, 13 March, 15 May, 19 June 1963; Aandstad, *Surrendering*, 87-94.

152 SLA DCS C18, memo 17 June 1963. Even before the Tokyo talks, the CMD expected that a settlement would lead to 'a large increase in profit (...) in Indonesia', SHA Conference minutes 17 April 1963.

153 SLA DCS C18, memo 17 June 1963; SHA Conference minutes, 19 June 1963 (quote from comments Lord Shawcross); Carlson, *Indonesia's Oil*, 14-8.

154 SHA Conference minutes 16 October, 20 November 1963; a call for volunteers resulted in seven Americans, five Germans, one Swiss, one Norwegian, three Venezuelans, five Frenchmen, one Belgian, two Dutchmen, two Italians, and one Dane offering to work in Kalimantan. Further British expatriates were being withdrawn in October 1964: SHA Conference minutes, 21 October 1964.

155 SLA DCS C18, CMD minutes 3 June 1965; SHA Conference minutes, 10 March 1965.

156 SHA Conference minutes, 15 April, 20 May 1964, 10 March 1965.

157 SLA DCS C18, CMD minutes 10, 24 August 1965.

158 Booth, *Indonesian Economy*, 73.

159 SHA Conference minutes 15 December 1965, 19 January 1966 with memo outlining the deal; SLA DCS C18, CMD minutes 6 January 1966.

160 SLA DCS C18, CMD minutes 19 November 1968, outlining a deal with Pertamina to ensure 'a substantial and continuing market for Shell lubricants and bitumen'; SHA Conference minutes 10 June 1970, 5.

161 SLA SC45/2, interview Loudon 17 January 1989, 15.

162 Booth, *Indonesian Economy*, 72-8.

163 Carlson, *Indonesia's Oil*, 23, 27.

164 SLA DCS C18, CMD minutes 20 September 1966; see also 7 October 1966, for another project.

165 SLA DCS C18, CMD minutes 22 January 1969.

166 SLA DCS C18, CMD minutes 1 April 1969

167 SLA DCS S10, CMD minutes 29 September and 4 October 1960. In its report at the end of October, the Group delegation noted that the relations with the Arab and Venezuelan representatives had been most friendly throughout. Shell had given both a cocktail party for 400 people and a dinner party for a smaller gathering, including the entire Venezuelan delegation.

168 SLA DCS S10, CMD minutes 1 February, 7 March 1961.

169 SLA GHC/ME/A13, G. G. Stockwell (BP) to Dowson and Woollcombe (Group), 29 December 1960.

170 SLA GHC/ME/A13, memo Piets, 15 November 1960; DCS S8, memo Piets January 1962; S10, CMD minutes 1 February 1961; SLA GHC/ME/A13, meeting of Loudon and Stephens with BP chairman Bridgeman, 9 March 1961.

171 SLA GHC/ME/A13, Searight to Berkin, 2 November 1960. Rodney Searight (1909-91) worked for Shell in the Middle East for many years. He loved the region, becoming an expert in its history and culture, and built up a large collection of 18th- and 19th-century watercolours and drawings of it. The Searight Collection was presented to the London Victoria and Albert Museum in 1985.

172 SLA DCS S10, CMD minutes 20 December 1960.

173 SLA DCS S10, CMD minutes 7 March 1961; SLA GHC/ME/A13, memo discussions between Shell, BP and Jersey, 7-8 March 1961.

174 Skeet, *OPEC*, 64; a vivid description of McCloy and his activities in Sampson, *Seven Sisters*, 254-9.

175 Cf. Skeet, *OPEC*, 35-7.

176 SHA Conference minutes 17 October 1962, 8, 12 December 1962, 10, 17 April 1963, 7, 18 September 1963, 12-3, 16 October 1963, 12-3, 15 January 1964, 9-10.

177 SHA Conference minutes 15 June 1966, 9.

178 SLA DCS S10, CMD minutes 7 March 1961; SLA GHC/ME/A13, meeting of Loudon and Stephens with Bridgeman, 9 March 1961.

179 SHA Conference minutes 11 October 1967, 5.

180 SHA Conference minutes 26 February 1969, 5.

181 SLA SC33/12, memoirs Carlisle; Howarth, *Century*, 300.

182 SHA Conference minutes 13 January 1971, 4-6; Yergin, *Prize*, 580.

183 Skeet, *OPEC*, 64-5; Sampson, *Seven Sisters*, 254-9.

184 Skeet, *OPEC*, 65.

185 SHA Conference minutes 13 January 1971, 4-6, 14 April 1971, 10, 12 May 1971, 10; Yergin, *Prize*, 580-3; Nouschi, *La France*, 248-9.

186 SHA Conference minutes 15 September 1971, 4-7.

187 SHA Conference minutes 13 October 1971, 6.

188 SHA Conference minutes 10 November 1971, 9.

189 SHA Conference minutes 10 November 1971, 10, 8 December 1971, 10.

190 Skeet, *OPEC*, 57. The Group started to consider the likelihood and consequences of government participation in 1964: SHA Conference minutes 15 January 1964, 8.

191 SHA Conference minutes 14 June 1972 (quote), 5-6, 12 July 1972, 3-4, 13 September 1972, 6-8, 11 October 1972, 7-9, 15 November 1972, 17-8, 28 February 1973, 8-9; Bamberg, *BP* III, 467-74; Yergin, *Prize*, 583-5.

192 SLA DCS S10, memo 24 January 1961.

Chapter 4

1 Jones, 'Control', 437-8.
2 Shell Oil Annual Reports, 1950-70.
3 Shell Oil Annual Reports, 1950-70.
4 Source for Figures 4.1 and 4.2, *Financial and Operational Information*, 1955-75; Shell Oil Annual Reports.
5 SHA 2/364, Burns to Legh-Jones, 7 November 1947.
6 Priest, 'Americanization', 194 (training Deutsche Shell marketers in 1957); SHA 2/1910, Percy to Van der Woude, 26 May 1952, Scholtens to Percy 6 November 1952 (reorganizations Indonesia and Burma).
7 SLA DCS S9, CMD minutes 1 October 1958; S10, reports E. E. de Bruyn, management advisory committee.
8 Shell Oil Houston, board minutes 22 November 1953, 23 March 1961, 10 December 1965, 23 December 1966, 21 December 1967.
9 SHA Conference minutes 9 December 1970, 8.
10 Shell Oil, Annual Report 1961.
11 SLA DCS C19, CMD minutes 15 December 1966.
12 SLA DCS C19, CMD minutes 15 December 1966. According to McCurdy, the pay-out time of the new system was now eight years, which he considered long but still economically justified.
13 The move was quickly considered a success, notably in bridging a perceived gap between operations and research: Shell Oil Houston, interview C. W. Humphreys, 71. There were also other advantages, such as uniting the company's patents department, located on the West Coast, with the licensing department from New York: 'I remember back when', *Shell News* Fall 1998, 26.
14 Shell Oil Houston, Historical Documents Room, interview McCurdy, 13, crediting Kemball-Cook with the reorganization into MTM.
15 SHA agendas Bataafsche, 15 September 1949, memo 6 September 1949.
16 Beaton, *Enterprise*, 658-9; Royal Dutch, Annual Report 1948, 22-3. See 'Shell Chemical Corporation 1945-55: becoming a national contender' for more detail.
17 SHA agendas Bataafsche, minutes Bataafsche board 22 May, 8 August, and 7 November 1946, and memo 20 November 1946; cf. 190A/179A, for a vigorous effort by the director of KSLA, Pfeiffer, to concentrate petrochemical production near the research facilities at Amsterdam. Homburg, Van Selm, and Vincken, 'Industrialisatie', 387; *Van Rotterdam Charlois naar Rotterdam Pernis*, 60.
18 SHA agendas Bataafsche, minutes 16 January 1947 (memo 20 November 1946), 10 September 1947 (memo 8 September 1947), 20 May 1948 (memo 4 May 1948), 20 October 1949 (memo 10 October 1949), 21 December 1950 (undated memo capital budget Pernis 1951), 21 December 1951 (memo capital budget Pernis 1952), 18 December 1952 (memo capital budget Pernis 1953); *Van Rotterdam Charlois naar Rotterdam Pernis*, 64; Homburg, Van Selm, and Vincken, 'Industrialisatie', 385.
19 SHA Royal Dutch minutes, 11 April 1957, report on the annual accounts for 1956; *Van Rotterdam Charlois naar Rotterdam Pernis*, 123; Hengstebeck, *Petroleum processing*, 24-5.
20 SHA 15/202, two- and three-party discussions; specifically on refineries in consumption countries; 49/664, Fenwick to Sluyterman van Loo, 13 July 1935 (joint resistance with AIOC to Italian Government's wish for a refinery); 15/181 talks and correspondence about siting refineries in consumer countries, 1934-38; 10/581, Kessler to New York, 1 May 1935 (joint action successful in thwarting refinery plans in Japan, Denmark and Ireland).
21 BEA G1/482, memo Group position, 3 December 1942, Annexes C (refining capacity) and D (volume sales), combined with crude supplies to Germany as proxy for refining capacity there.
22 SHA board agendas Bataafsche, 10 September 1947, memo Capital requirements of the oil industry, 8 September 1947.
23 SHA board agendas Bataafsche, minutes 16 November 1950, quarterly report production and exploration areas; idem 21 December 1950, undated memo Curaçao capital expenditure budget; idem 15 February 1951, memo 5 February. Through careful negotiations the restrictions were gradually somewhat eased.
24 SHA 190C/387 1-2, two booklets with manufacturing data and statistics, 1945-75, to which we have added the capital expenditure figures from the Shell Oil Annual Reports.
25 *One hundred years of Shell refining*, 3; *Financial and operational information*, 1955-64, 14-5.
26 SLA 141/9/1, board folders Shell Transport, Hopwood to board, 2 January 1952, DCS C35, CMD minutes 20 July 1971; Corley, *Burmah Oil* II, 313-5; Royal Dutch Annual Report 1955, 34.
27 SLA DCS C7, CMD minutes 10, 11, 17 and 24 September 1957, 20 and 22 January, 9 February, 3, 5 and 24 March, 12 May, 9 and 10 December 1959, 19 January and 31 May, 14 June, 20 and 27 July 1960, 6 March, 12, 18, 19 April, 9 May, 7, 20 November, 14 December 1961, 5 and 10 April 1962.
28 SLA DCS C7, CMD minutes 1 November 1962. Mattei had died on October 27, when his private plane crashed in bad

weather.

29 DCS S15, memo 21 May 1963.

30 Corley, *Burmah Oil* II, 313-5; SHA Royal Dutch agendas, board minutes 25 February 1959, Manufacturing Oil report December 1958-January 1959; idem, 23 September 1959, Manufacturing Oil report February-August 1959; Brouwer to Conference, 21 February 1962. The 30% stake in the Sri Racha refinery in Thailand was only acquired in the late 1960s.

31 Cf. SHA Conference 14 June 1967, Manufacturing Oil report July 1966-June 1967; cost breakdown 1964-67, year totals; idem 11 June 1969, Manufacturing Oil report July 1968-June 1969; idem 12 May 1971, Manufacturing Oil report 1970-71. By 1960, however, Group managers started to question the economics of joint refineries, since the administrative complications and technical compromises involved appeared to outweigh the benefits: SLA DCS S13, CMD minutes 24 March 1960. In 1968, Barran raised the same objections to the economics of joint refineries: SLA DCS 32, CMD minutes 19 November 1968.

32 SHA Bataafsche board minutes, 15 September 1949, memo 6 September 1949 on the rehabilitation of the Pladju reforming unit; idem 19 January 1950, memo 4 January 1950 on the building of a sharp fractionating plant at Pladju. The unit came on stream that same year: Royal Dutch Annual Report 1950, 27. US$ 64.8 Bataafsche minutes 17 January 1952, memo 7 January 1952 on the 1952 capital expenditure budget for Indonesia (restriction to maintenance only); idem 15 July 1954, memo 1 July 1954 on building the Edeleanu plant at Pernis rather than Balik Papan.

33 *Financial and operational information* 1965-74, 18-9.

34 SHA Conference 17 July 1963, Manufacturing Oil Annual Report; idem 15 January 1964, Manufacturing Oil Annual Report; Murray, *Go Well*, 164-5.

35 BEA G1/482, memo 3 December 1942, Annexes C and D.

36 Waller and Swain, 'Changing patterns', 148-9.

37 Waller and Swain, 'Changing patterns', 148.

38 SHA Royal Dutch minutes 26 March 1958, memo 6 March on the building of a refinery in the Upper Rhine region; Waller and Swain, 'Changing patterns', 149. The Marseilles pipeline opened in 1963, pre-empting by several years the ENI project, which was beset by continuing difficulties with planning authorities along its route.

39 Waller and Swain, 'Changing patterns', 149, 151-2.

40 *One hundred years of Shell refining*, 3, 6; *Financial and operational information*, successive issues for the years 1955-70.

41 *One hundred years of Shell refining*, 6.

42 SHA Conference minutes, McFadzean to Conference, 17 October 1962.

43 SHA Conference minutes 13 July 1963, 3.

44 Shell Oil Annual Report 1966, 10.

45 Debottlenecking first mentioned in Shell Oil Houston, board minutes 16 March 1950; SHA Conference minutes, Hoog to Conference, 19 September 1962 (refinery efficiency teams set up in 1959).

46 SHA Conference 19 September 1962, Manufacturing Oil report February-August 1962; idem 16 June 1965, report July 1964-June 1965; idem 11 June 1969, report July 1968- June 1969.

47 Interview Joost Jonker with Jan Verloop, 26 September 2005.

48 *Financial and operational information*, successive issues for the years 1955-70.

49 SHA Royal Dutch minutes, 25 June 1958, Manufacturing Oil report January-May 1958; idem 18 December 1958, Manufacturing Oil report June-November 1958; Conference 15 July 1964, Manufacturing Oil report January-June 1964. SLA DCS C7, CMD minutes 24 September 1957. In 1972, the Group possessed only 255,000 b/d of primary processing capacity in Italy, against 380,000 b/d in Germany and 630,000 b/d in France, whereas the countries had roughly similar population sizes and numbers of motor vehicles.

50 *One hundred years of Shell refining*, 6.

51 In 1962, unit cost at old refineries was US$64.8 per barrel, new refineries 55.9, Group total 64.6; three years later, old refineries were at 58.6, new ones 55.8, Group total 58.0. SHA Conference 16 June 1965, Manufacturing Oil report July 1964-June 1965; Conference 15 June 1966, Manufacturing Oil report July 1965-June 1966; *Van Rotterdam Charlois naar Rotterdam Pernis*, 119.

52 SLA DCS S32, CMD minutes 23 May 1967 and Shell Safety Bulletin Nos 1 and 2, January and July 1968; SHA Conference 12 June 1968, Manufacturing Oil, Report July 1967-June 1968. The accident rates reported in these sources show the figures for the Group outside North America stagnant for 1963, 1964 and 1965, but falling since 1966. Shell Oil's figures had actually risen slightly, so there was concern over that, too. During 1949, Shell Oil's frequency rate was 8.17 and the severity rate 0.64: *Shell News* October 1950, 6-7. There must have been some exchange of safety information during the 1950s, for in August 1957 *Olie* republished an article on safety helmets which first appeared in Shell Oil's *Shell News*, March 1954. In May 1959 the Group held its first industrial safety conference in The Hague: *Shell Spiegel* June-July 1959, 17.

53 SHA Conference 10 June 1970, Manufacturing Oil , Report July 1969-June 1970. Cf. SLA DCS S63, for safety statistics from from 1957, which show that the Group had already made considerable progress, the frequency rate for all operating companies dropping from 20.1 to 11.6 from 1957 to 1969.

54 Landau, 'Process', 148-9.

55 Beaton, *Enterprise*, 661; Spitz, *Petrochemicals*, 176-83; Hengstebeck, *Petroleum processing*, 179-207; Gary and Handwerk, *Petroleum refining*, 76-98. SHA Bataafsche minutes, 21 December 1950, undated memo capital expenditure Pernis; Shell Oil Annual Report 1955, 10.

56 *Shell News*, October 1952, 1.

57 SHA Conference 19 September 1962, Manufacturing Oil report February-August 1962; idem 10 June 1970, Manufacturing Oil report July 1969-June 1970.

58 SHA Conference 17 July 1963, Manufacturing Oil report February-June 1963.

59 Hengstebeck, *Petroleum processing*, 272-3; Gary and Handwerk, *Petroleum refining*, 130-2.

60 SHA Royal Dutch minutes 25 June 1958, Manufacturing Oil report January-May 1958.

61 SHA Royal Dutch minutes, 18 December 1958, Manufacturing Oil report, June-November 1958.

62 *Shell News* May-June 1966. The hydrocracking process was partially developed in collaboration with BP and Gulf: SHA Conference 17 July 1963, Manufacturing Oil report February-June 1963.

63 Shell Oil Annual Report 1967, 11.

64 Royal Dutch Annual Report 1974, 28.

65 For the passage on the Group's commitment to hydrogenation we are much indebted to information supplied by Jan Verloop. A survey of Shell research on hydrogenation in Schweppe, *Research*, 136-8, 188-92, and Oelderik, Sie, and Bode, 'Progress'.

66 SHA Conference 16 June 1965, Manufacturing Oil report July 1964-June 1965, 12 June 1968, July 1967-June 1968.

67 SHA Conference 20 September 1961, Manufacturing Oil report February-August 1961; idem 19 September 1962, Manufacturing Oil report February-August 1962; idem 10 June 1970, Manufacturing Oil report July 1969-June 1970.

68 SHA Conference 12 May 1971, Manufacturing Oil report mid 1970-mid 1971.

69 SHA Conference 10 June 1970, Manufacturing Oil report July 1969-June 1970; idem 14 June 1972, Manufacturing Oil report May 1972.

70 SHA Conference 10 June 1970, Manufacturing Oil report July 1969-June 1970.

71 SHA Conference minutes 14 June 1972, Manufacturing Oil report May 1972. Excess capacity was put at 530,000 b/d, which added US 6 cents to manufacturing costs: SLA DCS S53, CMD minutes 31 May 1972.

72 SLA, Chairman's address to the Shell Transport AGM of 1947 gives the British-flag figure of 632,000 dwt. Dutch-flag figures were La Corona, 184,220; NIT, 52,382; CSM, 10,514. Calculated from Wouters, *Shell Tankers* 100, 126-7.

73 Middlemiss, *Anglo-Saxon/Shell Tankers*, 145-7.

74 Ratcliffe M., *Liquid Gold Ships: History of the Oil Tanker*, 128.

75 SHA Conference minutes 17 April 1963, 10; 17 July 1963, 13; 10 March 1965, 13; 21 April 1965, 3-4.

76 SHA Conference minutes 16 November 1964, 8.

77 Wouters, *Shell Tankers*, 165.

78 SHA Conference minutes 17 February 1965, 4.

79 SLA MR 53/1-2, VLCC Tank Cleaning Safety, January 1969-July 1970 and July-December 1970, together give detailed records on the industry tests, which were successful. The most probable cause was established as electro-static ignition: the occurrence of a spark within the mist created by the cleaning water-jets inside the tanks. Tested solutions included altering the proportion of gas to air inside the tanks, making the mixture either too rich or too lean to explode, but the final answer was to flood the tanks with inert gas which simply could not explode.

80 SLA SC33/2, 'Middle East Crisis 1967': 'Competitors' Imports into Europe', memo 10 July 1967.

81 SLA SC33/1, 'Middle East Crisis 1967': 'Cape Movements', memo 10 July 1967.

82 SLA SC33/2, 'Middle East Crisis 1967': 'Middle East Contingency Study', memo 30 May 1967.

83 SHA Conference minutes 12 April 1967, 5, minutes 14 June 1967, 3.

84 SLA: SC33/1, 'Middle East Crisis 1967': 'Tonnage Position', memo 21 June 1967; SC33/2, 'Middle East Crisis 1967': 'Monthly Marine Highlights
June 1967', 6 July 1967; SHA Conference minutes 14 June 1967, 3; J. MacDonald, 'Why Tanker Rates are Rising Sharply', *Financial Times*, 16 June 1967.

85 SLA SC33/1, 'Middle East Crisis 1967', memo 21 June 1967.

86 SLA SC33/2, 'Middle East Crisis 1967', memo 14 June 1967.

87 SLA SC33/2, 'Middle East Crisis 1967': 'UK Price Surcharge', unsigned undated paper.

88 Interview Stephen Howarth with Barran, 27 October 1994.

89 SLA SC33/2, 'Middle East Crisis 1967': 'Middle East Contingency Study', memo 30 May 1967.

90 SLA: SC32/2, 'Middle East Crisis 1967', memo 18 July 1967; J. MacDonald, 'Why Tanker Rates are Rising Sharply', *Financial Times*, 16 June 1967.

91 Wouters, *Shell Tankers*, 160.

92 SHA Conference minutes 10 July 1968, 6.

93 In ships this is governed by the 'square/cube' rule. Take two cubes, cube A being two metres each way and cube B, four metres. The volume of cube A is 8 cubic metres (2 x 2 x 2), and the volume of cube B, 64 cubic metres (4 x 4 x 4); thus cube B has eight times the volume of cube A. But the surface area of cube A is 2 x 2 x 6, the number of sides, for a total of 24 square metres, whereas the surface area of cube B is 4 x 4 x 6, or 96 square metres just four times the surface area of cube A. In other words, multiplying the surface area by four times produces eight times the volume, and since in a ship the surface area broadly dictates the building cost, the much greater capacity (and therefore the earning power) of cube B is achieved at much less proportionate cost. Likewise with pipelines: a 20-inch diameter pipeline uses twice the amount of steel of a 10-inch, but has four times the capacity. Moreover, the larger flow encounters less friction and resistance, so is easier to pump and requires fewer pumping stations, and overall a fully-laden 20-inch pipeline moves oil at 45 per cent the cost per barrel of a fully-laden 10-inch pipeline.

94 Beaton, *Enterprise*, 789-90.

95 Beaton, *Enterprise*, 652-5.

96 *Shell Magazine*, 1958, 189; *London Shell*, 13 September 1961; SHA Conference minutes 10 March 1965, 4.

97 *London Shell*, 13 May 1959.

98 *Shell Magazine*, 1958, 357.

99 SHA Conference minutes 14 June 1967, 6.

100 SLA GHS/3H/1, memo 13 June 1956, Pipeline Survey in North Sea and Mediterranean, May-June 1956.

101 *Ibid*.

102 SHA2/299, Legh-Jones to Fowlie, 30 June 1947.

103 SLA St Helen's Court office telephone directory, January 1938.

104 SHA 2/364 , Marketing and Marketing Policy, June 1946-June 1958 US, Oostermeyer to Kessler, 17 September 1946.

105 SHA 2/96 Organization London Office, memo Hopwood to Heads of Departments, 24 September 1951.

106 SHA 2/96, Loudon to Davidson, 22 December 1953.

107 SLA, unnumbered folders with organization charts, 1953-60.

108 Unfortunately only one of the Organisation Study Unit's files has survived, showing its work in analysing and rearranging the Wax and Candles General Department: SLA GHS/3E/4.

109 SHA 190D/707-708, personnel files A. Hofland, notes on his career and circular Bloemgarten, 21 October 1955.

110 SHA 2/94, 3 December 1956, The Hague to Buenos Aires, 11 December 1956; idem 2/1910, Report of Technical Organization visit to Bataafsche Handelszaken, October 1957.

111 SHA 190C/274, Bataafse (Djakarta) to Shell International Petroleum, 9 July 1959.

112 Source for Figure 4.3, *One Hundred Years of*

Shell Refining.

113 Shell Oil Houston 163055274, competition analysis Jersey Standard, plus Shell Oil Annual Report 1968.

114 SHA Conference 11 June 1969, Marketing Report 1968.

115 SLA GHS/3G/1, Products general, GHS/3D/1, Marketing Prices and Competition.

116 SHA 2/364, Shell Oil aviation sales statistics 1949-53.

117 *Shell News*, July 1960, 4.

118 *London Shell*, 10 December 1958.

119 *Shell New,s* July 1960, 4, May 1961.

120 SHA Conference 15 June 1966, Marketing Report 1965; Royal Dutch Annual Report 1970, 5.

121 *Shell News*, July 1952 (introduced on the aviation market); July 1959 (in use with armed forces since 1951); Shell Oil Annual Report 1951, 8. The patent was finally granted in 1959.

122 SHA minutes Bataafsche, memo 22 July 1954 on building a cat cracker and reformer at Harburg (Germany); idem memo 23 August 1954, on building a cat cracker and alkylation plant at Cardón in Venezuela. Cf. Hengstebeck, *Petroleum processing*, 4, for an interesting graph tracing the development of octane numbers and engine compression ratios from 1925 to the late 1950s.

123 SHA 2/364, Burns to Hopwood, 7 December 1953; SLA 141/3/11, board minutes Shell Transport, 13 July 1954, 12 October 1954.

124 SHA Bataafsche minutes 20 May 1954, report Consolidated Oil Company, February 1954; idem 21 December 1954, report Consolidated September-October 1954, noting increased sales in all areas, notably South Africa, following the introduction of Shell ICA.

125 SHA Bataafsche minutes memo 22 July 1954 on building a cat cracker and reformer at Harburg (Germany); idem 21 October 1954, memo 13 October 1954, erection of a platformer at La Spezia. Shell Française took similar steps in 1957, SHA board minutes Royal Dutch 21 March 1957, memo 1 March 1957 on the building of a platformer at Petit-Couronne in the expected rise of octane numbers to 94 premium and 84 regular.

126 SHA 2/364, Burns to Hopwood, 19 January 1954.

127 SLA DCS S4, CMD minutes 1 October 1958.

128 SHA 2/364, Burns to Hopwood, 30 October 1953.

129 *London Shell*, June 1956.

130 SHA 2/364, Burns to Loudon, 16 May 1958.

131 SHA board minutes Royal Dutch, 22 February 1959, Marketing Report 1958, memo 1 February 1959.

132 Until 1951 Esso Petroleum Co. Ltd. (the British marketing arm of Jersey Standard) was known as Anglo-American Oil Co. Ltd.

133 Wall, *Growth*, 132-3; Esso magazine centenary issue, spring 1988, 40-1.

134 SHA Conference 12 June 1968, Marketing Report 1967.

135 Betjeman resigned in 1964 after a row about interference with the text of one of the guides: Hillier, *New fame, new love*, 66-79.

136 SLA Organization Chart 15 January 1959.

137 SHA Conference 16 June 1965, Marketing Report 1964.

138 SHA Conference minutes 16 June 1965, 6.

139 SHA Conference 11 June 1968, Marketing Report 1967.

140 SHA Conference minutes 20 February 1963, 6.

141 SLA SC55/13, Shell-Mex and BP Year Book 1954, 34-39.

142 'Monopolies Commission: Petrol - A *Report*, on the Supply of Petrol to Retailers in the United Kingdom' (HMSO: London, 1965), 19-22, 31.

143 Monopolies Commission, *Report*, 35, 130, 149-51, 153-6; Bamberg, *BP* III, 230-4.

144 Monopolies Commission, *Report*, 22-3.

145 SHA Conference minutes 16 June 1965, 7.

146 SHA Conference minutes 15 December 1965, 4-5.

147 Bamberg, *BP* III, 234-40.

148 *Shell Magazine*, 1959, 176-9 and *London Shell*, 11 March 1959.

149 *London Shell*, 11 March 1959.

150 *Financial Times*, 15 September 1959.

151 Shell Oil Annual Report 1971, 7.

152 SHA Conference minutes 12 December 1962, 6-7, 21 October 1964, 5, 18 November 1964, 6.

153 SHA Conference minutes 9 December 1964, 3-4.

154 SHA Conference minutes 21 April 1965, 5.

155 SHA Conference minutes , 16 October 1963, 6-9, 21 April 1965, 5-6, 8.

156 SHA Conference minutes 15 April 1964, 3-4, 6-7.

157 SHA Conference minutes 15 July 1964, 9; Shell Houston Historical Documents Room, interview Spaght, 31-2, 'It's impressive to put it this way, that we came from seventh to second. But what that represented was increasing one's market share from about 6% to 8%, this kind of thing. Because there's this grouping, I mean, everybody and his dog has between 8% and 5% if you know what I mean. And so while we looked very good in our league standing, it wasn't revolutionary in that respect'.

158 Shell Oil Houston 163055274, competition analysis Jersey Standard, overview of advertising budgets.

159 SHA Conference 20 January 1965, Memo Group Objectives 1966-70.

160 Beaton, *Enterprise*, 292-3, 399, 418; Shell Oil Annual Report 1966, 15, *Shell News*, November-December 1966, showing that even in the US, Shell Oil had to stock 15 types of air filters, 15 types of air filters, and 8 types of gasoline filters; considered in Europe, SLA DCS S44, CMD minutes 4 February 1969. For Shell Oil, the tyres took some time to realize and only arrived in 1969: *Shell News*, No. 3 1969, 12-3.

161 *Shell News*, October 1958 (first mention credit card, applications open), June 1961, 4 (application form for employees; cards also accepted by Conoco stations), December 1961, 13-4 (1.1 million issued first nine months 1961). Shell Oil's credit card was re-issued 1971. *Shell News*, May 1962, 1-4 (data centres at Tulsa, Oklahoma, and Menlo Park, California). In 1964, all operations were concentrated at Tulsa: *Shell News*, January-February 1967, 6-10. *Shell News*, May-June 1966 (nationwide car breakdown service).

162 Considered for Europe but decided not to take a lead because of the high costs involved, SLA DCS S32, CMD minutes 4 June 1968, SHA Conference 14 June 1967, Marketing Report 1966. The introduction of the Euroshell card in *Shell Spiegel*, March 1972.

163 Shell Oil Houston, board minutes 22 December 1949 for an outline of the scheme, 28 March 1956 and 24 August 1961 for changes in the general arrangement, 2 December 1964, $93 million financed under the scheme. Shell Oil stopped using the sale and lease-back construction in 1967, when market conditions changed and the Securities and Exchange Commission imposed new requirements for the disclosure of off-balance financing: SHA Conference minutes 8 March 1967, 4.

164 SHA Conference 11 June 1969, Marketing Report 1969.

165 SHA Conference 20 September 1961, Report US and Canada, second quarter 1961.

166 SHA Conference 12 June 1968, Marketing Report 1967.

167 SHA Conference minutes, 15 September 1965, 9, 14 May 1969, 7. On the lack of integration, Conference minutes 16 October 1963, 8.

168 SHA Conference 16 June 1965, Marketing Report 1964; idem 11 June 1969, Marketing Report 1968.

169 SHA Conference 16 June 1965, Marketing report 1964; idem Conference minutes, 14 December 1966, 3.

170 SHA Conference 12 June 1968, Marketing Report 1967; idem Conference 10 June 1970, Marketing Report 1969.

171 SHA Conference 14 June 1968, Marketing Report 1967.

172 SHA Conference 14 June 1968, Marketing Report 1967; idem, Conference 11 June 1969, Marketing Report 1968; interview authors with Bill Bentley, London 27 July 2005. Bentley was personal assistant to the Marketing Coordinator R.A. Meyjes at the time.

173 SHA Conference 9 June 1971, Marketing Report 1970.

174 SHA Conference 10 June 1970, Marketing Report 1969.

175 Shell Oil Annual Report, 1970, 7, 1972, 5.

176 SHA Conference 10 June 1970, Marketing Report 1969.

Chapter 5

1 Arora and Rosenberg, 'Chemicals', 94.

2 Shell Oil Houston, interview Humphreys, 40.

3 SLA SICC 173, Shell Staff Lecture series, 'Planning', Dr M. A. Matthews, Planning Coordinator, Chemical; SLA DCS S4, CMD minutes 20 October 1960.

4 SLA DCS S48, CMD minutes 30 March 1971. The same point had been raised earlier in the Conference, minutes 9-10 February 1971, 4.

5 Shell Oil Houston, interview Spaght, 40; the conversation with Kessler which Spaght describes probably took place during the early 1960s.

6 Spitz, *Petrochemicals*, 513-5; Homburg, Van Selm, and Vincken, 'Industrialisatie', 393-5.

7 Previously, coal tar had been a popular source of benzene, toluene, xylene, methanol and ammonia, but now 32 per cent of benzene production, 75 per cent of toluene, and 90 per cent of xylene, methanol and ammonia came from petrochemicals. Likewise, molasses had long been the favoured basis for producing ethyl alcohol, acetone and isopropyl alcohol, but now petro-chemistry provided 80 per cent of ethyl alcohol, 95 per cent of acetone, and 100 per cent of isopropyl alcohol. The fermentation of fusel oil as a source of the anaesthetic amyl alcohol had been 90 per cent replaced; wood distillation and fermentation were simply no longer used for making acetic acid, now entirely made through petrochemicals; and in the manufacture of ethyl chloride, even raw alcohol had been 100 per cent super-seded. SLA SC54, Faupel, *History of the Petroleum Industry*, 1958, 91-6.

8 Chapman, *International Petrochemical Industry*, 211.

9 Homburg, Van Selm, and Vincken, 'Industrialisatie', 391.

10 Chapman, *International Petrochemical Industry*, 84. SHA Conference minutes 20 April 1966, 7; SLA DCS S48, CMD minutes 31 March 1970.

11 Chapman, *Petrochemical Industry*, 209; *London Shell*, 11 March 1959; Murray, *Go Well*, 160-5.

12 Kennedy, *ICI*, 45-7.

13 Spitz, *Petrochemicals*, 345-50.

14 SLA DCS C6, CMD minutes 29 May, 18 June, 24 September, 16 December 1958.

15 Bamberg, *BP* III, 344-50; Larson, Knowlton and Popple, *New Horizons*, 766-70.

16 Wall, *Growth*, 177-85.

17 SHA Conference minutes 12 April 1972, Chemical report 1973, 3.

18 SLA DCS C6, CMD minutes 29 May, 18 June, 24 September, 16 December 1958; S4, CMD minutes 13 March 1961, 9 March 1963. By 1967, SES had run up a cumulative loss of £5.2 million: DCS C20, CMD minutes 31 October 1967.

19 Pradier, *Double anniversaire*, 30-5, 99-103.

20 *Shell Spiegel*, November 1953, 1-3; Abelshauser et al., *German Industry*, 441-51; SHA Conference minutes 12 April 1972, Chemical report 1971, 27.

21 SLA DCS C20, 11 January 1966, 16 February 1971.

22 SHA Conference minutes 15 April 1964, Chemical report 1963, Appendix 3, idem 20 April 1966, Chemical report 1965, Appendix 3A.

23 SLA DCS S4, memo 9 January 1959.

24 Homburg, Van Selm, and Vincken, 'Industrialisatie', 388. Shell and Montecatini also considered building a polypropylene plant in Argentina: SLA DCS S4, minutes 13 February 1962.

25 About the doubts SHA Conference minutes 11 December 1963, 2; poor results

and quarrels, Conference minutes 15 September 1965, 7.

26 The Group set aside £2 million to cover the associated administration, but otherwise suffered no great financial loss from the affair: Montecatini bought Shell's 50 per cent of MSP for £58 million, paid in nine equal half-yearly instalments without interest, and repaid its loan with interest. A condition of the sale and purchase was that the MSP name should immediately be changed so that it would not reflect any relationship with Shell. SHA Conference minutes 11 December 1963, 16 February 1966, 20 July 1966, 21 September 1966; Brouwer and Wagner, memo to boards of Bataafse and Shell Petroleum Company Ltd., 9 September 1966. Loudon later thought that the Group had been taken for a ride by Montecatini: SLA SC45/2, interview 17 January 1989, 41.

27 SLA DCS S4, CMD minutes 22 June, 13 July, 16 December 1960, 1 August 1962; S28, Chemical review 1967; SHA Conference minutes 21 February 1962, 2-3, 13 March 1963, 3-4, 16 November 1966, 13.

28 SLA DCS S4, CMD minutes 9 and 14 November 1961; SHA Conference minutes 21 April 1965, Chemical report 1964, 23. In 1959 the Group had started looking into the plastic pipe industry: SLA DCS S4, CMD minutes 28 January 1959. By 1975, Shell's stake in Wavin had risen to 59 per cent.

29 SLA DCS C15, CMD minutes 4 February 1964, C27, CMD minutes 16 February 1971.

30 Shell Oil Houston, board minutes 27 August 1964, 28 January 1965, 10 December 1965, 28 April 1966, 23 June 1966, 28 July 1967, 25 January 1968, 13 November 1968, 26 June 1969. In November 1969, a total of 13 agrochemical marketing companies were merged into a single organization, idem 25 November 1969.

31 SHA Conference agendas 17 April 1963, chemical review 1962, appendix V.

32 SHA Conference minutes 15 April 1964, 1.

33 SHA Conference minutes 12 April 1972, 4.

34 The division had operated as Shell Chemical Company until 1943, when it was absorbed into Shell Union after a lengthy debate about the interface between oil and chemicals: cf. SLA GHC/US/D9/2/2, merger 1939-43, memo 22 February 1943. Beaton, *Enterprise*, 676. Shell Oil Houston, interview Humphreys, 8.

35 SHA Bataafsche agendas, 25 September 1957, memo 9 September 1957 on setting up a separate corporate entity for Pernis; Royal Dutch agendas, 22 October 1958, Chemicals report January-June 1958; SLA, unnumbered file with London office organization charts, 1953-59.

36 SLA DCS S4, CMD memo 6 September 1961.

37 SLA SICC 173, Mitchell, 'Principal Facets of the Chemical Industry'.

38 Shell Oil Houston, interview McCurdy.

39 SHA Conference minutes 9 December 1970, 5.

40 Shell Oil Houston, interview Spaght, 23, interview Humphreys, 8-9, 13.

41 SLA 119/11/34/7, board memos Anglo Saxon, memo 27 June 1945. Methyl ethyl ketone, a very versatile solvent, was also used as a base for certain perfumes and for imparting a buttery taste to margarine.

42 SHA agendas Bataafsche 16 January 1947, memo 20 November 1946; SLA 119/11/34/7, board memos Anglo Saxon, memo 27 June 1945.

43 Homburg, Van Selm, and Vincken 'Industrialisatie', 387.

44 SHA 195B/95E, report 14 June 1945 (DDT pilot production at Amsterdam laboratory during the war), 11/16-2, Braybrook to Van Eck, 15 July 1944 (DDT in the US). Around 1950, Shell Oil explored the opportunities in vinylchloride and PVC, but decided against going ahead with it: Shell Oil Houston, interview Humphreys, 10.

45 Shell Oil Houston, interview Humphreys, 64.

46 Beaton, *Enterprise*, 676-80.

47 Beaton, *Enterprise*, 678. Spitz, *Petrochemicals*, 351, records that in 1953 a German manufacturer of synthetic rubber received an offer from a French company for the supply of alcohol produced from surplus red wine. The idea of an international red wine pipeline entertained many people, but the offer was declined.

48 Source for Figure 5.5: SHA agendas Bataafsche, 18 December 1958.

49 Source for Figure 5.6: SHA agendas Bataafsche, 18 December 1958, Conference agendas, chemical reviews 1960-74

50 Wall, *Growth*, 187.

51 SLA DCS C4, CMD minutes 13 November 1957.

52 SHA agendas Bataafsche 24 July 1957, memo 4 July, 23 April 1958, memo 9 March; *Chemical Age*, 1948, 482; 1953, 414, 6; 1955, 606. The history of Petrochemicals summarized in Tugendhat, 'Development'. NA Kew WO 195/11177, Visit of the Organic Chemistry Committee of HMG Munitions Chemistry Advisory Board, 3-4 July 1950. *Chemical Age*, 26 May 1956, 4 August 1956, 15 September 1956. Weizmann later became the first president of Israel, 1948-52.

53 SHA Conference 21 April 1965, Chemical review 1964, 22.

54 SLA *Shell Magazine*, April 1963, 111-3; Beaton, *Enterprise*, 678-79; Forbes and O'Beirne, *Technical Development*, 621.

55 Shell Oil Houston, board minutes 22 May 1952, interview Humphreys, 25-7.

56 Beaton, *Enterprise*, 681; *Chemical Age*, 3 October 1953, 712.

57 *Chemical Age*, 3 October 1953, 712.

58 SHA agendas Bataafsche, minutes 21 February 1952, memo 11 February 1952, 15 May 1952, memo 1 May 1952, 18 December 1952, memo capital budget Pernis 1953, 16 July 1953, memo 6 July 1953; *Van Rotterdam Charlois naar Rotterdam Pernis*, 92; Homburg, Van Selm, and Vincken, 'Industrialisatie', 387-8.

59 SHA Conference minutes 9 April 1969, Chemical review 1968, 3.

60 In 1964, the Group considered moving from epoxy resins into fibreglass, but relented after opposition from its French partner Saint-Gobain, which already had a strong position in the product: SLA DCS S15, CMD minutes 2 July, 6 and 20 August, 16 and 22 September 1967.

61 SHA Conference minutes, annual reviews chemical operations, 1960-68. For the allure of high technology products see 'Food and Health for Man's Future, *Shell News* September 1952, 10-5; 'This Glue has Glamour', *Shell News*, September-October 1963, 18-21.

62 Royal Dutch Annual Report 1969, 22.

63 SLA SICC 173, Shell Staff Lecture series, W. F. Tuson, 'Finance', 8-9 ('heroic'); cumulative figures, DCS S48, Werner to CMD, 28 March 1972.

64 SLA DCS S4, CMD minutes 19 September 1957.

65 *Ibid*.

66 SLA DCS S4, CMD minutes 20 October 1960.

67 Shell Oil Houston, interview Spaght, 20-1.

68 SHA Conference minutes, Brouwer to Conference, 16 February 1966, 11.

69 Spitz, *Petrochemicals*, 537-41.

70 SLA DCS S4, CMD minutes 4 July 1957 ('cautious' and 'rushing'); Jersey's 'fertilizer binge', Wall, *Growth*, 192-227.

71 Bamberg, *BP* III, 357, 375, 390. When BP moved into petrochemicals, Shell privately criticized it as 'rushing into the chemical business for prestige reasons without waiting to acquire petrochemical know-how and experience.' SLA DCS S4, CMD minutes 4 July 1957.

72 SLA DCS S4, CMD minutes 20 March 1961 for Mitchell's defence.

73 SLA DCS S4, memo 6 September 1961.

74 Shell Oil Houston, board minutes 27 April 1972; Shell Oil Annual Report 1972, 12.

75 SLA DCS S4, memo 2 March 1961.

76 SLA DCS S4, Schepers' note, 6 September 1961, 1.

77 Ibid., 12.

78 Ibid., 8.

79 The policy change related in SHA Conference minutes 12 April 1972, 4.

80 Shell Oil Houston, interview Humphreys, 44-5, 69; SHA Conference minutes 8 April 1970, 6 (competition between natural rubber and elastomers), 14 April 1971, 2 (alternatives for so many products).

81 Shell Oil Houston, interview Humphreys, 62-3.

82 Shell Oil Houston, interview Humphreys, 65-6.

83 Dumping was officially viewed as occurring if the domestic sales price of the product was higher than its export price. With petrochemicals, the practice had started in the middle 1950s, and remained habitual well into the 1960s: an irritated Shell note recorded that in 1954 the domestic US price of synthetic rubber had been between 20.7 and 23 cents a pound weight, against its export price of 18.5 cents, and that in 1964, while the domestic price remained unchanged, the export price had fallen to 17.7 cents a pound. SLA DCS S4, CMD minutes 9 November 1961; DCS S15, CMD minutes 23 January 1964, 14 July 1964.

84 SHA board agendas Royal Dutch, March-May 1958, chemical review 1957-58, 6, 27 April 1960, chemical review 1959, 3; Conference minutes 17 April 1963, review 1961, 5.

85 SHA Conference minutes 17 April 1963, Chemicals review for 1962, 1, and 15 April 1964, review 1963, 1.

86 Homburg and Rip, 'Chemische industrie', 405.

87 See SHA Conference minutes 12 April 1972, 4.

88 For a description Homburg, Van Selm, and Vincken, 'Industrialisatie', 389, 399-400.

89 SHA Conference minutes 9 April 1969, Chemicals review 1968; Grimshaw to Conference and Werner to Conference, 12 February 1969; Werner to Conference, 12 April 1972.

90 SHA Conference minutes 8 April 1970, 5.

91 SLA DCS S48, CMD minutes 31 March 1970.

92 SHA Conference minutes 12 April 1972, chemical review 1971, and 10 April 1974, review for 1973; SLA DCS S48,CMD minutes 28 March 1972.

93 SHA Conference 8 April 1970, 6.

94 SLA DCS S58, CMD minutes 27 March 1973.

95 SLA DCS S48, CMD minutes 30 March 1971.

96 SLA DCS S48, CMD minutes 30 March 1971.

97 This was not a uniquely low figure, but British Leyland's eventual fate showed what could happen to a good business handicapped by out-of-date plant, under-investment and a militant labour force.

98 SLA DCS S48, CMD minutes 30 March 1971; Shell Briefing Service, August 1971, 'Economics of Chemicals from Petroleum'.

99 SLA DCS S48, CMD minute, 30 March 1971; SHA Conference minutes, 14 April 1971.

100 SHA Conference minutes, 14 April 1971, 3. The original minutes reads indeed 'underestimated demand', whereas the secretary taking the minutes must have intended to write 'overestimated', for that was the crux of the problem.

101 Ibid., 4.

102 SLA DCS S58, minutes strategy meeting 1 October 1973.

103 Shell Oil Houston, interview Spaght, 40.

104 SLA DCS S10, memo 11 January 1962.

105 SLA SICC 173, Shell Staff Lecture series, 'Planning'.

106 SLA DCS S16, Spaght to Loudon, 21 November 1963.

107 NA Kew FCO 54/69, T317/1152.

108 SLA DCS S5, memo 2 July 1962.

109 SHA Conference minutes 17 October 1962, 5-6; cf. also minutes 21 November 1962, 3-5, 11 December 1963, 3. CSV also subscribed for substantial amounts to bonds backed by the Venezuelan state for housing and for public utilities: Conference minutes 16 January 1963, 5, 17 July 1963, 8.

110 SLA DCS S5, CMD minutes 16 July 1962, memo 14 November 1962.

111 SLA DCS S5, CMD minutes 11 December 1962.

112 SLA DCS S16, memo 4 January 1963.

113 SHA Conference minutes 13 January 1963, 5.

114 SLA DCS S16, CMD minutes 9 July 1963.

115 SLA DCS S22, memo 6 June 1966.

116 SLA DCS S16, CMD minutes 16 April 1963, memo 18 April 1963, memo 23 April 1963; SHA Conference minutes, 17 April 1963, 6-7, 12, 15 May 1963, 11-2, 21 October 1964, 10.

117 SLA DCS S16, CMD minutes 12 June 1963, also commented that Zwolsman himself was 'energetic and imaginative though somewhat unconventional', and that while Shell was satisfied as to his integrity and solvency, 'Some misgivings must nevertheless be felt in regard to his tendency to pursue rather ostentatious public relations methods. (...) No effort will be spared to impress on Zwolsman our expectation that he will modify his methods of dealing with the public and will be guided by the Group's practices in this regard'.

118 SLA DCS S16, memo 3 September 1963.

119 SHA Conference minutes 17 July 1963, 10-1.

120 SHA Conference minutes 17 July 1963, 11.

121 SHA Conference minutes 12 May 1971, 5; SLA DCS S65, CMD minutes, 4 May 1971; cf. also memo 24 September 1971, voicing a concern for the employment of future generations as inspiring diversification.

122 SLA DCS S16, Spaght to Loudon, 21 November 1963, memo 1 January 1964, CMD minutes 14 January 1964. Other projects at the back of Berkin's and Brouwer's minds at the time included iron- and -steel making in non-coal-producing areas, where oil in substantial quantities could perhaps be economically substituted for coal
a pilot plant had been built at Egham in the UK for this; the transportation of solids by pipeline, which Shell Pipeline Company was actively researching; and pharmaceuticals, Shell Oil already being active in veterinary medicines. Spaght himself was later doubtful about the

synergies between mining and metals, and oil: Shell Oil Houston, interview Spaght, 42-6.

123 SHA Conference minutes 21 October 1964, 9.

124 Cf. for instance SHA Conference minutes 17 June 1964, 10 April 1968, the latter reporting on a joint study with the Reactor Centrum Nederland into high-speed centrifuges for isotope separation.

125 SLA DCS S22, CMD minutes 30 June 1966.

126 SLA DCS S22, memo 4 January 1965.

127 SLA DCS S16, memos 12 and 21 August 1964, CMD minutes 9 September 1964.

128 SHA Conference minutes 8 October 1969, 6; Shell Oil Houston, interview Spaght, 46.

129 SHA Conference minutes 8 October 1969, 7.

130 SHA Conference minutes 8 October 1969, 8.

131 SLA DCS S65, Letter to Chief Executives of operating companies, 16 June 1969.

132 SLA DCS S65, CMD minutes 5 February 1969.

133 SLA DCS S65, CMD minutes 26 February, 3 June 1969.

134 SLA DCS S65, CMD minutes 5 February 1969.

135 Broersma, *Zaak*, 79-85; Murray, *Go Well*, 237; SHA Conference minutes, 10 June and 9 September 1970; SLA DCS S65, new enterprises annual review, April 1970-March 1971.

136 Dankers and Verheul, *Hoogovens* 279-88.

137 SLA DCS S65, memo 24 September 1971 and CMD minutes 28 September 1971.

138 SHA Conference minutes 13 May 1970, 8.

139 SHA Conference minutes 12 May 1971, 2-4.

140 SHA Conference minutes 15 September 1971, 8-12. Apparently Jersey Standard, BP, Gulf and AGIP all had some activities in tourism.

141 SHA Conference minutes, 10 November 1971, 10.

142 SHA Conference minutes 10 November 1971, 10, 9 February 1972, 9.

143 SLA DCS S48, CMD minutes 3 and 10 August, 13 September and 26 October 1971, memo 6 August 1971.

Chapter 6

1 Spitz, *Petrochemicals*, 87-9.

2 SHA agendas Bataafsche 1947, memo PVC plant IJmuiden, 25 October 1946, with appendix; Schweppe, *Research aan het IJ*, 123-4; Homburg, Small, and Vincken, 'Van carbo- naar petrochemie', 355; Homburg, Van Selm, and Vincken, 'Industrialisatie', 387. On the completion of the first Pernis cat cracker in 1951, the Group did switch to the conventional petrochemical process for making PVC.

3 For the preparation of this meeting, see SHA 2/387, memo discussions Bloemgarten-Davis, February 1946.

4 SHA agendas Bataafsche 1946, memo 4 June 1946.

5 SHA agendas Bataafsche 1946, memo 4 June 1946.

6 SHA agendas Bataafsche 1946, meeting 26 June 1946.

7 SHA board minutes Bataafsche 1946, memo 22 October 1946; SHA Research Reports, minutes Group technical and research survey 1959, 45. US laboratories tended to employ more scientists, 40 out of every 100 staff in 1962, as against only 20 in Europe: SHA Research Reports, minutes Group technical and research survey 17-18 May 1962, 32. SHA 2/387, Memo discussions with Mr. Alexander Fraser in London and The Hague, May 1950 (sharing research costs). The agreement also helped Shell Oil to defend the results-sharing to its outside shareholders, who might otherwise have had concerns about the Group getting unfair benefits. Information obtained from Pieter Folmer. Shell Oil and Bataafsche each owned half of Shell Development until 1953, when Shell Oil obtained another 15 per cent in return for a transfer of research assets to Bataafsche. Two years later Shell Oil became sole

owner: Beaton, *Enterprise* 691.

8 Beaton, *Enterprise*, 684. The need to coordinate between the new Houston centre and Amsterdam's E&P research activities was one of the reasons for setting up the first research meeting in April-May 1946: SHA 2/387, memo discussions Bloemgarten-Davis, February 1946. SHA 2/387-2, letters and memos concerning the 1948-49 Houston production engineering research conference 1948, 1949, and 1950. SHA 2/102, De Bruyn to American Department, London 31 August 1951. SHA 2/105, Bataafsche to Shell Development, 13 February 1952, announcing the first issue of a regular research bulletin. SHA Research Reports, minutes Group technical and research survey, 1957 and following years.

9 SHA 190D/708, file A. Hofland with a clipping from *Olie*, June 1959; SHA 190D/707, circular Bloemgarten 21 October 1955. The London Product Research and Development Department covered the UK operations alone: SHA 2/96, circular Hopwood 25 March 1952. In January 1949, the Delft and Amsterdam laboratories dropped the adjective 'Bataafsche' from their names and adopted the term 'Royal Dutch/Shell Laboratories' to underline their contribution to the Group as a whole, rather than to one of its operating companies alone: Schweppe, *Research aan het IJ*, 128-9; *ibidem* 134 for the appointment of J. H. Vermeulen as research coordinator in 1954. SHA Research Reports, Minutes Group Technical and Research Survey 17-8 May 1962, 12 (rare duplication of research).

10 SHA Conference minutes 21 November 1962, 8, 17 June 1964, 8.

11 SHA Conference minutes 9 October 1968, 5.

12 SHA agendas Bataafsche 1946, memos 26 June, 22 October 1946; Forbes and O'Beirne, *Technical development*, 460, quoting Kessler in 1928 emphasizing that research needed to earn its keep.

13 SHA Research Reports, Minutes 1st Shell Group Research Meeting, October 7-10, 1946; Minutes Group Technical and Research Survey 3, 4 and 6 May 1960, 61.

14 SHA Research Reports minutes Group research meeting 4-5 June 1957, 33.

15 SHA agendas Bataafsche 1949, memo 2 November 1949; idem 1950, memo 6 November 1950; idem 1951, memo 5 November 1951; Schweppe, *Research aan het IJ*, 124-5.

16 SHA Research Reports, minutes research meeting, 4-5 June 1957, 13; minutes Group technical and reseach survey, 14-5 May 1962, Part 1, 21 (close and constant cooperation); SHA Conference minutes 20 June 1962, 3 (contribution to profits).

17 SHA Conference 11 June 1969, Manufacturing Oil report July 1968-June 1969.

18 SHA Research Reports, minutes meeting oil products research, 1-2 July 1958, 6.

19 SHA agendas Bataafsche 1949, memo 2 November 1949.

20 SHA Research Reports, minutes research meeting, 4-5 June 1957, 13; idem 18-9 June 1959, 51 and Chart E1/d-e; idem 3, 4 and 6 May 1960, 61-2, emphasizing that the Group's effort on oil research trailed the competition and needed to be addressed; Data on Group Research, April 1965, giving a Group total spending of £13.9 million or 0.68 per cent of sales, against Jersey £16.3 million or 0.45 percent.

21 SHA agendas Bataafsche 1950, memo 6 November 1950; idem 1951, memo 5 November 1951; agendas Royal Dutch 1958, memo 8 September 1958; SHA Research Reports, Data on Group Research, April 1965, unnumbered pages.

22 SHA Research Reports, minutes meeting oil research, 17-8 November 1959.

23 SHA agendas Royal Dutch August-October 1958, memo 8 September 1958; *25 Jaar KSEPL*, a special issue of *De Drum*, 1982.

24 Schweppe, *Research aan het IJ*, 152-3.

25 SHA board agendas Royal Dutch September-October 1958, memo 8 September 1958, 24 June 1959, Research Report 1958-59; Schweppe, *Research aan het IJ*, 152-3; SHA Research Reports, minutes Group technical and research survey, 15-6 May 1957, 20, 29; idem 15-6 May 1958, 22; idem 18-9 June 1959, 41 and Chart B1 a/e; idem 17-8 May 1962, 5, 8.

26 SLA SC54/2, Royal Dutch/Shell, 1958, 45-6; *Shell Magazine* 1958, 169; DCS S15, memo 12 February 1963. Staff numbers: SHA board agendas Royal Dutch, September-October 1958, memo 8 September 1958.

27 SHA Research Reports, minutes Group technical and research survey 1961, 12-3.

28 SHA 2/387 shows the proliferation of such meetings, *ibidem*, Perquin to Cutting, 16 April 1948, on the desirability not to have a conference on internal combustion turbine fuels, since this would 'place our experts in the invidious position of sitting around a table discussing a subject, well knowing that they are not free to talk openly about the whole of their subject'.

29 SHA Research Reports, minutes Group technical and research survey 14-5 May 1962, 12. In the report, Gershinowitz estimated that the Group reported $44 million spent on R&D, but total costs were $65 million.

30 SHA Research Reports, minutes Group technical and research survey 14 and 15 May 1962, 12.

31 SHA Research Reports, minutes Group technical and research survey 3, 4 and 6 May 1960, 62.

32 Beaton, *Enterprise*, 683; Schweppe, *Research aan het IJ*, 124, 152-3; SHA agendas Royal Dutch 1958, memo 8 September 1958; idem 24 June 1959, Research report for 1958-59; SHA Research Reports, minutes Group technical and research survey 18-9 June 1959, Chart e1/g. The desire to draw more talent from Germany led to the establishment of a separate chemicals research lab near Bonn during 1959.

33 SHA Research Reports, report to Managing Directors, 26 September 1957, 2.

34 The Ferranti was replaced by a more powerful IBM in 1962, the capacity of which already became insufficient in 1966: Schweppe, *Research aan het IJ*, 150-2, 197, 203-4.

35 SHA agendas Royal Dutch 1958, memo 8 September 1958.

36 SHA Conference 11 June 1969, Marketing Report 1968.

37 SHA Research Reports, minutes Group technical and research survey 3-6 May 1960, 39.

38 SHA Research Reports, minutes Group technical and research survey, 15-6 May 1958, 22.

39 SHA Research Reports, minutes Group technical and research survey 1961, 6, 11, and Chart B1 a/a; in agricultural chemicals, only one in 600 synthesized compounds reached commercial testing stage, idem 14-15 May 1962, 10.

40 *Shell Magazine*, 2/1961, 'Fighting the Foam', 41-2; Van der Most et al., 'Nieuwe synthetische producten', 374-5.

41 NA Kew DSIR 13/155, E. L. Shepherd, Shell Chemical Ltd., to Dr. B. A. Southgate,

Water Pollution Research Laboratories, Stevenage, UK, 26 March 1965, Tarring to Southgate, 7 January 1964; and testing in the week of 2 November 1964; award announcement, Miss K. Hawkins to DSIR, 3 December 1964.

42 NA Kew DSIR 13/155, Shepherd to Southgate, 26 March 1965, DSIR 13/155. The Group sold a large volume of Teepol to Unilever. Van der Most et al, 'Nieuwe synthetische producten', 370-5; Howarth, *Century*, 273; Schweppe, *Research*, 190-1. Cf. Jones, *Tradition* 342, for Unilever's policy on biodegradable detergents.

43 SHA agendas Bataafsche, memo Teepol 21 June 1950; SHA Research Reports, minutes Group technical and research survey, 1959 39. The Group spent a total of £300,000 on research into biodegradable detergents during 1956-64; since the introduction of the first product to incorporate it into the UK in 1959 until 1965, Shell made a profit on it of £1.8 million: SHA Conference minutes 19 May 1965, Annual Report research, 6.

44 SHA Research Reports, minutes Group technical and research survey 3, 4 and 6 May 1960, 15; minutes meeting oil research 7-8 February 1961, II-3; idem highlights manufacturing oil's research programme 1966, 2.

45 SHA Conference 14 June 1972, Manufacturing Oil report May 1972.

46 Source for Figure 6.1: Shell, Intellectual Property Department, 2005. The figure shows the number of inventions and not the number of patent applications, because one invention will often lead to multiple applications.

47 SLA MK 72/1-2, Pohl to Davies, 18 July 1960.

48 SLA DCS C6, Schepers to Mumford, 14 April 1959, suggesting that the Technological Department act as an auditing and advisory body for pollution abatement at the chemical plant planned for Berre l'Étang.

49 SLA DCS S12, memo 8 November 1960, and MK72/2, Van Raalte, Internal Circular No. 1, 26 October 1964.

50 SHA Conference minutes 14 June 1967, manufacturing oil report July 1966-June 1967.

51 SLA MR 12, Environment and Conservation: Co-ordination Minutes and Meetings October 1971-July 1974 (Houston), notes these facts but no further details have been found in Group archives.

52 SLA DCS S63, CMD minutes 10 June 1969.

53 SLA DCS S63, memo 10 June 1969.

54 Ibid. Accordingly, the Royal Dutch Annual Report for 1969, 25, mentioned the formation of the conservation committee, the first time the Annual Report devoted a section to environmental conservation.

55 SLA DCS S63, memo 7 November 1969.

56 SLA DCS S63, CMD minutes 10 June 1969; SLA MK 009/70, marketing oil report 1970; SHA Conference minutes 14 June 1967, Manufacturing Oil report July 1966-June 1967 (first separate section on air and water pollution); SHA Research Reports, Manufacturing Oil research programme 1970, 1-3. In 1974, Shell initiated the foundation of the International Petroleum Industry Environmental Conservation Association, which had a similar purpose as CONCAWE: Royal Dutch Annual Report, 1974; Shell Oil Company, Annual Report 1978.

57 SLA MR 12, minutes and meetings October 1971-July 1974 (Houston).

58 SLA MK 009/70, Marketing Oil report for 1970.

59 SLA MK 009/70, marketing oil report for 1970; *Olie* June 1972, 165.

60 SHA 49/594, H. H. Buss to Bataafsche, 3 September 1937.

61 SHA 49/594, ibid., and Knoops to the mayor and aldermen, 9 February 1938.

62 SHA49/594, J. U. de Kempenaer to Bataafsche, 9 September 1937.

63 SHA49/594, Knoops to the mayor, 9 February 1938, M. Hardonk to Bataafsche, 14 July 1938.

64 Homburg, Van Selm, and Vincken, 'Industrialisatie', 400.

65 SHA board agendas Bataafsche, 16 June 1949, memo 2 June.

66 Forbes and O'Beirne, *Technical Development*, 517-8.

67 SHA board minutes Bataafsche, 5 December 1949.

68 SHA Research Reports, minutes meeting oil research 7-8 February 1961, II-3, III-4.

69 SLA DCS S10, memo 11 January 1962.

70 Shell Oil Annual Report 1958, 5, 1965, 5.

71 SHA Conference minutes 20 May 1964, 5-6; Loudon congratulated Kirby on this progress, which had been 'due largely to Mr Kirby's personal efforts.'

72 NA Kew DSIR 13/155, E. L. Shepherd of Shell Chemical Ltd. to Dr. B. A. Southgate, Water Pollution Research Labs, Stevenage, 26 March 1965.

73 SLA MR 75, Devices for containing floating oil, 1963.

74 SLA RS 34, Oil Pollution (Torrey Canyon) Biological Aspects, 14 March-9 October 1967.

75 SLA RS 17, Oil Pollution (Torrey Canyon), Sweeping and Sinking Methods, March-October 1967.

76 NA Kew FV12/481 Beach Protection: Shell Oil Herder chemical: report to DTI from [independent] Warren Spring Laboratory, Stevenage, Hertfordshire, 1967. A related product named Shell Dispersant LT was discovered later which it was proposed could be used in conjunction with the Herder (SLA: MR 12, Environment and Conservation: Co-ordination Minutes and Meetings October 1971-July 1974 (Houston)).

77 SLA MR 26, Pollution: accidental oil spillages, NW Europe and Baltic, September 1968-June 1971.

78 SHA Conference minutes, 8 July 1969, 6-7, and 8 July 1970, 6.

79 SLA MR 53/1-2, VLCC Tank Cleaning Safety, January 1969-July 1970 and July-December 1970, together give detailed records on the two-year inquiry at industry level into the accidents.

80 SHA Conference minutes, 11 February 1970, 13-4.

81 Homburg, Van Selm, and Vincken, 'Industrialisatie', 400-1; Van Zanden and Verstegen, *Groene geschiedenis*, 144-5; Cramer, *Groene golf*, 28-9; Boender, *Milieuprotest*, 129-40.

82 SHA Conference minutes 14 June 1967, Manufacturing Oil report for July 1966-June 1967, and 10 June 1970, Manufacturing Oil report for July 1969-June 1970.

83 SHA Conference minutes 10 June 1970, 8.

84 M. I. Rigter, Van CMRK tot DCMR. De vervuiling in de Rijnmond te lijf. MA thesis Social and Economic History, Utrecht University 1993, 54.

85 SHA board minutes Royal Dutch 18 December 1958, Manufacturing Oil report.

86 SHA board minutes Royal Dutch 25 February 1959, Manufacturing Oil report.

87 SHA Highlights of Manufacturing Oil's Research programme for 1968 (outside US), 3 April 1968, 3.

88 SHA Conference minutes 21 September 1966, 9.

89 SHA Conference minutes 16 November 1966, 8.

90 SHA Conference minutes 8 February 1967, 6-9, 14 June 1967, 7; SLA DCS S28, memo 29 June 1967.

91 SHA Conference minutes 14 June 1967, 7.

92 SHA Conference minutes 14 February 1968, 7.

93 SHA Conference minutes, 12 February 1969, 13-4.

94 Shell Transport Annual Report 1963, 16.

95 Markovitz and Rosner, *Deceit,* 108-19.

96 SLA DCS S28 has a major internal Group report from 1967 on pollution R&D comments on the 'intrinsic toxicity' of lead additives. In contrast, three years later Spaght claimed in the Conference that lead was 'not proved to be toxic in itself', SHA Conference minutes 25 February 1970, 7-8. Spaght presumably meant that it was no longer toxic when burnt, because its inherent toxicity as a substance was wellknown.

97 Contemporary interpretations of scientific data derived from research differed considerably then and also differed very much from later interpretations. For example, evidence gathered in the late 1960s was believed to indicate that carbon monoxide emissions were much more dangerous than emissions of carbon dioxide. In 1967, the US National Center for Atmospheric Research at Boulder, Colorado, conducted experiments on samples of ice cores from the North and South Poles that seemed to indicate that (partly as a consequence of CO emissions) pollution of the atmosphere could prevent the sun's rays from reaching the Northern Hemisphere to the extent that by 1980 it might suffer a drop in temperature of six degrees Centigrade. One eminent Shell consultant (Sir Graham Sutton FRS, a world-famous meteorologist) was asked by the Group 'to say whether the prediction of a dramatic, if not catastrophic, fall in temperature between now and 1980 is science fiction or a real possibility. He says the case is "not proven" but will not reject it.' Professor James Lovelock was another Shell consultant at the time and in contact with the National Center for Atmospheric Research (SLA DCS S28, Environmental Conservation 1967-68). He later conceived and in 1979 first published the Gaia theory of complete global environmental interdependence.

98 SLA RS 18/1-2, Joint Shell-BLMC agreement, August 1967-December 1968.

99 SLA MK 009/70, Marketing Oil report for 1970.

100 SLA DCS S28, pollution R&D report.

101 SLA MK 005/68, Marketing Oil report 1968, 20.

102 SLA DCS S63, memo 10 June 1969.

103 SLA DCS S28, pollution R&D report.

104 SLA MK 009/70, Marketing Oil report 1970.

105 SHA Conference minutes 9 September 1970, 8.

106 SHA Research Reports, highlights Manufacturing Oil research programme for 1970 18 March 1970, 1-3.

107 SLA MK 72/2, T. M. B. Marshall of Oil Products Development to Dr. C. G. Hunter of Shell's Tunstall laboratory, 11 May 1960; Hunter to W. A. Munday, Shell Research Ltd, 18 July 1960; C. B. Davies to Dr. H. G. S. van Raalte, 20 February 1961.

108 TTP, earlier known as tricresyl phosphate or TCP, was the additive first used in Shell X-100. Although very effective, its toxicity led in 1960 to its replacement by a non-metallic formulation containing triphenyl phosphate. The virtues of the new formulation were firstly that as before it was a detergent oil, capable of maintaining cleanliness in the crankcase, on piston skirts and ring grooves and other moving parts; but secondly as a non-ash oil, it did not leave the metallic ash of organo-metallic compound additives.

109 SLA MK72/2, J. H. Brookman, SIPC London, to Norske Shell, 2 January 1964; Brookman to Norske Shell, 5 May 1964.

110 SLA MK 72/2, Peter Hodges, Norske Shell, to SIPC London, 26 November 1963.

111 SLA MK 72/2, telegram SIPC London to Shell Verkoop, The Hague, 10 May 1965.

112 SLA MK 72/2, Hodges to SIPC, 26 November 1963.

113 SLA MK72/2, Jansen to Lacaille, July 1965.

114 *London Shell*, October 1957.

115 Ibid.

116 Examples of safety instructions in Shell Oil Houston MOR213679; the 'drins said to be more toxic than DDT in circular Shell Chemical Ltd., 11 February 1958.

117 For details on one such eradication programme, see Blu Buhs, *Fire Ant Wars*, 39-80. Lear, *Carson* 119-20, 305-6, 312-5; De Steiguer, *Age*, 33; National Institute for Occupational Safety and Health, 'Special occupational hazard review for aldrin/dieldrin', (Cincinnati, 1978), 82.

118 Van der Most et al., 'Nieuwe synthetische producten', 375; Lear, *Carson*, 320, 548.

119 *Chemical Age*, 12 February 1955.

120 Sheail, *Environmental History*, 237.

121 SHA board agendas Royal Dutch, 24 June 1959, research survey 1959-60.

122 *London Shell*, October 1957, January 1958, June 1958.

123 *Chemical Age*, 25 January 1958.

124 SLA DCS S4, memo 4 January 1962.

125 *London Shell*, 31 January 1962.

126 Lear, *Carson*, 320, 548.

127 Lear, *Carson*, 419-20, Sheail, *Environmental History*, 240.

128 M. Stoll, 'Rachel Carson's *Silent Spring* in Europe and America, a Comparative View of its Reception and Impact', Paper European Society for Environmental History, Prague 8 September 2003.

129 Shell Oil Houston MOR213710 contains a photocopy of a draft Nature Conservancy report on the same subject from 1967, which summarizes the earlier report. It was sent by Bataafse Internationale Chemie Maatschappij to Shell Chemical in Houston with the note 'Confidential. We are not supposed to have it nor refer to it'.

130 Breidenbach and Lichtenberg, 'DDT'.

131 SLA DCS S15, CMD minutes 27 February 1964.

132 Lear, *Carson*, 416-71.

133 Sheail, *Environmental History*, 237-8.

134 SLA SC54/2, 45-46; *Shell Magazine* 1958, 169; SLA DCS S15, memo 12 February 1963. Shell Oil Houston, interview Spaght, 61-2.

135 SLA DCS S63, memo 10 June 1969.

136 National Institute for Occupational Safety and Health, 'Special occupational hazard review for aldrin/dieldrin', (Cincinnati, 1978), 4.

137 H. G. Huckle, general manager Shell Chemical Company agricultural division, in *London Shell*, 12 July 1961. In 2000, the European Union banned Kotol for use in agriculture; Sheail, *Environmental History*, 239.

138 SHA Conference minutes 12 July 1967, 12-3.

139 SLA DCS S48, CMD minutes 30 March 1971 and memo 13 May 1971.

140 Sheail, *Environmental History*, 244-5.

141 SLA DCS S15, memo 12 February 1963.

142 SLA DCS S15, Searight to Loudon, 14 March 1963.

143 Shell Transport Annual Report 1962, 27;

Royal Dutch Annual Report 1962, 27-8. The film is mentioned by Starrenburg in SHA Conference minutes, 17 October 1962, 5, but there are no details of its method of distribution or audiences achieved.

144 SLA DCS S15, memo 12 February 1963.

145 SLA DCS S15, Searight to CMD, 14 March 1963.

146 SLA DCS S15, CMD minutes 21 May 1963.

147 Ibid.

148 SLA DCS S15, CMD minutes 31 December 1963.

149 SLA DCS S15, CMD minutes 27 February 1964.

150 SLA DCS S15, memo 27 February 1964.

151 SLA DCS S15, CMD to Shawcross and Rothschild, 19 March 1964.

152 Sheail, *Environmental History*, 244-5.

154 Jager, *Aldrin*.

154 An overview of the hearings in National Institute for Occupational Safety and Health, 'Special occupational hazard review for aldrin/dieldrin', (Cincinnati, 1978). Shell's attitude in Shell Oil Houston, MOR213705.

155 SLA Publication 217, Petroleum and Chemistry, Bataafsche 1957. Group production at the Arsenal continued until December 1982.

156 SLA PA 12, Reviews of Public Affairs 1968-85, December 1983, 5, and December 1984, 6.

157 SHA Conference minutes 11 June 1969, Manufacturing Oil report July 1968-June 1969.

158 SHA Conference minutes 10 July 1968, 4. Outsiders considered the Trade Relation department's name quaint. In 1968 the *Financial Times* noted that Shell called its department Trade Relations 'to avoid the pejorative association of what the rest of the world calls Public Relations', SLA DCS S34, CMD minutes 14 November 1968. In the Conference, the coordinator defended this name by saying that 'The promotion of the company's commercial interests was an essential part of Trade Relations, and to that extent a certain amount of "special pleading" was inevitable; however, it seemed that Shell had succeeded in establishing a reputation for presenting its interests in an objective way, without over-pressing its case. (It was this kind of consideration which had led to the rejection of the title "Public Relation" some years ago.)', SHA Conference minutes 10 July 1968, 4.

159 SHA Conference minutes 13 January 1971, 10.

160 SLA MK 009/70, Marketing Oil report 1970, 9-10, 12, 22.

161 SLA MR 29, Holdsworth to Kirby, 16 April 1969.

List of tables and figures

Chapter 1

The Rectiflow process 13

The principle of the alkylation process 13

The Shell-designed jet combustion chamber 21

1.1 Crude oil production indices, World, Jersey Standard, and the Group, 1930-1950 34

1.2 The Group's main areas of production in percentages of total production, 1930-1950 35

1.3 The Group's production by main areas in metric tons, 1930-1950 35

The formula for butadiene 74

Chapter 2

2.1 The Group's top managerial structure in 1959 144

2.2 The Group's corporate structure in 1959 145

2.3 The Group's staff numbers and staff costs as a percentage of sales, 1955-1975 156

Chapter 3

3.1 Table: The Group's main financial data, 1951-1970 180

3.1 The Group's crude supplies by region, 1950-1975 185

3.2 The Group's upstream and downstream investments, 1950-1970 188

3.2 Table: The Group's reserves, expected discoveries, and projected E&P spending, 1966 and 1967-1971 189

Chapter 4

4.1 Manufacturing productivity at Shell Oil and the rest of the Group, 1955-1970 253

4.2 Net revenue per staff member for Shell Oil and the rest of the Group, 1955-1970 255

Manufacturing oil flow chart 260

Petrochemical production flow chart 261

4.3 The Group's cut-of-the-barrel, 1945-1975 298

Chapter 5

Petrochemical products derived from refinery gases 336

Petrochemical products flow chart 337

5.1 The growth of Group revenues, 1955-1975 344

5.2 Group chemicals sales, 1954-1975 344

5.3 Group investment in manufacturing, 1955-1975 345

5.4 Oil and chemicals investment as a percentage of the manufacturing total, 1955-1975 345

5.5 Group chemical sales revenue in Europe, by product segment, 1952 349

5.6 Group chemical sales revenue by product segment, 1955-1972 350

Chapter 6

6.1 The number of patents obtained by the Group and research costs, 1950-1975 400

Abbreviations

A

AG — Aktien Gesellschaft (joint-stock company)
AGIP — Azienda Generali Italiana di Petroli
AGM — annual general meeting of shareholders
AGNS — Allied General Nuclear Services
AIOC — Anglo-Iranian Oil Company
AN CAMT — Archives Nationaux, Centre des Archives du Monde du Travail (Roubaix, France)
API — American Petroleum Institute
APOC — Anglo-Persian Oil Company
Avgas — aviation fuel

B

B/d, bpd — barrels per day
BASF — Badische Anilin- und Soda-Fabriken
BB RWM — Bundesarchiv Berlin, Reichswirtschaftsministerium
BB — Bundesarchiv Berlin
BEA — Bank of England Archives, London (UK)
BEF — British Expeditionary Force
BHP — Broken Hill Proprietary Company
BIM — Bataafsche Import Maatschappij
BNOC — British National Oil Corporation
BP — British Petroleum
BPC — Basrah Petroleum Company
BPM — Bataafsche Petroleum Maatschappij

C

CBE — Commander of the Order of the British Empire
CEI — Compagnie d'Esthétique Industrielle
CEO — chief executive officer
CEP — Current Estimated Potential
CERA — Cambridge Energy Research Associates
CFCs — chlorofluorocarbons
CFO — Chief Financial Officer
CFP — Compagnie Française des Pétroles
CIF — cost, insurance, freight
CMD — Committee of Managing Directors
CNOOC — China National Offshore Oil Corporation
CONCAWE — Conservation of Clean Air and Water, Western Europe
COT — Curaçao Oil Terminal
CPIM — Curaçaosche Petroleum Industrie Maatschappij
CPMR — Pipeline Mozambique Rhodesia Company
CSM — Curaçaosche Scheepvaart Maatschappij
CSV — Compañia Shell de Venezuela

D

DAPG — Deutsch-Amerikanische Petroleum Gesellschaft
DCC — Director of Coordination Chemical
DCO — Director of Coordination Oil
DEA — Deutsche Erdöl Aktiengesellschaft
DNB — Nederlandsche Bank, Amsterdam (Netherlands)
DSM — Dutch State Mines
DWT — deadweight tonnes

E

E&P — exploration and production
EC — European Community
EEC — European Economic Community
EGM — extraordinary general meeting
Elf — Essence et Lubrifiants français
ENI — Ente Nazionali Idrocarburi
EP, E&P — Exploration and Production
EPU — Europäische Petroleum Union
ERAP — Entreprise de Recherches et d'Activités Pétrolières
Expro — Exploration and Production

F

FCE — Fletcher Challenge Energy
FIH — free in harbour
FOB — free on board
FSA — Financial Services Authority
FTC — Federal Trade Commission

G

GRT — Gross Registered Tonnes
GTL — Gas to Liquids

H

HDNP — Historisch Documentatiecentrum voor het Nederlands Protestantisme (Vrije Universiteit, Amsterdam, the Netherlands)
HMG — His/Her Majesty's Government
HR — Human Resources
HTGR — High Temperature Gas-cooled Reactor

I

ICC — International Chamber of Commerce
IHECC — International Hydrogenation Engineering and Chemical Company
IHP — International Hydrogenation Patents Company
IMCO — Intergovernmental Maritime Consultative Organisation
INOC — Iraq National Oil Company
IPC — Iraq Petroleum Company

K

KBE Knight Commander of the Order of the British Empire
KNPM Koninklijke Nederlandse Petroleum Maatschappij
KOC Kuwait Oil Company
KPM Koninklijke Paketvaart Maatschappij
KSLA Koninklijke Shell Laboratorium Amsterdam

L

LEAP Leadership and Performance
LNG Liquid Natural Gas
LPG Liquified Petroleum Gas
LWR Light Water Reactor

M

Mekog Maatschappij tot Exploitatie van Kooks-Oven Gassen
MFH Ministry of Finance, The Hague (the Netherlands)
Mogas automobile fuel
Mori Market & Opinion Research International
Mosop Movement for the Survival of the Ogoni People

N

NA Kew National Archives, Kew (UK)
NA The Hague National Archives, The Hague (the Netherlands)
NAM Nederlandse Aardolie Maatschappij
NGO Non-governmental organization
NHM Nederlandsche Handel-Maatschappij
NIIHM Nederlandsch-Indische Industrie- en Handel-Maatschappij
NIOC National Iranian Oil Company
NIOD Nederlands Instituut voor Oorlogsdocumentatie (Amsterdam, the Netherlands)
NIT Nederlandsch-Indische Tankstoomboot Maatschappij
NKPM Nederlandsche Koloniale Petroleum Maatschappij
NNPC Nigerian National Petroleum Corporation
NSB Nationaal Socialistische Beweging
NT New Technology Ventures Division
NTB Non-Traditional Business
NV Naamloze Vennootschap (joint-stock company)
NVD New Venture Divisions

O

OAPEC Organization of Arab Petroleum Exporting Countries
OBE Officer of the Order of the British Empire
OECD Organization for Economic Cooperation and Development
OPC Oil Price Collapse
OPEC Organizaton of Petroleum Exporting Countries
OVA Overhead Value Analysis

P

PA Public Affairs
PDO Petroleum Development Oman
PDVSA Petróleos de Venezuela S.A.
Pemex Petróleos Mexicanos
PET Polyethylene Terephthalate
PPAG Petroleum Produkte Aktien Gesellschaft
PVC Polyvinyl Chloride

R

R&D Research and Development
RA Alkmaar Regionaal Archief Alkmaar (the Netherlands)
RAF Royal Air Force
RD Royal Dutch
RD/S Royal Dutch/Shell
RDS Royal Dutch Shell plc
RIS Republik Indonesia Serikat
RM Reichsmark
ROACE Return on Average Capital Employed
RTZ Rio Tinto-Zinc Corporation
RVI Retail Visual Identity
RWM Reichswirtschaftsministerium

S

SAPREF Shell and BP South African Petroleum Refineries
SASOL South African Synthetic Oil Ltd
SBR Styrene butadiene rubber
SCORE Service Companies Operations Review Exercise
SEC Securities and Exchange Commission
SHA Shell Archives, The Hague (the Netherlands)
SHAC Shell High Activity Catalyst
SHOP Shell Higher Olefins Process
SIEP Shell International E&P
Sietco Shell International Eastern Trading Company
SIMEX Singapore International Monetary Exchange
Sinopec China Petroleum & Chemical Corporation
SIPC Shell International Petroleum Company Ltd.
SIPM Shell Internationale Petroleum Maatschappij NV
SIS Shell International Shipping
Sitco Shell International Trading Company
SLA Shell Archives, London (UK)
SM/PO Styrene Monomer/Propylene Oxide
SMBP Shell-Mex & BP Ltd.
SMDS Shell Middle Distillate Synthesis
SOC Standard Oil Company
SOCAR State Oil Company of Azerbaijan Republic
Socony Standard Oil Company of New York
Stanvac Standard-Vacuum Oil Company

T

TBA tyres, batteries, and accessories
TCP tri-chresyl phospate
TEL tetra-ethyl lead
TINA There Is No Alternative
TNT tri-nitro toluene
TPC Turkish Petroleum Company
TVP true vapour phase

U

UMWA United Mine Workers of America
UNC United Nuclear Corporation
UOP Universal Oil Products

V

VLCC very large crude carrier
VOC Venezuelan Oil Concessions Ltd.
VU HDNP Historisch Documentatiecentrum voor het Nederlands Protestantisme (Vrije Universiteit, Amsterdam, the Netherlands)

W

WOCANA World outside the Communist area and North America
WTI West Texas Intermediate

Y

YPF Yacimientos Petroliferos Fiscales

Bibliography

Aandstad, S. A., *Surrendering to Symbols, United States Policy Towards Indonesia 1961-1965* http://www.lulu.com/content/322967.

Abelshauser, W., **W. von Hippel**, **J. A. Johnson**, **R. G. Stokes**, *German Industry and Global Enterprise, BASF: the History of a Company* (Cambridge University Press: Cambridge 2004).

Allen, R. S., *Wartime with Shell, the Autobiography of R.S. Allen*, ed. **N. L. Middlemiss** (Shield Publications: Newcastle 1996).

Anderson Jr., I. H., *The Standard-Vacuum Oil Company and United States East Asian Policy, 1933-1941* (Princeton University Press: Princeton 1975).

Anderson, R. O., *Fundamentals of the Petroleum Industry* (Weidenfeld & Nicolson: London 1984).

Arora, A. and **N. Rosenberg**, 'Chemicals: a US Success Story', in: **A. Arora**, **R. Landau** and **N. Rosenberg**, eds., *Chemicals and Long-Term Economic Growth, Insights from the Chemical Industry* (Wiley: New York 1998) 71-102.

Bakker, A., **M. M. P. van Lent**, *Pieter Lieftinck 1902-1989, een leven in vogelvlucht* (Veen: Utrecht 1989).

Bamberg, J. H., *The History of the British Petroleum Company*, II, *The Anglo-Iranian Years, 1928-1954* (Cambridge University Press: Cambridge 1994); III, *British Petroleum and Global Oil, 1950-1975, the Challenge of Nationalism* (Cambridge University Press: Cambridge 2000).

Bartlett, A. G., **R. J. Barton**, **J. C. Bartlett**, **G. A. Fowler**, and **C. F. Hays**, *Pertamina, Indonesian National Oil* (Amerasian: Djakarta 1972).

Birkenfeld, W., *Der synthetische Treibstoff, ein Beitrag zur nationalsozialistischen Wirtschafts- und Rüstungspolitik* (Musterschmidt: Göttingen 1964).

Blanken, I. J., *Geschiedenis van Philips Electronics NV*, IV, *Onder Duits beheer* (Europese Bibliotheek: Zaltbommel 1997).

Blu Buhs, J., *The Fire Ant Wars, Nature, Science, and Public Policy in Twentieth-entury America* (University of Chicago Press: Chicago 2004).

Boender, K., *Milieuprotest in Rijnmond, sociologische analyse van milieusolidariteit onder elites en publiek* (Sijthoff: Rijswijk 1985).

Boer, J., *Koninklijke Olie in Indië, de prijs voor het vloeibare goud 1939-1953* (Bonneville: Bergen 1997).

Booth, A., *The Indonesian Economy in the Nineteenth and Twentieth Centuries, a History of Missed Opportunities* (Macmillan: London 1998).

Borghuis, G. J., *Veertig jaar NAM, de geschiedenis van de Nederlandse Aardolie Maatschappij 1947-1987* (NAM: Assen 1987).

Breidenbach, A. W., **J. J. Lichtenberg**, 'DDT and Dieldrin in Rivers: A Report of the National Water Quality Network', *Science* 141 (1963) 899-901.

Bridges, H., *The Americanization of Shell, the Beginnings and Early Years of Shell Oil Company in the United States* (Newcomen Society: New York 1972).

Broersma, K. E., *Eene zaak van regt en billijkheid. Enkele episoden uit de enerverende Billiton-geschiedenis 1860-1985* (Billiton: Leidschendam 1985).

Brouwer, G. C., **M. J. Coenen**, *Nederland = aardgasland* (Roelofs van Goor: Amersfoort 1968).

Burleson, C. W., *Deep Challenge, the True Epic Story of our Quest for Energy beneath the Sea* (Gulf Publishing: Houston 1999).

Carlson, S., *Indonesia's Oil* (Westview Press: Boulder 1977).

Chandler, A. D., *Strategy and Structure, Chapters in the History of the American Industrial Enterprise* (MIT: Harvard 1962).

Chandler, A. D., *Scale and Scope, the Dynamics of Industrial Capitalism* (Harvard University Press: Harvard 1990).

Chandler, A. D., **T. Hikino**, and **D. C. Mowery**, 'The Evolution of Corporate Capabilities and Corporate Strategy and Structure within the World's Largest Chemical Firms: The Twentieth Century in Perspective', in: **A. Arora**, **R. Landau** and **N. Rosenberg**, eds., *Chemicals and Long-Term Economic Growth, Insights from the Chemical Industry* (Wiley: New York 1998) 436-39.

Chapman, K., *The International Petrochemical Industry: Evolution and Location* (Blackwell: Oxford 1991).

Cohen, R. B., *Japan's Economy in War and Reconstruction* (University of Minnesota Press: Minneapolis 1949).

Cooke, R. C., R. Conyers Nesbit, *Target: Hitler's Oil, Allied Attacks on German Oil Supplies 1939-1945* (William Kimber: London 1985).

Corley, T. A. B., *A History of the Burmah Oil Company*, Vol. II, *1924-1966* (Heinemann: London 1988).

Correljé, A., *Hollands Welvaren, de geschiedenis van een Nederlandse bodemschat* (Teleac: s.l., 1998).

Correljé, A., C. van der Linde, T. Westerwoudt, *Natural Gas in the Netherlands, from Cooperation to Competition?* (Oranje-Nassau Groep: s.l. 2003).

Cramer, J., *De Groene Golf, geschiedenis en toekomst van de milieubeweging* (Van Arkel: Utrecht 1989).

Dankers, J. J., J. Verheul, *Hoogovens 1945-1993. Van staalbedrijf tot twee-metalenconcern. Een studie in industriële strategie* (SDU: The Hague 1993).

DiNardo, R. L., A. Bay, 'Horse-drawn Transport in the German Army', in: *Journal of Contemporary History* 23 (1988) 129-42.

Engle, A., 'Organisation and Management Planning in the Royal Dutch/Shell Group', in: **R. S. Edwards, H. Townsend**, *Business Enterprise, its Growth and Organisation* (Macmillan: London 1957) 342-52.

Faassen, M. van, A. E. Kersten, *Documenten betreffende de buitenlandse politiek van Nederland 1919-1945, Periode C*, VI, *16 December 1942-1930 juni 1943* (ING: The Hague 1996).

Faassen, M. van, R. J. J. Stevens, *Documenten betreffende de buitenlandse politiek van Nederland 1919-1945, Periode C*, VIII, *1 juli 1944-14 augustus 1945* (ING: The Hague 2004).

Fabricius, J, *Brandende aarde, de vernieling en evacuatie van de olieterreinen in Nederlandsch-Indië* (Bataafsche Petroleum Maatschappij: The Hague 1949), also published as *East Indies Episode, an Account of the Demolitions carried out and of some Experiences of the Staff in the East Indies Oil Areas of the Royal Dutch-Shell Group during 1941 and 1942* (Shell Petroleum Company: London 1949).

Gabriëls, H., *Koninklijke Olie: de eerste honderd jaar 1890-1990* (Shell Internationale Petroleum Maatschappij: The Hague, 1990).

Gales, B. P. A., J .P. Smits, 'Olie en gas', in: **H. W. Lintsen**, ed., *Techniek in Nederland in de twintigste eeuw*, II, *Delfstoffen, energie, chemie* (Walburg Pers: Zuthpen 2000) 67-89.

Gary, J. H., G. E. Handwerk, *Petroleum Refining: Technology and Economics* (Dekker: New York 1984).

Green, W., *The Warplanes of the Third Reich* (Macdonald and Jane's: London 1970).

Hillier, B., *John Betjeman, New fame, New Love* (Murray: London 2002).

Homburg, E., A. Rip, 'De chemische industrie in de twintigste eeuw', in: **H. W. Lintsen et al.**, ed., *Techniek in Nederland in de twintigste eeuw*, II, *Delfstoffen, energie, chemie* (Walburg Pers: Zutphen 2000) 403-7.

Homburg, E., A. van Selm, P. Vincken, 'Industrialisatie en industriecomplexen, de chemische industrie tussen overheid, technologie en markt', in: **H. W. Lintsen et al.**, ed., *Ttechniek in Nederland in de twintigste eeuw*, II, *Delfstoffen, energie, chemie* (Walburg Pers: Zutphen 2000) 377-401.

Hope, S., *Tanker Fleet: The War Story of the Shell Tankers and the Men who manned them* (London: The Anglo-Saxon Petroleum Company, 1948).

Jager, K. W., *Aldrin, dieldrin, endrin and telodrin, an epidemiological and toxicological study of long-term occupational exposure* (Elsevier: Amsterdam 1970).

Jones, G. G., 'Control, Performance, and Knowledge Transfers in Large Multinationals, Unilever in the United States, 1945-80', *Business History Review* 76 (2002) 435-78.

Jones, G. G., *Renewing Unilever, Transformation and Tradition* (Oxford University Press: Oxford 2005).

Jong, L. de, *Het Koninkrijk der Nederlanden in de Tweede Wereldoorlog*, Vol II, *Neutraal* (Nijhoff: Leiden 1969); Vol. III, *Mei 1940* (Nijhoff: Leiden 1970); Vol XIa, XIb, XIc, *Indië* (Nijhoff: Leiden 1984-1986).

Kamp, A. F., *De standvastige tinnen soldaat, 1860-1960, NV Billiton Maatschappij 's Gravenhage* (S.n., s.l., 1960).

Karlsch, R., and R. G. Stokes, *Faktor Öl, die Mineralölwirtschaft in Deutschland 1859-1974* (C.H. Beck: Munich 2003).

Kennedy, C., *ICI: The Company that Changed our Lives* (Chapman: London 1993).

Kerr, G. P., *Time's Forelock: A Record of Shell's Contribution to Aviation in the Second World War* (Shell Petroleum Company: London 1948).

Kipping, M., 'Consultancies, institutions and the diffusion of Taylorism in Britain, Germany and France, 1920s-1950s', *Business History* 39(1997) 67-84.

Kipping, M., 'American Management Consulting Companies in Western Europe, 1920 to 1990, Products, Reputation, and Relationships', *Business History Review* 73 (1999) 190-220.

Kipping, M., L. Engwall, ed., *Management Consulting, Emergence and Dynamics of a Knowledge Industry* (Oxford University Press: Oxford 2002).

Kruisinga, H. J., *The Balance between Centralization and Decentralization in Managerial Control* (Stenfert Kroese: Leiden 1954).

Kruisinga, H. J., *Vraagstukken van directievoering in geografisch gedecentraliseerde bedrijven* (Stenfert Kroese: Leiden 1956).

Landau, R., 'The Process of Innovation in the Chemical Industry', in **A. Arora, R.Landau** and **N. Rosenberg**, eds., *Chemicals and Long-Term Economic Growth: Insights from the Chemical Industry* (Wiley: New York 1998) 139-80.

Lear, L., *Rachel Carson, Witness for Nature* (Holt & Co.: New York 1997).

Lucas, W. E., *Eagle Fleet : the Story of a Tanker Fleet in Peace and War* (Weidenfeld & Nicholson: London 1955).

Manning, A. F., A. E. Kersten, *Documenten betreffende de buitenlandse politiek van Nederland 1919-1945, Periode C*, I, *10 mei-31 oktober 1940* (Nijhoff: The Hague 1976); III, *1 juni-7 December 1941* (Nijhoff: The Hague 1980); IV, *8 December 1941-30 juni 1942* (Nijhoff: The Hague 1984); V, *1 juli-15 December1942* (Nijhoff: The Hague 1987).

Markowitz, G., D. Rosner, *Deceit and Denial, the Deadly Politics of Industrial Pollution* (University of California Press: Berkeley 2002).

McKenna, C. D., *The World's Newest Profession, Management Consulting in the Twentieth Century* (Cambridge University Press: Cambridge 2006).

Meijer, H., *Den Haag-Djakarta, de Nederlands-Indonesische betrekkingen 1950-1962* (Het Spectrum: Utrecht 1994).

Meijer, H., *Indische rekening, Indië, Nederland en de backpay-kwestie 1945-2005* (Boom: Amsterdam 2005).

Micheels, P., *De vatenman, Bernard van Leer (1883-1958)* (Contact: Amsterdam 2002).

Middlemiss, N. L., *The Anglo-Saxon/Shell Tankers* (Shield Publications: Newcastle upon Tyne 1990).

Monopolies Commission, *A Report on the Supply of Petrol to Retailers in the United Kingdom* (HMSO: London 1965).

Mook, H. J. van, *Nederlandsch-Indië en Japan: hun betrekkingen in 1940-1941* (Netherland Publishing Company: London 1945), also published as *The Netherlands Indies and Japan: their relations, 1940-1941* (Allen & Unwin: London 1944).

Most, F. van der, E. Homburg, P. Hooghoff, A. van der Selm, 'Nieuwe synthetische producten', in: **H. W. Lintsen** et al., ed., *Techniek in Nederland in de twintigste eeuw*, II, *Delfstoffen, energie, chemie* (Walburg Pers: Zutphen 2000) 358-75.

Muller, H., *The search for the qualities essential to advancement in a large industrial group: an exploratory study* (S.n., The Hague 1980)

Murray, R., *Go Well: one hundred years of Shell in Australia* (S.n., s.l. 2001).

Nockolds, H., *The Engineers, a Record of the Work done by Shell Engineers in the Second World War* (Shell Petroleum, London 1949).

Nouschi, A., *La France et le Pétrole de 1924 à nos Jours* (Picard: Paris 2001).

Oelderik, J. M., S. T. Sie, D. Bode, 'Progress in the Catalyses of the Upgrading of Petroleum Residue: A Review of 25 Years of R&D on Shell's Residue Hydroconversion Technology', *Applied Catalysis*, 47(1989) 1-24.

Overy, R. J., *Why the Allies Won* (Jonathan Cape: London 1995).

Payton-Smith, D. J., *Oil: A Study of War-time Policy and Administration* (HMSO: London 1971).

Penrose, E. T., *The International Petroleum Industry* (Allen & Unwin: London 1968).

Philip, G., *Oil and Politics in Latin America: Nationalist Movements and State Companies* (Cambridge University Press: Cambridge, 1982).

Pradier, H., *Shell France, une double Anniversaire, 1919-1989, 1948-1988* (Shell France: Paris 1989).

Priest, T., 'The Americanization of Shell Oil', in: **G. G. Jones, L. Galvez-Muñoz**, ed., *Foreign Multinationals in the United States, Management and Performance* (Routledge: London 2002) 188-206.

Rabe, S. G., *The Road to OPEC: United States Relations with Venezuela, 1919-1976* (University of Texas Press: Austin, 1982).

Ratcliffe, M., *Liquid Gold Ships: a History of the Tanker 1859-1984* (Lloyds of London: London 1985).

Sampson, A., *The Anatomy of Britain Today* (Hodder & Stoughton: London, 1965).

Schippers, J. L., G. P. J. Verbong, 'De revolutie van Slochteren', in: **H. W. Lintsen** et al., ed., *Techniek in Nederland in de twintigste eeuw*, II, *Delfstoffen, energie, chemie* (Walburg Pers: Zuthpen 2000) 203-291.

Schweppe, J., *Research aan het IJ LBPMA 1914-KSLA 1989 : de geschiedenis van het 'Lab Amsterdam'* (Shell Research: Amsterdam 1989).

Sheail, J., *An Environmental History of Twentieth-Century Britain* (Palgrave: Basingstoke 2002).

Skeet, I., *OPEC: Twenty-five Years of Prices and Politics* (Cambridge University Press: Cambridge, 1988).

Spitz, P. H., *Petrochemicals, the Rise of an Industry* (Wiley: New York 1988).

Steiguer, J. E. de, *The Age of Environmentalism* (McGraw Hill: New York 1997).

Stheeman, H. A., A. A. Tiadens, 'A History of the Exploration for Hydrocarbons within the territorial boundaries of the Netherlands', in: **P. Hepple**, ed., *The Exploration for Petroleum in Europe and North Africa* (Institute of Petroleum: London 1969) 259-70.

Taselaar, A., *De Nederlandse koloniale lobby, ondernemers en de Indische politiek 1914-1940* (CNWS: Leiden 1998).

Tielhof, M. van, *Banken in bezettingstijd, de voorgangers van ABN AMRO tijdens de Tweede Wereldoorlog en de periode van rechtsherstel* (Contact: Amsterdam, 2003).

Tugendhat, G.,'Development of Manchester Oil Refinery Limited and Petrochemicals Limited', in: **R. S. Edwards, H. Townsend**, *Business Enterprise, its Growth and Organisation* (Macmillan: London 1957) 112-22.

Van der Wee, H., M. Verbreyt, *Oorlog en monetaire politiek, de Nationale Bank van België, de Emissiebank te Brussel en de Belgische Regering, 1939-1945* (Nationale Bank van België: Brussels 2005).

Van Rotterdam Charlois naar Rotterdam Pernis : 1902-1977: jubileumuitgave Onder de Vlam, bedrijfskrant van Shell Nederland (Shell Nederland: Rotterdam 1977).

Veldman, J. L., *De voorziening van aardolieproducten in Nederland gedurende de Tweede Wereldoorlog* (Rijksbureau Aardolieproducten: The Hague 1949).

Visser, W. A., J. I .S. Zonneveld, A. J. van Loon, *Seventy-five Years of Geology and Mining in the Netherlands (1912-1987)* (Royal Geological and Mining Society of the Netherlands: The Hague 1987).

Vuurde, R. van, *Los Países Bajos, el petróleo y la Revolución Mexicana, 1900-1950* (THELA: Amsterdam 1997).

Wall, B. H., *Growth in a Changing Environment: A History of Standard Oil Company (New Jersey) 1950-1972 and Exxon Corporation 1972-1975* (McGraw-Hill: New York, 1988).

Waller, P. P., H. S. Swain, 'Changing Patterns of Oil Transportation and Refining in Germany', *Economic Geography* 43(1967) 143-56.

Weenink, W. H., *Bankier van de wereld, bouwer van Europa, Johan Willem Beyen 1897-1976* (Prometheus: Amsterdam 2005).

Wesseling, L., *Fuelling the war, revealing an oil company's role in Vietnam* (Tauris: London 2000).

Whittle, F., *Jet: The Story of a Pioneer* (Frederick Muller: London 1953).

Wouters, W., *Shell tankers, van koninklijke afkomst* (Shell Tankers: Rotterdam 1984).

Wubs, B., *Unilever between Reich and Empire, 1939-1945, International Business and National War Interests* (PhD Rotterdam 2006).

Wielenga, F., 'James Marnix de Booy', in *Biografisch Woordenboek van Nederland* IV, 53-4.

Zanden, J. L. van, S. W. Verstegen, *Groene geschiedenis van Nederland* (Spectrum: Utrecht 1993).

Illustration credits

The publisher has made every effort to contact all those with ownership rights pertaining to the illustrations. Nonetheless, should you believe that your rights have not been respected, please contact Boom Publishers, Amsterdam.

IISG: International Institute of Social History, Amsterdam
IWM: Imperial War Museum, London
KITLV: Royal Netherlands Institute of Southeast Asian and Caribbean Studies, Leiden
MAI: Maria Austria Institute, Amsterdam
NFM: Dutch Photo Museum, Rotterdam
SHA: Shell Archive, The Hague
SLA: Shell Archive, London
SRTCA: Shell Research and Technology Centre, Amsterdam

Introduction

1 Beaulieu National Motor Museum/Shell Advertising Art Collection/Shell Studio, 1952

Chapter 1

1 Stadsarchief Amsterdam
2 Getty Images/Horace Abrahams
3 SHA, 190D/730
4 Shell Photographic Services/Shell Int. Ltd.
5 Shell Photographic Services/Shell Int. Ltd.
6 From: Forbes, *Technical development*, 376
7 Corbis/E.O. Hoppe
8 SLA, from: *Petroleum at war*
9 Getty Images
10 Corbis
11 Hollandse Hoogte/Magnum/Robert Capa
12 Hollandse Hoogte/Magnum/W. Eugene Smith
13 Getty Images/Margaret Bourke-White
14 Getty Images
15 Corbis/Philip Wallick
16 From: *Time's Forelock*, 71
17 Corbis
18 Corbis
19 Getty Images/Davis
20 Corbis
21 Corbis
22 Corbis
23 Corbis
24 Getty Images
25 KITLV
26 SHA, 11/78
27 SHA, 190C/5A
28 SHA, 190C/5A
29 Gemeentearchief Den Haag
30 SHA, 12/169
31 Shell Oil Houston, AVRP0002331
32 Shell Photographic Services/Shell Int. Ltd.
33 Corbis
34 SHA, 190F/204
35 SHA, 190F/204
36 SHA, 190F/204
37 Stadsarchief Amsterdam
38 Stadsarchief Amsterdam
39 Stadsarchief Amsterdam
40 Gemeentearchief Rotterdam
41 Shell Photographic Services/Shell Int. Ltd.
42 From: *Shell France, une double anniversaire*, 29
43 Deutsches Historisches Museum, Berlin
44 Getty Images/Russell Lee
45 IWM, HU39714
46 Corbis
47 Corbis
48 Arjen van Susteren
49 From: *Oil for victory*
50 Corbis
51 Corbis
52 Corbis
53 IWM, T31
54 Arjen van Susteren, based on: *Time's forelock*
55 IWM, T30
56 SHA, Shell Tankerarchive 849/365
57 SHA, Shell Tankerarchive 658
58 SHA, Shell Tankerarchive 433
59 KITLV
60 Shell Photographic Services/Shell Int. Ltd.
61 SHA, Shell Tankerarchive 192
62 SHA, Shell Tankerarchive 852/643
63 SHA, Shell Tankerarchive 850/523
64 SHA, Shell Tankerarchive 433
65 SHA, Shell Tankerarchive 433
66 SHA, Shell Tankerarchive 1404
67 SHA, Shell Tankerarchive 1404
68 Shell Photographic Services/Shell Int. Ltd.
69 SHA, Shell Tankerarchive 1446
70 SHA, Shell Tankerarchive 1401
71 Corbis
72 Getty Images
73 SHA,190F/30
74 SHA, 190F/30
75 SHA, 190F/32
76 SHA, Shell Tankerarchive 138077
SHA, Shell Tankerarchive 845/91

78 SHA, Shell Tankerarchive 845/92
79 Corbis
80 Shell Photographic Services/Shell Int. Ltd.
81 Bildarchiv Preussischer Kulturbesitz, 30010554
82 SHA, 190F/Box 4-United States
83 Getty Images/Dmitri Kessel
84 Shell Photographic Services/Shell Int. Ltd.
85 Shell Oil Houston, AVRP0002596
86 Getty Images
87 IWM, A22693
88 IWM, NYP15163
89 Shell Deutschland Oil LmbH, from: *Der Ring*, Vol 14, March/April 1941
90 Gemeentearchief Den Haag
91 SHA, 8/1861
92 SHA, 190D/736
93 SHA,190D/736
94 SHA, 190C/96
95 SHA, Verwaltung 16/88
96 SRTCA
97 SRTCA
98 SRTCA
99 Getty Images/A. J. O'Brien
100 MAI/Ad Windig
101 Gemeentearchief Den Haag
102 Shell Photographic Services/Shell Int. Ltd.
103 Shell Photographic Services/Shell Int. Ltd.
104 Shell Photographic Services/Shell Int. Ltd.
105 SHA,190F/Portraits
106 SHA,190F/Portraits
107 SHA,190F/Portraits
108 Nationaal Archief/Anefo
109 Shell Australia
110 SHA, 190F/KON 47
111 Shell Australia
112 Shell Australia
113 SHA, 190F/KON 382W
114 SHA, 190F/KON 556
115 From: Gabriëls, *Koninklijke olie*, 138
116 SHA, 2/3
117 MAI/Ad Windig
118 SHA, Shell Tankerarchive 851/599
119 SHA, Shell Tankerarchive 846/207
120 Shell Photographic Services/Shell Int. Ltd.

Chapter 2

1 SHA, 190F/KON 350
2 SHA, 190F/204
3 Shell Photographic Services/Shell Int. Ltd.
4 Shell Photographic Services/Shell Int. Ltd.
5 SHA, 190D/760
6 SHA, 190D/760
7 Getty Images
8 Corbis
9 SHA, 190F/Portraits
10 Nationaal Archief/Anefo
11 SHA, 190C/393A-2
12 SHA, 190C/393A-1
13 SHA, 190C/393A-1
14 SHA, 190C/393A-2
15 SHA, 190C/393A-2
16 SHA, 190C/393A-1
17 SHA, 190C/393A-2
18 SHA, 190C/393A-2
19 From: *Shell News*, May/June 1965
20 SHA, 190C/393A-2
21 SHA, 190C/393A-Enclosure
22 Hollandse Hoogte/Transworld/Tim Hethrington
23 Shell Photographic Services/Shell Int. Ltd.
24 Shell Oil Houston, AVRP0002594
25 SHA, 2/579
26 SHA, 2/579
27 SHA, Shell Tankerarchive 469
28 SHA, 190F/107
29 SHA, 190F/107
30 SHA, 2/16-1
31 SHA, 2/632
32 Shell Photographic Services/Shell Int. Ltd.
33 Shell Photographic Services/Shell Int. Ltd.
34 Shell Photographic Services/Shell Int. Ltd.
35 SHA, 2/653
36 SHA, 190F/143
37 SHA, 190F/Box 4-Venezuela
38 SHA, 2/16-1
39 Shell Photographic Services/Shell Int. Ltd.
40 Getty Images
41 Getty Images
42 Shell Photographic Services/Shell Int. Ltd.
43 Corbis
44 Corbis/David Paterson
45 Shell Photographic Services/Shell Int. Ltd.
46 Shell Photographic Services/Shell Int. Ltd.
47 Shell Photographic Services/Shell Int. Ltd.
48 SHA, From: *Shell Magazine*, Februar 1953
49 From: *Shell Oil Company 1965 Annual Report*
50 Shell Oil Houston, from: *Shell News*, November/December 1963
51 Shell Australia
52 From: *Shell News*, April 1963
53 Getty Images
54 SHA, 190F/Portraits
55 SHA, 190F/4-1
56 SHA, 190F/Portraits
57 Shell Photographic Services/Shell Int. Ltd.
58 From: *Shell News*, April1969
59 From: *Shell News*, April 1969
60 From: *Shell News*, July/August 1964
61 From: *Shell Magazine*, September 1961
62 Shell Photographic Services/Shell Int. Ltd.
63 From: *Shell News*, August 1968
64 SHA, 190F/70
65 Getty Images
66 SRTCA

Chapter 3

1 Getty Images/Andreas Feininger
2 Spaarnestad Photo
3 Spaarnestad Photo/Black Star/Groebli
4 Spaarnestad Photo
5 Getty Images
6 Getty Images/Harry Kerr
7 Corbis
8 Spaarnestad Photo
9 Corbis/David Lees
10 Getty Images/Andreas Feininger
11 Spaarnestad Photo
12 Getty Images/Ed Clark
13 Shell Photographic Services/Shell Int. Ltd.
14 Arjen van Susteren
15 SHA, 190F/70
16 SHA, 190F/70
17 SHA, 190F/70
18 SHA, 190F/70
19 SHA, 190F/70
20 Shell Photographic Services/Shell Int. Ltd.
21 SHA, 190F/Box 4-Venezuela
22 Spaarnestad Photo
23 Spaarnestad Photo
24 Getty Images/George Silk
25 Spaarnestad Photo
26 Spaarnestad Photo
27 Shell Photographic Services/Shell Int. Ltd.
28 Shell Photographic Services/Shell Int. Ltd.
29 Arjen van Susteren, based on: *Shell Magazine*, 1961
30 Shell Photographic Services/Shell Int. Ltd.
31 From: *Long-term Exploration and Production Development Review*, 1964
32 Arjen van Susteren, based on: *Shell News*, 1/1959
33 NFM/Aart Klein
34 Spaarnestad Photo
35 Spaarnestad Photo
36 SHA, 190F/9
37 SHA, 190F/9
38 SHA, 69/74
39 NFM/Aart Klein
40 NFM/Aart Klein
41 Joost Guntenaar
42 Arjen van Susteren, based on: *Aardgas, energie zonder grenzen*
43 Arjen van Susteren
44 NFM/Aart Klein
45 NFM/Aart Klein
46 NFM/Aart Klein
47 Shell Photographic Services/Shell Int. Ltd.
48 Shell Photographic Services/Shell Int. Ltd.
49 Getty Images
50 Getty Images/Raymond Kleboe
51 Hollandse Hoogte/Magnum/Bruno Barbey

52 Spaarnestad Photo/ANP
53 Spaarnestad Photo
54 Getty Images/Bert Hardy
55 Spaarnestad Photo
56 SHA, Shell Tankerarchive 849/433
57 SHA, 190F/131
58 Spaarnestad Photo
59 Getty Images
60 Getty Images
61 Getty Images
62 Getty Images
63 SHA, 190F/131
64 SHA, 2/1948
65 SHA, 2/1948
66 SHA, 2./1948
67 SHA, 2/1948
68 Spaarnestad Photo
69 Getty Images/John Dominis
70 Spaarnestad Photo
71 Spaarnestad Photo
72 Spaarnestad Photo
73 Spaarnestad Photo/ANP
74 Spaarnestad Photo/ANP
75 Getty Images
76 Corbis
77 SHA, from: *Shell Venster*, 7/1977
78 OPEC
79 SHA, from: *Shell Venster*, 7/1977
80 SHA, from: *Shell Venster*, 7/1977
81 Shell Photographic Services/Shell Int. Ltd.
82 Shell Deutschland Oil GmbH, E246-44
83 Shell Photographic Services/Shell Int. Ltd.
84 SHA, 190F/78
85 Spaarnestad Photo/United Press International

Chapter 4

1 Spaarnestad Photo/Artica Press
2 Collection Minneapolis Institute of Arts/Walker Evans
3 Getty Images/Ernst Haas
4 From: *Shell Oil Company 1964 Annual Report*
5 Shell Oil Houston, from: *Shell News*, Januar 1966
6 Shell Oil Houston, from: *Shell News*, Januar 1966
7 Getty Images
8 From: Beaton, *Enterprise*
9 Getty Images
10 ANP
11 Shell Pernis
12 Collectie Ton Kuyer/Photo: Hans van den Boogaard
13 ANP
14 SHA, 190Y/1030
15 NFM/Aart Klein
16 Spaarnestad Photo/Artica Press
17 Spaarnestad Photo/Artica Press
18 Spaarnestad Photo/Artica Press
19 SHA, 2/621
20 SHA, 2/621
21 SHA, 2/609
22 SHA, 2/624
23 SHA, 190F/Box 4-Singapore
24 SHA, 190F/Box 4-Singapore
25 Shell Oil Houston, AVRP0002261
26 Spaarnestad Photo/Anefo
27 Shell Pernis
28 Shell Photographic Services/Shell Int. Ltd.
29 SHA, Shell Tankerarchive 669
30 Shell Photographic Services/Shell Int. Ltd.
31 SHA, Shell Tankerarchive 1533
32 SHA, Shell Tankerarchive 1533
33 Shell Photographic Services/Shell Int. Ltd.
34 SHA, Shell Tankerarchive 531
35 SHA, Shell Tankerarchive 531
36 SHA, Shell Tankerarchive 469
37 SHA, Shell Tankerarchive 469
38 SHA, Shell Tankerarchive 433
39 Shell Photographic Services/Shell Int. Ltd.
40 SHA, Shell Tankerarchive 469
41 SHA, Shell Tankerarchive 469
42 Shell Deutschland Oil GmbH/WW6016
43 SHA, 190F/131
44 SHA, 190F/10
45 Arjen van Susteren, based on: Howarth, *Century* 277
46 SHA, 190F/10
47 Shell Deutschland Oil GmbH
48 Collection Ton Kuyer/Photo: Hans van den Boogaard
49 Shell Photographic Services/Shell Int. Ltd.
50 Advertising Archives
51 IISG
52 Spaarnestad Photo
53 Print Room Leiden/Emmy Andriesse
54 SHA, Advertisements/green album, 1960
55 Shell Australia
56 Shell Oil Houston, from: *Shell News*, September/October 1963
57 SHA, 2/364
58 SHA, 2/364
59 Advertising Archives
60 Collection Ton Kuyer/ Photo: Hans van den Boogaard
61 Collection Ton Kuyer/ Photo: Hans van den Boogaard
62 Advertising Archives
63 From: Alistair Cooke, *American journey*
64 SHA, 2/364
65 Corbis/Walker Evans
66 Collection Ton Kuyer/Photo: Hans van den Boogaard
67 Corbis/Michael Buselle
68 SHA/190F/205
69 Collection Ton Kuyer/Photo: Hans van den Boogaard
70 Shell Australia
71 MAI/Kees Scherer
72 Shell Deutschland Oil GmbH/WW12540
73 Collection Tim van Kooten
74 Shell Oil Houston, from: *Shell News*, July/August 1968
75 Shell Oil Houston, from: *Shell News*, July/August 1968
76 Shell Oil Houston, from: *Shell News*, July/August 1968
77 Reclamearsenaal
78 Reclamearsenaal
79 Recalmearsenaal
80 Reclamearsenaal
81 Collection Joost Jonker
82 From: *Shell France, une double anniversaire*
83 Beaulieu National Motor Museum/Shell Advertising Art Collection
84 Shell Photographic Services/Shell Int. Ltd.
85 Shell Deutschland Oil GmbH
86 Rue des Archives
87 Reclamearsenaal

Chapter 5

1 Spaarnestad Photo
2 Getty Images/W. Eugene Smith
3 Getty Images/Gjon Mill
4 Getty Images
5 Getty Images
6 Corbis
7 Getty Images/Ralph Morse
8 Shell Photographic Services/Shell Int. Ltd.
9 Shell Photographic Services/Shell Int. Ltd.
10 Shell Photographic Services/Shell Int. Ltd.
11 Editions Images Plurielles/Abed Abidat
12 From: *De Journalist*, 15 December 1965
13 From: *De Journalist*, 1958
14 Shell Photographic Services/Shell Int. Ltd.
15 SHA, 190F/Portraits
16 SHA, 190F/Portraits
17 Collection Ton Kuyer/Photo: Hans van den Boogaard
18 Shell Photographic Services/Shell Int. Ltd.
19 Spaarnestad Photo
20 Getty Images/George Silk
21 Getty Images/George Freston
22 Shell Photographic Services/Shell Int. Ltd.
23 Shell Photographic Services/Shell Int. Ltd.
24 Shell Photographic Services/Shell Int. Ltd.
25 Shell Australia
26 Shell Film Unit, London
27 Shell Photographic Services/Shell Int. Ltd.
28 Shell Photographic Services/Shell Int. Ltd.
29 Shell Photographic Services/Shell Int. Ltd.
30 Advertising Archives
31 Advertising Archives

32 SHA, 190F/ 4-1
33 SHA, 190F/ 4-1
34 Shell Photographic Services/Shell Int. Ltd.
35 Getty Images
36 Corbis
37 Shell Photographic Services/Shell Int. Ltd.
38 Shell Oil Houston, from: *Shell News*, September/October 1963
39 Spaarnestad Photo
40 Spaarnestad Photo
41 Spaarnestad Photo
42 Spaarnestad Photo
43 Spaarnestad Photo
44 Shell Photographic Services/Shell Int. Ltd.
45 Shell Photographic Services/Shell Int. Ltd.
46 Shell Photographic Services/Shell Int. Ltd.
47 SHA, 190F/3.81.04.018
48 From: *Shell Oil Company 1968 Annual Report*
49 SHA, 190F/205
50 Spaarnestad Photo
51 SHA, 190D/812
52 Spaarnestad Photo
53 MAI/Eli van Zachten
54 MAI/Eli van Zachten
55 MAI/Eli van Zachten
56 Spaarnestad Photo
57 Spaarnestad Photo
58 Spaarnestad Photo
59 Spaarnestad Photo
60 Collection Minneapolis Institute of Arts/W. Eugene Smith

Chapter 6

1 Spaarnestad Photo
2 SRTCA
3 MAI/Carel Blazer
4 SHA, 190F/KON 329W
5 MAI/Carel Blazer
6 MAI/Carel Blazer
7 SHA, 190F/89
8 SHA, 190F/91
9 Shell Photographic Services/Shell Int. Ltd.
10 Shell Oil Houston, AVRP0002534
11 Shell Photographic Services/Shell Int. Ltd.
12 Shell Photographic Services/Shell Int. Ltd.
13 Spaarnestad Photo
14 SRTCA
15 SHA, 2/657
16 Spaarnestad Photo
17 Shell Photographic Services/Shell Int. Ltd.
18 Shell Oil Deutschland GmbH, WW14933
19 Shell Deutschland, WW24045
20 Corbis
21 Shell Oil Deutschland GmbH, WW9112
22 Corbis
23 Shell Oil Deutschland GmbH, WW10362/70
24 Shell Oil Deutschland GmbH
25 SHA, Shell Tankerarchive 846/207
26 SHA, Shell Tankerarchive 846/207
27 SHA, Shell Tankerarchive 846/214
28 Shell Oil Deutschland GmbH
29 Spaarnestad Photo/Anefo
30 SHA, 190F/205
31 SHA, 190F/205
32 Corbis
33 Corbis
34 Getty Images/Ted Russell
35 Getty Images/Andreas Feininger
36 Corbis
37 Getty Images
38 Getty Images/Gerald R. Brimacombe
39 Collection Ton Kuyer/Photo: Hans van den Boogaard
40 Collection Ton Kuyer/Photo: Hans van den Boogaard
41 Getty Images/Joe Munroe
42 Getty Images/Howard Sochurek
43 Corbis
44 Getty Images
45 Corbis
46 Cor Jaring
47 IISG/Ben van Meerendonk
48 Cor Jaring
49 Hollandse Hoogte/Magnum/Bruno Barbey
50 Corbis/Sygma/James Andanson
51 Collection Ton Kuyer/Photo: Hans van den Boogaard
52 From: *Shell Oil Company 1968 Annual Report*
53 Rue des Archives/Gerald Bloncourt

Tins: Collection Ton Kuyer/Photo: Hans van den Boogaard

Index

Page references in grey refer to illustrations or maps. Page numbers followed by a slash and a number refer to notes. Names of oil platforms, ships, newspapers and periodicals are in *italics*.

A

'A' Agreement 5

Abadan (Iran) 29, 265

Abrahams, Mark 270

Abu Dhabi 207

accidents, *see also* safety 139, 276, 285, 406, 410, 415, 469/52, 477/79

 shipping accidents, *see also* pollution 285, 406

  Kong Haakon VII 285

  Mactra 285, 413, 470/79

  Marpessa 285, 413

  Torrey Canyon 406, 408, 410, 410-1

Accra (Ghana) 131

Aceh (Sumatra) 227-8

Achnacarry Agreement (As-Is Agreement) 6, 14, 172

acquisitions, take-overs, *see also* joint ventures, mergers 92, 189, 232, 348-9, 370-3, 376, 379, 381, 394

Adjustment Agreement 5, 11

advertising / advertisements and posters, *see also* branding, marketing, publicity 78, 91, 119, 212, 301-9, 299, 306-7, 306-10, 309, 320-2, 324, 328, 343, 347

 Shellubrication 301

Africa 14, 136, 166, 202, 221, 265, 268

 West Africa 189, 189

AGIP (Azienda Generali Italiana di Petroli), *see also* ENI/AGIP 78, 176, 241, 268, 475/140

Agnew, Sir Andrew 14, 17, 63, 87-8, 91, 97, 457/163

agrochemicals / agricultural chemicals 261, 337, 339, 342, 348, 351, 360, 394, 399, 435-6, 441, 473/30, 476/39

 fertilizers 76, 332, 346, 348, 351, 361, 393, 394-7, 402, 415, 436

 pesticides, herbicides, fungicides and insecticides 76, 80, 176, 261, 332, 337, 339, 342, 346, 348-52, 351-3, 360, 362, 383, 393-5, 399, 401, 404, 427-31, 428-31, 433, 435-9, 436, 440, 441, 445, 456/137

  5% DDT Dust 427

  aldrin 350-1, 399, 427, 431, 433, 435, 438

  Aldrin Soil Pest Killer 427

  Ant Doom 427, 429

  bidrin 427

  Coppicide 427

  DDT 80, 93, 346, 427, 429, 430-1, 431, 438, 478/116

  DDT Dust 427

  Derris Dust 427

  dieldrin 350-1, 352, 399, 427, 429, 431, 435, 438

  Dieldrin Garden Pest Killer 427

  'drins' 351, 353, 427-39, 441, 446, 478/116

  endrin 350, 399, 427, 431

  heptachlor 431, 435

  Kotol 435, 437, 478/137

  Netelex 429-30

  phosdrin 427, 430

  Pillakiller 427

  Proponex 429

  Shell Liquid Derris 427

  Shell Tomato-Set 427

  Shell Weedkill for Lawns 427

  Shelltox 427

  Slug Doom 430

  Slug Kill 430

  Sulficide 427

  Telodrin 430

  Universal DNC Fruit Tree Wash 427

  Vapona 353

Agricultural Products Committee 436

AIOC *see* Anglo-Iranian Oil Company (AIOC)

Aioi (Japan) 288-9

air pollution, *see* pollution

Alabama (US) 51

Alaska (US) 196, 205, 464/54

Albatros fertilizer works 402

Albion-Cuthbertson Water Buffalo 191

Albion Motors Ltd. 191

Alexandria (Egypt) 57, 176, 222

Algemene Bank Nederland 166

Algeria 219, 244, 246-7

Allegheny Mountains (US) 52

Alsace (France) 454/58

Alusuisse 376

Amerada 220

American government, *see also* Washington, DC 39, 63, 75, 172

 Department of Agriculture 428

 Department of Health, Education and Welfare 422

 Department of Justice 150, 243-4

 Environmental Protection Agency (EPA) 438

 Maritime Commission 281

 Securities Exchange Commission (SEC) 112

 State Department 17, 41, 63, 239, 243, 455/104

American Institute of Chemical Engineers 385
American Journal, The 120
American Petroleum Institute (API) 276, 385
American Public Health Service 423
Amsterdam (Netherlands) 9, 12, 42-5, 47, 72, 78, 80, 82-3, 85, 98, 100, 114, 149, 154, 159, 168, 316, 346, 384, 386, 385-6, 391, 393, 400, 428, 456/149, 468/17, 476/8, 476/9
Amsterdam Stock Exchange 113
Anacortes, WA (US) 279
Andlinger, G. R. 162
Anglo-American Oil Co. Ltd. 471/132
Anglo-Egyptian Oilfields Ltd. 37, 220-1, 454/67
Anglo-Iranian Oil Company (AIOC), later British Petroleum (BP) 19, 29, 37, 57, 172, 191, 219, 265, 336, 451/8, 454/67, 468/20
Anglo-Saxon Petroleum Company (A-S), *see also* advertising, brands and branding, collaboration, concessions, corporate governance, exploration and production, finance, management and organisation, manufacturing, marketing and distribution, mergers and acquisitions, offices, research 5, 5, 11-2, 28, 31-2, 53, 57, 62, 75-8, 88, 91-2, 94-7, 100, 113, 122, 281, 282, 299, 391, 451/11, 453/42, 455/84, 456/132, 457/180
Antilles (Caribbean) 412-3
Antwerp (Belgium) 10, 366
API, *see* American Petroleum Institute
Arab League 179
Arab Oil Congress 179
Arab Petroleum Congress 241
Arabian American Oil Company (Aramco) 172, 173, 175, 247
Aramco, *see* Arabian American Oil Company
ARCO (Atlantic Richfield Company) 205
Argentina, *see also* Buenos Aires, Diadema 33, 37, 76, 94, 99, 122, 131, 140, 299, 454/59, 456/137, 462/171, 472/24
Ariërverklaring 84
Arif, Abdul Salam 220
Armour & Company 361
Aruba 14, 29, 40
As-Is Agreement, *see* Achnacarry Agreement
Asia 14, 17, 29, 57, 63, 66, 69, 99, 130, 136, 144, 228, 231, 265, 268, 270, 272, 274, 455/85
Asiatic Petroleum Company Ltd. 5, 5, 11, 19, 32, 56, 88-9, 91-2, 95, 122, 137, 163, 443, 458/2
Asiatic Petroleum Corporation 34, 112, 150, 154-5, 253, 464/67
asphalt, *see* bitumen
Association of British Manufacturers of Agricultural Chemicals 436
Astley-Bell, L. A. 232
Astra Romana 31, 34-5, 69, 78, 86, 453/52, 455/116
Aswan High Dam (Egypt) 220
Atlanta (US) 164
Atmospheric Pollution Committee 401
Atomic Power Lubricants *see* APL
Atomium (Brussels, Belgium) 343
Atoom, Het 374-5
Australasia 144, 189, 189
Australia 91, 122, 136, 153, 164, 270, 274, 305, 317, 334, 349, 350, 375
Austria, *see also* Graz 32, 69, 78, 281
aviation, aircraft 14, 16, 19, 19, 23, 60, 61, 62, 66, 69, 76, 302, 322
  Armstrong Whitworth AW-38 Whitley [monoplane] 22
  Boeing 707 jet airliner 302
  Douglas DC-8 jet airliner 302
  Fairey Swordfish torpedo bomber [biplane] 60-1, 62
  Heinkel He-162 'Völksjäger' jet fighter 19, 22-3
  Messerschmitt Me-262 'Schwalbe' jet fighter 19, 20-1
  transatlantic flights 62, 149, 302
  Vickers Viscount turboprop airliner 302
aviation engines, jet engines 19, 21, 75, 149, 452/29
  Whittle Turbine Project 452/30
aviation fuel / aviation gasoline / aircraft fuel / aircraft gasoline 12, 14-5, 16, 17, 36, 63, 70-2, 75, 76-7, 81, 298, 302, 304, 391
  jet fuel 183, 261, 302, 304
Azienda Generali Italiana di Petroli, *see* AGIP
B
'B' Agreement 5
Bader, Sir Douglas 19
Badische Anilin- und Soda Fabriken, *see* BASF
Bahamas 122
Baku (Azerbaijan) 84
Balik Papan (Borneo, Dutch East Indies / Indonesia) 26, 63, 65, 66-7, 130, 226, 258, 264, 270, 455/107, 469/32
Ballard, Dr. S. A. 154, 154, 400
Bangkok (Thailand) 429
Barran, Sir David 154, 154, 163-4, 169, 244, 286, 371, 379, 402, 463/198, 466/115, 469/31
BASF (Badische Anilin- und Soda-Fabriken) 335, 339
Basrah (Iraq) 223
Bataafsche / Bataafse Petroleum Maatschappij (BPM), *passim*, *see* advertising, brands and branding, collaboration, concessions, corporate governance, exploration and production, finance, management and organisation, manufacturing, marketing and distribution, mergers and acquisitions, offices, research
Bataafse Handelszaken 147
Bataafse Internationale Chemie Maatschappij 145, 478/129
Bataafse Internationale Petroleum Maatschappij 145
Batavia, Djakarta (Dutch East Indies) 17, 26, 27, 29, 63, 452/23, 453/45, 453/49
Baton Rouge, LA (US) 51
Battle of Britain 48
Battle of the Atlantic 49, 62
Bavaria (Germany) 274, 295
Bayer 335, 362
Bearsted, 1st Viscount, *see* Samuel, Sir Marcus
Bearsted, 2nd Viscount, *see* Samuel, Walter Horace
Beaumont, TX (US) 51
Beirut (Lebanon) 241
Belgian Shell 156
Belgium, *see also* Antwerp, Brussels, River Meuse, River Scheldt 10, 28, 48, 120, 298, 351, 453/45, 467/154
Bénard, A. 379
Bendul (Indonesia) 226
Benelux nations 298, 321, 341, 354
Benson, F. P. 141
Bentley, Bill 472/172
Benton, IL (US) 191
Benzolverband 454/52
Berkin, John 141, 375, 474/122
Berkshire (UK) 28
Berlin (Germany) 81
Bernhard, Prince 108
Berre-l'Étang (France) 263, 274, 337, 339, 339-41, 349, 477/48
Betjeman, John 309, 471/135
Beyen, J. W. 101
Billiton Maatschappij 376, 377-8, 379
biodegradability 399, 409, 477/43
Birkenhead (UK) 283-4
Birmingham (UK) 84, 162, 167
Birmingham University 167

Bishopsgate (London, UK) 139
bitumen / asphalt 36, 71, 122, 240, 260-1, 268, 298, 300, 336, 346, 350, 373, 391, 438, 454/62
Black, Eugene 153, 234, 466/115, 467/142
Blazer, Carel 388-9
Blitzkrieg 10, 23
Bloemgarten, Henk 128, 354, 391
Bolivia 39
Bombay (Mumbai) (India) 226, 265
Bonn (Germany) 476/32
Bonny (Nigeria) 295
Boots 162
Booy, James M. de 31, 33, 87, 89-94, 89, 451/2, 453/48, 453/49, 454/53, 457/164
Bordeaux (France) 48
Borneo, *see also* Miri, Sarawak, Tarakan 65, 200
Böttcher, Prof. Dr. Frits 7
Boulder, COL (US) 478/97
Bournemouth (UK) 28
BP, *see* British Petroleum
BPM, *see* Bataafse Petroleum Maatschappij
brands and branding, *see also* advertising, marketing 5-6, 14, 152, 185, 240, 251, 301, 306-7, 309, 313, 317, 319-20, 325, 329, 445
Shell brand / Pecten logo 5-6, 78, 123-4, 152, 254, 287, 300, 306, 323, 326, 461/138
individual brands, see separately agrochemicals
Aeroshell 302, 303
AeroShell Compound '7' 426
AeroShell Fluid '4' 426
Aeroshell Fluid '12' 427
Dobane JN 399
Epikote / Epon resins 261, 337-9, 337, 349-50, 350, 355
Essence B 339
Esso Extra 306-7
Euroshell Card 320
Kraton rubbers 261, 337
Neosol 348
Shell APL (Atomic Power Lubricants) 302
Shell ETR (Extreme Temperature Range) grease 302
Shell [premium fuel] with ICA (Ignition Control Additive) 299, 304, 304-5, 306, 306, 471/124
Shell [premium fuel] with TCP (tricresyl phosphate) 304, 305-6, 306-7, 399, 478/108
Shell X-100 motor oil 301, 301-2, 395, 478/108
Shellzone antifreeze 426
Super Shell 306
Teepol detergent 76, 258, 332, 337, 346, 346, 348, 399, 404, 477/42
Tide detergent 347
Brazil 99, 140, 462/171
Briejèr, C. J. 429-30
Brigitta 211, 465/92
Brindisi (Italy) 339
Brink, J. R. M. van den 264
Bristol (UK) 157
Bristol University 167
British Association for the Advancement of Science 399
British Borneo 135
British Empire 86, 91
British government 12, 39, 63, 87, 91, 93, 96-7, 101, 172, 175, 365, 399, 401, 409, 431, 435, 437, 451/9, 452/20, 453/51, 457/183
Advisory Committee on Poisonous Substances used in Agriculture and Food Storage 431
Air Ministry 12, 28, 70, 451/11, 451/17, 456/123
Chancellor of the Exchequer 101, 157
Clean Air Act 401, 420-1
Department of the Environment 409
Department of Trade and Industry 409
Directorate of Scientific and Industrial Research 399
Foreign Office 17, 39, 63, 87, 94, 97, 162, 221, 286, 464/31, 465/105
Inspectorate for Schools, HM 167
Minister of Agriculture 437
Ministry of Labour and National Service 141
Ministry of Transport 405
Monopolies Commission 313
Treasury 31, 87, 93, 95-6, 101, 453/50, 457/161, 457/173, 457/180
Treasury Agreement 96-7, 101, 457/161, 457/183
War Office 167
British Honduras Shell Petroleum & Development Company Ltd. 122
British Hydrocarbon Chemicals 361
British Leyland Motor Corporation 369, 423, 474/97
British Motor Corporation 162
British Petroleum (BP), *see also* Anglo-Iranian Oil Company (AIOC) 19, 72, 158, 161, 166, 172, 175, 180, 187-9, 196, 202, 204, 205, 213, 220-1, 243, 268, 286, 295, 297, 313, 317, 326, 342, 361, 371, 401, 406, 415, 452/32, 463/4, 469/62, 474/71, 475/140
Brouwer, L. E. J. 148, 169, 230, 244, 359, 474/122
Brunei, *see also* Panaga Club, Seria 34-5, 132, 167, 186, 219, 243, 270, 285
Brussels (Belgium) 343
Bruyn, Dirk de 403
Buenos Aires (Argentina) 142
Burma 255
Burmah-Shell 265, 268, 460/64
Burns, Max 149, 202, 255, 307, 400
C
Cactus ordnance works 72
Caernarvon (UK) 124
Cairo (Egypt) 176, 179, 233, 241
Calais (France) 54-5
Calder Hall (UK) 302
California (US) 17, 36, 74, 88-9, 126-7, 149, 181, 195-6, 255, 348, 350, 386, 393, 403, 409, 427, 471/161
Caltex 226, 233-5, 239, 265, 268, 297, 415, 466/131
Cambridge (UK) 163
Cameroon 465/78
Canada, *see also* Jumping Pound, Oakville, Totonto 98, 158, 189, 234, 255, 274, 298-300, 321, 342, 373, 401, 461/136, 465/88
Canadian Shell 115, 231, 466/137
Canadian Shell Overseas Ltd. 466/137
capacity, *see* manufacturing capacity
Cape of Good Hope (South Africa) 37, 176, 285-6, 290, 290
Caracas (Venezuela) 108, 140, 244
Cardón (Venezuela) 40-1, 41, 107, 132, 263-5, 270, 301, 422, 471/122
Caribbean 43, 144, 265, 286
Caribbean Petroleum Company 89, 94
Carrington (UK) 295, 348
cars, *see* motor cars
Carson, Rachel 430-1, 430, 436
cartels (oil cartels), *see also* Achnacarry Agreement, Seven Sisters 78, 176, 179, 226, 242-3
Castle, John 301
catalysts, *see* manufacturing
Caucasus 84, 86

CBI, *see* Confederation of British Industry
Celebes, *see also* Makasar 234
Central America 144, 189, 265, 300
Central College of Shellmanship 151
Ceram (Indonesia) 455/107
Ceylon 268
CFP, *see* Compagnie Française des Pétroles
CFPCS, *see* Compagnie Française des Produits Chimiques Shell
Chamberlain, Arthur Neville 10, 11
Chandler, Alfred 139
Channel / English Channel 54-5, 291
Chase Manhattan Bank 114
Chemical Age 429
chemicals, *see* petrochemicals
Chevron 172
Chicago, IL (US) 361
Chile 39
China 15, 17, 17, 56, 57, 60, 135, 412-3
Choufoer, J. H. 300
Churchill, Winston Leonard Spencer 10, 11, 102, 457/173
Cities Service 306
City of London College 167
Clark Oil 306
Cleveland, OH (US) 313
Club Miramar 107
Club of Rome 439
    The Limits to Growth (by Dennis L. Meadows, 1972) 439
Clyde (Australia) 274, 349, 350
CMD (Committee of Managing Directors), *see* management and organisation
coal, *see also* hydrogenation 80, 82, 334-5, 379, 421, 423, 465/71, 474/122
    coal gasification 80, 280
    coal to liquids process 17
    coal versus fuel oil 176, 443
Coca-Cola 326
Cochimé 339
Coevorden (Netherlands) 37, 207, 210
Cold War 172, 375
Colijn, Hendrikus 137, 444
Cologne (Germany) 274, 339
collaboration, *see also* cartels, joint ventures 14, 26, 82, 176, 202, 243, 249, 254, 268, 279, 339, 410, 415, 423, 433, 469/62
Colombia 39, 108, 149, 205
Colón Development Company 94
colonialism, *see also* imperialism 26, 35, 103, 130, 225-6, 265
Colorado (US) 350, 438, 478/97
Columbia University 166
Committee of Managing Directors (CMD), *see* management and organisation
communism, *see also* Cold War 6, 150, 162, 172, 239, 240, 375, 466/115
Compagnie de l'Esthétique Industrielle 323
Compagny Française des Pétroles (CFP) 172, 175, 189, 225, 244, 268, 406
Compagnie Française des Produits Chimiques Shell (CFPCS) 339
Compañia Shell de Venezuela (CSV) 140, 162, 167, 195, 195-6, 282, 372, 474/109
computer technology, computers, data centres 165, 255-6, 256-7, 320-1, 395
Con Edison 422
concessions, *see also* nationalization, participation 39-40, 69, 175, 178, 195, 197, 199, 220, 244, 255, 454/73, 464/46, 466/130
    locations:
    Dutch East Indies 5, 90
    Egypt 221-2
    Indonesia 229-30, 232, 234-5, 239
    Iraq 223
    Kuwait 187, 197, 200, 202, 220
    Middle East 99, 176, 178, 200, 219
    Netherlands 211
    Venezuela 265, 371
    Vietnam 465/77
Conch International Methane 214
Confederation of British Industry (CBI) 158
Conoco 195, 471/161
Conservation of Clear Air and Water, Western Europe (CONCAWE) 401, 410, 415, 477/56
Consolidated 221
Constock International Methane Ltd. 214
Continental Oil Company 195, 205
Continental Shelf 199, 211, 213
Cooper Pedy (Australia) 317
Cook, Alistair 311
Cornwall (UK) 408, 410
corporate culture 169, 444
corporate governance 92, 443
corporate organization 139, 155
Corredor, Ruben 40
County Hall (London) 146
Courtaulds 158
Crena de Iongh, Daan 87
Cressier (Switzerland) 275
crude oil, *see also* exploration and production 6, 48, 51, 63, 70, 72, 86, 171-2, 176, 178, 183, 187-9, 191, 199, 202, 204, 249, 251, 254, 258, 260-1, 262-3, 269-70, 274-6, 275, 278-80, 290, 295, 318-9, 329, 336, 339, 345, 391, 403, 405-6, 409, 422, 441, 443, 454/60, 455/101, 468/21
    low sulphur crude oil 325, 417, 419, 422-3, 441
    prices 176, 178, 188, 202, 247, 268
    production, supply and trade 22-3, 34-41, 49, 51, 51-2, 63, 66, 69-70, 73, 99, 150, 172, 178, 178, 180, 183, 185-9, 191, 204, 229, 232, 239, 247, 251, 254, 269, 274, 276, 298, 319, 339, 423
    purchases 34-41, 187, 202, 205, 220, 265
    Crude Oil Washing 414
    crude position Royal Dutch Shell 34-5, 70, 187, 249, 463/29
CSM *see* Curaçaosche Scheepvaart Maatschappij
CSV *see* Compañia Shell de Venezuela
Cuba 131, 207, 462/171
Curaçao, *see also* Rio Canario, Schottegat, Suffisantdorp, Willemstad 14, 29, 29-30, 31, 34, 36-7, 40-1, 53, 56-7, 60, 62, 62, 71-2, 87, 94-7, 103, 107, 128, 131-2, 136, 142, 263-5, 269-71, 270, 301, 403, 403, 422, 453/45, 453/47, 453/48, 453/51, 453/52, 455/85, 456/123, 468/23
Curaçaosche Scheepvaart Maatschappij (CSM) 53, 57, 60, 281, 288-9
CUSS consortium 195-6
Cyprus 167
Czechoslovakia, *see also* Sudetenland, Waldheusel 24-5, 78
D
D-Day 54-5
DAF (motor car) 424
Dalton, Hugh 101
damage, *see* war-damage
Dansk Shell 80, 156
decolonisation, *see also* Dutch-Indonesian relations 265
Deep Rock 306
Delft (Netherlands) 80, 385-6, 393, 451/11, 476/9

Denmark, *see also* Fredericia, Odense 10, 156-7, 213-4, 276, 285, 290, 359, 467/154, 468/20
Denver, COL (US) 350, 438
Department of Plant Pathology 429
Deterding, Henri 5-6, 9-10, 86, 88, 99, 106, 125, 137, 149, 169, 204, 385, 443, 452/20, 459/56
Detergents and Wildlife 409
Deutsche Shell 149, 255, 306, 318, 322
Deutsche Shell Chemie 365
Deutsche Shell Tankers GmbH 122
Deutsche Werft 291
Devoe & Raynolds 349
Diadema Argentina 33, 34, 94
diesel 261, 336, 339
diesel engines 164
Distillers Company 336
diversification, *see also* agrochemicals, coal, metals, nuclear energy, petrochemicals, pharmaceuticals 6-7, 75, 152, 189, 191, 244, 261, 320, 331-42, 336-7, 345, 352, 359, 370-3, 375-6, 379, 381, 383, 445-6, 474/121
Diversification Studies Division (Shell) 372
dividends, *see* finance
Djakarta (Indonesia) 131, 228-9, 230, 231, 233, 236, 239, 300
Dominguez, CAL (US) 73, 73, 76
Dominion (oil company) 313, 317
Doolittle, James (Jimmy) 14, 16, 357, 361, 451/17
downstream activities, downstream business 7, 41, 43, 150, 183, 188, 205, 247, 249, 251, 257, 280, 329, 444-5, 460/96
Doyle, William 108
Dresden (Germany) 70
drilling 138, 193, 195-7, 199, 203, 209, 464/54
drilling equipment, drilling techniques, drilling rigs, *see* exploration and production
DSM (Dutch State Mines / De Staatsmijnen) 211
Dubai 242
Duke of Edinburgh, H. R. H. Prince Philip 437
Dungeness (UK) 54-5
Dunsterville, G. C. K. 156, 462/156
Dupont & Co, Francis I. 75, 119-20, 336
Durban (South Africa) 279
Dutch East Indies, *see also* Batavia, Indonesia (including places in Indonesia) 5-6, 17, 26, 28-9, 31-2, 34-5, 41, 43, 63, 65-6, 69, 86-7, 89-90, 93-4, 96, 99, 103, 106, 224, 455/85
Dutch government 17, 39, 63, 89, 93-4, 97, 101, 211, 226, 230-1, 415, 429, 443, 452-3/41, 453/45, 455/104, 457/173, 457/183
  Department of Plant Pathology 429
  minister for Foreign Affairs 97
  minister for Overseas Territories 97
  Ministry of Finance 7, 101
  Mobilization Council (Staatsmobilisatieraad) 28, 65, 452/37, 452/40
Dutch-Indonesian relations 224, 225-41, 228, 230-2, 235-8, 248, 249
Dutch Round Table Agreements 130
Dutch Royal Air Force (Koninklijke Luchtmacht) 415
E
E&P, *see* Exploration and Production
Eagle hood (invention) 60
Eagle Oil and Shipping Company 53, 57, 60, 62, 282, 299, 455/84
East Asia 144
East Coast (US) 255
Eastern Bloc countries 301
Eastern Hemisphere 300
Eastern Seaboard 464/67
Eck, Jan Carel, Baron van Panthaléon van 10, 17, 28, 29, 31, 33, 60, 63, 86, 89-90, 89, 93-4, 97, 100, 109, 451/1, 453/48, 453/49, 454/53, 457/164
Ecopetrol 205
Ecuador, *see also* Mera 39, 108, 134, 205
Edinburgh (UK) 214
Egham (UK) 393, 474/122
Egypt, *see also* Alexandria, Aswan High Dam, Cairo, Hurghada, Luxor, Port Said, Port Suez, Ras Gharib, Suez, Suez Canal, Suez Canal crises 37, 60, 175-6, 176-7, 220-3, 454/67, 466/115
Einar, Dr. J. F. E. 378
Eire, *see* Ireland
Eisenhower, Dwight David ('Ike') 22
El Aguila oil company (Mexican Eagle) 39-41, 106
El Paso 154, 371, 461/141
embargo 17, 39, 63, 175-6
Emeryville, CA (US) 72-3, 75, 279, 385-6, 386, 393
emissions, *see also* air pollution, gasoline: leaded vs. unleaded, low-sulphur fuel 418, 423-5, 424-5
  carbon dioxide 478/97
  carbon monoxide 478/97
  hydrogen sulphide 402, 404
  lead 423-5
  nitrogen oxide, nitrous oxides 415
  sulphur 419-23, 435
  sulphur dioxide 401, 415, 417, 419, 422-3, 433
Empire Oil Company 294
employees, *see* staff
ENI (Ente Nazionali Indrocarburi), *see also* ENI/AGIP 247, 274, 342, 469/38
ENI/AGIP 176, 178, 241, 268
environment, environmental concerns, *see also* agrochemicals, biodegradability, emissions, pollution, toxic substances, toxicological research 7, 139, 205, 247, 280, 325, 343, 383, 401-4, 410, 415, 417, 417, 419, 424, 425, 430-1, 433, 438-9, 439, 441, 446, 477/54, 477/56, 478/97
environmental conservation, *see also* CONCAWE, Shell Committee for Environmental Conservation, International Petroleum Industry Environmental Conservation Association, pollution: Load-on-Top system, TOVALOP 280, 325, 401-2, 437-8, 477/54
environmental movement 425, 430, 433
  Silent Spring (by Rachel Carson, 1963) 430-1, 433, 436-7
environmental regulations 280, 410, 426
ERAP (Entreprise de Recherches et d'Activités Pétrolières) 225
Erb, Josef Theodore 461/139
Esso Petroleum Company, *see also* Standard Oil Company of New Jersey (Jersey Standard), Exxon 18, 158, 211, 213, 268, 276, 295-7, 302, 306-7, 307, 313, 325, 348, 471/132
Ethiopia 136
Eureka (exploration ship) 196
Europe 14, 28, 31, 34, 43, 49, 63, 66, 69, 78, 90, 91, 97, 99-100, 112, 114-5, 119, 131, 135, 144, 148-50, 154, 157, 162, 166-7, 172, 176, 183, 186, 189, 202, 207, 209, 228, 253, 258, 262-4, 274, 276, 277, 278, 280-1, 285-6, 287, 290, 295, 298, 300, 308-9, 313, 317, 320-1, 334-7, 339, 342, 346, 348-9, 349, 359, 361, 365, 367, 373, 381, 394, 399, 401, 422, 434-5, 451/12, 460/96, 475/7
  Western Europe 337, 373, 399, 401
European Economic Union 254
European Union 478/137
Europoort Botlek Belangen, Stichting 415
expansion of organisation / operations 5-6, 77, 99-101, 110, 119, 130, 154, 159, 180, 183, 185, 189, 202, 226, 251, 258, 262, 264, 272, 275, 278-9, 310, 332, 342, 344, 346, 348, 357, 371, 385, 393,

441, 443
expansion of production / production capacity 22, 35, 43, 275, 278, 281, 298, 332, 366, 417
exploration and production (in general), *see also* concessions, crude oil, expansion of organisation, expansion of production, Exploration and Production (E&P), geology, geophysics 7, 37, 39, 134, 140, 171, 188-9, 191, 202, 211, 220, 222, 225, 239, 247, 268, 274, 310, 342, 376, 391, 393, 404, 461/136, 465/88
derricks, rigs 181-2, 184, 194-5, 195, 209, 214, 216-7, 409
drilling equipment and drilling techniques 138, 190, 192, 195-7, 199, 198-9, 203, 209, 225
geological surveys, *see also* geology 225
geophysical surveys, *see also* geophysics 197, 213
'nodding donkeys' 206-7
offshore exploration and production 194-7, 195-7, 199, 200-1, 203, 211
production by region: *see* fields
seismic surveys 192-3, 196-7, 197, 246, 465/95
supply pattern 185-9, 191
upstream activities, upstream investments 150, 188, 188, 202, 204, 247, 249, 329
Exploration and Production (E&P) – *business sector of Royal Dutch Shell Group*; *see also* exploration and production (in general) 5, 36, 80, 132, 137, 142, 144, 152, 155, 166-7, 171, 188-9, 196-7, 199, 202, 204-5, 207, 220, 229, 240-1, 247, 249, 329, 371, 444-5, 461/136, 476/8
Expo 1958 343
Exxon 167, 326
F
Far East 68, 166, 186, 224
Ferranti 257, 395, 476/34
Ferrara (Italy) 339
Festival of Britain Exhibition 147
fields: oil and/or gas fields 6, 26, 28, 35-7, 36, 38-9, 65, 172, 178, 185, 187, 205, 210, 213, 214, 225, 229, 229, 232, 290, 294, 332, 393, 395, 435
Brent (North Sea) 199
Burgan (Kuwait) 219
Coalinga (California, US) 195
Jambi (Sumatra) 41
Jumping Pound (Canada) 189
North Rumaila (Iraq) 223, 466/125
Poza Rica (Mexico) 39
Prudhoe Bay (Alaska, US) 196, 205, 464/54
Ras Gharib (Egypt) 37
Schoonebeek (Netherlands) 37, 159, 191, 206-7, 207, 208-9, 210, 258, 262, 454/70
Signal Hill (California, US) 195
Slochteren (Netherlands) 210-1, 445
South Pass Block 24 field (Gulf of Mexico) 195
Tandjung (Kalimantan) 232-4, 466/139
TXL field (Texas, US) 182
finance 75, 78, 101, 110, 113, 115, 136, 150, 164-5, 240-1, 320-1, 323, 332, 367, 372-3, 466/125, 471/163
accounting 113, 178, 249, 255, 257, 310, 313, 395
dividends 47, 96 97, 110, 119, 156, 430, 456/146, 457/172
investments 19, 26, 47, 96, 119, 119, 150, 183, 188, 197, 199, 213-4, 220-1, 228, 232, 234, 249, 262, 270, 276, 280, 304, 319, 322-3, 334, 341-2, 345, 353-5, 357, 359, 362, 366-7, 369-73, 375, 379, 381, 387, 400, 402, 410, 444, 474/115
profits 41, 43, 47, 76, 120-1, 141, 147, 154, 180, 185, 204, 211, 223, 230, 249, 255, 331, 345, 351-2, 359, 361, 365, 370, 379, 385, 391, 399, 433, 445
Finance / Finance Administration – *function of Royal Dutch Shell Group* 112-3, 138, 144, 166, 255, 300, 343, 355
Financial Times 479/158
Finland 122
Fisons 158
Food or Famine 351
For Your Information: Chemicals and Our Food 436
Formica 355
Formosa 68
France, *see also* Alsace, Berre-l'Étang, Bordeaux, Calais, Grasse, Lorraine, Marseilles, Mericourt Dam, Normandy, Paris, Pauillac, Penhoet, Petit-Couronne, Reichstett-Vendenheim, River Seine, Rouen, Saint-Gobain, Sedan, Strasbourg 10, 15, 26, 28, 43, 47-8, 54-5, 69, 73, 80, 101, 112, 120-2, 136, 160, 167, 175-6, 219, 221, 225, 244, 264, 268, 274-6, 282, 285, 322, 323, 326, 336-7, 338-9, 339, 342, 351, 373, 394, 398-9, 401, 408, 440, 452/20, 453/43, 454/58, 462/155, 467/154, 469/49, 473/47, 473/60
Franks Report 164
Franks, Lord 164
Fraser, Alexander 112, 149
Fredericia (Denmark) 276
Freital (Germany) 70
Friends of the Earth 424
fuel-cells 424
fuel oil 19, 22, 36, 49, 71, 80, 183, 253, 260-1, 262-3, 275, 298, 300, 336, 419, 441, 464/67
low-sulphur fuel 325, 419, 422-3, 441
G
Gabon 136, 205, 422
Gaia theory 478/97
gas, *see* natural gas
gas fields, *see* fields
gas reserves, *see* reserves
gas stations, *see* service stations
gasification, *see* coal
gasoline, *see also* aviation fuel, branding, manufacturing, marketing 15, 19, 34, 36-7, 36, 48, 55, 57, 58, 69, 71-2, 75, 77, 80, 84, 102, 183, 185, 189, 210, 253-4, 260-1, 262, 275, 277, 278-9, 300, 304, 306-7, 309-10, 313, 319, 325, 334, 336, 339, 387, 391, 393, 399-401, 423-5, 433, 435, 441, 444, 459/33, 471/160
additives in gasoline 19, 75, 279, 301, 304, 306, 325, 348, 399-400, 423, 425-7, 478/96, 478/108
high-octane gasoline, engine knock 15, 17, 19, 63, 69, 71-2, 75, 262, 278-9, 304, 306, 348, 399, 400
iso-octane 12, 14-5, 19, 71, 451/17
leaded vs. unleaded / low-lead / lead-free gasoline, *see also* emissions 279, 325, 400-1, 423-5, 424-5, 433, 441, 478/96
rationing 298, 313
synthetic gasoline, *see also* hydrogenation 17, 19, 70, 339, 452/22, 452/24
Gasunie 211
Geelong (Australia) 334
Geismar (US) 364
Gelsenberg oil company 295, 297
General Electric 424
General Motors 180, 424
Geneva Convention 26, 65-6
Genoa (Italy) 274, 296-7
geology, *see also* exploration and production 86, 172, 195-6, 213, 219, 225
Geopotes VII (oil clearing ship) 411
geophysics, *see also* exploration and production 195, 197, 213, 393
Georgia (US) 51
German occupation (of the Netherlands) 31-2, 37, 66, 78-87, 79, 99, 207, 452/38, 454/58, 455/107

'Verwalter' (caretaker manager), *see also* Klass, Eckhardt von; Lochner, Hans 22, 31-2, 48, 69, 78, 80-2, 80-1, 86, 453-4/52, 456-7/158
Germany, *see also* Hitler government, Luftwaffe, national-socialism: Nazi regime, Nazi's; *see also* Bavaria, Berlin, Bonn, Cologne, Dresden, Freital, Godorf, Hamburg, Harburg, Ingolstadt, Karlsruhe, Leipzig, Monheim, Munich, Nürburgring, Pölitz, Regensburg, Reisholz, River Rhine, Ruhrgebiet, Wesseling, Wilhelmshaven 6, 10, 12, 17, 18, 19, 22-3, 24-5, 26, 28-9, 31-2, 34, 36-7, 39-40, 47-9, 62, 63, 69-70, 70-1, 75, 78, 80-2, 84, 86-7, 94, 95, 97, 137, 167, 207, 211, 213, 254, 264, 274-6, 291, 295, 318, 318, 323, 327, 334-5, 339, 341, 351, 359, 394, 401, 426, 452/22, 452/29, 452/34, 453/45, 453/52, 454/53, 454/58, 455/115, 456/155, 466/115, 467/154, 468/21, 469/49, 471/122, 471/125, 473/47, 476/32
Reichswirtschaftsministerium 22
West Germany 401
Gershinowitz, Dr. Harold 154, 155, 387, 391, 394, 422, 476/29
Getty Oil 156
Gewerkschaft Elwerath 80
Ghana 122, 131
Gist-Brocades 379
Glenmoore, PA (US) 53
Godber, Sir Frederick, 1st Baron Godber of Mayfield 10-1, 11, 31, 86, 88-9, 88, 91-3, 95, 97, 100, 108, 148-9, 169, 451/1, 457/163, 457/173, 463/197
Godorf (Germany) 274-5, 279
Goering, *see* Göring, Hermann Wilhelm
Golden Lands, The 351
Göring, Hermann Wilhelm 19
government participation, *see* participation
Govindachari, Ray Bahadur 460/64
Graan, W. de 78
Grasse (France) 338
Graz (Austria) 281
Great Britain, *see* United Kingdom
Greece 81, 136
Greensboro, NC (US) 51
Groningen (Netherlands) 210
Group, *see* Royal Dutch Shell Group
Grozny (Chechnya) 84
Guépin, F. A. C. 299, 343, 344
Gulbenkian, Calouste 172, 189
Gulf of Mexico 195
Gulf Oil Corporation (GOC) 40, 49, 99, 172, 180, 183, 187, 196, 202, 204, 207, 213, 219-20, 258, 265, 285-6, 306, 319, 415, 419, 465/105, 469/62, 475/140
Gulf States 244, 247
Gusto Shipyard 197
H
Haensel, Vladimir 278
Hague, The (Netherlands), *see also* offices 10-1, 26, 28-9, 28, 31-2, 79, 82, 86-7, 89, 97, 97, 131, 136-8, 140-2, 145, 147-8, 152, 157, 160, 163, 166, 229-31, 299, 343, 354, 401-2, 443, 454/53, 460/79, 461/136
Haifa (Israel) 37, 223
Hamburg (Germany) 70, 274, 291, 297
Harburg (Germany) 70-1, 102, 274, 279, 471/122, 471/125
Harden, Orville 209
Harris, Frank 89
Harvard University 166, 462/156
Hasselt, Barthold Th. W. van 41, 94, 94, 97, 99, 101, 106, 108-9, 112, 114, 209, 264, 452/23, 454/73, 457/183, 458/5
Hatta, Mohammed 225
Hawaii (US) 17, 17, 64, 347, 356
health hazards, *see also* environment, pollution 279, 394, 422, 426-7, 426-7, 436, 478/96, 478/108
Heinkel aircraft works 19
Hereford (UK) 429
Heysham (UK) 12, 12-3
Hinde, J. F. K. 370
Hiroshima (Japan) 374-5
Hitler, Adolf 6, 12, 17, 49, 78
Hoechst 335
Hofland, A. 299, 309
Hollands Diep (Netherlands) 295
Hollywood, CA (US) 342
Holmes, Sir Peter 109, 136
Hong Kong 56
Hoogovens 376
Hook of Holland (Netherlands) 411
Hopwood, Hon. Sir Frank (Lord Southborough) 95, 109, 299, 307, 372-3
Hornet, USS 16
Houston, TX (US) 15, 51, 72, 73, 75-6, 99, 161, 256, 258, 258, 279, 294, 346, 348-50, 362, 386, 393, 402, 439, 461/130, 476/8, 478/129
Houston Data Service Center 256
Huiskamp, G. B. 112
Humble Oil Company 205, 348
Hungary 69, 78, 81
Hurghada (Egypt) 37
hydrogen 15, 73, 77, 279-80, 443
hydrogen sulfide 189, 402, 404
hydrogenation, *see* manufacturing
Hyman & Company, Julius 350, 399, 438
I
IBM, *see* International Business Machines Corporation
ICI, *see* Imperial Chemical Industries
Ickes, Harold 15
Idd el Shargi (Qatar) 290
IG Farben (Interessen-Gemeinschaft Farbenindustrie A.G.) 17, 19, 40, 334-5
IJmuiden (Netherlands) 80, 346, 385
Illinois (US) 51, 191, 294
Imperial Chemical Industries (ICI) 12, 158, 335-6
imperialism, *see also* colonialism 136
India 135, 265, 339, 460/75
Indian Ocean 62
Indiana (US) 294
Indochina 63
Indonesia, *see also* Aceh, Balik Papan, Bendul, Borneo, Celebes, Ceram, concessions, Djakarta, Dutch East Indies, Jambi, Java, Makasar, Medan, Miri, Palembang, Pankalan Brandan, Pladju, Praboemoetih, Probolingo, Sarawak, Semarang, Sumatra, Tandjong Priok, Tarakan, Tjepa, Wonokromo 35, 65, 106, 128, 130-2, 147, 186, 225-32, 234-5, 234, 236-9, 239-41, 248, 255, 258, 265, 269-70, 274, 299-300, 365, 376, 378, 444, 453/49, 462/171, 466/126, 466/130, 466/131, 466/135, 466/139, 466/140, 467/150, 467/152, 469/32
Ingolstadt (Germany) 274-5, 295, 296-7
innovation 6-7, 12, 41, 60, 70, 75-6, 93, 147, 162, 167, 195-7, 255, 257, 371, 376, 385, 404, 444-5
INOC, *see* Iraq National Oil Company
Institute of Petroleum Research Group 426
International Business Machines Corporation (IBM) 166, 476/34
International Petroleum Industry Environmental Conservation Association 477/56
IPC *see* Iraq Petroleum Company (IPC)

Iran, *see also* Abadan, Teheran 39, 172, 174-5, 175, 179, 186, 207, 244, 247, 274, 350, 427
Iranian Oil Exploration and producing Company 175
Iranian Oil Refining Company 175
Iraq, *see also* Basrah, fields (North Rumaila), Nahr Umr, Zubair 37, 179, 207, 220, 220, 223, 225, 244
Iraq National Oil Company (INOC) 223, 225
Iraq Petroleum Company (IPC) 37, 158, 160-2, 166, 172, 189, 223, 225
Irbid (Jordan) 223
Ireland (Eire) 144, 276, 468/20
Ishikawajima Harima Heavy Industries 288-9
Isle of Wight (UK) 54-5
Israel, *see also* Haifa 37, 175, 222-3, 223, 225
Italy, *see also* Brindisi, Ferrara, Genoa, La Spezia, Po Valley, Rome, Trieste 15, 22, 23, 26, 34, 37, 39, 63, 69, 122, 176, 178, 221, 268, 276, 295, 298, 319, 339, 341-2, 351, 359, 361, 394, 400-1, 467/154, 468/20, 469/49
J
Japan, *see also* Hiroshima, Nagasaki, Osaka, Sodegaura, Tokyo 15, 16-7, 17, 22, 26, 34, 39, 49, 62-3, 62, 64, 65-6, 69, 77, 87, 132, 136, 176, 178, 218-9, 219, 225, 270, 274, 282, 285, 287-9, 299, 339, 341, 374-5, 401, 452/22, 453/49, 455/101, 455/106, 468/20
Japanese occupation (of the Dutch East Indies) 34, 66, 90, 225
Java 27
Jersey Enterprises 375
Jersey Standard, *see* Standard Oil Company of New Jersey
joint ventures, *see also* acquisitions, cartels, mergers 41, 75, 80, 149, 155, 172, 175, 178, 195, 205, 207, 211, 213, 219-20, 222-3, 230, 268-9, 274-5, 294, 337, 339, 359, 367, 373, 376
  Shell and Anglo-Iranian 19
  Shell and BASF 339
  Shell and BP 202, 221
  Shell and Esso 213
  Shell, ICI and Trinidad Leaseholds: Trimpell 12
  Shell and Jersey Standard 207, 209
  Shell, National Distillers and Chemical Corporation (Shorko) 341
  Shell and NIAM 41, 230-2
  Shell and Union Carbide 334
Jongh, A. de 137
Jordan 223
Journalist, De 343
Jupiter, SA des Pétroles 43, 47, 80, 122
K
Kalimantan (Borneo) 232, 240-1, 467/154
Kansas (US) 36
Karachi (Pakistan) 268-9, 276
Karlsruhe (Germany) 274
Kay, Fred 17, 63
Kemball-Cook, Dennis 155, 468/14
Kennedy, John Fitzgerald 239, 430, 439
Kent (UK) 54-5, 360, 393, 437
Kentucky (US) 398-9
kerosene 183, 245, 260, 262, 279, 304, 336
Kessler (jr.), Jean Baptist August (Guus) 10, 11, 28, 32, 33, 73, 87-95, 89, 94, 99, 101, 106, 125, 137, 148, 298, 334, 357, 370, 371, 385, 387, 391, 451/1, 453/48, 454/53, 457/164, 457/173, 458/2, 458/3, 459/19, 463/197, 468/20, 472/5, 476/12
Khrushchev, Nikita Sergeyevich 220, 233
Kirkpatrick Award for Chemical Engineering Achievement 385
Klass, Eckhardt von 32, 78, 80-2, 80-1, 84, 86, 453-4/52
KNMI (Koninklijk Nederlands Meteorologisch Instituut: Royal Dutch Meteorological Institute) 415
KOC *see* Kuwait Oil Company
Kok, Johan Egbert Frederik (Frits) de 6, 28, 31-2, 32, 86-8, 87, 90, 454/53
Kontinentale Öl 69, 84, 86
Korea, *see* South Korea
Krimpen, J. L. van 343, 344, 354
Kruisinga, Dr. H. J. 7, 141, 372, 461/111
Krupp 12
Kuwait 99, 172, 175-6, 179, 183, 187, 197, 200, 202, 204, 207, 219-20, 258, 406, 411, 419, 441
Kuwait Oil Company (KOC) 172, 219
L
La Corona, NV Petroleum Maatschappij 53, 58, 62, 62, 452/41, 470/72
La Spezia (Italy) 276, 400, 471/125
laboratories, *see* research
Lagos (Nigeria) 142
Laguinillas (Venezuela) 195-6
Lake Maracaibo, *see* Maracaibo (Venezuela)
Lancashire (UK) 12, 12
Land Must Provide, The 351
Land's End (UK) 62
Lankro 349
Larson, J. O. 145, 461/122
Lavies, Jan 328
laws, *see* legislation
Lebanon, *see also* Beirut 37
Leer, Bernard van 466/126
Legh-Jones, Sir George 11, 86, 88-9, 88, 91-3, 112, 149, 298, 385, 391, 463/197
legislation / laws, *see also* litigation / lawsuits 14-5, 29, 31, 57, 84, 90, 100, 112, 122, 130, 150, 175, 230, 232, 240, 247, 348, 401, 405, 410, 453/45, 454/73, 464/54
  anti-trust legislation 150, 175, 244, 461/132
  British Companies Act 112
  Clean Air Act 401, 420-1
  company laws 31, 112, 125
  environmental legislations 325, 400, 419, 423, 425-6, 441
  Hazardous Substances Labelling Act 426
  Mining Rights Bill (Indonesia) 233
  National Recovery Act 15
  petroleum law 40-1, 227, 232, 234
  Securities Exchange Act 112
  tax legislation 202, 251, 444
Leipzig (Germany) 400
Lelyveld, E. 345
Liberia 123
Liberty ships 49, 50-1, 281-2
Libya, *see also* Tripoli 122, 178, 183, 207, 220, 244, 245, 247
Libya Shell NV 122
licences, *see* patents and licences
Lieftinck, Pieter (Piet) 101
Life 307
Limburg (Netherlands) 281
Linden, NJ (US) 51
liquefied natural gas, *see* LNG
litigation / lawsuits 40, 149-50, 154-5, 268, 350, 426, 438
  Halpern vs Barran (US court case) 154, 204
  Rocky Mountain Arsenal / Julius Hyman & Company 350, 393, 399, 438, 479/155
Lloyd's of London 226
LNG (liquefied natural gas), *see also* natural gas 167, 174, 214, 218-9, 219, 290, 349, 423, 445
Load-on-Top system 405-6, 406, 414
Lochner, Hans 453-4/52, 456-7/158

Loewy, Raymond 287, 323, 326-7
Lombok 226
London (UK), *see also* offices 10-1, 28, 31-2, 34, 81, 88-9, 92, 94-5, 97, 101, 109, 112-4, 119, 125, 131, 136-8, 138-9, 140-2, 143, 145, 147-50, 147, 152, 152, 157, 159-60, 159, 163, 166-7, 176, 202, 225, 228-9, 231, 255, 257, 299, 309, 385-6, 401-2, 420-1, 427, 443, 453/45, 454/53, 455/113, 460/79, 467/171, 476/9
London Airport (UK) 302
London Business School 164
London Policy Group 244
London Press Exchange 115
London Shell 459/27
Long Beach, CA (US) 126-7
Longview, TX (US) 51
Lorraine (France) 454/58
Los Angeles, CA (US) 36, 75, 421
Los Angeles Data Service Center 256
Los Angeles Times 64
Loudon, Jhr. Hugo 106
Loudon, Jhr. Dr. J. H. 100-1, 105-6, 107-8, 108-10, 109, 113, 114, 120, 125, 130, 140-2, 147, 150, 154, 156, 161, 169, 175, 179, 202, 231, 240, 255, 307, 458/11, 459/56, 460/96, 461/113, 461/141, 463/4, 463/198, 464/60, 465/92, 466/115, 473/26, 477/71
Louisiana (US) 191, 195, 202, 275, 364
Lovelock, James 478/97
LPG (liquefied petroleum gas) 166, 260, 372
lubricants, lube oils, *see also* brands and branding 12, 13, 24-5, 70, 78, 80, 240, 258, 260-2, 268, 270, 279, 298, 300-2, 302, 313, 323, 325-6, 385, 391, 403, 426, 438
  additives in lubricants 310, 426-7
  Shell APL (Atomic Power Lubricants) 302
  Shell ETR (Extreme Temperature Range) grease 302
Luftwaffe 19, 23, 70
Lumina SA 122
Luns, Joseph Antoine Marie Hubert 236
Luxor (Egypt) 220
Lynn Nelson, W. 60
Lyons & Co, J. 162
M
Maat, Johannes 80
Maatschap Groningen 211
Maatschappij tot Exploitatie van Kooksovengassen (MEKOG) 80, 346, 361, 385
MAC *see* Merchant Aircraft Carrier
Magellan Strait 288-9
Makasar (Celebes / Sulawesi) 234
Malaysia / Federation of Malaysia 65, 131, 135, 186, 240, 270
management and organization, *passim*, *see also* corporate culture, corporate governance, corporate organization, staff, 50:50 ownership (Dutch-British) 87, 101, 109, 121, 149, 264, 443, 457/173
  Anglo-Dutch relations 6-7, 86-97, 169, 443, 463/197, 463/198
  centralization 5, 125, 141, 169, 255, 299, 309, 320, 383
  Committee of Managing Directors (CMD) 7, 144, 318, 332, 334, 341-3, 354-5, 361, 365-7, 372-3, 375-6, 379, 381, 383, 387, 394, 401-2, 437, 443, 445, 462/188, 464/64, 465/71, 467/152
  Conference 7, 105, 147-8, 188, 243-4, 247, 341, 345, 359, 366-7, 369-70, 372-3, 375-6, 379, 387, 391, 401, 422-3, 435, 438-9, 443, 445
  decentralization 141-2, 147, 155, 169, 228, 276, 321
  delegation of authority, power, responsibilities etc. 5, 92, 137-8, 147, 169, 242, 401, 444
  matrix structure 137-55, 329, 444-5
  top management 10, 144, 169, 400
  transfer of Royal Dutch's legal seat to Curaçao 29, 29-30, 31-2, 34
Manchester (UK) 166, 349
Manchester Business School 164
Manhattan, New York, NY (US) 148, 421
manufacturing, *passim*, *see also* downstream operations, expansion of organisation, expansion of production, petrochemicals, refining
  additives, *see* gasoline, lubricants, health hazards 19, 75, 279, 301, 304, 306, 325, 348, 399-400, 423, 425-7, 478/96, 478/108
    TTP (tritolyl phosphate) 426, 478/108
  alkylation 12, 13, 63, 69, 71-2, 75, 260, 451/8, 456/123, 471/122
  barrel cut 183, 253, 298, 319, 393
  blending 72, 258, 262, 270, 301, 344, 387, 399, 422-3
  cracking (in general) 71, 80, 86, 258, 260, 278, 304, 339, 366, 385
    catalysts 71, 274, 277, 278-80
    catalytic cracking (cat cracking) 71, 73, 99, 258, 260, 269-70, 270-1, 274, 278, 304, 332, 471/122, 471/125, 475/2
    Catarole cracking process 349
    desulphurization 260, 279, 419, 422-3
    Dubbs crackers 37, 48, 69, 71, 86
    hydrocracking 260, 279, 281, 469/62
    Shell Trickle Hydrodesulphurization process 279, 419
    thermal cracking 71-2, 260, 349
  distilling, distillates 13, 72, 258, 258-61, 278-80, 336, 344, 393, 472/7
    atmospheric distilling 258, 260
    fractionating 36, 71-2, 258, 270, 280, 469/32
    high-vacuum distilling 69, 71, 258, 260, 269
    middle distillates 183, 253, 275, 279, 298, 300, 393
    Trumble process 12, 69
  extraction 12, 13, 36, 72-3, 189, 402, 419
    Duo-Sol extraction unit 12
    Edeleanu process 13, 72-3, 270, 469/32
    Rectiflow (backwash) process 12, 13
  hydrogenation 15, 17, 22, 70-1, 73, 73, 75, 278-9, 339, 452/22, 470/65
    Base Stock Hydrogenation process 71
    Bergius process 279
    dehydrogenation 12, 75, 278
  hydrotreating 260, 279-80
  platforming, platformers 258, 262, 277-80, 277-8, 471/125
  polymerization 271, 336
  refining 65, 109, 263-5, 268-70, 280, 298, 344
  reforming 37, 69, 270, 304, 339
    catalytic 260, 269, 277, 278-9, 304
    thermal 278-9
  secondary processing 260-1, 278-9, 345, 441, 443
  secondary recovery techniques 191, 210, 210, 464/46
  'whitening the barrel' process 300
manufacturing capacity 10, 15, 22, 29, 36-7, 40-1, 43, 48, 63, 69-71, 265, 268-70, 274-6, 278, 280-1, 294, 298, 304, 306, 318, 329, 339, 346, 349, 357, 359, 365, 369, 401, 443, 451/17, 468/21, 469/49, 470/71
Manufacturing Chemicals – *business sector of Royal Dutch Shell Group* 402
Manufacturing Oil – *business sector of Royal Dutch Shell Group* 275-6, 280, 300, 391, 402
Manufacturing, Transport and Supplies, and Marketing (MTM) –

division of Shell Oil 256-7
Maracaibo / Lake Maracaibo (Venezuela) 140, 194-6, 195
market share, *see* marketing and distribution
marketing and distribution, *passim*, *see also* advertising, brands, downstream operations, expansion of organization, expansion of production, service stations
Handelszaken 93, 130, 147, 229
market share 6, 14, 172, 183, 187-9, 220, 229-30, 265, 268, 300-1, 304, 313, 317-9, 332-3, 351, 359, 435, 445, 471/157
marketing organization 89, 255, 298, 323, 345
marketing policy 298-325
sponsoring 71, 141, 152, 152-3, 428, 436
TBA (tyres, batteries and accessories) 313, 319-20
trade, trading 14-5, 22, 34, 97, 150, 185, 202, 204, 269, 282, 292, 298, 301, 323, 329, 453/49, 455/85
Marketing-Chemical – *business sector of Royal Dutch Shell Group* 309
Marketing-Oil – *business sector of Royal Dutch Shell Group* 309-10, 322-3, 325
Marseilles (France) 274-5, 337
Marshall Islands (Pacific Ocean) 77
Martinez, CA (US) 15, 76, 275, 279, 386, 400
Massachusetts Institute of Technology (MIT) 166
matrix structure, *see* management and organization
Mattei, Enrico 176, 178, 241-2, 268, 296-7, 468/28
McCloy, John 243
McCurdy, R. C. (Dick) 255, 468/12
McKinsey & Co 125, 139-40, 141-2, 144-5, 144, 147, 156-7, 162, 255, 309, 343, 460/96, 461/113
McKinsey review 121, 140-8, 342
Meadows, Dennis L. 439
Medan (Sumatra) 300
Mediterranean 37, 244, 295
MEKOG, *see* Maatschappij tot Exploitatie van Kooksovengassen
Melbourne (Australia) 142
Menlo Park, CA (US) 471/161
Mera (Ecuador) 134
Merchant Aircraft Carrier (MAC) 60, 60-1
mergers, *see also* acquisitions, joint ventures 5, 77-8, 119, 339, 341, 345, 361, 370-1, 460/96, 473/30
Mericourt Dam (France) 440
metals 279, 373, 376, 445, 474-5/122
light metals: aluminium (incl. bauxite: aluminium ore), magnesium, titanium 376, 378
non-ferrous metals 376
precious metals 277, 278
platinum 277-8, 280
tin 376, 378
vanadium 73
Mexican Eagle, *see* El Aguila oil company
Mexico 6, 34-5, 39-41, 86-7, 100, 130, 140, 229, 232, 239, 275
Mexico City (Mexico) 106
Meyjes, R. A. 472/172
Middle East 35, 99, 144, 171-2, 175-6, 178, 183, 185-6, 188-9, 189, 200, 202, 207, 214, 219, 225, 243-4, 262-3, 270, 279, 286, 290, 295, 298, 329, 444-5, 463/19, 467/171
Middle Eastern Crude Project (MEC) 258, 279
Minnesota (US) 425
Miri (Sarawak, Borneo) 290
Mirza, MV 56
Misr Petroleum 222
Mississippi (US) 195, 294, 364
Mitchell, Sir William Foot 343-4, 345, 354-5, 357, 361, 365, 369
Mobil 172, 180, 242, 268, 297
Modesto, CA (US) 393-4, 395
Møller, A. P. 213
Moerdijk (Netherlands), Shell Moerdijk 366-7, 366-7, 417
Monheim (Germany) 70, 254, 274
monopoly, *see also*: British government: Monopolies Commission 5-6, 14, 313
Montecatini 339, 341, 359, 472/24, 473/26
Monteshell 359
Monteshell Agricola 339
Monteshell Petrochimica 339, 341
Mosaddiq, Dr. Mohammad 175, 175, 265
Moss, Stirling 301
motor cars, automobiles *see also* gasoline 19, 126-7, 170, 176, 299, 302, 310, 400, 420-1, 424-5, 433
racing 301, 318, 322
Mr Pollution 401
Mullock, D. W. 163
Mumbai (Bombay) (India) 265, 269
Munich (Germany) 275, 451/17
Muscat 243
Mussolini, Benito Amilcare Andrea 15, 78
N
Nafta Italiana 78, 122
Nagasaki (Japan) 374-5
Nahr Umr (Iraq) 223
Nairobi (Kenya) 142
NAM, *see* Nederlandsche Aardolie Maatschappij
naphtha 253, 260, 278, 302, 304, 336, 366
Nasser, Gamal Abdel (Jamal Abd al-Nasser) 175, 176, 179, 220, 220, 222-3, 233, 466/115
National Benzole 313, 317
National Bulk Carriers 285
National Center for Atmospheric Research, US 478/97
National Farmers' Union 437
National Institute for Occupational Safety and Health 479/154
National Iranian Oil Company (NIOC) 205, 247
national socialism, *see also* German occupation, NSB (Dutch Nazi party).
Nazis 6, 22, 24-5, 32, 69, 78-86, 79-82, 456/158
Nazi regime 6, 17, 22, 32, 34, 69, 78, 79, 81, 82, 86, 339, 452/20
nationalism, nationalists 6, 89, 171, 227, 244, 249, 438-9
Arabian nationalism 175, 221
Egypt 220-2, 220
Indonesia 35, 224, 225-41
Iraq 220, 223-5
OPEC (economic nationalism) 241, 445
nationalization, *see also* nationalism 6, 34-5, 39-40, 81, 175, 176, 186, 220, 222, 225, 230, 232-5, 239, 244, 265, 376
natural gas, *see also* gas fields, LNG (liquefied natural gas) 171, 185, 189, 199, 210, 213, 214, 253, 261, 335, 337, 395, 465/88
Nature Conservancy (UK environmental organisation) 431, 478/129
Nazis, Nazi regime, *see* national-socialism
Nederlandsche Aardolie Maatschappij (NAM) 196, 207, 209-11, 465/85, 465/88
Nederlandsche Bank, De (DNB) 100
Nederlandsche Handel-Maatschappij (NHM) 114
Nederlandsche Koloniale Petroleum Maatschappij (NKPM) 26, 66, 226
Nederlandsch-Indische Aardolie Maatschappij (NIAM) 41, 230-2
Nederlandsch-Indische Tank-Stoomboot Maatschappij (NIT) 53, 62
Netherlands, *see also* Amsterdam, Coevorden, concessions, Delft, fields (Schoonebeek, Slochteren), Groningen, The Hague,

Hollands Diep, Hook of Holland, IJmuiden, Limburg, Moerdijk, Pernis, Plaspoelpolder, Rijswijk, Rotterdam, Rotterdam-Charlois, Scheveningen, Schiphol Airport, Schoonebeek, Slochteren, Utrecht, Vlaardingen, Wassenaar 7, 10, 15, 27-8, 28-9, 32, 37, 39, 47-8, 53, 57, 62, 63, 69, 79, 80, 82, 86-7, 89, 94-7, 99, 100, 101, 104, 108-10, 112, 120-1, 125, 128, 130-2, 135, 140, 156, 158-60, 166-7, 169, 171, 175, 191, 196-7, 206-7, 207, 210-1, 211-2, 213, 223, 225-6, 230-1, 233, 235, 236-8, 248, 262, 276, 281-2, 284, 285, 290, 295, 301-5, 320, 341, 343, 361, 365-6, 366-7, 373, 385, 394, 398, 401, 407, 415, 419, 431, 432-3, 438, 443, 445, 452/41, 453/45, 453/47, 453/51, 453/52, 454/57, 455/84, 455/85, 455/104, 457/165, 457/173, 457/183, 458/3, 459/33, 461/133, 463/198, 466/115, 466/135, 466/140, 467/154, 470/72
Netherlands, oversees territories 29
Neuchâtel (Switzerland) 275, 280
New Guinea 231, 235
New Jersey (US) 51
New Mexico (US) 294
New York, NY (US) 28-9, 34, 51, 60, 89-90, 93-4, 101, 112, 114, 118, 119-20, 140, 142, 149-50, 152, 154, 169, 170, 209-10, 209, 252, 255-6, 400-1, 422, 453/48, 453/49, 459/19, 468/13
New York Stock Exchange (NYSE) 6, 105, 110, 111, 112-5, 113-4, 121, 133, 185, 255, 444
New York Times 29
New York World Telegram 65
New Yorker 430
New Zealand 270
Newbury (UK) 28
NIAM, *see* Nederlandsch-Indische Aardolie Maatschappij
Niger Delta 190, 191, 192-3, 295
Nigeria, *see also* Bonny, Lagos, Oloibiri, Port Harcourt 165, 183, 186, 190, 191, 192-3, 202, 244, 295, 295, 422
NIOC, *see* National Iranian Oil Company
NIT *see* Nederlandsch-Indische Tank-Stoomboot Maatschappij
NKPM *see* Nederlandsche Koloniale Petroleum Maatschappij (NKPM)
Non-Traditional Business (NTB) – *business sector of Royal Dutch Shell Group* 379
Noortwijk, A. van 453/43
Norco, LA (US) 275, 275, 279
Normandy (France) 47, 54-5
Norris City, IL (US) 51
Norske Shell 80, 426-7
North Africa 144, 188, 189
North Carolina (US) 51
North Sea 148, 186, 196-7, 199, 207, 211, 213-4, 214-7, 291, 295, 376, 465/92, 465/95
Norway, *see also* Værnes 10, 177, 285, 409, 467/154
NSB (Nationaal-Socialistische Beweging: Dutch Nazi party) 69, 84, 453/52, 454/57
    Dutch Nazis 32, 69
nuclear energy 302, 375-6, 375, 379
Nürburgring (Germany) 318
O
Oakville (Canada) 274
Odense (Denmark) 285
Odessa, TX (US) 182
offices 42, 86, 134, 136, 139, 141, 255-6, 299, 309-10, 320, 343, 373
    BPM head office Batavia / Djakarta 27, 228-9, 231
    central office The Hague (Carel van Bylandtlaan) 10-1, 28, 31-2, 79-80, 97, 99, 136, 148, 354, 401
    central office London (St Helen's Court) 11, 28, 31, 34, 97, 125, 136-7, 138-9, 140, 143, 145, 146-7, 299, 302, 401
    central offices 131, 136-8, 140, 142, 144-5, 147, 149-50, 154-6, 162, 169, 27-9, 255-7, 310, 323, 344-5, 372, 379, 401-2, 460/96
    dual office structure 148
    Shell Center 147-9, 257, 309-10, 322-3, 461/123, 463/198
offshore exploration and production, *see* exploration and production; *see also* fields, platforms
offshore platforms, *see* platforms
Oil Board 12
Oil Control Board 14
oil fields, *see* fields
oil prices, *see* prices of oil and oil products
oil reserves, *see* reserves
oil tankers, *see* tankers
Oklahoma (US) 255, 294, 307, 471/161
Olie 469/52
Oloibiri (Nigeria) 295, 295
Oman 189, 243
Oosten, W. H. 28
Oostermeyer, Jan 149, 298
OPEC (Organization of Petroleum Exporting Countries) 171, 179, 188-9, 197, 227, 241-4, 241-2, 245, 247, 247, 329, 444-6
Oppenheim, A. S. 29, 94, 451/2, 452/38, 453/45, 457/183
Oregon (US) 50-1, 281, 361, 459/33
Organization of Petroleum Exporting Countries *see* OPEC
Os, Koen van 301
Osaka (Japan) 423
Oxford (UK) 164, 167
O/Y Kamex A/B 122
P
Pahlavi, Mohammad Reza, *see* Shah of Iran
Painter, Carl W. 112, 153, 459/19
Pakistan, *see also* Karachi 268, 460/75
Palembang (Sumatra) 300
Palestine 37
Panaga Club (Brunei) 132
Pangkalan Brandan (Sumatra) 63, 65
Panhandle (US) 72
Papua New Guinea / West New Guinea 205, 230-1, 235, 235-6
Paraguana (Venezuela) 41
Paris (France) 5, 142, 323, 434-5
Parker, Hugh 141, 460/96, 461/110, 461/113
participation (shared ownership) 22, 133, 163, 166, 247, 268, 465/85
    government participation 176, 178, 223, 247, 268, 460/75, 467/190
Partington (UK) 337, 349, 394
Pasadena, TX (US) 422
patents and licences 19, 30, 71, 80, 213, 278-9, 301, 304, 306, 335, 339, 349-51, 362, 385, 391, 394, 399-400, 400, 431, 435, 441, 451/8, 468/13, 471/121, 477/46
Patton, George Smith 22
Pauillac (France) 48, 337
Pearl Harbor, Hawaii (US) 17, 17, 49, 64, 65
Peet, E. Chester 113, 113, 255, 355, 459/22, 459/24
Penhoët (France) 282
Pennsylvania (US) 53
Permigan 239
Permindo 230, 235
Pernis (Netherlands), Shell Pernis 12, 26, 46, 48, 73, 76, 80, 86, 99, 159, 220, 250, 254, 258, 262-5, 262-7, 270, 274, 276, 277-8, 278-80, 297, 301, 337, 346, 349, 351, 354, 363, 365, 385, 402, 415, 415-8, 417, 435, 438, 454/78, 469/32, 473/35, 475/2
personnel, *see* staff
Personnel – *function of Royal Dutch Shell Group* 132, 144, 157, 163

Pertamina 467/160
Perth (Australia) 305
Peru 140, 205
Petit-Couronne (France) 15, 48, 84, 301, 337, 471/125
petrochemicals 6-7, 260-1, 331, 334-5, 336-7, 337, 346-70, 351, 353, 381, 427, 429, 443-5, 472/7, 474/83 (and *passim*)
acetone 336, 337, 339, 472/7
ammonia (synthetic) 12, 63, 395, 472/7 394-5
benzene 13, 72, 279, 335-6, 337, 339, 349, 472/7
butadiene 74, 75-6, 332, 332, 335-6, 336, 339, 348, 365, 385
derivatives 339, 427
detergents 76, 332, 337, 347, 357, 364, 375, 398-9, 399, 406, 409, 425, 435, 477/42, 477/43
ethanol / ethyl alcohol 346, 348, 472/7
ethylene 86, 261, 335, 36, 336-7, 339, 346, 348-9, 348-9
fertilizers, *see* agrochemicals
glycerine (synthetic) 75, 258, 258, 261, 337, 346, 349, 362, 365, 385
methanol 424, 472/7
isopropyl alcohol 336, 337, 472/7
pesticides, herbicides, fungicides and insecticides, *see* agrochemicals
polymers 76, 261, 271, 336-7, 400
nylon / polyamide 333, 350, 355
plastics 80, 176, 331, 339, 341-2, 348-9, 351-2, 353, 355, 358-9, 359, 362, 366-7, 368, 369, 375, 382, 383, 385, 393, 441, 445, 473/28
polyethylene 339, 349, 355-6
polypropylene 261, 336-7, 339, 341, 346, 355, 407, 472/24
polystyrene 261, 336-7, 339, 348, 355
PVC (polyvinyl chloride) 80, 336, 341-2, 346, 348, 355, 373, 385, 473/44, 475/2
resins (synthetic) 76, 332, 339, 348-9, 350, 352, 445
epoxy resins, *see also* brands: Epikote, Epon 261, 331, 336-7, 348-50, 350, 351-2, 355, 375, 473/60
rubber (synthetic), *see also* rubber (natural) 40, 73, 74, 75, 332, 332, 334-5, 336, 336, 339, 348-9, 351-2, 354, 357, 362, 365, 393, 441, 445, 473/47, 474/83
thermoplastics 339
slack wax 12, 258, 451/10
solvents 13, 76, 261, 332, 337, 339, 346, 348, 350-2, 473/41
toluene (synthetic), *see also* toluene (natural) 72-3, 75, 279, 349, 385, 472/7
wax (synthetic) 137, 258, 261, 279, 337
Petrochemicals Ltd. 349, 354, 365, 394
petrol, *see* gasoline
petrol stations, *see* service stations
Petroleum Board 14, 88, 298, 451/13
Petroleum Development (Oman) Ltd. (PD(O)) 189
Pfeiffer, H. 468/17
pharmaceuticals 162-3, 373, 379, 436, 474-5/122
Philippines 131-2, 136
Philips 29, 31, 82, 453/45
Philips, August 97
Philips-Duphar 379
Phillips Petroleum 302, 336, 339, 342
Phoenixville, PA (US) 53
Pijzel, *see* Pyzel
pipelines 37, 49, 51, 53, 54-5, 65, 75-6, 122, 187, 200, 211, 215, 223, 223, 225, 226, 229, 245, 254, 274-5, 290-1, 292-5, 294-5, 298, 349, 366, 375, 456/134, 470/93, 473/47, 474-5/122
Basin Pipeline system 294
Big Inch and Little Big Inch Pipelines 51-2, 52-3
Marseilles Pipeline 469/38
Ozark system 294
Plantation Pipeline 52
PLUTO (Pipeline Under The Ocean) 54-5
Rancho system 294, 294
Rotterdam-Rhine Pipeline 274, 295
Southern European Pipeline 274
Transalpine Pipeline 295, 296-7
West Texas-Houston Rancho Pipeline 294
Pladju (Sumatra) 17, 63, 66, 226, 226, 270, 469/32
platforming, *see* manufacturing
platforms: offshore oil and gas production platforms 195-7, 199-200, 215-8
individual platforms
Orient Explorer 200-1
Seashell 197, 198-9
Triton 196
Plaspoelpolder, Rijswijk (Netherlands) 390
Platt, J. W. 299
Pleyte, C. M. 137, 444
Ploesti (Romania) 18, 69, 70, 455/116
PLUTO (Pipeline Under The Ocean), *see* pipelines
Plymouth (UK) 28
Po Valley (Italy) 178
Poland 10
Pölitz (Germany) 22, 70
Pollard's Rock (UK) 408
pollution, *see also* accidents, emissions, environment, environmental concerns, environmental conservation, gasoline: leaded vs. unleaded, health hazards, toxic substances 280, 295, 325, 383, 387, 400-2, 403, 409-10, 415, 417, 419, 422-3, 425, 433, 438-9, 441
air pollution, atmospheric pollution (in general), smog, *see also* Atmospheric Pollution Committee, emissions 267, 280, 400-1, 404, 415, 415, 419-21, 419-22, 423-5, 424-5
by oil spills 405-6, 409-10, 405, 407-14
by production and consumption of oil, oil products and chemicals 382, 398, 403
by shipping accidents, *see* accidents: shipping accidents
by tanker cleaning 405
Load-on-Top system 405, 406, 414
light pollution 267
protests against pollution 414, 433, 439-40
soil pollution 401, 438
techniques against pollution 407, 409-11
TOVALOP (the Tank Owner's Voluntary Agreement concerning Liability for Oil Pollution) 410
water pollution, marine pollution 280, 410, 405-6, 407-9, 409-10, 433
Port Harcourt (Nigeria) 142, 295, 295
Port Said (Egypt) 177
Port Suez (Egypt) 222
Portland, OR (US) 50-1
Power Petroleum 313, 317
Praboemoelih (Indonesia) 462/171
Pratt's (Esso Brand) 314-5
preparations for war 12-5, 17, 19, 22-3, 26, 29, 47, 79, 452/20, 453/45
prices of oil and oil products, 5-6, 77, 178-9, 183, 185, 188, 202, 204, 221, 228, 244, 247, 253, 265, 268, 275, 285, 304, 309, 313, 318-9, 329, 336, 345, 354, 359, 362, 365, 369, 379, 422, 433, 445
marked prices 154, 178-9, 204, 464/67
posted prices 178-9, 202, 241-4, 249
Princess Beatrix dock 62
Probolingo (Indonesia) 238

profits, *see* finance
Progress and Prospects 436
public opinion 115, 410
  protests 60, 325, 422, 425
publicity and press / media, *see also* advertising, brands and branding, public opinion, Trade Relations 27, 77, 107, 115, 119, 128, 156, 167, 258, 268, 305, 309, 317, 322, 402, 415, 433, 435-6, 439
Puerto Rico 136
Pulau Bukom (Singapore) 270, 272-3
Pure Oil Company 195
Pyzel, Dr. Daniel (*a.k.a.* Pijzel, Dan) 403
Q
Qatar, *see also* Idd el Shargi 189, 196, 198-9, 199, 207, 290
R
Raalte, Dr. H. G. S. van 401
RAF, *see* Royal Air Force
Rathbone, Jack 359
Raymond, C. L. 436
Reactor Centrum Nederland (RCN) 475/124
Red Line Area, Red Line Agreement 172, 173
refineries, *passim see various geographical locations*
refining, *see* manufacturing
refining capacity, *see* manufacturing capacity
Regensburg (Germany) 70
regulations and government control, *see also* legislation, nationalization 15,31, 60, 175, 234, 280, 401, 419, 426, 435, 439
Reichstett-Vendenheim (France) 275
Reign of Chemistry, The 380
Reisholz (Germany) 70
Republik Indonesia Serikat (RIS) *see* Indonesia
research 385-400
  budget 122, 385-7, 391, 393-4, 402
  commercial dimensions 332, 334-7, 339, 341-5, 381, 385-7, 391, 393-5, 399-400, 441
  laboratories (in general) 75, 166-7, 199, 256, 349, 362, 386, 388-9, 391, 394, 415, 415, 419, 441, 443-4, 475/7
  laboratories: individual locations, *see also* Delft, Egham, Emeryville, Houston, Martinez, Modesto, Partington, Rijswijk, Rocky Mountain Arsenal (Denver), Salida, Sittingbourne, Thornton, Wilmington, Wood River, Woodstock.
    Amsterdam (Netherlands): Bataafsche laboratory / Royal Shell Laboratory Amsterdam (Netherlands) / Shell Research Centre / Shell Research Laboratory (KSLA) 12, 42-3, 45, 47, 72, 78, 80, 82, 154, 159, 168, 306, 346, 384-6, 391, 395, 400, 428-9, 456/149, 476/9, 468/17
  mobile laboratory 415, 415
  outsourcing 310, 394, 441
Research and Development (R&D) – *function of Royal Dutch Shell Group, see also* research 476/9, 476/29
Research Planning Conference 387
reserves
  coal 379
  crude oil 39, 172, 187-9, 189, 197, 214, 232, 244, 464/60, 466/139
  deepwater reserves 244
  natural gas 188, 210-1, 244, 464/60
Rheinische Olefinwerke Wesseling (ROW) 339, 353
Rhenania-Ossag 6, 12, 22, 43, 78, 78, 97, 137
Rhodesia 131, 356
Richmond (US) 51
Riemens, Dr. H. 101
Rijckevorsel, Jhr. J. M. van 130
Rijswijk (Netherlands) 390, 392, 393, 407
Ring, Der 78
Rio Canario (Curaçao) 136
Rio de Janeiro (Brazil) 142
Rio Tinto mining company 376
RIS (Republik Indonesia Serikat), *see* Indonesia
Rising Sun Oil Company 15
River Meuse (Belgium) 366
River Rhine (Germany) 274
River Scheldt (Belgium) 366
River Seine (France) 398-9
River Thames (UK) 14, 48, 147, 148, 159, 321
Rock Hill, PA (US) 53
Rocky Mountain Arsenal, military base and laboratory, Denver, COL (US) 350, 393, 438, 479/155
Romania 18, 19, 31-2, 35, 69, 70, 86, 100, 130, 186
Rome (Italy) 276
Roosevelt, Franklin Delano (FDR) 15, 39
Roque, Graterol 40
Rost van Tonningen, J. H. W. 69, 453/52
Rost van Tonningen, M. M. 453/52
Rothschild family 5, 121, 443
Rothschild, Nathaniel Mayer Victor, third Baron Rothschild 422
Rotterdam (Netherlands) 12, 47, 62, 86, 122, 258, 291, 295, 297, 337, 366, 378, 415
Rotterdam-Charlois (Netherlands) 262
Rotterdam-Rijn Pijpleiding Maatschappij, NV 122, 297
Rotterdamse Polyolefinen Maatschappij 339
Rouen (France) 48
Roxana Petroleum Company 88
Royal Air Force (RAF) 19, 19, 21, 406, 451/17, 462/183
Royal Dutch, *see* Royal Dutch Company for the Exploitation of Petroleum Wells in the Dutch East Indies
Royal Dutch Company for the Exploitation of Petroleum Wells in the Dutch East Indies *a.k.a.* Royal Dutch (RD) (*Dutch name:* Koninklijke Nederlandsche Maatschappij tot Exploitatie van Petroleumbronnen in Nederlandsch-Indië, *later* Koninklijke Nederlandse Petroleum Maatschappij), *passim*, *see* advertising, brands and branding, collaboration, concessions, corporate governance, exploration and production, finance, management and organisation, manufacturing, marketing and distribution, mergers and acquisitions, offices, research
Royal Dutch Petroleum Company, *see* Royal Dutch Company for the Exploitation of Petroleum Wells in the Dutch East Indies
Royal Dutch Navy 60, 62
Royal Dutch Shell Group (RD/S), *passim*, *see also* advertising, brands and branding, collaboration, concessions, corporate governance, exploration and production, finance, management and organisation, manufacturing, marketing and distribution, mergers and acquisitions, offices, research, Shell Oil Company, staff
  main business sectors and functions, see separate entries:
    Chemicals, *see* Shell Chemicals
    Exploration and Production (E&P)
    Finance / Finance Administration
    Manufacturing Chemicals
    Manufacturing Oil
    Marketing-Chemical
    Marketing-Oil
    Personnel, later Human Resources (HR)
    Research and Development (R&D)
    Shell Chemicals
    Trade Relations
  Group structure in 1907 5

Royal Society for the Protection of Birds 409
royalties
on oil production 40-1, 175, 225, 391
on patents 279, 419
rubber (natural), *see also* polymers: rubber (synthetic) 74, 75, 362, 474/80
Ruhrchemie 295
Ruhrgebiet (Ruhr region, Germany) 122, 274, 295
Ruhröl 295
Russia, *see also* Soviet Union 88
S
SA des Pétroles Jupiter, *see* Jupiter
safety, safety regulations, *see also* accidents 276, 402, 404, 430-1, 433, 437, 469/52
Saigon (Vietnam) 465/77
Saint-Gobain 337, 339, 473/60
Salida (US) 386
Salomon Brothers & Hutzler 118
Samuel, Sir Marcus, 1st Viscount Bearsted 88, 169, 272
Samuel, Walter Horace, 2nd Viscount Bearsted 87-8, 88, 91-3, 95, 97, 106, 272
San Francisco, CA (US) 108, 400
Santa Barbara, CA (US) 409
Sarawak (*or* British Borneo) 34-5, 65, 197, 199, 290
Saudi Arabia 172, 175-6, 178-9, 219, 244, 247, 444
Saudi Aramco 173
SBM, *see* Single-Buoy Mooring
Scandinavia 276, 354
Scheffer, B. 231, 466/136
Schepers, Lijkle 'Skippy' 355, 361-2, 365, 369, 371
Scheveningen (Netherlands) 31
Schiedam (Netherlands) 197, 198
Schiphol Airport (Netherlands) 303, 374-5
Schoonebeek (Netherlands) 37, 159, 191, 206-7, 207, 208-9, 210, 258, 262, 454/70
Schottegat (Curaçao) 403
Science 431
Science Advisory Committee 430
Scientific Development and Experimental Effort 325
Scotland (UK) 149
Searight, Rodney 467/171
Sedan (France) 10
SEDCO 445 (drilling ship) 197
Semarang (Indonesia) 300
Senegal 205
Seoul (South Korea) 431
Seria (Brunei) 132, 142
service stations 22, 183, 253, 254, 255, 299-300, 299, 309, 310, 311, 314-7, 317, 319-20, 319, 323, 325, 326, 373, 459/33
self-service / self-serve stations 317, 318, 327
solus trading 310, 311-4, 313, 317, 317
Seven Sisters, *see also* cartels 172, 176, 268, 326
Seven Stones reef (UK) 408
Seyss-Inquart, Arthur 31
Shah of Iran (Pahlavi, Mohammad Reza) 175, 178
shareholders / stockholders 29, 31, 37, 40, 47, 90, 92, 95-6, 97, 110, 113, 114-5, 115-9, 119-21, 130, 154-6, 169, 173, 204, 223, 232, 274, 317, 370, 373, 375, 381, 444, 446, 465/88, 475/7
Annual General Meetings (AGMs) 114, 117, 121, 135
minority shareholders 150, 154-5
shares 22, 28, 47, 78, 81, 94, 95, 97, 101, 104, 104, 110, 112, 114-5, 118, 119-21, 133, 150, 172, 185, 231-2, 234, 339, 342-3, 373, 444, 454/76, 457/172, 460/75
dividends, *see* finance
preference shares 114, 453/44, 457/172
share price 110, 120
market capitalization 110, 185
Shaw, USS 64
Shell, *passim*, *see also* Royal Dutch Shell Group
Shell Agricultural Research Centre 360
Shell Aircraft Ltd. 122
Shell Argentina 300
Shell Australia 91
Shell Australia Securities Ltd. 122
Shell Automotive Retail 323
Shell Aviation 19, 298
Shell (Bermuda) Overseas Ltd. 283
Shell Berre 322
Shell Boutiques 323
Shell-BP Nigeria 191
Shell-BP Petroleum Development Company of Nigeria Limited *see* Shell-BP Nigeria
Shell-BP Trade School 165
Shell Cambodge 460/75
Shell Canada 155, 166, 189, 202, 204, 298, 402, 466/137
Shell Candles 298
Shell Caribbean Petroleum Company 101, 119, 459/19
Shell Center, *see* offices
Shell Chemical Company Ltd. 73, 75-7, 258, 295, 298, 348, 343, 346, 348-9, 356, 365, 399, 473/34, 478/129
Shell Chemical Corporation 76, 149, 343-5, 350
Shell Chemical Ltd. 478/116
Shell Chemicals – *business sector of Royal Dutch Shell Group* 157, 221
Shell Chemicals UK 369
Shell International Chemical Company 145
Shell Chimie 339
Shell Committee for Environmental Conservation 325
Shell Company of Egypt 221
Shell Company of Ghana 122, 131
Shell Company of Thailand 131
Shell Company of the Bahamas 122
Shell Company of the Philippines 131
Shell Deutschland 7
Shell Development Company 72, 75, 149-50, 349, 385, 391, 393, 400-1, 475/7
Shell Enterprises 202, 204
Shell Expro, *see* Shell UK Exploration and Production Ltd.
Shell Française 122, 149, 459/48, 460/91, 471/125
Shell Group, *see* Royal Dutch Shell Group
Shell Guides 309, 310
Shell Haven (UK) 46, 264-5, 346, 361, 365
Shell Indonesia 233-5, 239-40
Shell International Petroleum Company (SIPC) 145, 166, 414, 426-7
Shell International Research 401
Shell Internationale Research Maatschappij 145
Shell Italiana 122, 306
Shell Lodge 159, 159, 162, 165
Shell Magazine 458/3
Shell Market Development 309
Shell-Mex 14, 19, 138, 148, 161, 166, 295, 317
Shell-Mex & BP (SMBP) 302, 309, 313, 313, 317, 321
Shell Moerdijk, *see* Moerdijk (Netherlands)
Shell Nederland 373
Shell Nederland Chemie 363, 366-7
Shell News 306, 357
Shell Oil Canada 461/136
Shell Oil Company (Shell Oil), US, *see also* 'Manufacturing,

Transport and Supplies, and Marketing' (MTM) 16, 36, 76-7, 88, 108, 112, 115, 122, 125, 149-50, 152-5, 157-8, 160, 166, 169, 180, 182, 187-8, 191, 195-7, 202, 204-5, 232, 253-7, 253, 255-6, 260-1, 275-6, 279, 299, 302, 317, 319-21, 323, 325, 332, 342-6, 348, 350, 357, 361-2, 365, 371, 386-7, 393, 400, 402-4, 425-6, 438, 441, 459/33, 459/56, 460/52, 460/96, 461/130, 461/131, 461/136, 461/141, 462/142, 462/164, 463/1, 464/54, 464/64, 464/67, 465/71, 465/77, 465/78, 466/135, 468/24, 471/113, 471/160, 471/161, 471/163, 473/44, 474/122, 475/7
Shell Overseas Exploration Ltd. 231
Shell Pernis, *see* Pernis (Netherlands)
Shell Petroleum Company 95, 97, 100, 113, 122, 141, 145, 145, 299, 460/82
Shell Petroleum NV 125
Shell Pipe Line Corporation 294
Shell Pipeline Company 474/122
Shell Point 403
Shell Production of Argentina Ltd. 122
Shell Research Centre / Shell Research Laboratory (KSLA), *see* research: laboratories
Shell Research Ltd. 391, 409
Shell Saint-Gobain 337, 339, 339
Shell Shops 323
Shell Societa Italiana 122
Shell Switzerland 122
Shell Tankers BV 285
Shell Tankers Ltd. 125
Shell Tankers Ltd. (UK) 285
Shell Tankers NV 125
Shell Travel Services 379
Shell Transport, *see* Shell Transport & Trading Co.
Shell Transport & Trading Co. (STT, *also* Shell Transport), *passim*, *see also* advertising, brands and branding, collaboration, concessions, corporate governance, exploration and production, finance, management and organization, manufacturing, marketing and distribution, mergers and acquisitions, offices, research
Shell UK Exploration and Production Ltd. (Shell Expro) 213-4
Shell Union Oil Corporation 6, 14-5, 16, 36, 43, 47, 69-73, 75, 89, 90, 94, 101, 103, 122, 149, 202, 453/49, 454/60, 459/56, 461/131, 473/34
Shell Wax and Candles 470/108
Shellman's Snakes and Ladders (STATO) 151
Shellstar fertilizer plant 361
Sheridan, TX (US) 36
shipping / transport / transportation, *see also* pipelines, tankers 5, 7, 23, 29, 37, 43, 48-9, 51, 53, 63, 68, 69, 86, 91, 96, 202, 214, 219, 225, 251, 254, 262, 274-5, 281-2, 284, 290-1, 297, 301, 309, 329, 348, 375-6, 379, 405, 453/45, 474/122
shipping accidents, *see* accidents
Signal Hill, CA (US) 74, 195, 181
Simplex Agreement 150, 386
Simplex game (Simulation Planning by Executives), *see* staff: training
Sinclair 294
Singapore 137, 142, 236, 270, 272, 274, 301, 466/126
Single-Buoy Mooring (SBM) 290-1
SIPC, *see* Shell International Petroleum Company
Sittingbourne (UK) 393, 429, 437
Slochteren (Netherlands) 210-1, 445
Slotboom, H. W. 391
SMBP, *see* Shell-Mex & BP
Smit, J. E. G. 141
Smith, Sydney 75
Smith, W. Eugene 380
Socal, *see* Standard Oil Company of California
Societa Edison 341
Société des Elastomères de Synthèse (SES) 339
Société du Pipe-Line Sud-Européen 274
Socony, *see* Standard Oil Company of New York
Socony Vacuum 220
Sodegaura (Japan) 218-9
soil pollution, *see* pollution
Solvay (rival PVC producer) 342
Somalia 205
South Africa, *see also* Cape of Good Hope, Durban 304, 310, 471/124
South America 39, 136, 144, 189, 299-300
South Korea, *see also* Seoul 205, 268, 431
Southborough, Lord, *see* Hopwood, Hon. Sir Frank
Southern Alaska 205
Soviet Union / USSR, *see also* Russia 6, 69, 81, 84, 172, 178, 220, 221, 374-5, 466/125
Spaght, Monty 147, 153-4, 154, 156, 204-5, 357, 361, 371, 373, 400, 422, 459/56, 461/130, 461/141, 461/150, 463/29, 464/54, 464/64, 471/157, 472/5, 474/122
Spain 339
St Helen's, OR (US) 361
St Ives (UK) 410
staff / employees / Personnel – *function* 156-8
  career development, career planning 133, 135, 157, 462/166
  employment of local people, *see also* staff: regionalization 66, 100, 132, 262, 455/109
  employment opportunities, terms and conditions 66, 159, 444, 462/188
  expatriates (expats) 125, 128, 130-3, 130-6, 135, 158, 166-7, 221, 231, 460/79, 466/135, 467/154
  internationalization 105, 128, 130-1, 136, 158, 231
  nationalities 33, 34, 106, 128, 129, 130-1, 135-6, 165, 231, 466/135
  numbers 66, 80, 82, 131-3, 231, 280, 369, 393-4, 455/109
  pay / salaries 60, 99, 125, 157-8
    bonuses 158
    Group Basic Salary (GBS) 125
    Local Current Salary (LCS) 125
    pensions 125, 156, 458/189, 462/155
    retirement bonus 125, 156
  recruitment 99, 131, 136, 157-8, 160, 166, 169, 461/131
  redundancies, redundancy payments 133, 156, 391
  regional coordinators 343, 460/89, 461/122
  regionalization ('local for local'), *see also* staff: employment of local people 99-100, 105, 128, 130-3, 136, 159, 164, 166-7, 221, 226, 228-9, 230, 240, 444
  retirement 100, 125, 156-7
  training, trainees 65, 99, 103, 105, 130, 135-6, 149-50, 159-167, 159-65, 168, 169, 199, 230-1, 255-6, 404, 436, 444, 462/171
    Lensbury Club, Teddington 28, 31, 159, 161, 165
    Simplex game (Simulation Planning by Executives) 162, 162-3
Standard Oil Company of California (Socal) 172, 173, 180, 189, 279
Standard Oil Company of Kentucky 455/83
Standard Oil Company of New Jersey / Jersey Standard (JS), *see also* Esso 5, 14, 19, 29, 34-7, 34, 39-40, 57, 70, 73, 75, 145, 172, 173, 179-80, 183, 185-9, 205, 207, 208-9, 209-11, 213, 220, 223, 242-3, 306-7, 313, 319, 336-7, 339, 342, 348, 359, 361, 375, 391, 406, 415, 419, 424, 454/60, 454/83, 455/84, 463/19, 471/113, 471/132, 471/158, 475/140

Standard Oil Company of New York (Socony) 172, 173
Standard Oil Trust 5, 335
Standard Vacuum Oil Company (Stanvac) 15, 17, 26, 63, 66, 226, 233-5, 240, 265, 268
Stanford University 166, 400
Stanlow (UK) 12, 73, 76, 99, 258, 259, 263-5, 279, 281, 295, 301, 337, 346, 349, 451/11
Stanvac, *see* Standard Vacuum Oil Company
Stephens, Frederick J. (Fred) 108, 110, 144, 200, 230, 462/166
Stewart, Jackie 318
Stockholm (Sweden) 439
Strasbourg (France) 274-5
Studebaker 327
Sudetenland (Czechoslovakia) 24-5
Suez (Egypt) 37, 220
Suez Canal (Egypt) 175-6, 176-7, 200-1, 221, 222, 281, 282, 285-6, 290, 405
Suez Canal crisis 177, 223, 274
Suffisantdorp (Curaçao) 131-2
Suharto, General Thojib 239, 241
Sukarno, President Achmad 224, 225, 233, 234, 234-5, 239
Sulawesi 234
sulphur, *see also* emissions, manufacturing: desulphurization 189, 279, 337, 402-3, 403, 417, 419-23, 435, 441
Sumatra, *see also* Aceh, Jambi, Medan, Palembang, Pankalan Brandan, Pladju 224, 225, 226, 228-30, 232, 264, 300
Superior Oil 195
supply pattern, *see* exploration and production
Surabaya 68
Surinam 32, 377-8
Sutton, Sir Graham, FRS 478/97
Swart, Karel 376
Sweden, *see also* Stockholm 282, 291, 319, 401, 417, 424, 431
Switzerland, *see also* Cressier, Neuchâtel 101, 112, 120-2, 240, 275-6, 280, 341, 373, 401, 467/154
Sydney (Australia) 142
Syria 37, 176, 223, 225, 245
T
Tabaksblat, Morris 463/198
take-overs, *see* acquisitions
Tandjong Priok (Indonesia) 68
Tank Syndicate 270
tankers, *see also* pollution, accidents 9, 12, 14, 28, 49, 53, 57, 60, 62-3, 69, 125, 176, 177, 199, 274, 281-2, 283-5, 285-6, 287-9, 290-1, 291-3, 301, 405-6, 409-10, 452-3/41, 455/84, 455/85, 455/101
D-class tankers 282, 291
L-class tankers 282
M-class tankers 285, 288-9
N-class tankers 285, 291
S-class tankers 284
T2 tankers 49, 281-2, 281
Ultra-Large Crude Carriers 129
V-class tankers 283
Very Large Crude Carriers (VLCCs) 285, 290-1, 290, 410, 412-3
Z-class tankers 282
individual tankers
Angelina 68
Drupa 291
Eulota 62
Gadila 60-1
Kong Haakon VII 285
Metula 129
Macoma 60
Mactra 285, 412-3, 470/79
Marinula 285
Marisa 59, 290
Marpessa 100, 285, 412-3
Methane Pioneer 214
Methane Princess 214
Methane Progress 214
Metula 129, 288-9, 292-3
Naticina 291
Ordina 62
Ovula 57, 226
Paula 68
Niso 285, 287
Sepia 282, 284
Serenia 282
Sitala 282
Sivella 282
Solen 282
Sunetta 58-9
Thelidomus 281
Tomocyclus 281
Torrey Canyon 406, 408, 410, 410-1
Universe Apollo 282
Vexilla 282, 283
Tanzania 136
tar sands 244
Tarakan (Borneo) 26, 65-6
taxes, taxation, *see also* legislation: tax legislation 6, 26, 41, 47, 125, 148, 155-6, 158, 175-6, 179, 183, 189, 202, 204, 211, 230, 234, 244, 334, 424, 445, 453/47, 466/130
Te Werve (country club, Rijswijk) 94
Teagle, Walter 35
Teddington (UK) 28, 31, 159
Teepol 76, 258, 332, 337, 346, 346, 348, 399, 404, 477/42
Tehran (Iran) 175, 244, 247
Tennessee (US) 51
Texaco 140, 172, 173, 180, 189, 255, 279, 294, 460/96
Texas (US) 36, 36, 51, 72, 178, 182, 206-7, 294, 422
West Texas (US) 294
Texas Instruments 460/96
Texas Pipe Line Company 294
Thailand, *see also* Bangkok 65, 131, 268
Thames Haven (UK) 46, 54
The Hague, *see* Hague, The (Netherlands)
Thornton (UK) 385-6, 393, 424, 451/11, 456/123
Threat in the Water, The 351
Thune, Trygve 409
Time magazine 109, 109, 148, 307
Tjepu (Indonesia) 63, 65-6
Tokyo (Japan) 16, 239, 467/152
toluene, *see also* petrochemicals: toluene (synthetic) 71-3, 472/7
Toronto (Canada) 150
Torrance, CA (US) 75, 348
TOVALOP, *see* pollution
toxic substances, toxicity 342, 399, 401, 409, 423, 426-30, 429, 438, 446, 478/96, 478/108 478/116
toxicology, toxicological research 387, 394, 399, 401-2, 429, 436
trade, trading, *see* marketing and distribution
Trade Relations – *function of Royal Dutch Shell Group* 144, 401-2, 438, 441, 479/158
Trans-Volta-Togoland 131
Transalpine Pipeline 295, 298-9
transport, *see* shipping.
Trieste (Italy) 295
Trinidad 12, 37, 38-9, 136

Trinidad Leaseholds Ltd. 12
Tripoli (Libya) 37, 244, 245, 247
Truman, President Harry S. 175
Tulsa, OK (US) 471/161
Turkey 122
Turkse Shell NV 122
Tuyll van Serooskerken, Baroness Marie Cornelie van 108
Tyne (UK) 282
U
U Thant, Maha Thray Sithu 236
Unie van Kunstmestfabrieken (UKF) 361
Unified Planning and Control Machinery (UPM) 138
Unilever 29, 31, 82, 158, 162, 166, 462/142, 463/198, 477/42
Union Carbide 334, 335
Union Kraftstoff 339
Union Oil 195
United Arab Emirates 241
United British Oilfields of Trinidad 37
United Kingdom (UK), *see also* British Empire, British government; *see also* Berkshire, Birkenhead, Birmingham, Bournemouth, Bristol, Caernavon, Calder Hall, Cambridge, Carrington, Cornwall, Dungeness, Edinburgh, Egham, Hereford, Heysham, Isle of Wight, Kent, Lancashire, Land's End, London, Manchester, Newbury, Oxford, Partington, Plymouth, Pollard's Rock, River Thames, Scotland, Seven Stones Reef, Shell Haven, Sittingbourne, St Ives, Stanlow, Teddington, Thames Haven, Thornton, Tyne, Wales, Woodstock 5, 7, 10, 12, 14-5, 19, 28-9, 28-9, 31, 39, 46, 47-9, 51, 53, 54-5, 57, 62-3, 62, 69, 72-3, 77, 84, 86-7, 91-2, 94-7, 99, 101, 108-9, 120-2, 131-2, 135-6, 140-1, 144, 148, 156, 158-60, 162, 164, 166, 169, 172, 175-6, 176, 191, 200, 213-4, 219-23, 225, 231, 233, 240, 258, 259, 264-5, 270, 276, 282, 283-4, 285, 291, 295, 298, 301, 302, 307, 309-10, 310, 312, 313, 314-5, 317, 318, 323, 324, 333, 334-5, 337, 339, 341, 348, 349, 351, 353, 360, 362, 365, 367, 369, 371, 374-5, 375, 385-6, 393, 394, 399, 401, 406, 408, 409, 419, 420-1, 424-5, 429, 431, 435-7, 443, 451/17, 452/20, 452/40, 452/41, 453/50, 453/51, 454/78, 455/84, 456/123, 457/165, 457/183, 458/3, 461/133, 462/183, 463/198, 467/154, 470/72, 471/132, 474/122, 476/9, 477/43
United Nations 236
United Nations Conference on the Human Environment 439
United Nations Food and Agriculture Organization 350, 401
United States of America (US), *see also* American government; *see also* Alabama, Alaska, Allegheny Mountains, Anacortes, Atlanta, Baton Rouge, Beaumont, Benton, Boulder, California, Chicago, Cleveland, Colorado, Denver, Dominguez, East Coast, Emeryville, fields (Coalinga, Prudhoe Bay, Signal Hill, TXL field), Geismar, Georgia, Glenmoore, Greensboro, Hawaii, Hollywood, Houston, Illinois, Indiana, Kansas, Kentucky, Linden, Long Beach, Longview, Los Angeles, Louisiana, Manhattan, Martinez, Menlo Park, Minnesota, Mississippi, Modesto, New Jersey, New Mexico, New York, Norco, Norris City, North Carolina, Odessa, Oklahoma, Oregon, Panhandle, Pasadena, Pearl Harbor, Pennsylvania, Phoenixville, Portland, Richmond, Rock Hill, Rocky Mountain Arsenal, Salida, San Francisco, Santa Barbara, Sheridan, Signal Hill, St Helen's, Tennessee, Texas, Torrance, Tulsa, Utah, Ventura, Virginia, Washington DC, Wilmington, Wood River 6, 14-5, 17, 17, 23, 29, 31, 34, 36-7, 39-41, 43, 49, 49-52, 62-3, 65, 70-2, 73, 75, 76, 77, 88-90, 96, 99, 101, 103, 106, 110, 112-3, 115, 115-7, 119-22, 130, 135, 139-42, 149-50, 152, 154-6, 158, 162, 162-3, 166-7, 172, 175-6, 178, 180, 183, 184, 185-9, 202, 204-5, 205, 211, 214, 220, 225, 225, 233, 239, 241, 243-4, 247, 249, 251, 253-8, 254, 262, 264-5, 268, 270, 276, 277, 279-80, 285, 294-5, 299-302, 304, 310, 311-2, 318-22, 321, 325, 326, 333, 334-7, 339, 341-2, 348-9, 350, 351, 352, 357, 360, 365, 370, 373, 374-5, 381, 385-6, 394, 399-402, 406, 419, 422-3, 425-6, 428, 430-1, 433, 436-8, 443-4, 451/17, 455/104, 459/19, 460/96, 461/132, 461/136, 462/142, 462/171, 465/71, 465/88, 467/150, 467/154, 474/83, 475/7
Universal Oil Products 278
University Institute of Education, Oxford (UK) 167
upstream activities, upsteam investments, *see* exploration and production 150, 188, 188, 202, 204, 247, 249, 329
US Army 438
US Army Air Force / US Air Force 16, 302, 304
USS Shaw 64
Utah (US) 357
Utrecht (Netherlands) 89
V
Værnes (Norway) 426
Veer, Jeroen van der 463/198
Velsicol 350, 431
Venezuela, *see also* Caracas, Cardón, Laguinillas, Maracaibo, Paraguana 36, 39-41, 40-1, 53, 60, 73, 92, 96, 103, 106, 107, 108-9, 130, 135, 136, 140-1, 149, 156-7, 162, 167, 175, 179, 183, 186, 189, 191, 194, 202, 229, 231, 239, 241, 244, 253, 255, 258, 262, 264-5, 270, 279, 282, 306, 371-2, 419, 422, 441, 452/23, 454/73, 455/85, 460/96, 462/156, 463/19, 464/67, 467/154, 467/167, 471/122, 474/109
Ventura, CA (US) 348, 394
Verenigde Kunstmestfabrieken Mekog-Albatros (VKF) 361
Vermeulen, J. H. 387
Very Large Crude Carriers (VLCCs), *see* tankers
Victoria and Albert Museum (London) 467/171
Vietnam 465/77
Virginia (US) 51
Vlaardingen (Netherlands) 47, 402
Voltol (lube oil works) 70
W
Wagner, Gerrit A. 154-5, 154, 183, 465/80
Walden, George 63
Waldheusel (Czechoslovakia) 24-5
Wales (UK) 124, 302
Waley Cohen, Sir Robert 88, 88, 91-2, 97, 451/1, 457/163
Wall Bake, J. W. M. van den 108
Wall Street, *see* New York Stock Exchange
Wall Street Journal 121
Walton, C. L. 141
war, *see also* Cold War, D-Day, Dutch-Indonesian Crisis, Suez Canal crisis
  First World War (1914-18) 5-6, 22, 26, 35, 43, 72, 137, 373, 443-4, 455/84
  Nigerian Civil War / Biafran War (1967-70) 191
  Second Sino-Japanese War (1937-45) 15, 17, 17
  Second World War (1939-45), *see also* Battle of Britain, Battle of the Atlantic, Blitzkrieg, German occupation, Japanese occupation, preparations for war, war-damage 7, 100-3, 24-5, 49, 61-2, 64-5, 68, 76-7, 79-85, 95, 98, 114, 123, 128, 137, 143, 150, 152-3, 171-2, 191, 207, 226, 255, 304, 374-5, 385, 405, 429, 438, 443, 455/84, 461/131
  Six Day War (1967) 222, 222-3, 225
war damage
  to oil fields, buildings and installations 8, 26, 28, 42-6, 43, 47-8, 65-6, 66-7, 69-70, 70-1, 103, 222-3, 225, 227, 262, 335, 454/78, 455/106, 455/116
  to ships / tankers 49, 51, 53, 56, 57, 58-9,60, 62, 62, 64-5, 65, 68
Washington, DC (US) 17, 39, 41, 88, 279
Wassenaar (Netherlands) 167
Wavin (pipe manufacturer) 342, 373, 473/28

Weizmann, Dr. Chaim 349
Welling, A. H. S. 262
Werner, E. G. G. 7, 367, 369, 379
Wesseling (Germany) 339, 349
West New Guinea, *see* Papua New Guinea
Westinghouse 424
White House Garage (New York) 252
Whittle, Sir Frank 21
Wickes, Forsyth 459/56
Wieringa, W. G. 119, 119
Wijk, N. van 32
Wilhelmshaven (Germany) 274, 295, 297
Wilkinson, Sir Harold 112, 140, 209, 209, 461/132
Wilkinson, Lady 284
Willemstad (Curaçao) 403
Wilmington, CA (US) 15, 71, 258, 279, 386
Wishart, I. S. 141
Wonderful World of Golf, The 152-3
Wonokromo (Indonesia) 65, 226
Wood River, IL (US) 15, 71, 73, 73, 258, 275, 279, 304, 386
Woodstock (UK) 360, 362, 393-4, 399, 436
World Bank 153, 234, 467/142
World Health Organization 401
World Wildlife Fund 437
Wright, Myron A. 210
Y
Yugoslavia 69, 81
Z
Ziegler patents 349, 394
Zimbabwe 131
Zubair (Iraq) 223
Zuiderkruis 237
Zulver, Cornelis 452/41
Zwolsman, Reinder 474/117
Zwolsman Group 373

Colophon

Book design
Marise Knegtmans, Amsterdam
Picture research
Nienke Huizinga, Amsterdam
Picture production
Karin Creemers, Nijmegen
Lithography
Colorset, Amsterdam
Map design
Arjen van Susteren, Schiedam
Technical illustrations
Paul Maas & Eric van Rootselaar, Tilburg
Photography cover
Kees Rutten, Amsterdam
Slipcase
Ruimtelijke Zaken, Eindhoven
Print
Drukkerij Wilco, Amersfoort
Binding
Binderij Callenbach, Nijkerk
Index
Hans van der Pauw & Aida van Gelderen

Typeface
DTL Argo
Paper
Absolut mat, Proost & Brandt
Paper cover
Tom & Otto, MoDo van Gelder

Uitgeverij Boom
Geert van der Meulen, Aranka van der Borgh, Ton van Lierop, Max Dumoulin
Oxford University Press
David Musson, Matthew Derbyshire, Tanya Dean

With many thanks to
Julia Bate and colleagues (Shell Photographic Services), Wim Blom Communicatie, Tom Chandler, Veronica Davies (Shell London), Jack Doherty, Matthew Green, Piet Holleman & Nico Rozendaal (Abottroom, SRTCA), Rob Lawa (Shell The Hague), Marjolein van der Tweel